Advances in Mathematics Education

For further volumes:
www.springer.com/series/8392

Helen Forgasz • Ferdinand Rivera

Editors

Towards Equity in Mathematics Education

Gender, Culture, and Diversity

 Springer

Editors
Helen Forgasz
Faculty of Education
Monash University
Melbourne, Victoria, Australia

Ferdinand Rivera
Department of Mathematics
San Jose State University
San Jose, CA, USA

ISSN 1869-4918   e-ISSN 1869-4926
Advances in Mathematics Education
ISBN 978-3-642-27701-6   e-ISBN 978-3-642-27702-3
DOI 10.1007/978-3-642-27702-3
Springer Heidelberg Dordrecht London New York

Library of Congress Control Number: 2012934455

Printed on acid-free paper

Springer is part of Springer Science+Business Media (www.springer.com)

Series Preface

The third volume in the series *Advances in Mathematics Education* deals with current equity issues in mathematics education. It addresses 21st century issues involving minorities whose lives are complexly intertwined within/against the dominant discourse. Further, due to the blurring of gender lines and the porosity of upward mobility, referring to the possibility of development in the social ladder, research studies on differences between groups have also become increasingly nuanced. Thus, the volume concentrates on perspectives that involve gender, culture, and diversity. Educational systems have been anchored in the Western Anglo-Saxon system of patriarchal hegemony and the under-appreciation of females drawn from various historical, cultural, and social forces, which should lead us to consider more encompassing concerns drawn from recent discussions in philosophy, feminist theory, and social justice.

While the volume contains many new papers, however, it preserves the special flavor of the series, namely, its close ties to *ZDM—The International Journal on Mathematics Education*—that was the basis of many of the topical and highly influential papers. The first section of the volume, which involves gender, begins with the groundbreaking paper of Leone Burton, *Moving Towards a Feminist Epistemology of Mathematics*, which was first published in 1995. Burton's article sets the scene with an embracing new perspective on mathematics from a radical feminist perspective followed by several chapters that involve an empirical methodology. These chapters underscore the view that a feminist perspective on mathematics and its epistemology is overdue. The next section broadens the debate by bringing to bear issues involving culture, race, ethnicity, and indigeneity. The third section focuses on equity and curriculum diversity and narrows down the debates on schooling and its influences on equity or institutionalized inequity. The final section attempts to thread a new direction in matters involving equity by discussing aspects of biology, one of the hot potatoes of the debate on gender in the 1970s, and the emerging topic of neuroscience. These approaches seem capable of overcoming the classical borderline, that is, those paralyzing distinctions that characterize the different strands of the debate.

The voices in this volume represent a truly international and secular reach across cultures and institutions. Both of us knew Leone Burton personally, and we are certain that she would have been very proud of this volume.

Hamburg, Germany   Gabriele Kaiser
Missoula, MT, USA   Bharath Sriraman

Contents

Introduction

Ferdinand Rivera and Helen Forgasz

The issue of equity remains one of the most difficult and persistent problematics in the theory and practice of mathematics education. In the broadest sense, *equity* encompasses matters involving regimes of inclusions and exclusions that both enable and constrain the lived conditions—by gender, cultural identity, and diversity—of individuals in complex societies. Undoubtedly, this is why the presumption of *access* is foregrounded by Bishop and Forgasz (2007) in their synthesis of research on equity in mathematics education. They claimed that "without *access* to mathematics education there can be no equity" (p. 1146). This view offers a powerful counterdiscourse to *inequity* which, as a concept, is represented by marked (negative) differences in outcomes among groups within a given context or setting. Inequities are manifest in educational situations that involve achievement, enrolment, and attitudes. While differential outcomes implicitly convey inequity of some kind, they are distinguished from *diversity*, a concept that relates to the complex of interacting ways in which groups of individuals are identified and categorized. In this book, diversity encompasses: gender; race, ethnicity and culture; nationality and language background; socioeconomic status; exceptionality; and physical and learning disabilities.

In this volume, we have assembled two categories of research articles that address various aspects of equity. Direct reproductions of earlier published ZDM articles comprise the first category of papers. To bring them up-to-date, we asked the authors of these articles to write a short Preface. In light of Leone Burton's passing, we invited Diana Erchick to write the Preface to the re-printed Burton article. New chapters encompass the second category of papers. We invited authors who could

F. Rivera (✉)
Department of Mathematics, San Jose State University, Washington Square 1, San Jose, CA 95192, USA
e-mail: ferdinand.rivera@sjsu.edu

H. Forgasz
Faculty of Education, Monash University, Wellington Road, Melbourne, Victoria 3800, Australia
e-mail: helen.forgasz@monash.edu

H. Forgasz, F. Rivera (eds.), *Towards Equity in Mathematics Education,*
Advances in Mathematics Education,
DOI 10.1007/978-3-642-27702-3_1, © Springer-Verlag Berlin Heidelberg 2012

assist in addressing gaps in the field that we believed had not been sufficiently covered by the first category of papers to prepare new chapters. We carefully selected authors because we felt they might also bring with them "new" methodological and theoretical perspectives that would benefit and advance the field. Each new chapter is also accompanied by two commentaries. Commentators were selected who were well-known, experienced and knowledgeable researchers in the field, or less experienced but promising, emerging researchers. Each commentator was asked to comment on the current state of the field based on the views raised in the chapters assigned to them. We also encouraged them to develop views that would provoke forward, productive, and different ways of thinking, including ways in which the work could potentially influence changes in policy and institutional practices.

Overall, while the book aims to both broaden and expand the complicated conversations concerning equity in mathematics education, it should be clear that the primary intent, which we share with equity scholars within and outside of mathematics education (e.g., Luke et al. 2010; Nasir and Cobb 2007), is to develop ways in which up-to-date research knowledge in the field can assist in eliminating the various *equity gaps* that are so pervasive in the mathematics education literature.

The remaining nineteen major chapters in this volume have been structured around four parts. Each major chapter is accompanied by either an updated Preface or two follow-up chapter commentaries. We caution readers that while part labels helped us in organizing the chapters, they are nonetheless subjective and arguable as there are considerable overlaps across dimensions of equity. The six chapters comprising Part I deal with issues associated with equity and gender. Part II consists of six chapters in which issues pertinent to equity and culture, ethnicity, race, and indigeneity are discussed. The four chapters included in Part III tackle issues of equity and matters that pertain to curriculum. The final three chapters found in Part IV address equity and matters of a biological nature. Each part commences with brief introductory comments from the editors of the volume.

Across the nineteen chapters, a strong and mutually determining relationship between theory and empirical evidence can be discerned. We share Lukas and Beresford's (2010) view concerning the significance of emerging theories on equity in mathematics education that can inform the development of equitable practices and, more generally, policy:

> (B)y itself, empirical research offers little to social analysts and policymakers; theory is essential for drawing proper inferences from the research. Yet the wide set of plausible theories, and strategies of analysis that are not designed to eliminate nonviable theories, can ultimately render social science evidence of little value to policy. (p. 26)

We are just as concerned about efforts that seek to translate evidence to norms, but we hope the chapters in the volume generate further productive discussions in terms of how equity gaps can be eliminated so that all learners, in whatever category of diversity they are located, are able to achieve success in their encounters with and engagement in the learning of mathematics.

Finally, consistent with the structure of the two earlier volumes in the Advances in Mathematics Education series, one important feature of this book is the international dimension of authorship and mathematics learning contexts. This has enabled us to pursue matters involving equity in mathematics education from several different, and interesting, lenses, angles, and perspectives. We were impressed with the scholarship of the researchers whose locations and contexts were from across the globe. They have provided thought-provoking views and empirical evidence that should further enrich both the theory and praxis of equity in mathematics education well into the 21st century.

Acknowledgements We acknowledge the assistance of the range of funding agencies around the world that may have supported the research reported in these pages. We also thank Gabriele Kaiser (handling series editor) and members of the Editorial Board of *Advances in Mathematics Education*, and the Springer team who have supported us in the development and preparation of this volume. We also acknowledge the helpful and insightful reviews provided by the following scholars (in alphabetical order):

Alan Bishop	Steven Khan	Swee Fong Ng
Allan Bernardo	Richard Kitchen	Nuria Planas
Nerida Ellerton	Robert M. Klein	Joanne Rossi Becker
Margaret Flores	Evelyn Kroesbergen	Wolfgang Schlöglmann
John Francisco	Gilah Leder	Bharath Sriraman
Peter Galbraith	Roza Leikin	Olof Bjorg Steinthorsdottir
Vasilis Gialamas	Stephen Lerman	Catherine Vistro-Yu
Merrilyn Goos	Swapna Mukhopadhyay	David Wagner

Finally, our sincere gratitude and thanks to the contributing authors and commentators for their patience and impressive work.

References

Bishop, A. & Forgasz, H. (2007). Issues in access and equity in mathematics education. In F. Lester (Ed.), *Second handbook of research on mathematics teaching and learning* (Vol. 2, pp. 1145–1167). Reston: National Council of Teachers of Mathematics.

Lukas, S. & Beresford, L. (2010). Naming and classifying: Theory, evidence, and equity in education. In A. Luke, J. Green, & G. Kelly (Eds.), *Review of Research in Education, Volume 34 (What counts as evidence in educational settings? Rethinking equity, diversity, and reform in the 21st century)* (pp. 25–84). Washington: American Educational Research Association.

Luke, A., Green, J., & Kelly, G. (2010). What counts as evidence and equity? In A. Luke, J. Green, & G. Kelly (Eds.), *Review of Research in Education: Volume 34 (What counts as evidence in educational settings? Rethinking equity, diversity, and reform in the 21st century)* (pp. vii–xvii). Washington: American Educational Research Association.

Nasir, N. & Cobb, P. (2007). *Improving access to mathematics: Diversity and equity in the classroom*. Columbia: Teachers College Press.

Part I
Equity and Gender

Helen Forgasz and Ferdinand Rivera

Six chapters are included in Part I of the book. The first three are reproduced articles from ZDM. With the exception of the first chapter, authored by the late Leone Burton, each is accompanied by a preface written by the original authors. The preface to the Burton chapter has been thoughtfully crafted by Diana Erchick. The remaining three chapters in Part I are newly authored. Two commentaries by various authors with expertise in the pertinent research area follow each new contribution.

A major focus of each chapter in this part of the book is *gender equity*. We decided on gender equity for the opening part of the book as, historically, gender was the initial dimension of equity researched widely, and later served as the springboard for emphases on, or in combination with, other dimensions of equity within the field of mathematics education research.

Leone Burton's chapter, *Moving towards a feminist epistemology of mathematics*, was a seminal contribution to feminist theoretical discussions on addressing gender inequities in mathematics learning beyond the liberal feminist approaches which had underpinned early work in the field. As identified by Diana Erchick in her preface, Burton "argued for a feminist epistemology that is parallel to the epistemologies of other STEM fields, but a case unto itself because of the nature of mathematics". Burton proposed five categories for defining "knowing mathematics" based on her reading of philosophical, pedagogical and feminist literature: "its person- and cultural/social-relatedness; the aesthetics of mathematical thinking it invokes; its nurturing of intuition and insight; its recognition and celebration of different approaches particularly in styles of thinking; the globality of its applications". Erchick honed in on Burton's expressed hopes for the future, pointing out that some progress had been made in meeting Burton's vision, but more was needed: "we are as a field moving in the right direction to enrich the discipline by nurturing learners whose epistemology of mathematics is less and less objectivist".

In her chapter, *Equity in mathematics education: Unions and intersections of feminist and social justice literature*, **Laura Jacobsen** (formerly Spielman) conceived of gender equity as socially constructed and inextricably intertwined with social class and race/ethnicity, with women's ways of knowing mathematics recognized as different from men's. For a sustainable impact on equity and for the common good, traditional mathematics, it was argued, was in need of an overhaul. Support was needed for mathematics education to include mathematical literacy, critical literacy, and community literacy. In her preface, **Jacobsen** describes ensuing research and curricular developments for pre-service teachers based in the theoretical perspectives she put forward in the original article.

In her chapter, *Adolescent girls' construction of moral discourses and appropriation of primary identity in a mathematics classroom*, **Jae Hoon Lim** presents findings from a qualitative study of three American girls from different socio-economic and ethnic circumstances. Within their traditional mathematics classrooms, she traces their developing social and academic identities. The dynamics and culture of the classroom, peer group pressures, and personal class and ethnic backgrounds were interacting influences on their identity formations. The girls, she found, made choices that lead them into mathematics pathways that limited their future mathematical options. In the pursuit of "a more equitable intellectual pursuit and joyful learning in school mathematics across diverse groups of students", **Lim** argued for a systematic change to the US public school system "directed toward various groups of underserved students". In the preface to the chapter, she provides further research evidence to support her contention that "social and cultural disconnection between working class minority girls and their teachers has contributed to existing gender, racial/ethnic, and class inequities in mathematics education". She called for the re-conceptualizing of mathematics teaching "in light of social activism". She lamented that this perspective seemed "far removed from the dominant research discourse in mathematics education which has focused on the cognitive process of individual learners to explain how authentic mathematics learning may occur".

Paul Dowling and **Jeremy Burke** provoke readers to contemplate addressing issues of social inequity with sound mathematical content within the mathematics classroom in their chapter entitled, *Shall we do politics or learn some maths today? Representing and interrogating social inequality*. They reflect on historical portrayals and the inferences drawn of gendered and class inequities manifest in photos and in the images found in mathematics textbooks, and present an overview of pertinent research findings. They introduce two variables for examining images: whether the image expresses tacit (connotative) or explicit (denotative) social inequity, and then whether the image is consonant or dissonant with that representation. They conclude that it is possible to ensure that students learn appropriate mathematics within the classroom, and they strongly endorse the imperative to remove representations reflecting patterns of social inequity. Yet, they claim, serious understandings of social injustice are not possible in the mathematics classroom "because the public domain settings that we construct will always be mathematically motivated distortions of the alliances that we want to destabilize". Privileging "political motivations over our mathematical ones" by switching activities is needed. That is, "we can be both mathematics educators and political activists, just not at the same time".

Gabriele Kaiser, **Maren Hoffstall**, and **Anna Orschulik** present recent empirical findings on the gendered perceptions of mathematics held by two age cohorts (11–12 year-olds, and 15–17 year-olds) of German secondary students in their chapter, *Gender role stereotypes in the perception of mathematics: An empirical study with secondary students in Germany*. They administered a modified version of the *Who and mathematics* instrument developed by Leder and Forgasz (2002) and included an additional open-ended question. They found that overall, mathematics continued to be viewed as a male domain, with the older group holding stronger traditionally stereotyped views than the younger group who were generally more egalitarian. The reasons students provided for males or females being considered the higher achieving group differed. The three most frequently cited reasons for males to be considered more successful mathematically were: interest in mathematics, their ability to reason logically, and career. For girls to be considered the higher achieving group, effort, concentration, and ambition, were the three most prevalent response categories. The authors implored teachers and parents to "to strengthen girls' mathematical self-concepts and make them believe in their own achievements".

In the chapter, *Students' attitudes, engagement and confidence in mathematics and statistics learning: ICT, gender, and equity dimensions*, **Anastasios Barkatsas** reports findings from three studies in which the *Mathematics and Technology Attitudes Scale* [MATS] were used to gauge students' attitudes towards technology use for mathematics, and a further two studies in which the *Survey of Attitudes Toward Statistics* Scale [SATS] were administered. The MATS was completed by secondary school students and the SATS by university students. The participants were from Australia and Greece. Replicating previous findings, boys expressed greater confidence than girls in mathematics and in technology, greater confidence in using computers and CAS calculators, and had more positive attitudes to learning mathematics with computers and CAS calculators. At the tertiary level, males also demonstrated more positive attitudes than females toward statistics. In one of the studies, cluster analysis was conducted to examine the inter-relationships between the factors included in the MATS scale. The findings revealed that girls were more likely than boys to be found in some clusters, while boys were more likely than girls to be found in others. The commentators on this chapter both identified this analysis as of particular interest in that the model **Barkatsas** had described as underpinning the development of the MATS instrument appeared to have been challenged by the findings. It was suggested that gender might be a variable to be included in the model.

References

Leder, G., & Forgasz, H. (2002). *Two new instruments to probe attitudes about gender and mathematics*. ERIC, Resources in Education (RIE). [ERIC document number: ED463312]

Preface to "Moving Towards a Feminist Epistemology of Mathematics"

Diana B. Erchick

In *"Moving towards a feminist epistemology of mathematics,"* Leone Burton (1995/2008) argued for a feminist epistemology that is parallel to the epistemologies of other STEM fields, but a case unto itself because of the nature of mathematics. Her proposal defined knowing in mathematics in relation to five categories including social relatedness, the aesthetics of mathematical thinking, intuition and insight, recognition and celebration of different approaches and ways of thinking, and global applications. These categories incorporate processes mathematicians employ as well as the experience of the learner in constructing mathematical meaning. Her conjecture for the future was that:

> [T]he scientism and technocentrism which dominate much thinking in and about mathematics, and constrain many mathematics classrooms, would no longer be sustainable. Mathematics could then be re-perceived as humane, responsive, negotiable and creative. One expected product of such a change would be in the constituency of learners who were attracted to study mathematics but I would also expect changes in the perception of what is mathematics and of how mathematics is studied and learned. (p. 527)

Burton's discussion of this epistemology and its consequence for the discipline and for mathematics education involves a cultural view, meanings of knowing mathematics and being a mathematician, and the challenge of an inclusive epistemology. The interactions among topics of Burton's argument suggest that a (re)consideration of concepts in terms of how they relate to each other could better serve us in raising new questions for furthering this work. For discussion, I suggest associating concepts in terms of conceptual clusters—groupings of concepts or constructs that are interrelated. This short preface cannot do justice to the few clusters I consider; nor does it allow introduction of more groupings. But these clusters can help us think about connected concepts in terms of their influences upon each other and what that has to offer. I also expect that they will become examples of how fluidly we can organize concepts to consider new perspectives and generate new questions, and a

D.B. Erchick (✉)
The Ohio State University, Newark, Newark, USA
e-mail: erchick.1@osu.edu

H. Forgasz, F. Rivera (eds.), *Towards Equity in Mathematics Education,*
Advances in Mathematics Education,
DOI 10.1007/978-3-642-27702-3_2, © Springer-Verlag Berlin Heidelberg 2012

lens for reading a feminist epistemology. I comment on three conceptual clusters: Feminism/Gender/Connected Social Constructs; Mathematics/Equity/Social Justice Pedagogies; and Instruction/Perspectives on Mathematics/Testing.

1 Feminism/Gender/Connected Social Constructs

Thinking about this cluster, consider how the feminism grounding Burton's perspective is defined. First, Burton's perspective includes that feminism and gender are analogous, defined in terms of woman and girls, their perspectives and experiences. As Burton prepared to introduce her five categories relating to knowing mathematics, she pointed to the "need to accept the personal experience of women as a valid component of experimental observations" (p. 526). Reference to Rosser's "women's experience of knowing and doing science" (p. 523) grounded her argument. The assumed connection between feminism and gender was further exemplified by the explanation that "[D]espite many reports calling for curriculum reform in mathematics and science... the reforms suggested do not take feminist concerns into account; in fact, in the case of mathematics they tend to put added emphasis on curricular areas in which young women regularly perform less well than their male counterparts" (Damarin 1991, p. 108, cited in Burton 1995/2008, pp. 519–520). More recently, Damarin (2008) noted how in mathematics education research, "use of the term 'gender' still signals attention to women and serves as both code and container for attention to women, girls, females, and the feminine" (p. 105). Although Burton did suggest a broadening of the conception of gender when she wrote "more women (*and men*) might feel comfortable with an epistemology of mathematics described in this way [of relating to Burton's five categories]" (italics added, p. 526), but more is needed.

Limiting feminist epistemology discussions to gender is problematic at the very least in the short term. As scholarship around sex/gender develops, the complexity encourages scholars to call for a clear meaning for the word *gender*.

> Glasser and Smith (2008) call the attention of the educational research community to the persistence of "vagueness" in the meaning of the term *gender* as it is used in educational research and writing and argue for more careful and accountable uses of the term.... [W]e note that mathematics is involved in many understandings of what is meant by gender and that it is, therefore, important that mathematics education researchers take up these issues and address the use of gender specifically in mathematics education literature. (Damarin and Erchick 2010, p. 310)

When interrogating the term gender leads to discussion about a sex/gender system, masculinities and femininities, and intersexuality, how will the new meanings influence an evolving feminist epistemology?

Even if there were more clarity on the meaning of the term gender, it is not a new understanding that gender intersects with other social constructs in meaningful ways. Alcoff and Potter (1993) explained:

> Growing awareness of the many ways in which political relationships... are implicated in theories of knowledge has led to the conclusion that gender hierarchies are not the only

ones that influence the production of knowledge. ...Moreover, developments in feminist theory have demonstrated that gender as a category of analysis cannot be abstracted from a particular context while other factors are held stable; gender can never be observed as a 'pure' or solitary influence.... Thus, because gender as an abstract universal is not a useful analytical category and because research has revealed a plethora of oppressions at work in productions of knowledge, feminist epistemology is emerging as a research program with multiple dimensions. (p. 3)

Alcoff and Potter (1993) not only make the case that gender is integrally connected to its context, but their points also suggest that the need we have to define gender, as already noted, may be served well by considering the "multiple dimensions" of gender and its "complex interrelationships with other systems of identification and hierarchy" (p. 3).

Feminism *began* with work toward better lives for women and girls, and that foundation—both in terms of the work done and the development of the field—frames current perspectives. It makes sense for us to begin to think about this work as *centering* on gender and including additional dimensions. Burton acknowledged a social/cultural element to person-ness, and black feminist scholars (e.g., Collins 1989, 1991, 2008; hooks 1989, 2000a, 2000b) have theorized around the connections among race, gender, class, and sexuality, as well as the (dis)connections between feminist scholarship and black feminist thought. Damarin (2008) noted "that the plural framing of masculinities invites and requires attention to issues of race and class when considering boys and men in educational settings; this stands in sharp contrast to the absence of attention to racial, ethnic, or economic diversity in most literature on gender, coded as female, and mathematics" (p. 105). The need is clear and the time is right for enriching feminism and its epistemologies with clearer and enriched meanings of gender.

2 Mathematics/Equity/Social Justice Pedagogies

If we center our considerations of feminist epistemology on our understandings of gender, and also work toward an epistemology resulting from analysis of related social hierarchies, the work would inform a Mathematics/Equity/Social Justice Pedagogies conceptual cluster. It is important, as pointed out by Burton, that these discussions be distinctly different for mathematics. So, just as feminism may focus on gender, acknowledge other social constructs, but realize gender is a distinctive, central case, this is also the case for mathematics and equity and social justice perspectives.

In working toward equity, as Burton (1995/2008) wrote, "Re-telling mathematics, both in terms of context and person-ness, would consequently demystify and therefore seem to offer opportunities for greater inclusivity" (p. 522); and, I would say, also of both mathematical ideas and human involvement. Although she may have intended this for women and girls, her statement clearly can speak to the intricacies of gender's relationships with race, class, and sexuality, and working toward "greater inclusivity" with equity pedagogy. For example, concepts such as voice and agency, beyond feminism and gender discussions to include sexuality and race (Blackburn 2004; Curtis-Tweed 2003), work a way not only into general equity agendas, but also for the growing body of work in social justice in mathematics education. One of the challenges in teaching mathematics is determining focus in negotiating the equity and social justice/mathematics terrain (Erchick and Tyson 2008). Are we teaching mathematics from a social justice perspective or teaching social justice through mathematics? How does an epistemology of mathematics become different when the *agenda* is as much or more about equity and social justice as it is about the mathematics? Current work in equity and social justice no doubt addresses the inclusion of the learner; but is that changing mathematics and assuming so, how?

3 Instruction/Perspectives on Mathematics/Testing

Perspectives on mathematics influence instructional decisions and thus learning. When this perspective is absolutist, and the teaching perpetuates the perspective, we find the system nurturing the very perspective we propose to change. Burton (1995/2008) wrote how it is "exceedingly difficult to dismantle the beliefs which have been integral to our learning experiences of mathematics and almost impossible to construct in our imaginations alternatives to the processes which we have been taught and with which we have gained 'success' " (p. 522). With the belief system so difficult to change, and perspectives deeply engrained, one might wonder how, or if, we can make progress with a feminist epistemology. Given Burton's five categories noted earlier in this preface, in the ways in which mathematics education (e.g., Boaler and Staples 2008) presents findings aligned with these categories, we may indeed be making progress. Burton continued, though, with a suggestion that sounds today like a caution, in that if we:

> scratch a pedagogical or philosophical constructivist... underneath you are likely to expose an absolutist. In other words, it might be acceptable to negotiate a curriculum or introduce a collaborative, language-rich environment within which to make the learning of mathematics more accessible, but the mathematics itself is considered non-negotiable. (p. 522)

Which is the stronger influence in learning: the message of the pedagogy, or the message of the mathematics? Perhaps it is as much the conflicting message as it is anything to do with mathematics or culture or gender or feminism that is in the way of learning and (re)defining a feminist epistemology.

Burton (1995/2008) discusses another conflict where:

> Adopting an objectivist stance within mathematical philosophy means accepting that mathematical 'truths' exist and the purpose of education is to convey them into the heads of the learners. This leads to conflicts both in the understanding of what constitutes knowing, and of how that knowing is to be achieved through didactic situations. (p. 520)

Burton noted that the "philosophical myth" of objectivity in knowing "continues to exercise enormous power over mathematics both in curricular and in methodological terms" (p. 520). And the power remains at our current reading of her work.

What does this mean for teaching methodology? In the US, many teachers enter programs with limited mathematics understanding and with objectivist perspectives developed in their own learning experiences. How can they then be expected to incorporate a feminist epistemology of mathematics in their practice, one that rejects an objectivist stance? To teach mathematics constructively, one needs to learn mathematics constructively (Schifter and Fosnot 1993). I have no reason to believe it would be any different for teaching a feminist epistemology of mathematics.

Add to this relationship between mathematics and teaching the drive for accountability, national core standards, testing for school students, and more focus on accountability at the college level (Alexander 2000; Huisman and Currie 2004), and one cannot help but wonder what the impact will be. It is not testing, but its impact on pedagogy that is the problem. If the absolutist is under the skin of even the constructivists, the pressure to teach for the test as opposed to teaching for learning is likely to break the skin, with absolutism becoming the guiding influence.

Burton (1995/2008) ended her paper with the hope that her work questioned the nature of mathematics, resulting in a mathematics that is more open to "the experience and the influence of members of as many different communities as possible, thereby... not only enriching the individuals but also the discipline" (p. 527). It seems with the kind of research coming out of mathematics education internationally, we are as a field moving in the right direction to enrich the discipline by nurturing learners whose epistemology of mathematics is less and less objectivist. It is a start for learners as young children, as young adults, as mature adults. We need not be intimidated by the magnitude of the work still to be done. What we must do is continue to work systemically at every level of teaching and learning mathematics. Five-year old children, teachers of all levels, adolescent students making decisions about college, or graduate students choosing careers, will need to make informed decisions about participating in a discipline that is as Burton envisioned. The systemic view also allows for research and theorizing around related clusters of concepts, as I have noted here, so that a broader, more integrated focus can inform feminist epistemologies in mathematics.

References

Alcoff, L., & Potter, E. (Eds.) (1993). *Feminist epistemologies*. New York: Routledge, Chapman and Hall, Inc.

Alexander, F. K. (2000). The changing face of accountability: Monitoring and assessing institutional performance in higher education. *Higher Education, 48*(4), 411–431.

Blackburn, M. V. (2004). Understanding agency beyond school-sanctioned activities. *Theory Into Practice, 43*(2), 103–110.

Boaler, J., & Staples, M. (2008). Creating mathematical futures through an equitable teaching approach: The case of Railside school. *Teachers College Record, 110*(3), 608–645.

Burton, L. (1995/2008). Moving towards a feminist epistemology of mathematics. *ZDM The International Journal on Mathematics Education, 40*, 519–528.

Collins, P. H. (1989). The social construction of black feminist thought. *Signs, 14*(4), 745–773.

Collins, P. H. (1991). *Black feminist thought: Knowledge, consciousness, and the politics of empowerment*. New York: Routledge.

Collins, P. H. (2008). *Black feminist thought: Knowledge, consciousness, and the politics of empowerment* (2nd ed.). New York: Routledge.

Curtis-Tweed, P. (2003). Experiences of African–American empowerment: A Jamesian perspective on agency. *Journal of Moral Education, 32*(4), 397–409.

Damarin, S. (1991). Rethinking science and mathematics curriculum and instruction: Feminist perspectives in the computer era. *Journal of Education, 173*(1), 107–123

Damarin, S. (2008). Toward thinking feminism and mathematics together. *Signs: Journal of Women in Culture and Society, 34*(1), 101–123.

Damarin, S. K., & Erchick, D. B. (2010). Toward clarifying the meanings of "gender" in mathematics education. *Journal for Research in Mathematics Education, 41*(4), 310–323.

Erchick, D. B., & Tyson, C. (2008). Teaching social justice by the numbers: Toward a pedagogy of social justice in the MCP. In D. B. Erchick (Chair), *Keeping the focus on social justice in a K-6 mathematics coaching program.* Symposium presented at the National Council of Teachers of Mathematics annual research pre-session, Salt Lake City, Utah.

Glasser, H. M., & Smith, J. P., III. (2008). On the vague meaning of "gender" in educational research: The problem, its sources, and recommendations for practice. *Educational Researcher, 37*, 343–350.

hooks, b. (1989). *Talking back.* Boston: South End Press.

hooks, b. (2000a). *Feminist theory: From margin to center.* London: Pluto Press.

hooks, b. (2000b). *Feminism is for everybody.* Cambridge: South End Press.

Huisman, J., & Currie, J. (2004). Accountability in higher education: Bridge over troubled waters? *Higher Education, 48*(4), 529–551.

Schifter, D., & Fosnot, C. T. (1993). *Reconstructing mathematics education: Stories of teachers meeting the challenge of reform.* New York: Teachers College Press.

Moving Towards a Feminist Epistemology of Mathematics

Leone Burton

Abstract There is, now, an extensive critical literature on gender and the nature of science, three aspects of which, philosophy, pedagogy and epistemology, seem to be pertinent to a discussion of gender and mathematics. Although untangling the inter-relationships between these three is no simple matter, they make effective starting points in order to ask similar questions of mathematics to those asked by our colleagues in science. In the process of asking such questions, a major difference between the empirical approach of the sciences, and the analytic nature of mathematics, is exposed and leads towards the definition of a new epistemological position in mathematics.

1 Introduction

Received science has been criticised on three grounds from a gender perspective. The first is its reductionism and its claim to be objective and value-free (e.g. Harding 1986, 1991; Keller 1985; Rose and Rose 1980). Second, the conventional style of learning and teaching in science, its pedagogy, has been challenged. It is suggested that enquiry methods used by scientists are often intrusive and mechanistic, separating observer and observed, and reinforcing competition. Further, these methods are presented not only as 'correct' but also as the only way possible (e.g. Kelly 1987; Whyte 1985). Third, having rejected objectivity as an untenable criterion for judging science, a new scientific epistemology was required and has been derived (see Rosser 1990) by examining the connections between the discipline and those who

This is a version of a paper first presented at the ICME7 theme group of the International Organisation on Women and Mathematics Education, Quebec, 1992. Its present content owes much to discussion with and comments from members of that network. In addition, I would particularly like to thank Mary Barnes, Leonie Daws, Stephen Lerman and the anonymous reviewers for challenging and provoking re-working of the ideas.

Reprinted from Educational Studies in Mathematics (1995) 28(3), 275–291 with permission from the publisher.

This chapter is a reprint of an article published in ZDM—The International Journal on Mathematics Education (2008) 40(4), 519–528. DOI 10.1007/s11858-008-0109-9.

H. Forgasz, F. Rivera (eds.), *Towards Equity in Mathematics Education*,
Advances in Mathematics Education,
DOI 10.1007/978-3-642-27702-3_3, © Springer-Verlag Berlin Heidelberg 2012

use it, and the society within which it develops. This line of reasoning is consistent with a broad range of thinking in the sociology of science.

The old certainties about science, the old belief in its cultural uniqueness and the old landmarks of sociological interpretation have all gone (Mulkay 1981, p. vii).

Mathematics and mathematics education have been subjected to a similar challenge from within on philosophical, pedagogic and epistemological grounds. The philosophical arguments for a rejection of absolutism in mathematics have been explored elsewhere (see Ernest 1991). Lakatos (1976, 1983), Bloor (1976, 1991) and Davis and Hersh (1983) have all made similar philosophical and epistemological criticisms on those outlined in the science literature with respect to the so-called objectivity of mathematics.

Likewise, a gender critique similar to that found in science has been made of mathematical pedagogy (see, for example, Burton 1986, 1990a; Fennema and Leder 1990; Leder and Sampson 1989).

> Despite many reports calling for curriculum reform in mathematics and science... the reforms suggested do not take feminist concerns into account; in fact, in the case of mathematics they tend to put added emphasis on curricular areas in which young women regularly perform less well than their male counterparts (Damarin 1991, p. 108)

Mathematics tends to be taught with a heavy reliance upon written texts which remove its conjectural nature, presenting it as inert information, which should not be questioned. Predominant patterns of teaching focus on the individual learner and induce competition among learners. Language is pre-digested in the text, assuming that meaning is communicated and is non-negotiable. In Hull's (1985, pp. 45–50) terms this defines:

> knowledge as an object and so equates knowing, and coming to know, with its possession; it effaces the crucial distinction between the learner's subjective experience of moving towards knowledge and the objectifying of a knowledge finally achieved

Like science, therefore, mathematics is perceived by many students and some teachers as "a body of established knowledge accessible only to a few extraordinary individuals" (Rosser 1990, p. 89). Indeed, the supposed 'objectivity' of the discipline, a cause for questioning and concern by some of those within it, is often perceived by non-mathematician curriculum theorists as inevitable (see, for example, Hirst 1965, 1974 and, for a critique expanding the points being made here, Kelly 1986). But

> the processes of knowing (and so also of science) in no way resemble an impersonal achievement of detached objectivity. They are rooted throughout... in personal acts of tacit integration. They are not grounded on explicit operations of logic. Scientific inquiry is accordingly a dynamic exercise of the imagination and is rooted in commitments and beliefs about the nature of things (Polanyi and Prosch 1975, p. 63)

Adopting an objectivist stance within mathematical philosophy means accepting that mathematical 'truths' exist and the purpose of education is to convey them into the heads of the learners. This leads to conflicts both in the understanding of what constitutes knowing, and of how that knowing is to be achieved through didactic situations. For example, such conflicts can be found between the UK mathematics national curriculum, expressed in terms of a hierarchy of mathematical truth statements, and the support documentation given to teachers which includes such relativistic statements as:

> Each person's "map" of the network and of the pathways connecting different mathematical ideas is different, thus people understand mathematics in different ways (Non-statutory Guidance to the Mathematics National Curriculum, para. 2.1, p. C1)

> The teacher's job is to organise and provide the sorts of experiences which enable pupils to construct and develop their own understanding of mathematics, rather than simply communicate the ways in which they themselves understand the subject (Non-statutory Guidance to the Mathematics National Curriculum, para. 2.2, p. C2).

Although "the ideal of pure objectivity in knowing and in science has been shown to be a myth" (Polanyi and Prosch 1975, p. 63), it is a philosophical myth which continues to exercise enormous power over mathematics both in curricular and in methodological terms.

Proposed as an alternative, social constructivism is a philosophical position which emphasises the interaction between individuals, society and knowledge out of which mathematical meaning is created. It has profound implications for pedagogy. Classroom behaviours, forms of organisation, and roles, rights and responsibilities have to be re-thought in a classroom which places the learner, rather than the knowledge, at the centre. Epistemology, too, requires reconsideration from a theoretical position of knowledge as given, as absolute, to a theory of knowledge, or perhaps better, of knowing, as subjectively contextualised and within which meaning is negotiated.

With respect to science, Rosser stated:

> If science is socially constructed, then attracting a more heterogeneous group of scientists would result in different questions being asked, approaches and experimental subjects used, and theories and conclusions drawn from the data (Rosser 1990, p. 33)

How might including many of those currently outside the mainstream of mathematical development influence its conjectures, its methods of enquiry and the interpretation of its results? In turn, how might any changes which resulted from a philosophical shift, affect the pedagogy and epistemology of the discipline? In particular, what are the epistemological questions which are sharpened by bringing a feminist critique to bear on the discipline of mathematics? These issues are the focus of this paper.

2 Adopting a Cultural View of Mathematics

In writing about mathematics, Harding drew attention to its cultural dependency:

> Physics and chemistry, mathematics and logic, bear the fingerprints of their distinctive cultural creators no less than do anthropology and history. A maximally objective science, natural or social, will be one that includes a self-conscious and critical examination of the relationship between the social experience of its creators and kinds of cognitive structures favoured in this inquiry... whatever the moral and political values and interests responsible for selecting problems, theories, methods, and interpretations of research, they reappear at the other end of the inquiry as the moral and political universe that science projects as natural and thereby helps to legitimate (Harding 1986, pp. 250–251).

Despite the stance taken by many mathematicians on the objectivity and value-free nature of the discipline, Bloor convincingly argued from a historical perspective that it is possible to conceive of alternative mathematics differently derived at different periods:

> Seeing how people decide what is inside or outside mathematics is part of the problem confronting the sociology of knowledge, and the alternative ways of doing this constitute alternative conceptions of mathematics. The boundary (between mathematics and meta-mathematics) cannot just be taken for granted in the way that the critics do. One of the reasons why there appears to be no alternative to our mathematics is because we routinely disallow it. We push the possibility aside, rendering it invisible or defining it as error or as nonmathematics (Bloor, 1991, pp. 179–80).

More recently, Harding (1991) has pushed the argument further to locate mathematics firmly within its interpretative context despite its overtly comparable formalistic expression:

> There can appear to be no social values in results of research that are expressed in formal symbols; however, formalisation does not guarantee the absence of social values. For one thing, historians have argued that the history of mathematics and logic is not merely an external history about who discovered what when. They claim that the general social interests and preoccupations of a culture can appear in the forms of quantification and logic that its mathematics uses. Distinguished mathematicians have concluded that the ultimate test of the adequacy of mathematics is a pragmatic one: does it work to do what it was intended to do? Moreover, formal statements require interpretation in order to be meaningful... Without decisions about their referents and meanings, they cannot be used to make predictions, for example, or to stimulate future research (Harding 1991, p. 84)

In his discussion of mathematical epistemology, Joseph (1993) drew attention to two major philosophical presuppositions which underlie Western (European) mathematics. These are, first, that mathematics is a body of absolute truths which are, second, argued (or 'proved') within a formal, deductive system. However, he pointed out that dependence upon an axiomatically deduced system of proof was a late nineteenth century development which was pre-dated by 'proofs' closer in style to that of non-European mathematicians:

> The Indian (or, for that matter, the Chinese) epistemological position on the nature of mathematics is very different. The aim is not to build up an imposing edifice on a few self-evident axioms but to validate a result by any method, including visual demonstration (Joseph 1993, p. 9)

Joseph further stated that:

> None of the major schools of Western thought... gives a satisfactory account of what indeed is the nature of objects (such as numbers) and how they are related to (other) objects in everyday life. It is an arguable point... that the Indian view of such objects... may lead to some interesting insights on the nature of mathematical knowledge and its validation. Irrespective of whether this point can be substantiated or not, a more balanced discussion of different epistemological approaches to mathematics would be invaluable. However, a different insight into some of the foundational aspects of the subject is hindered by the prevalence of the Eurocentric view on the historical development of mathematics (pp. 11–12)

Joseph is criticising the dominance of a Eurocentric (and male) mathematical hegemony which has created a judgemental situation within the discipline whereby, for example, deciding what constitutes powerful mathematics, or when a proof proves and what form a rigorous argument takes, is dictated and reinforced by those in influential positions. How often do we hear statements, often made about a geometric proof, dismissing it as 'merely a demonstration' or the suggestion that computer-assisted proofs are not quite as 'good' as those developed without a computer? How frequently are students encouraged to believe that the mathematico-scientific and technological development of the West has been made independently of a systematic knowledge and resource exploitation of the rest of the world? The colonisation of mathematics has been so successful that the history of their own mathematical culture and its contribution to knowledge is often unknown to students in Africa, Asia and Latin America. Such bias is increasingly under attack (see, for example, Needham 1959; Zaslavsky 1973; van Sertima 1986; Joseph 1991; Nelson et al. 1993) as researchers uncover the richness and power of mathematical and scientific development in the non-European world which has been obscured by the rewriting of history from a European perspective. If the body of knowledge known as mathematics can be shown to have been derived in a manner which excluded non-Europeans and their mathematical knowledge, why not conjecture that the perceived male-ness of mathematics is equally an artefact of its production and its producers?

Since I am arguing that mathematics is socio-cultural in nature, the conditions under which it is produced are factors in determining the products. "Important" mathematical areas are identified, value is accorded to some results rather than others, decisions are taken on what should or should not be published in a society determined by power relationships, one of which is gender. Mathematical products can then be seen as the outcome of the influence of a particular 'reading' of events at a given time/place. Such readings are referred to by Sal Restivo as

> stories about commercial revolutions and mathematical activity, as in Japan, or about the 'mathematics of survival' that is a universal feature of the ancient civilizations. And they can be stories about how conflict and social change shape and reflect mathematical developments (1992, p. 20)

In a Plenary lecture given at the 1994 American Educational Research Association Conference, Jerome Bruner (1994) pointed out that explanation as causal is a post-nineteenth century phenomenon. A longer history can be found for interpretation whose objective is understanding and not explanation. He made out a case for understanding being viewed as both contextualising and systematising and he advocated a route to contextualising in a disciplined way through narrative. From this perspective, codified mathematics can be viewed as reified narrative and it no longer seems so absurd to ask how different narratives, or stories, might constitute alternative mathematics (in the plural). Mathematics as a particular form of story about the world *feels*, to me, very different from mathematics as a powerful explanation or tool. Re-telling mathematics, both in terms of context and person-ness, would consequently demystify and therefore seem to offer opportunities for greater inclusivity.

3 Knowing Science and Mathematics

The feminist literature on the philosophy of science I find very valuable for the clarity with which it has sharpened the critical debate on the nature of knowledge in science and how that knowledge is derived. However, it is noticeable that the content criticisms of science are rooted in the empirical disciplines. For example, female primatologists such as Goodall (1971), Fossey (1983) and Hrdy (1986) challenged conceptions of interactive behaviour by refusing to accept the prevailing (male) views on dominance and hierarchy in sexual selection. Keller (1985) highlighted McClintock's approach to her study of maize as a symbiotic relationship between the plant and its environment which was distinctively different to the more usual 'objective' investigation undertaken by botanists. Carson (1962) is frequently cited for her early work on ecology and the broad view that she took about the environmental effects of pesticides. In all these cases, the results of the science were different from what had, formerly, been expected because different questions were asked about what was being observed and different methods were used to make the observations.

However, criticisms of, for example, nuclear physics are more likely to focus upon the social effects of the science, rather than the science itself (see, for example, Easlea 1983). This is not to diminish the importance of developing models of scientific use and abuse which criticise the purposes, products and implications of scientific developments. But, as with mathematics, it is difficult to confront the abstractions which are the substance and tools of the discipline and the methods used in their derivation, especially where these are analytic and non-observational in order to ask what differences a female perspective would make to them.

In what ways might the questions, or the styles of enquiry or the mathematical products differ if mathematics were to be admitted to be a socio-cultural construct? Part of the difficulty in responding to this question resides in the highly successful socialisation experiences through which we all go in order to achieve success at mathematics. It is exceedingly difficult to dismantle the beliefs which have been integral to our learning experiences of mathematics and almost impossible to construct in our imaginations alternatives to the processes which we have been taught and with which we have gained 'success'. Hence, scratch a pedagogical or philosophical constructivist and underneath you are likely to expose an absolutist. In other words, it might be acceptable to negotiate a curriculum or introduce a collaborative, language-rich environment within which to make the learning of mathematics more accessible, but the mathematics itself is considered non-negotiable. However, to be consistent in our critiques, we cannot avoid addressing the nature of knowing mathematics along with the philosophy and pedagogy of the discipline.

Knowing mathematics, and science, has traditionally required entry into a community of knowers who accord the status of 'objective', "in some sense eternal and independent of the flux of history and culture" (Restivo 1992, p. 3), to the knowledge items as well as to the means by which these items are derived. However, objectivity is a variable; it is a function of the generality of social interests. Aesthetic and truth motives exist in the realm of ideas, but they are grounded in individual and social interests ranging from making one's way in the world (literally, surviving) to exercising control over natural and cultural environments (Restivo 1992, p. 135).

A consequence of this is that

a mathematical object… like a hammer or a screwdriver, is conceived, constructed, and put to use through a social process of collective representation and collective elaboration (Restivo 1992, p. 137)

If we are to argue for a different conception of mathematical knowing from that traditionally accepted, we must address the meaning which is to be understood by 'objectivity' since the truth status accorded to mathematical objects underpins the pervading epistemology. Criticising scientific 'objectivity' along similar lines, Harding (1991) called for an epistemology of the sciences which requires a more robust standard than that currently in use. This would include the critical examination, "*within scientific research*" (original italics, p. 146)

of historical values and interests that may be so shared within the scientific community, so invested in by the very constitution of this or that field of study, that they will not show up as a cultural bias between experimenters or between research communities (Harding 1991, p. 147).

And she further noted that:

the difficulty of providing (such) an analysis in physics or chemistry (and, I would add, mathematics) does not signify that the question is an absurd one for knowledge-seeking in general, or that there are no reasonable answers for those sciences too (p. 157)

Rosser, in her book *Female-Friendly Science* (1990), used women's experience of knowing and doing science to draw out differences from what she called the conventional androcentric approaches. Amongst many of the inclusionary methods she listed are:

- expanding the kinds of observations beyond those traditionally carried out;
- increasing the numbers of observations and remaining longer in the observational stage of the scientific method;
- accepting the personal experience of women as a valid component of experimental observation;
- being more likely to undertake research which explores questions of social concern than those likely to have applications of direct benefit to the military;
- working within research areas formerly considered unworthy of investigation because of links to devalued areas;
- formulating hypotheses which focus on gender as an integral part;
- defining investigations holistically.

This list, useful as it is for science, does not generalise easily to mathematics although the links to the history, philosophy and pedagogy of mathematics are more obvious. But help appears to be at hand.

4 Being a Mathematician

In *The Emperor's (sic) New Mind*, Penrose (1990) arguing from the powerful position of a research mathematician at the top of his profession, claimed that the mathematician's "consciousness" is a necessary ingredient to the comprehension of the mathematics. He said:

> We must 'see' the truth of a mathematical argument to be convinced of its validity. This 'seeing' is the very essence of consciousness. It must be present whenever (original emphasis) we directly perceive mathematical truth. When we convince ourselves of the validity of Gödel's theorem we not only 'see' it, but by so doing we reveal the very non-algorithmic nature of the 'seeing' process itself (p. 541)

Elsewhere in his book, and in contradiction to the above, Penrose supported a Platonic approach to mathematics in that he propounded a discovery, rather than an invented, perspective on the discipline. That is, mathematics is out there waiting to be uncovered rather than within the head (and possibly the heart?) of the mathematician. And yet, Penrose himself admitted that 'seeing' the validity of a mathematical argument must be a personal experience and one which, it seems reasonable to me to assert, can be assumed to differ between individuals. By arguing that 'seeing' is non-algorithmic, Penrose permitted the personalisation of the process. He reinforced this with the statement:

> There seem to be many different ways in which different people think and even in which different mathematicians think about their mathematics (Penrose 1990, p. 552)

However, for me, far from accepting that the outcomes of mathematical thinking are discovered mathematical 'truths', the inevitable conclusion of his statement is that there are potentially many different mathematics.

The contradiction would appear to lie in a different perspective on the mathematician than on mathematics itself. Penrose viewed a mathematical statement, once articulated, as being absolute, that is either right or wrong, and its status verifiable by any interested party. But he said, in the

> conveying of mathematics, one is *not* simply communicating *facts*. For a string of (contingent) facts to be communicated from one person to another, it is necessary that the facts be carefully enunciated by the first, and that the second should take them in individually... the *factual* content is small. Mathematical statements are necessary truths... and even if the first mathematician's statement represents merely a groping for such a necessary truth, it will be that truth itself which gets conveyed to the second mathematician... The second's mental images may differ in detail from those of the first, and their verbal description may differ, but the relevant mathematical idea will have passed between them (original emphasis, p. 553).

Despite the assumed personal nature of the communication and the expectation of differences in human images and descriptions, there is an assumption that the 'mathematics', the essential 'truth' of the statement, can and will be the same for all. This is repeatedly refuted by the message of many of the anecdotes which are recounted by and about mathematicians. For example, how is it possible to interpret the kind of intuitive insights which Penrose himself, and other mathematicians such as Poincaré, Hadamard, Thom, claim to have had and which have led to their finding particular, personal resolutions of a mathematical problem? Given that it is reasonable to expect that any one problem might be amenable to a number of different routes for solution, an individual is likely to fall on the one which matches her or his experience, approach, preferences, possibly making the mathematical outcome different from that which would be offered by another individual. Of course, once articulated, the internal consistency of the mathematical argument is claimed to be verifiable. However, the most recent attempt to prove of Fermat's Last Theorem provided an example of the unverifiability, by most mathematicians, of the claims being made and, consequently, both the potential non-uniqueness and fragility of their status. And, even if the internal consistency is substantiated, this does not additionally encompass any objective status or any implication of uniqueness, it seems to me. The social context within which the mathematics is placed does, however, offer one explanation for apparent uniqueness, or at least convergence of 'solutions', given that it describes and constrains the 'possible'. Thus, a piece of mathematics is both contributory to, and defined by, the context within which it is derived.

A belief in the world of mathematical concepts existing independently of those who develop or work with them is attached to embracing the 'objective' truths of mathematics. An image of 'variable' truth, that is degrees of correctness, or solutions responsive to different conditions, is unacceptable to many within the discipline despite the support from the history of mathematics that understandings change over time as the foci and the current state of knowledge change. The social context of a mathematical statement, the impact upon it of the interests, drives and needs of the person deriving and then communicating it, are dismissed by many mathematicians as inappropriate to the product. Thus, the distinction is made between the person who is working at the mathematics, and the mathematics itself. But I believe that Penrose failed to sustain this distinction particularly in his discussion of intuition, insight and the aesthetic qualities of mathematical thinking. He underlined person-ness by reiterating an argument (see, for example, Thom 1973, pp. 202–206) that:

> the importance of aesthetic criteria applies not only to the instantaneous judgements of inspiration, but also to the much more frequent judgements that we make all the time in mathematical (or scientific) work. Rigorous argument is usually the *last* step! (original emphasis, Penrose, p. 545)

In drawing a close analogy between mathematical thought and intuition and inspiration in the arts, Penrose added:

> The globality of inspirational thought is particularly remarkable in Mozart's quotation (from Hadamard 1945) 'It does not come to me successively... but in its entirety' and also in Poincaré's 'I did not verify the idea; I should not have had time' (p. 347)

Any de-personalisation of the mathematical process and reification of the product pushes mathematics back into the absolutist position by objectivising the 'truths'. However, accepting a mathematics which is not absolute, is culturally defined and influenced by individual and social differences is not only of great interest to those who have argued for an inclusive mathematics but challenges the discipline epistemologically as well as philosophically and pedagogically.

It does not seem untimely to suggest a theory of knowing that draws attention to the knower's responsibility for what the knower constructs (von Glasersfeld 1990, p. 28)

Once we re-focus from knowing *that* a particular mathematical outcome exists to knowing *why* that outcome is likely under particular circumstances, we are distinguishing between the 'objective' knowledge of the outcome and the 'subjective' knowing which underlies how to achieve that outcome. This begins to be familiar as the old debate between product and process. However, by attempting to construct a theory of knowing, I am moving past the false dichotomy of product/process which polarised the how and the what, towards a re-conceptualisation and integration of the how with the what. The value to pedagogues of such an approach is obvious. As teachers, we can recognise when learners mimic a piece of mathematical behaviour rather than acquiring it as their own. The articulation of an epistemological position on knowing mathematics which is predicated on mathematical enquiry, rather than receptivity, challenges teacher behaviour. Rather than demanding evidence of the acquisition of mathematical objects by students, it assumes that mathematical behaviours and the changes in behaviour that might signify learning are products of, and responsive to, the community within which the learning is situated. Recounting different narratives, speculating about their similarities and differences, querying their derivations and applications, denies 'objectivity' and reinstates the person and the community in the mathematics. Such re-consideration of the characteristics of science and mathematics has underpinned much of the feminist work in the philosophy of science already referenced and is exemplified in the work of Damarin. She presented a table of generalised descriptors

not as a definitive description of feminist science, but rather as defining a tentative framework for examining whether and how the teaching of science might be made more consistent with feminist conceptions of science (Damarin 1991, p. 112).

As stated above, the philosophical challenge, while not necessarily acceptable to a large number of mathematicians, has been well formulated. (Reference has already been made to the work of Bloor, Davis and Hersh, Harding, Lakatos and Restivo.) Gadamer (1975) added his argument that:

all human understanding is contextual, perspectival, prejudiced, that is hermeneutic (and) fundamentally challenges the conception of science as it has been articulated since the Enlightenment (cited in Hekman 1990, p. 107)

as did Fee (1981) to:

attack the objectivity that is part of the 'mythology' of science... (and)... re-admit the human subject into the production of scientific knowledge (also cited in Hekman 1990, p. 130)

Much of the pedagogic challenge is focused on the dysfunctional nature of the continuum between an absolutist philosophy of mathematics and a transmissive pedagogy and the poverty of the product/process distinction:

On the one hand, authors and publishers produce textbooks that do not have to be read before doing the exercises; on the other hand, teachers acquiesce by agreeing that this is the way mathematics ought to be taught... the real importance lies not in the students' ability to conceptualize, but rather in their ability to compute. Teachers tend to underscore this by their rapt attention to correctness, completeness, and procedure. Students comply with the grand scheme by establishing as their local goal the correct completion of a given assignment and as their global goal receiving their desired grade in the course. For most, once it's over, it's over (Gopen and Smith 1990, p. 5).

Compare this with a student-centred problem-solving approach:

An instructor should promote and encourage the development for each individual within his/her class of a repertoire of powerful mathematical constructions for posing, constructing, exploring, solving and justifying mathematical problems and concepts and should seek to develop in students the capacity to reflect on and evaluate the quality of their constructions (Confrey 1990, p. 112)

Researchers have argued that creative mathematicians are more likely to develop by encountering and learning mathematics in a classroom climate which supports individuals within social groupings; that the negotiation of meaning both within the group and between the group and conventional social understandings needs to be encouraged (see for example Davis et al. 1990). These philosophical and pedagogical critiques, in my view, would be strengthened by the focus, structure and consistency which are gained from an epistemological stance, that is, a formulation of the nature of knowing mathematics.

5 The Epistemological Challenge

I believe that we can discern the outline of an epistemological challenge to mathematics which, potentially, incorporates approaches consistent with and familiar to broader constituencies than European, middle-class males. These approaches are inclusive, rather than exclusive, accessible rather than mystifying, encompassing of as wide a range of styles of understanding and doing mathematics as possible rather than reducible to those styles currently validated by the powerful. I am claiming that knowing, in mathematics, cannot be differentiated from the knower even though the knowns ultimately become public property and are subject to public interrogation within the mathematical community. Knowing, however:

involves encouraging rebellious spirits to blossom with free rein to the imagination, preserving a certain nimbleness of mind while affording it the means of being creative. The 'training' procedures, as we conceive them and ordinarily practice them, hardly lend themselves, one must admit, to that kind of enticement, since they more often emphasize the transmission of acquired knowledge and apprenticeship in proven methods. And considering that those procedures resemble an obstacle course where the competition is tighter and tighter, this hardly encourages departing from the beaten path (Flato 1992, p. 75).

I am speculating that five categories, drawn from the work already cited and consistent with the above critique, might distinguish the ways in which (creative) mathematicians come to know mathematics and that, in their choice of mathematical areas to pursue, more women (and men) might feel comfortable with an epistemology of mathematics described in this way. The assumption is that such an epistemology would displace dualisms such as the relativist/absolutist dichotomy and expectations of a value-free mathematics with an hermeneutic and pluralist approach. It would open the way towards an inclusive perspective on mathematics by challenging our understanding of what constitutes knowing in mathematics.

I propose defining knowing in mathematics in relation to the following five categories derived from the reading reviewed above in the philosophical, pedagogical and feminist literature:

- its person-and cultural/social-relatedness;
- the aesthetics of mathematical thinking it invokes;
- its nurturing of intuition and insight;
- its recognition and celebration of different approaches particularly in styles of thinking;
- the globality of its applications.

Knowing mathematics would, under this definition, be a function of who is claiming to know, related to which community, how that knowing is presented, what explanations are given for how that knowing was achieved, and the connections demonstrated between it and other knowings (applications). What evidence we have, usually sited in the learning and assessing of school mathematics, suggests that inviting students to define and describe their knowing in mathematics in these ways does have gender implications (for example, see Burton 1990b; Forgasz 1994; Stobart et al. 1992).

The similarities with Rosser's (1990) and Damarin's (1991) lists of the differences between male-and female-friendly science are encouraging. For example, both refer to the expansion of the kinds of observations carried out, the recognition of and concern for personal responsibility and the consequences of actions. I have listed a valuing of intuition and insight and the recognition and celebration of different approaches. Globality, or in both Rosser's and Damarin's terms holism, is a feature. A need to accept the personal experience of women as a valid component of experimental observations is acknowledged where I have pointed to person-relatedness which is important to knowing mathematics. Hekman's (1990) analysis of the relationship between gender and post-modernism was also supportive of this approach both in drawing out the similarities in argument between feminists and post-modernists as well as pointing out the pervading influence of absolutism in affecting these stances. In Rose's words:

> A feminist epistemology... transcends dichotomies, insists on the scientific validity of the subjective, on the need to unite cognitive and affective domains; it emphasises holism, harmony, and complexity rather than reductionism, domination and linearity (1986, p. 72)

The next step is to open a dialogue with practising mathematicians with a view to discussing the appropriateness of my description to their understanding of the nature of knowing in mathematics. This would be done in a style which would be rich in ethnographic data, encouraging the expression of feelings, aesthetics, intuitions, and insights. It would also attempt to challenge the effects of socialisation into the mathematical culture in order to untangle differences from cultural similarities. Outcomes which are supportive of the suggested epistemological framework, especially where these emphasise impact on gender inclusivity, would provide a strong argument in favour of a re-perception and re-presentation of mathematics. The resulting narrative would have an internal consistency which should please all mathematicians.

Such anecdotal approaches as have already been made confirm the validity of the five categories in describing how mathematicians come to know. Arguments in favour of humanising and demystifying the mathematics curriculum in schools have long been made with an implication that such attempts change perceptions of mathematics, and subsequent performance by formerly under-represented groups. However, these suggestions are rarely connected to epistemological frameworks of the discipline more frequently relating either to constructivist philosophy or empowering pedagogy. And Suzanne Damarin criticises curriculum reformers for their

> reliance on the models of expertise and information processing, which are popular in current research on the cognitive bases of teaching and learning of science and mathematics (and) appear to be diametrically opposed to first-order implications of feminist pedagogical research (Damarin 1991, p. 108).

If the nature of knowing mathematics were to be confirmed as matching the description given in this paper, the scientism and technocentrism which dominate much thinking in and about mathematics, and constrain many mathematics classrooms, would no longer be sustainable. Mathematics could then be re-perceived as humane, responsive, negotiable and creative. One expected product of such a change would be in the constituency of learners who were attracted to study mathematics but I would also expect changes in the perception of what is mathematics and of how mathematics is studied and learned. That such a possibility, in schools, is not outside the realms of possibility is suggested in Boaler (1993). We can also learn from experiences in other disciplines. English, for example, attracts predominantly female constituencies of learners at the undergraduate level, many of whom have been successful in developing academic careers.

> English was constructed as a liberal humanist discipline which demanded personal and thoughtful response... The most important characteristic of English, in the view of students and staff, is its individualism: the possibility of holding different views from other people (Thomas 1990, p. 173).

Providing a new epistemological context would enable the questioning of what mathematics is taught, how it is learned and assessed within a consistent treatment.

> By adopting an epistemological view of mathematical knowledge that stresses change, development, and its social foundations generally, and by consciously relating this to the curriculum process, the result would be to make the subject more open in its nature and more easily accessible (Nickson 1992, p. 131).

My aim in attempting this work is to question the nature of the discipline in such a way that the result of such questioning is to open mathematics to the experience and the influence of members of as many different communities as possible, thereby, I hope, not only enriching the individuals but also the discipline.

References

Bloor, D. (1976, 1991). *Knowledge and social imagery* (2nd ed.). London: University of Chicago Press.

Boaler, J. (1993). Encouraging the transfer of 'school' mathematics to the 'real world' through the integration of process and content, context and culture. *Educational Studies in Mathematics, 25*(4), 341–373. doi:10.1007/BF01273906.

Bruner, J. (1994). *The humanly and interpretively possible.* Plenary address given to the AERA Annual Meeting, New Orleans.

Burton, L. (Ed.) (1986). *Girls into maths can go.* London: Cassell.

Burton, L. (1990a). *Gender and mathematics: An international perspective.* London: Cassell.

Burton, L. (1990b). Passing through the mathematical critical filter–implications for students, courses and institutions. *Journal of Access Studies, 5*(1), 5–17.

Carson, R. (1962). *Silent spring.* New York: Fawcett Press.

Confrey, J. (1990). What constructivism implies for teaching. In R. B. Davis, C. A. Maher, & N. Noddings (Eds.), *Constructivist view on the teaching and learning of mathematics, Journal for Research in Mathematics Education, Monograph* No. 4. Reston: NCTM.

Damarin, S. (1991). Rethinking science and mathematics curriculum and instruction: Feminist perspectives in the computer era. *Journal of Education, 173*(1), 107–123.

Davis, E., & Hersh, R. (1983). *The mathematical experience.* Harmondsworth: Penguin.

Davis, R. B., Maher, C. A., & Noddings, N. (Eds.) (1990). *Constructivist views on the teaching and learning of mathematics, Journal for Research in Mathematics Education, Monograph* No. 4. Reston: NCTM.

Department of Education and Science (1989). *Mathematics in the national curriculum.* London: HMSO.

Easlea, B. (1983). *Fathering the unthinkable: Masculinity scientists and the nuclear arms race.* London: Pluto Press.

Ernest, P. (1991). *The philosophy of mathematics education.* Basing-stoke: Falmer Press.

Fee, E. (1981). A feminist critique of scientific objectivity. *Science for the People, 14*, 30–33.

Fennema, E., & Leder, G. (1990). *Mathematics and gender.* New York: Teachers College Press.

Flato, M. (1992). *The power of mathematics.* London: McGraw-Hill.

Forgasz, H. (1994). *Society and gender equity in mathematics education.* Geelong: Deakin University Press.

Fossey, D. (1983). *Gorillas in the mist.* Boston: Houghton Mifflin.

Gadamer, H.-G. (1975). *Truth and method.* New York: Continuum.

Goodall, J. (1971). *In the shadow of man.* Boston: Houghton Mifflin.

Gopen, G. D., & Smith, D. A. (1990). What's an assignment like you doing in a course like this? Writing to learn mathematics. *The College Mathematics Journal, 21*(1), 2–19. doi:10.2307/2686716.

Hadamard, J. (1945). *The psychology of invention in the mathematical field.* Princeton: Princeton University Press.

Harding, S. (1986). *The science question in feminism.* Milton Keynes: Open University Press.

Harding, S. (1991). *Whose science? Whose knowledge?* Milton Keynes: Open University Press.

Hekman, S. (1990). *Gender and knowledge—elements of a postmodern feminism.* Boston: Northeastern University Press.

Hirst, P. H. (1965). Liberal education and the nature of knowledge. In R. D. Archambault (Ed.), *Philosophical analysis and education.* London: Routledge and Kegan Paul.

Hirst, P. H. (1974). *Knowledge and the curriculum.* London: Routledge & Kegan Paul.

Hrdy, S. B. (1986). Empathy, polyandry, and the myth of the coy female. In R. Bleier (Ed.), *Feminist approaches to science.* Oxford: Pergamon Press.

Hull, R. (1985). *The language gap.* London: Methuen.

Joseph, G. G. (1993). A rationale for a multicultural approach to mathematics. In D. Nelson, G. G. Joseph & J. Williams (Eds.), *Multicultural mathematics.* Oxford: Oxford University Press.

Joseph, G. G. (1991). *The crest of the peacock*. London: Tauris & Co.

Keller, E. F. (1983). *A feeling for the organism: The life and work of Barbara McClintock*. New York: W.H. Freeman.

Keller, E. F. (1985). *Reflections on gender and science*. New Haven: Yale University Press.

Kelly, A. (Ed.). (1987). *Science for girls?* Milton Keynes: Open University Press.

Kelly, A. V. (1986). *Knowledge and curriculum planning*. London: Harper and Row.

Lakatos, I. (1976). *Proofs and refutations*. Cambridge: Cambridge University Press.

Lakatos, I. (1983). *Mathematics*. Cambridge: Science and Epistemology, Cambridge University Press.

Leder, G., & Sampson, S.N. (Eds.). (1989). *Educating girls*. Sydney: Allen & Unwin.

Mulkay, M. (1981). Preface. In A. Brannigan (Ed.), *The social basis of scientific discoveries*. Cambridge: Cambridge University Press.

Needham, J. (1959). *Science and civilisation in China*. Cambridge: Cambridge University Press.

Nelson, D., Joseph, G. G., & Williams, J. (1993). *Multicultural mathematics*. Oxford: Oxford University Press.

Nickson, M. (1992). Towards a multi-cultural mathematics curriculum. In M. Nickson & S. Lerman (Eds.), *The social context of mathematics education: Theory and practice*. London: South Bank Press.

Penrose, R. (1990). *The emperor's new mind*. London: Vintage.

Polanyi, M., & Prosch, H. (1975) *Meaning*. London: University of Chicago Press.

Restivo, S. (1992). *Mathematics in society and history*. Dordrecht: Kluwer.

Rose, H. (1986). Beyond masculinist realities: A feminist epistemology for the sciences. In R. Bleier (Ed.), *Feminist approaches to sciences*. Oxford: Pergamon.

Rose, H., & Rose, S. (1980). The myth of the neutrality of science. In R. Arditti, P. Brennan & S. Cavrak (Eds.), *Science and liberation*. Boston: South End Press.

Rosser, S. V. (1990). *Female-friendly science*. New York: Pergamon.

Stobart, G., Elwood, J., & Quinlan, M. (1992). Gender bias in examinations: How equal are the opportunities? *British Educational Research Journal, 18*(3), 261–276. doi:10.1080/0141192920180304.

Thom, R. (1973). Modem mathematics: Does it exist? In A. G. Howson (Ed.), *Developments in mathematics education*. Cambridge: Cambridge University Press.

Thomas, K. (1990). *Gender and subject in higher education*. Buckingham: SRHE & Open University Press.

van Sertima, I. (Ed.) (1986). *Blacks in science: Ancient and modern*. New Brunswick: Transaction.

von Glasersfeld, E. (1990). An exposition of constructivism: Why some like it radical. In R. B. Davis, C. A. Maher & N. Noddings (Eds.), *Constructivist views on the teaching and learning of mathematics journal for research in mathematics education*, Monograph No. 4. Reston: NCTM.

Whyte, J. (1985). *Girl friendly schooling*. London: Methuen.

Zaslavsky, C. (1973). *Africa counts*. New York: Lawrence Hill Books.

Preface to "Equity in Mathematics Education: Unions and Intersections of Feminist and Social Justice Literature"

Laura Jacobsen

> The evidence is clear that groups of girls continue to be disadvantaged by previously identified dimensions—the mathematics curriculum, classroom practices, and assessment practices—as well as with respect to newer aspects including the adoption of technology into mathematics pedagogy. (Leder and Forgasz 2008, p. 518)

In recent years, research examining results from tests of mathematics performance has generally documented small or reduced gaps between male and female students (e.g., Else-Quest et al. 2010; Hyde et al. 2008; McGraw et al. 2006). However, this is not always the case. For example, research exploring the gender gap among high-achieving high school students, using data from the American Mathematics Competitions, has indicated that the gender gap widens dramatically at very high percentiles and that the highest-achieving girls are concentrated in a very small set of elite schools (Ellison and Swanson 2010). A study by Fryer and Levitt (2010) documented the emergence of a substantial mathematics gender gap in the early years of schooling in the United States, documenting that girls lose more than two-tenths of a standard deviation relative to boys over the first six years of school, across every strata of society.

Analyses of disparities in students' mathematical achievement on standardized tests have sometimes been referred to as "gap gazing" and have been used to assess whether schools offer equitable opportunities by race, gender, income, or other demographics (Lubienski 2008). Such analyses have sometimes shaped public opinion and informed educational policy. However, despite their routine use as tools for educational analyses, as Gutiérrez (2008) indicated, an achievement gap focus offers only a static picture of inequities and constructs some groups as "failures" relative to other groups. Describing the popular use of the "racial achievement gap" in mainstream mathematics education research, policy, and practice, Martin (2009) similarly indicated that attempts to compare population groups serve to sort students

L. Jacobsen (✉)
Radford University, Radford, VA, USA
e-mail: ljacobsen@radford.edu

Formerly Laura Spielman.

H. Forgasz, F. Rivera (eds.), *Towards Equity in Mathematics Education*,
Advances in Mathematics Education,
DOI 10.1007/978-3-642-27702-3_4, © Springer-Verlag Berlin Heidelberg 2012

into those who know, and those who do not know, mathematics and treats race as a causal variable for mathematics achievement. Students' racialized forms of experience are not addressed through mainstream mathematics education efforts.

Similar arguments can be made for why achievement gap studies in mathematics education by gender are likewise inadequate. Leder and Forgasz (2008) suggested that such studies sometimes overlook explanations for why scores vary little by gender. For example, Leder and Forgasz explained how by focusing heavily on lower cognitive level items, tests of mathematics performance may provide a limited database for analysis. The United Nations Educational, Scientific and Cultural Organization (UNESCO) reviewed international approaches to analysis of gender-disaggregated data (Huyer and Westholm 2007). Huyer and Westholm indicated that women's participation in science, engineering and technology contributed to increasing enrollments of up to 20–25% in many countries, but cautioned that the numbers have declined by approximately 10–15% since 2000. Given the serious impact this may have internationally, especially on developing countries (Huyer and Westholm 2007), clearly a broadened database beyond test scores and continued attention are needed to promote gender equity, as well as other forms of equity, in mathematics education.

In my original publication, "Equity in mathematics education: Unions and intersections of feminist and social justice literature" (Spielman 2008), I situated gender equity in mathematics within broader international equity concerns. I described connections between gender inequities and differences also by social class and race/ethnicity, and I also related gender inequities in mathematics to overarching inequities with respect to economic participation, economic opportunity, political empowerment, educational attainment, and health and well-being (Lopez-Claros and Zahidi 2006). I suggested the need for mathematics education to transform its largely decontextualized and impersonal traditions by fostering new goals supporting equity and serving broader public interest goals both in the U.S. and abroad. I argued that to secure a transformative and sustainable impact on equity, mathematics needs to be treated as an integral component of a larger system producing educated citizens, making the case that mathematics as a discipline must be reconstructed, beginning with the question, "What is the purpose of schooling?" Longstanding disciplinary boundaries must be torn down in favor of a newly constituted mathematics that rewrites traditional practices in the field in favor of a new "common sense" that integrates the learning of mathematics with social critique and with community relations and actions.

Since publication of Spielman (2008), the focus of my research efforts has been in the development of the *Mathematics Education in the Public Interest* (MEPI) project[1] with my colleague, Jean Mistele. The MEPI project aims to put into practice the theoretical framework described in that publication, and has three overarching goals: (1) To support equity and social justice in mathematics education, (2) To

---

[1]MEPI is supported by the National Science Foundation, award number 0837467. Any opinions, findings, and conclusions or recommendations expressed in this paper are those of the author(s) and do not necessarily reflect the views of the National Science Foundation.

diversify student interest and participation in mathematics, and (3) To broaden and enrich the ways mathematics is viewed as a discipline. Gender equity is included as one aspect within the project's emphasis on equity.

Thus far, the focus of the project has been in elementary and middle school teacher education. Our primary project objective involves curriculum and course development for a new junior-level course for preservice teachers (PSTs), *Elementary and Middle Grades Mathematics for Social Analysis* (*Math for Social Analysis*), offered through the Department of Mathematics and Statistics at Radford University. Our secondary objective involves conducting and disseminating related mathematics education research examining PSTs' learning and experiences in *Math for Social Analysis*. We have also begun work to create an online MEPI activities and resources center. Because nearly all PSTs in the elementary education program, in particular, are female, the MEPI project has primarily involved female PSTs as participants. In the remainder of this preface I describe progress on our two main objectives.

1 Introduction to Math for Social Analysis

Math for Social Analysis focuses foremost on mathematics content, but dualistically aims: (1) to have the learning of interdisciplinary applications and important social issues to strengthen and reinforce mathematical understandings and (2) to have the mathematical activities and projects to reinforce and strengthen understandings of the interdisciplinary applications and important social issues. Interdisciplinary content and choice of social issues vary depending on current events, student interest, and text selection. Interdisciplinary content always includes many diverse relationships to science, social studies, and language arts. Course content includes discussions of political, social, and economic challenges and implications associated with understanding and even changing the world using mathematics.

Curriculum units from the course have included: (1) A global economy unit addressing topics such as poverty and the distribution of wealth and sweatshop labor, (2) An environmental unit addressing topics such as water and energy conservation, mountaintop removal coal mining in Appalachia, and rain forest depletion, and (3) A gender equity unit addressing topics such as salary distribution in the workforce, international differences in mathematics participation and achievement by gender and other demographics, gendered media images, and stereotyped roles and expectations for women. We aim to help PSTs learn about and conduct research on these issues while simultaneously deepening their understanding of the mathematics needed for critical analysis of these issues. As instructors, we determine some mathematical concepts to be addressed within these social issues, and we also offer opportunities for PSTs to pose and try to answer their own questions about these issues, using mathematics. Although we face a constant struggle, and a wide range of successes and failures in doing so, we maintain a goal of creating and responding to questions for which citizens would care about the answer, rather than creating textbook-style, artificial word problems.

2 Introduction to MEPI Research

We have documented some of the issues and struggles PSTs face as they learn MEPI ideas in *Math for Social Analysis*. For example, when PSTs created their own MEPI lesson plans as part of a semester-long project, they struggled to balance emphases on mathematics, reform-based pedagogy, and social issues (Jacobsen and Mistele 2010). Some PSTs' projects *applied* various mathematical concepts, but failed to *teach* any of the mathematics being used; others had interesting projects, but used primarily traditional and/or non-challenging mathematics. Some PSTs focused on the mathematics, but gave only cursory attention to the meaningful aspects of the social issues; others created lessons with disconnect, or artificial connections, between social issues and the mathematics.

Despite these many challenges, on the whole, we have been very pleased with student feedback from *Math for Social Analysis*, especially feedback from students who described previous struggles with or dislike for mathematics. Many PSTs enter *Math for Social Analysis* describing high levels of mathematics anxiety. In Mistele and Spielman (2009), we communicated the course's overall success at reducing PSTs' mathematics anxiety and generating positive attitudes toward mathematics and toward mathematics teaching by increasing the utility of mathematics, redirecting attention away from anxiety, and building confidence to teach. PSTs universally label this course as their first extended experience with learning mathematics in connection with multiple meaningful real-world applications and social issues. Class discussions about the social issue units are often rich with enthusiasm, critique, and questions. For example, in analyzing salary data across many professions during the gender equity unit described above, PSTs have raised and debated questions regarding why teachers' salaries are low. They have discussed the role of societal gender inequity in producing teaching as a second-tier status profession in many nations.

One course strength has been that PSTs develop greater understanding of how to integrate social issues and mathematics, and most describe their interest in incorporating social issues into their mathematics classrooms in the future. For example, one PST, Allie, explained:

> When I become a teacher I will be able to look back on this course and remember how easy it is to teach about a social issue and also about a math lesson. Teaching a social issue will engage the student to be interested in the math lesson.... My students will not only be excellent math whizzes but they will also know what is going on with the world. I will encourage them to use the newspaper, book, Internet, and current issues in their learning experience. I hope to also show other teachers at my future elementary school the different ways they can help and teach their students. I now hope that all teachers will use current issues to teach the children.

Based on MEPI research, we have also reported on survey results providing evidence that PSTs' views about mathematics and about mathematics teaching changed over the semester (Spielman 2009). They came to see mathematics as increasingly useful for understanding and engaging with important issues and increasingly connected to home and community experiences. Further, we concluded that interwoven mechanisms supporting PSTs' engagement with and reframing of mathematics included: (1) Learning the relevance of mathematics to something they care about;

(2) Developing interest in mathematical applications and in supporting their future students' interest and learning in mathematics; and (3) Shifting their perspectives on mathematics by changing prior assumptions and instructional goals. As PSTs increasingly saw mathematics as relevant and important in social issues, they developed new teaching goals to help students integrate mathematics with other subjects and the world outside of school. Finally, they developed a new sense of agency to create mathematical learning opportunities that students will find interesting and relevant.

The majority of our PSTs describe support for having the mathematics situated within relevant social, economic and political contexts—such as in using mathematics to analyze gender equity. However, some PSTs do not support the course's emphases on mathematics in connection with social issues, and reject the new ideas outright or make comments such as, "I think that social issues are mostly for a political science and social science class to take care of" (Annie). This may be due largely to their prior experiences in mathematics classrooms having traditional classroom norms that do not include a public interest emphasis. In addition, some students— particularly those who hope to teach in the lower elementary grades—raise concerns about the potential relevance of social issues in their future classrooms. Others question whether an interdisciplinary approach to teaching mathematics is feasible given time constraints for lesson planning and implementation. They additionally have concerns regarding whether teaching mathematics using social issues is beneficial to students' academic performance on mandatory state standardized tests in mathematics. We continually adapt course design to address the concerns raised by our students.

3 Theoretical and Practical Challenges

We can represent the various practical challenges that we, as mathematics teacher educators, face in course design and implementation as three types of balancing acts, balancing emphases on: (1) Mathematics content and social critique; (2) Mathematics content and pedagogy; and (3) In-class and out-of-class experiences and learning.

Balancing Mathematics Content and Social Critique The greatest challenge we face as teacher educators is finding appropriate balance in depth and breadth of emphases on helping PSTs learn mathematics while helping them critically engage with social issues. Nolan (2009) explained how the "statistics and figures" content approach that takes mathematics as usual and appends social justice concepts will not be enough (p. 207). *Math for Social Analysis* addresses all five NCTM content strands, yet number and operations and statistical concepts generally align more easily with social issues than others. We struggle to balance time spent examining mathematical concepts for deep understanding with time spent engaging PSTs with social issues that does not trivialize either or produce artificial connections between them.

Balancing Mathematics Content and Pedagogy We dualistically aim to balance PSTs' personal experiences with learning rich mathematics within the context of social issues, while: (1) critiquing their own understanding of mathematics as a discipline in relation to the democratic purposes of schooling and (2) building new pedagogical content knowledge embedded in MEPI principles. Finding an appropriate balance has been a constant challenge.

Balancing In-Class and Out-of-Class Experiences and Learning Our challenges extend outside of the classroom as we strive to balance the in-class and out-of-class experience through mathematics service learning for PSTs in several after school programs, which requires that PSTs create and implement MEPI activities with children. We strive to balance the needs of our service learning partner and our course goals. Our service learning partner takes responsibility for the scheduling and assessment of our PSTs, and their feedback becomes a component of the PSTs' course grade. PSTs complete reflective journals periodically throughout the semester, and they submit their activity write-ups and reflections to their course instructor.

4 Final Thoughts

The theoretical outline for a new mathematics education posited in Spielman (2008) has been, and will continue to be, challenging but invigorating to explore and research at a practical level in the mathematics classroom. In the MEPI project, we have aimed to address gender equity issues as a component of broader equity and social justice considerations both in the U.S. and globally. We have aimed to invite participation, interest, and success in mathematics through changing the face of the mathematics itself and what it means to do mathematics in the classroom. Thus far, we have primarily done this by incorporating relevant and contextualized social issue mathematics units for PSTs into the *Math for Social Analysis* classroom and by having small group and whole class discussions about equity and social justice issues—including gender equity issues. Clearly, this is just a starting point for the project, and we hope to use our research to continue to develop our course and project in ways that further, and even more effectively, promote equity both in and outside of mathematics education.

It can be intimidating for teacher educators and researchers, including myself, to communicate practical attempts to apply theoretical principles, given that our idealistic notions for what is needed in mathematics education is extraordinarily challenging to visualize and actualize in the classroom. No attempts will be perfect, but each of us engaged in this sort of work must continue weaving and meshing theory and practice so that we may transform mathematics education to better support equity and social justice. We must further make clear that such work is not simply about making the curriculum more relevant, but instead, doing something important with our curriculum toward supporting democratic citizenship that the individual disciplines themselves would not be able to accomplish.

References

Ellison, G. & Swanson, A. (2010). The gender gap in secondary school mathematics at high achievement levels: Evidence from the American Mathematics Competitions. *Journal of Economic Perspectives, 24*(2), 109–128.

Else-Quest, N., Hyde, J. S., & Linn, M. C. (2010). Cross-national patterns of gender differences in mathematics: A meta-analysis. *Psychological Bulletin, 136*(1), 103–127.

Fryer, R. G., & Levitt, S. D. (2010). An empirical analysis of the gender gap in mathematics. *American Economic Journal, 2*(2), 210–240.

Gates, P., & Jorgensen, R. (2009). Foregrounding social justice in mathematics teacher education. *Journal of Mathematics Teacher Education, 12*(3), 161–170.

Gutiérrez, R. (2008). A "gap-gazing" fetish in mathematics education? Problematizing research on the achievement gap. *Journal for Research in Mathematics Education, 39*, 357–364.

Huyer, S., & Westholm, G. (2007). *Gender indicators in science, engineering and technology.* Paris: United Nations Educational, Scientific and Cultural Organization.

Hyde, J. S., Lindberg, S. M., Linn, M. C., Ellis, A. B., & Williams, C. C. (2008). Gender similarities characterize math performance. *Science, 321*, 494–495.

Jacobsen, L. J., & Mistele, J. M. (2010). Please don't do "Connect the dots": Mathematics lessons with social issues. *Science Education and Civic Engagement: An International Journal, 2*(2), 9–15.

Leder, G., & Forgasz, H. (2008). Mathematics education: New perspectives on gender. *ZDM—The International Journal on Mathematics Education, 40*(5), 513–518.

Lopez-Claros, A., & Zahidi, S. (2006). *Women's empowerment: Measuring the global gender gap.* World Economic Forum, Geneva, Switzerland.

Lubienski, S. T. (2008). On "gap gazing" in mathematics education: The need for gaps analyses. *Journal for Research in Mathematics Education, 39*(4), 350–356.

Martin, D. (2009). Little black boys and little black girls: How do mathematics education and research treat them? In S. L. Swars, D. W. Stinson, & S. Lemons-Smith (Eds.), *Proceedings of the 31st annual meeting of the North American chapter of the International Group for the Psychology of Mathematics Education* (pp. 22–41). Atlanta: Georgia State University.

McGraw, R., Lubienski, S. T., & Strutchens, M. E. (2006). A closer look at gender in NAEP mathematics achievement and affect data: Intersections with achievement, race/ethnicity, and socioeconomic status. *Journal for Research in Mathematics Education, 37*(2), 129–150.

Mistele, J., & Spielman, L. J. (2009). The impact of "Math for social analysis" on math anxiety in elementary preservice teachers. *Academic Exchange Quarterly, 12*(4), 93–97.

Nolan, K. (2009). Mathematics in and through social justice: Another misunderstood marriage. *Journal of Mathematics Teacher Education, 12*(3), 205–216.

Spielman, L. J. (2008). Equity in mathematics education: Unions and intersections of feminist and social justice literature. *ZDM—The International Journal on Mathematics Education, 40*(5), 647–657.

Spielman, L. J. (2009). Mathematics education in the public interest: Preservice teachers' engagement with and reframing of mathematics. In S. L. Swars, D. W. Stinson, & S. Lemons-Smith (Eds.), *Proceedings of the 31st annual meeting of the North American Chapter for the International Group for the Psychology of Mathematics Education* (pp. 408–415). Atlanta: Georgia State University.

Equity in Mathematics Education: Unions and Intersections of Feminist and Social Justice Literature

Laura Jacobsen Spielman

Abstract Traditional models of gender equity incorporating deficit frameworks and creating norms based on male experiences have been challenged by models emphasizing the social construction of gender and positing that women may come to know things in different ways from men. This paper draws on the latter form of feminist theory while treating gender equity in mathematics as intimately interconnected with equity issues by social class and ethnicity. I integrate feminist and social justice literature in mathematics education and argue that to secure a transformative, sustainable impact on equity, we must treat mathematics as an integral component of a larger system producing educated citizens. I argue the need for a mathematics education with tri-fold support for mathematical literacy, critical literacy, and community literacy. Respectively, emphases are on mathematics, social critique, and community relations and actions. Currently, the integration of these three literacies is extremely limited in mathematics.

Keywords Equity · Social Justice · Gender

Abbreviations

BLS	Bureau of Labor Statistics
NAEP	National Assessment of Educational Progress
NCTM	National Council of Teachers of Mathematics
NSF	National Science Foundation
SES	Socioeconomic status
WEF	World Economic Forum

L.J. Spielman (✉)
Radford University, Radford, VA, USA
e-mail: lspielman@radford.edu
url: http://www.radford.edu/lspielman

This chapter is a reprint of an article published in ZDM—The International Journal on Mathematics Education (2008) 40(4), 647–657. DOI 10.1007/s11858-008-0113-0.

H. Forgasz, F. Rivera (eds.), *Towards Equity in Mathematics Education*,
Advances in Mathematics Education,
DOI 10.1007/978-3-642-27702-3_5, © Springer-Verlag Berlin Heidelberg 2012

1 Introduction

"The public purpose of schooling in this nation—a nation founded on principles of freedom, justice and measures of happiness for all—is to educate the citizenry in understanding and abiding by these principles." (Goodlad 2004, pp. 14–17)

The question, "What is the purpose of schooling?," almost invariably leads to responses involving support for democratic ideals, or civic and global responsibility, or fairness and justice, or building moral character, or developing the whole person, or gaining knowledge useful for real life and for economic opportunity. Schools ideally help us to develop multiple forms of literacy—for personal growth, community livelihood, the workforce, and responsible citizenship. Yet historically and internationally, school mathematics is isolated from other subjects and from students' lives and interests outside of school. Mathematics is treated as independent from important social, political, and economic issues facing our communities and our world. This paper constructs unions and intersections between feminist and social justice literature in mathematics education. I make the case that to secure a transformative, sustainable impact on mathematics equity we must begin by promoting excellence and a challenging, inclusive mathematics curriculum for all students. But further, we must treat mathematics as an integral component of a larger social system producing educated citizens. Students must learn the relevance of mathematics for understanding and even remedying local, national, and global injustices—both gender related and otherwise.

We can rethink common sense in mathematics education when we revisit education's purpose and put principles such as "freedom, justice and measures of happiness for all" (Goodlad 2004, pp. 14–17) at the forefront. Drawing on Eric Gutstein's (2006) recommendations for mathematics curriculum, I use a broad range of existing research to argue the need for theoretical and practical frameworks in mathematics education that offer tri-fold support for students' mathematical literacy, critical literacy, and community literacy (described later). Respectively, emphases are on mathematics, social critique, and community relations and actions. Currently, the integration of these three literacies is extremely limited in mathematics.

This paper situates gender equity in mathematics within broader equity considerations both in the U.S., where I live, and across the globe. I first address gender equity in mathematics, then broaden the discussion to situate gender equity in mathematics within broader equity concerns, then shift and broaden discussion again by addressing the global picture of gender equity. This leads to specific suggestions for a new construction of mathematics education premised on these contextualizations, conceived as *mathematics education in the public interest*.

2 Gender Differences in Mathematics

Gender differences in mathematics achievement and attitudes are complex and changing over time. Since the early 1900s, a wide range of international research has reported gender inequities in mathematics, favoring males (e.g., Keitel 1998;

Leder 1992; Sriraman 2007). By about 1980, consistent findings from research on gender and mathematics showed that fewer females than males elected to study mathematics when it was optional in secondary schools; young women indicated mathematics as not particularly useful and tended to express less confidence in their ability to learn mathematics; mathematics was stereotyped as a male domain; and societal influences tended to suggest mathematical learning as not particularly appropriate for girls (Damarin 1995; Fennema 2000; Leder 1992).

Gender differences in mathematics have narrowed substantively over time and by some measures have even been eliminated. For example, in the U.S., sex differences in high school mathematics coursetaking no longer exist, and females earn approximately half (47%) of bachelor's degrees in mathematics [National Science Foundation (NSF), 2006]. In recent years, gender attention has increasingly shifted to include concerns about boys' educational needs and the problems boys experience (e.g., Forgasz and Leder 2001; Lingard et al. 2002; Weiner et al. 1997).

Interventions designed to address gender differences in mathematics have been classified by program type, time, school calendar, targeted population, education focus, strategy, elements of success, creation of new organizations, and teaching and learning strategies (Leder et al. 1996). Traditional models of gender equity created student norms based on male experiences and treated differences in personal characteristics as deficits on the part of females, or as the "girl problem" in mathematics (Campbell 1995). Deficit model assumptions that male behavior and outcomes are the desirable norm to which women should strive have underpinned previous policy, much research, and even many intervention programs (Forgasz and Leder 2001).

These traditional deficit models explaining gender differences in mathematics have been challenged by feminist models emphasizing the social construction of gender and positing that women may come to know things in different ways from men (e.g., Baxter Magolda 1992; Becker 1995; Belenky et al. 1986/1997; Brew 2001; Damarin 1995; Kaiser and Rogers 1995). Related interventions have sometimes attributed gender differences to pedagogical or assessment practices that are discriminatory toward females; others have explained inequities in mathematics in terms of the design, content, and structure of the mathematics curriculum (Goodell and Parker 2001).

3 Situating Gender Equity Within Broader Equity Concerns

Mathematics is "often regarded as the most abstract subject removed from responsibilities of cultural or social awareness" (Boaler and Staples 2005, p. 32). Mathematics has further been associated with the stratification of learning opportunities across race, ethnicity, gender, and social class. For example, Stinson (2004) referred to mathematics as "(re)produc[ing] and regulat[ing] racial, ethnic, gender, and class divisions" (p. 9). Among other nations, in the U.S., disparities and unequal access to mathematics course taking, achievement, and career fields remain a serious problem (Oakes et al. 2004; Secada 1992). Secada summarized:

Along a broad range of indicators, from initial achievement in mathematics and course taking to postsecondary degrees and later careers in mathematics-related fields, disparities can be found between Whites and Asian Americans on the one hand and African Americans, Hispanics, and American Indians on the other; between males and females; among groups based on their English language proficiency; and among groups based on social class. (p. 623)

Referring to U.S. education as a whole, a 2007 report by the Jack Kent Cooke Foundation showed lower income students disproportionately fall out of the high-achieving group during elementary and high school, rarely rise into the ranks of high achievers during those periods, and far too infrequently ever graduate from college or go on to graduate school (Wyner et al. 2007, p. 4). Sirin (2005) conducted a meta-analysis on socioeconomic status [SES] and academic achievement in journal articles published between 1990 and 2000; results showed a medium to strong SES-achievement relation. According to the long-term National Assessment of Educational Progress [NAEP] assessments, White students continue to outperform Black and Hispanic students in both reading and mathematics (National Center for Education Statistics 2007).

Gender gaps in mathematics achievement and participation have closed over time but do still exist and are interconnected with differences by social class and race/ethnicity. In Australia, Lamb (1996, 1997) suggested social class differences in mathematics participation at the senior level of schooling were substantial for both girls and boys. He described girls as less likely than boys to take advanced mathematics, but suggested girls from professional family origins experienced far less of a gender gap in mathematics participation, confidence, and interest than girls from skilled manual backgrounds.

Interestingly, research intersecting gender with socioeconomic status does not always yield consistent findings. In the U.S., analyzing relationships among achievement and mathematical content, student proficiency and percentile levels, race, and socioeconomic status, the gender performance gap favoring males is generally small but did not diminish from 1990 to 2003 (McGraw et al. 2006). McGraw et al. reported gender gaps were most consistent for White, higher socio-economic groups of students, and were non-existent for Black students. Further, female students' mathematics attitudes and self-concepts continued to be more negative than those of male students. Female students were less likely than male students to indicate they like or believe they are good at mathematics; yet females had similar views to males about their beliefs about their level of understanding of what goes on in mathematics class.

4 Global Picture of Gender Equity

The issue of gender inequity in mathematics must be considered not only in connection with inequities by SES and race/ethnicity, but also in connection with cultures and with gender inequities in and outside of education as a whole (Rogers and Kaiser 1995). The World Economic Forum (WEF) conducted a study to assess the current size of the gender gap internationally by measuring the extent to which women in 58 countries have achieved equality with men in five critical areas: economic participation, economic opportunity, political empowerment, educational attainment, and health and well-being (Lopez-Claros and Zahidi 2006). The U.S. ranked 17 among these countries with regard to gender gap rankings across these five measurable areas. The study concluded, "Even in light of heightened international awareness of gender issues, it is a disturbing reality that no country has yet managed to eliminate the gender gap" (p. 1).

Nordic countries lead the way in providing women with a quality of life almost equal to that of men (Lopez-Claros and Zahidi 2006). However, some other countries show wide variation, and some across all five dimensions. The WEF suggested *educational attainment* is "without doubt, the most fundamental prerequisite for empowering women in all spheres of society" (p. 5). Yet women represent more than two-thirds of the world's illiterate adults. According to the WEF, an obvious gender gap in education tends to appear early in most countries and on average to grow more severe with each year of education. The WEF makes a critical point that simply making literacy and education accessible to women will not be enough. To close the gender gap, the content of the curriculum and attitudes of teachers must also change so as not to reinforce prevalent stereotypes and injustices.

5 Reconstructing Mathematics: Mathematics Education in the Public Interest

"When we write a thesis or a paper, we learn that the first thing to do is to latch it on to the discipline at some point. This may be by showing how it is a problem within an existing theoretical and conceptual framework. The boundaries of inquiry are thus set within the framework of what is already established." (Smith 1974)

In 1974, sociologist Dorothy Smith questioned the taken-for-granted assumptions of traditional sociological thought—its methods, conceptual schemes, and theories. Smith began a longstanding effort to develop a sociology for women/people that takes issue with the disjunction that at times exists between women's lived experiences in the world and the theoretical schemes available to think about it. She explained:

> Our experience of the world is of one which is largely incomprehensible beyond the limits of what is known in a common sense. No amount of observation or face-to-face relations, no amount of analysis of commonsense knowledge of everyday life, will take us beyond our essential ignorance of how it is put together. (p. 13)

Smith argued that supplementing traditional male notions of sociology with components relevant to women's worlds, such as by addressing omitted and overlooked conversations, only serves to produce women's sociology as an addendum while still maintaining existing sociological thought and procedures and also extending their authority. Related arguments can be made regarding the transformation of mathematics education.

5.1 Setting New Goals in Mathematics Education

Putting aside deficit models popular in the past, current feminist perspectives generally posit the problem in mathematics is not with women's ability, but instead with mathematics as currently taught and constituted (Jacobs 1994; Kaiser and Rogers 1995). Feminist perspectives often suggest women tend to be *connected knowers* and men separate knowers (Belenky et al. 1986/1997; Becker 1995; Clinchy 1989; Gilligan 1982; Jacobs 1994). *Connected knowing* suggests knowledge is contextualized and built on personal or shared experiences; *separate knowing* suggests impersonal procedures and abstractions help establish truths.

Traditional approaches to mathematics instruction— stressing certainty, deduction, logic, argumentation, algorithms, structure, and formality—may be particularly incompatible with women's ways of learning (Becker 1995; Becker and Jacobs 2001; Jacobs and Becker 1997). To the extent girls perceive school mathematics and word problems as unrealistic or meaningless, school mathematics must not focus only on mathematical skills, but must also address problems in contexts that respect and develop self-confidence, awareness, and independence (Gellert et al. 2001). Gellert et al. argue that schools must foster students' ability to study and possibly change the local environment and to critique mathematical applications and the use to which mathematics and science are put in society.

Certain differences in reasoning and knowing patterns may exist between genders, depending on contexts (e.g., Galotti et al. 1999; Galotti et al. 2001; Knight et al. 1995). Feminist recommendations to mathematics educators typically encourage content and pedagogical change supporting *connected teaching*, where, for example, teachers and students problem solve and discover mathematics together in a supportive environment where alternate solution methods are encouraged (e.g., Becker 1995; Jacobs and Becker 1997). These recommendations have much in common with recommendations in the National Council of Teachers of Mathematics (NCTM) *Standards* (1989, 2000) guiding reform in the U.S. The *Standards* support instruction where students conjecture, test, and build mathematical arguments and also learn to value mathematics and to become confident in their ability to do mathematics.

More important than possible gendered differences in mathematics, however, is that many differences undoubtedly exist between individuals in mathematics in general. Any mathematics education aimed at reaching a broad audience must therefore accommodate the various individual, gendered, socioeconomic, racial/ethnic, cultural, and other differences, or otherwise risk excluding many learners. Given the important roles of mathematics in helping promote literacy and in opening doors to career possibilities, among other roles, exclusion of any individual or group is unacceptable. For this reason, mathematics education must necessarily transform its largely decontextualized and impersonal traditions. Goals to support equity and to diversify student interest and participation in mathematics must take center stage. These goals must also resonate with broader public interest goals to improve educational and social conditions both in the U.S. and abroad.

Fig. 1 Integrated literacies

Similar to Smith's development of a feminist sociology, we can argue that appending components of social justice (or other) theoretical or pedagogical viewpoints to certain accepted mathematics disciplinary constructions and assumptions might further sanction and privilege mainstream thought. To secure a transformative and sustainable impact on mathematics equity and bring additional new students into the world of mathematics, rather than appending "radical" concepts to the mainstream, we need to engage in "rethinking mathematics" (Gutstein and Peterson 2005). What would this new mathematics look like?

Mathematics educators and others have already done much to answer this question, paving the way for fundamental change. A range of issues contribute to inequities in mathematics education, including policy and social factors, curriculum choices and implementation, and institutional practices (Bishop and Forgasz 2007). For mathematics curriculum to help close gaps in achievement and participation— whether they are associated with gender, race/ethnicity, or social class—will require not only educating individuals with traditional mathematics knowledge, but also rewriting learning objectives to necessarily include feminist perspectives, culturally relevant content, and social justice emphases that help students understand and challenge dominant power relations.

A wide range of empirical and theoretical research introduced in the next several sections points to the urgent need for an integrated *mathematical literacy*, *critical literacy*, and *community literacy* to narrow the existing and perpetuating gaps in mathematics interest, participation, and achievement among different gender, racial/ethnic, and social class groups (see Fig. 1; Table 1 describes these forms of literacy). Respectively, these forms of literacy emphasize mathematics, social critique, and community relations and actions. Gutstein (2006) perhaps most closely identified and described the need for integrated literacies by proposing an exploratory orientation toward building mathematics curriculum with integrated components of community knowledge, critical knowledge, and classical knowledge. The 12 characteristics of the Connected, Equitable Mathematics Classroom proposed by Goodell and Parker (2001) also support similar emphases in the rethinking of mathematics.

Integrating these literacies will necessarily imply that mathematics curriculum and instruction be fundamentally redesigned with overlapping objectives that, for example: (1) incorporate feminist connected teaching approaches, (2) are more culturally responsive, (3) make use of individuals' and groups' funds of knowledge, (4) engage learners' more fully, more meaningfully, and more responsibly with their communities, and (5) explicitly aim to achieve social justice locally and globally. This may be a lot to ask of mathematics educators, but successful advocacy for positive social change requires also organizing and implementing revised structures and approaches in mathematics curriculum and instruction.

Table 1 Mathematical literacy, critical literacy, and community literacy

Mathematical literacy	*Mathematical literacy* suggests all students should be problem solvers who can communicate and reason mathematically; students should learn to value mathematics and have confidence in their ability to do mathematics. Instruction promoting mathematical literacy provides learners with cooperative opportunities for exploration, for problem solving and problem posing, and for using and justifying multiple solution methods in a supportive community of learners (cf., Becker 1995; Jacobs and Becker 1997; NCTM 1989, 2000)
Critical literacy	*Critical literacy* suggests students should learn to question "power relations, discourses, and identities in a world not yet finished, just, or humane.... [Critical literacy] connects the political and the personal, the public and the private, the global and the local, the economic and the pedagogical, for rethinking our lives and for promoting justice in place of inequity" (Shor 1999)
	Critical mathematical literacy concerns have to do with both mathematics research and practice, and include concerns for equity and social justice (Skovsmose 2004). Skovsmose suggested mathematics education can "contribute to the creation of a critical citizenship and support democratic ideals" (p. 1)
Community literacy	*Community literacy* engages students in the complex of social relations and actions to making and communicating meaning around issues of common concern throughout the community (Bishop and Bruce 2001). Premises include:
	• Individuals have and produce knowledge about their communities, including mathematical knowledge (e.g., Lave 1988; Moll and González 2004). This knowledge is integral to learning and must be valued and included in instruction
	• Teachers should incorporate culturally relevant curriculum to build on students' prior knowledge and experiences (e.g., Ladson-Billings 1995). "The goal of multicultural education is to teach students to know, to care, and to act to promote democracy in the public interest" (Banks 2006, p. 145)
	• Community service allows young people to deepen and demonstrate their learning while also becoming more civic minded (D. Hart et al. 2007; National Commission on Service Learning 2002)

Gutstein (2006) proposed building mathematics curriculum with integrated community knowledge, critical knowledge, and classical knowledge. MEPI's descriptions of literacy forms were informed by these classifications.

In making use of these overlapping ideas and reconstructing mathematics by starting from education's broader objectives and purposes to produce an educated citizenship, the need for mathematical literacy, critical literacy, and community literacy becomes "common sense." These forms of literacy overlap substantively. For example, Steen's (1997) description of quantitative literacy (including reading and reasoning, writing and calculating, problem solving and technology, practices and knowledge, and procedures and contexts), and his description of the economic and social consequences of innumeracy, contains certain elements of all three forms of literacy. I only distinguish these literacies to help clarify the aims of a mathematics education established to serve the public interest. Following justification of emphases on each of mathematical literacy, critical literacy, and community literacy, I describe how I view this combination as a useful merger of feminist and social justice perspectives.

5.2 Why Mathematical Literacy?

Regarding student learning, research has yielded largely positive support for reforms incorporating connected mathematics teaching and relatedly, NCTM *Standards*-based (1989, 2000) reforms. For example, evidence exists that connected teaching approaches can improve both success and attitudes in mathematics among young women (Becker 1996; Buerk 1996; Morrow 1996). Looking across genders, in an extensive, 3-year comparative study of two schools in England, Boaler (1998) suggested that students who receive project-based instruction learn more, and different mathematics than students receiving traditional skills-based instruction. Relatively consistent evidence also exists that students using reform-based curricula perform equally well on tests of mathematical skills and procedures as comparison students using traditional curricula, and perform better on tests involving mathematical concepts and problem solving (Schoenfeld 2002; Senk and Thompson 2003). Schoenfeld further explained, "Reform appears to work when it is implemented as part of a coherent systemic effort in which curriculum, assessment, and professional development are aligned. Not only do many more students do well, but the racial performance gap diminishes substantially" (p. 17). Also, both male students and female students in reform-based school programs in the U.S. outperformed their counterparts in traditional programs; and for female students, all performance differences by program were statistically significant (Riordan and Noyce 2001).

All students can learn, and must be supported to learn, challenging mathematics. Bob Moses, activist and founder of the *Algebra Project*, further argued mathematical literacy to be a civil right (Moses and Cobb 2001). Moses and Cobb argued mathematics education's role in the ongoing struggle for citizenship and equality for the poor and for people of color. They suggested as a floor for all middle school students that they be ready for the college preparatory sequence in high school, and for all high school students, that they be ready to engage in college curricula in mathematics and science.

Though many different conceptualizations of mathematics exist, including the one promoted in this paper, regardless of the definition used, access to challenging mathematics and emphases on problem solving are indeed civil rights. It is essential that all students develop mathematical literacy. In the U.S., the Bureau of Labor Statistics (BLS) projected differential growth of the labor force from 2002 to 2012, with much of the difference attributable to strong growth in mathematics and computer-science related occupations—occupations where women, African Americans, Hispanics, and other populations remain underrepresented (BLS 2004; NSF 2006). Outside of the workplace, the need for a strong understanding of mathematics is equally important. For example, Steen (1997) argued that in today's society, a strong tendency exists to reduce complex information to numbers, with these numbers also helping to shape public policy. He suggested citizens lacking strong quantitative reasoning skills are made increasingly vulnerable by the quantification of public policy issues.

5.3 Why Critical Literacy?

Goals to support equity in mathematics education are very important. Recommendations for how to achieve equity goals almost always include requirements for setting high expectations and providing strong support for all students (e.g., Moses and Cobb 2001; NCTM 2000). In the U.S., despite many strengths, reform documents such as the NCTM *Standards* (1989, 2000) do not go far enough. Gutstein (2006) indicated the *Standards* embody a relatively narrow perspective on equity, discussing equity in terms of opportunity to learn, but not critiquing societal inequities behind the lack of those opportunities for many segments of the population both in the U.S. and abroad. Apple (1992) similarly explained:

> One searches in vain among the specifics of what teachers should know for a substantive sense of social criticism and for a more detailed understanding of the complex and contradictory roles that mathematical knowledge may play in an unequal society. (p. 425)

Based on my dissertation research in one elementary teacher education program in the U.S., I argued that *Standards*-based teaching philosophies sometimes have limited, place-specific relevance in schools (Spielman 2006). Specifically, I argued the limited relevance unacceptably tended to favor middle-class White children and to marginalize urban or diverse schools and classrooms, or schools having more limited resources, as viable places to engage in teacher education program-recommended practices for good teaching.

A critical mathematical literacy is needed to help students—also citizens—to clarify issues, to understand the structure of society, and to justify or refute opinions (cf., Frankenstein 1989). Critical mathematical literacy increases learners' capacity to understand and also challenge oppressive social structures and power relations that perpetuate over time and across the globe. Paolo Freire's (1970/2004) work rejecting a class-based society and multiple forms of oppression provides guidance to views on critical literacy. Freire advocated problem-posing education, suggesting:

> In problem-posing education, people develop their power to perceive critically *the way they exist* in the world *with which* and *in which* they find themselves; they come to see the world not as a static reality, but as a reality in process, in transformation. (p. 83)

In a world where educational inequities and other inequities exist and persist, the treatment of school mathematics as abstract, as independent of students' lived experiences, and as independent of moral and social obligations is short-sighted. We can do better.

In recent years, an increasing number of mathematics educators have begun to ground mathematical investigations in meaningful personal and social contexts. A small group of teachers and researchers, primarily in the U.S., have begun to document students' experiences and learning from this process, as well as their own experiences and learning. Turner and Strawhun (2005) described New York City middle school students' experiences with mathematically investigating overcrowding at their school, concluding, "Not only did opportunities to engage in responsive action support students' sense of themselves as people who can and do make a difference, but using mathematics as a tool to support their actions challenged students' view of the discipline" (p. 86).

Any educator seeking equity and social justice must consider how mathematics is not only a tool to produce literate citizens who understand the discipline and its meanings and applications in sociocultural contexts, but also a powerful tool to identify and rectify injustices across the globe. As Gutstein and Peterson (2005) suggest, "Math has the power to help us understand and potentially change the world" (p. 5). Teaching in a middle school classroom in a diverse Chicago school, Gutstein's class included mathematical studies of the distribution of the world's wealth, possible racism in housing data and mortgage loans, and random drug testing (pp. 117–120). Based on his research, Gutstein (2007) suggested, "Students learned mathematics and began to develop sociopolitical awareness and see themselves as possible actors in society through using mathematics to understand social injustices" (p. 420).

Students must experience opportunities to be part of the change needed in this world by experiencing first-hand the many ways that mathematical knowledge makes us more informed, and more powerful, citizens. Particularly given disparities in educational and social opportunities and conditions across gender, race/ethnicity, and social class, an emphasis on critical literacy in the mathematics classroom is sorely needed. Examinations of how mathematical applications can be deliberately and explicitly used to challenge gender inequities such as those identified by the WEF (Lopez-Claros and Zahidi 2006), among other local, national, and global inequities, are likewise sorely needed.

5.4 Why Community Literacy?

People have and produce valuable knowledge about their communities and their lives. Studies on everyday cognition, in-school and out-of-school mathematics, and ethnomathematics are among those examining connections, and sometimes conflicts, between the school world and everyday and cultural practices (e.g., Civil 2002; D'Ambrosio 2006; Lave 1988; Nunes et al. 1993). A number of educators have developed teaching innovations that build on children's and their families' backgrounds and experiences. For example, Civil (2007) describes how in the Funds of Knowledge for Teaching project (González 1995; Moll 1992; Moll et al. 1992), efforts are made to build on cultural aspects of students' communities and to implement such innovations with an eye on the mathematics. A basic conjecture is that when educators view all learners as creators of knowledge and tap into their personal and community "funds of knowledge," this helps learners to produce important connections between school and non-school aspects of their lives (Moll and González 2004). Currently, mathematics curriculum too often has little personal or cultural relevance to children's lives outside of school. Gellert et al. (2001) explained the detrimental impact this can have:

> The culturally alienating effect of mathematics education in school is extremely destructive in nonindustrialized countries that have imported mathematics and science curricula.... People in developing countries may adopt, uncritically, Western thinking in mathematics

and science and thereby abandon some of their cultural identity, or they may come to regard themselves as lacking in mathematical ability. (p. 63)

The need and transformative potential of a mathematics curriculum that connects with children's lived experiences and draws upon their existing personal and community-based knowledge and cultural practices has begun to be documented for its significance. In Lim's (2004) study of girls' experiences in learning school mathematics, the authoritative and competitive culture of the mathematics classroom was found to be a primary source of pervasive anxiety or self-alienation among participants. Lim suggested the teaching authority in mathematics classrooms rarely listens to or respects the voices of learners themselves; she offered that this classroom structure can threaten girls' feelings of self-worth and can undetermine their rights as individuals and their freedom to learn. Ladson-Billings (1997) argued the existence of a correspondence between the nature of school mathematics teaching and middle class norms. She explained that given how middle class culture demands "efficiency, consensus, abstraction, and rationality," the traditional teaching of mathematics with emphasis on "repetition; drill; convergent, right-answer thinking; and predictability" (p. 669) may be most compatible with experiences and understandings of one segment of society—most notably, White middle-class male students. Ladson-Billings suggested that different forms of cultural expression are neither reinforced nor represented in school mathematics. This must change.

Service learning in mathematics education may also be important. The National Commission on Service-Learning defined service-learning as "a teaching method that combines meaningful service to the community with curriculum-based learning" (2002, p. 3) and described the inextricable linkage between individuals' well being and the well being of their local, national, and worldwide communities. In a study of community service among high school students, using the National Educational Longitudinal Study database, both voluntary *and* school-required community service were strong predictors of adult voting and volunteering (D. Hart et al. 2007). Another study compared undergraduate preservice teachers participating in a literacy tutoring service-learning experience with preservice teachers engaged in self-selected and independently directed tutoring sessions. Service-learning was found to positively influence student academic achievement (S. Hart and King 2007). Hadlock's (2005) edited text further illustrated numerous ways that mathematics students and communities have benefited from service learning.

At Salome Ureña Middle Academies, situated in a low-income, working-class neighborhood of New York City, and with a primary population of first-or second-generation immigrants from the Dominican Republic, sixth-grade students learned mathematics by teaching 4- and 5-year olds and planning for their needs (Clinkscales and Zaslavsky 1997). This approach was based on a view that community service empowers students to be productive members of their community. Clinkscales and Zaslavsky explained that middle school students must be actively engaged to build a knowledge base, and that they must understand how academic experiences affect the quality of daily life. They reported that the sixth-graders developed confidence in presenting their ideas to the public, used mathematical terms with increasing ease, and came to understand the connections among mathematics, the humanities, and the real world. Further, students gained experiences working with different age groups and with people in the community of diverse racial/ethnic backgrounds.

5.5 *Unions and Intersections of Feminist and Social Justice Literature*

In this paper, I have described many specific studies in feminist and social justice literature; still needed is a summary of how the proposal for integrated literacies represents the union and intersection of these knowledge bases. To begin, feminist recommendations for content and pedagogical change supporting connected teaching excel at promoting a mathematical literacy likely to support the learning of all students—female and male alike. Breaking from long-standing traditions in mathematics education, feminist connected teaching pedagogies stress student engagement, opportunities for firsthand experience, teacher/student dialogue, and a supportive learning environment (Becker and Jacobs 2001). These types of goals appropriately alter the confines of mathematics as a discipline. Another great strength of current feminist literature in mathematics education is that it stresses the significant role of shared experiences and cultural environments in shaping how we come to know things. This emphasis promotes certain components of community literacy, as the term is used in this paper.

Under-examined in feminist literature and in international mathematics education literature on the whole are critical literacy approaches to mathematics curriculum and instruction (Keitel 1998; Valero 2001). Explicit attempts in mathematics education curriculum to analyze issues such as poverty, or racism, or gender injustices, or capitalism and the global economy are sparse. Additional texts and resources such as *Relearning Mathematics* (Frankenstein 1989) and *Rethinking Mathematics* (Gutstein and Peterson 2005) are sorely needed, as is related research on the effectiveness of these materials and related instructional approaches. Although important, it is not enough for a broader audience of learners simply to learn mathematics better; we must also alter the global balance of power and resources to achieve lasting improvements in individuals' social, economic, political, and educational conditions worldwide. We must achieve a critical literacy, including a critical mathematical literacy, to actualize this goal.

Social justice literature comes in various forms that tend toward advocating one of either critical literacy or community literacy. Forms more supportive of critical literacy and citizen activism are exceptional at reframing mathematics in applications with transformative potential to understand and change society as we know it. A weakness of related literature is that it largely fails to temper its sometimes argumentative, or by some standards radical, recommendations with the personal and interpersonal feminist or multicultural recommendations that can help demonstrate the common-sense nature of the new version of mathematics being advanced.

Forms of social justice literature more supportive of community literacy demonstrate the significance of individuals' and communities' funds of knowledge in shaping all of what we experience. These forms are interconnected with and help to further situate connected teaching approaches in sociocultural contexts and democratic terms. Both forms overall are supportive of approaches to achieving mathematical literacy as advocated by feminist and NCTM *Standards*-based (1989, 2000) perspectives, though such approaches in social justice literature are typically couched within social justice applications rather than described for their possible inherent advantages. Neither feminist nor social justice literature tends to address possible advantages of service learning in mathematics to help achieve community literacy.

In describing unions and intersections, I aimed not to express all the nuances of various bodies of literature, but simply to help reveal the fruitful potential to integrate wide-ranging perspectives. As previously noted, authors such as Goodell and Parker (2001) and Gutstein (2006) have made similar recommendations. These recommendations help construct a narrative for thinking about for mathematics education in the public interest. However, the narrative is thus far incomplete.

Considerations for how to best implement reform are likely at least as important as the nature of the reforms themselves. For example, among other constraints, high stakes testing environments can and do undermine reform efforts, defining the framework of what is important in schools by controlling what is tested. It is advantageous for change to take shape through grassroots efforts that enlist support and involvement from individuals and communities. For example, the largely successful Algebra Project employs experiential strategies, ongoing teacher education, and grassroots community leadership to increase student achievement in math and prepare students to succeed in college-prep mathematics and science courses at the high school level. As Boaler (1998) described, even successful reform-based systems can be dismantled if they are not well understood or supported by parents. It will remain a challenge for future educators and community leaders to consider how to best organize and support grassroots efforts of the magnitude required to effect real change.

6 Summary

To promote gender, racial/ethnic, social class, and global equity, mathematics education needs fundamental change. However, the mathematics education community will be unable to produce the much-needed change by maintaining current structures and assumptions. Many mathematics educators and researchers have already set the stage and made the case for transformative, sustainable change. A space is produced for change in mathematics education, and emphases on multiple forms of literacies are easily legitimized, when we reconstruct mathematics through a lens of producing an educated and capable citizenship and when we come to understand the relevance of mathematics as a tool to learn about and even to remedy local, national, and global injustice—both gender related and otherwise. Together, we can reconstruct common sense in mathematics education by tearing down tired disciplinary boundaries and building up a new mathematics education in the public interest that can inspire hope and change.

References

Apple, M. W. (1992). Do the *Standards* go far enough? Power, policy, and practice in mathematics education. *Journal for Research in Mathematics Education, 23*(5), 412–431.

Banks, J. A. (2006). Democracy, diversity, and social justice: Educating citizens for the public interest in a global age. In G. Ladson-Billings & W. F. Tate (Eds.), *Education research in the public interest: Social justice, action, and policy* (pp. 141–157). New York: Teachers College Press.

Baxter Magolda, M. (1992). *Knowing and reasoning in college: Gender-related patterns in students' intellectual development*. San Francisco: Jossey-Bass.

Becker, J. R. (1995). Women's ways of knowing in mathematics. In P. Rogers & G. Kaiser (Eds.), *Equity in mathematics education: Influences of feminism and culture* (pp. 163–174). London: Falmer Press.

Becker, J. R. (1996). Research on gender and mathematics: One feminist perspective. *Focus on Learning Problems in Mathematics, 18*(1, 2, & 3), 19–25.

Becker, J. R., & Jacobs, J. E. (2001). Introduction. In W. G. Secada, J. E. Jacobs, J. R. Becker & G. F. Gilmer (Eds.), *Changing the faces of mathematics: Perspectives on gender* (pp. 1–7). Reston: NCTM.

Belenky, M. F., Clinchy, B. M., Goldberger, N. R., & Tarule, J. M. (1986/1997). *Women's ways of knowing: The development of self, voice, and mind* (2nd ed.). New York: Basic Books.

Bishop, A. P., & Bruce, B. C. (2001). *Fostering community literacy*. International Conference on Researching Literacy. Gevirtz Graduate School of Education, University of California Santa Barbara, Santa Barbara, CA, USA.

Bishop, A. J., & Forgasz, H. J. (2007). Issues in access and equity in mathematics education. In F. K. Lester Jr. (Ed.), *Second handbook of research on mathematics teaching and learning* (pp. 1145–1167). Charlotte: Information Age Publishing.

Boaler, J. (1998). *Experiencing school mathematics*. Philadelphia: Open University Press.

Boaler, J., & Staples, M. (2005). Transforming students' lives through an equitable mathematics approach: The case of Railside School. Retrieved June 20, 2005, from http://www.stanford.edu/~joboaler/pubs.html.

Brew, C. (2001). Women, mathematics, and epistemology: An integrated framework. *International Journal of Inclusive Education, 5*(1), 15–32.

Buerk, D. (1996). Our open ears can open minds: Listening to women's metaphors for mathematics. *Focus on Learning Problems in Mathematics, 18*(1, 2, & 3), 26–31.

Bureau of Labor Statistics (2004). *National industry-occupation employment projections 2002–2012*. Washington: U.S. Department of Labor.

Campbell, P. B. (1995). Redefining the "girl problem in mathematics". In W. G. Secada, E. Fennema & L. B. Adajian (Eds.), *New directions for equity in mathematics education* (pp. 225–241). New York: Cambridge University Press.

Civil, M. (2002). Culture and mathematics: A community approach. *Journal of Intercultural Studies, 23*(2), 133–148.

Civil, M. (2007). Building on community knowledge: An avenue to equity in mathematics education. In N. Nasir & P. Cobb (Eds.), *Improving access to mathematics: Diversity and equity in the classroom* (pp. 105–117). New York: Teachers College Press.

Clinchy, B. (1989). On critical thinking and connected knowing. *Liberal Education, 75*(5), 14–19.

Clinkscales, G., & Zaslavsky, C. (1997). Integrating mathematics with community service. In J. Trentacosta & M. J. Kenney (Eds.), *Multicultural and gender equity in the mathematics classroom: The gift of diversity, 1997 yearbook* (pp. 107–114). Reston: National Council of Teachers of Mathematics.

Damarin, S. K. (1995). Gender and mathematics from a feminist standpoint. In W. G. Secada, E. Fennema & L. B. Adajian (Eds.), *New directions for equity in mathematics education* (pp. 242–257). New York: Cambridge University Press.

D'Ambrosio, U. (2006). The program Ethnomathematics: A theoretical basis of the dynamics of intra-cultural encounters. *Journal of Mathematics and Culture, 1*(1). Retrieved November 12, 2007, from http://www.ccd.rpi.edu/Eglash/nasgem/jmc/issue_1_1.html.

Fennema, E. (2000). *Gender and mathematics: What is known and what do I wish was known?* Paper presented at the 5th Annual Forum of the National Institute for Science Education, Detroit, MI, USA.

Forgasz, H., & Leder, G. (2001). "A + for girls, B for boys": Changing perspectives on gender equity and mathematics. In B. Atweh, H. Forgasz & B. Nebres (Eds.), *Sociocultural research on mathematics education* (pp. 347–366). Mahwah: Erlbaum.

Frankenstein, M. (1989). *Relearning mathematics: A different third R—radical maths.* London: Free Association Books.

Freire, P. (1970/2004). *Pedagogy of the oppressed* (30th Anniversary ed.). New York: Continuum.

Galotti, K. M., Clinchy, B. M., Ainsworth, K. H., Lavin, B., & Mansfield, A. F. (1999). A new way of assessing ways of knowing: The attitudes toward thinking and learning survey (ATTLS). *Sex Roles, 40*(9/10), 745–766.

Galotti, K. M., Drebus, D. W., & Reimer, R. L. (2001). Ways of knowing as learning styles: Learning MAGIC with a partner. *Sex Roles, 44*(7/8), 419–436.

Gellert, U., Jablonka, E., & Keitel, C. (2001). Mathematical literacy and common sense in mathematics education. In B. Atweh, H. Forgasz & B. Nebres (Eds.), *Sociocultural research on mathematics education* (pp. 57–73). Mahwah: Erlbaum.

Gilligan, C. (1982). *In a different voice: Psychology theory and women's development.* Cambridge: Harvard University Press.

González, N. (1995). Educational innovation: Learning from households. *Practicing Anthropology, 17*(3), 3–24.

Goodell, J. E., & Parker, L. H. (2001). Creating a connected, equitable mathematics classroom: Facilitating gender equity. In B. Atweh, H. Forgasz & B. Nebres (Eds.), *Sociocultural research on mathematics education* (pp. 411–431). Mahwah: Erlbaum.

Goodlad, J. (2004). Fulfilling the public purpose of schooling. *School Administrator, 61*(5), 14–17. Retrieved November 17, 2007, from http://findarticles.com/p/articles/mi_m0JSD/is_5_61/ai_n6046265.

Gutstein, E. (2006). *Reading and writing the world with mathematics: Toward a pedagogy for social justice.* New York: Routledge.

Gutstein, E. (2007). "And that's just how it starts": Teaching mathematics and developing student agency. *Teachers College Record, 109*(2), 420–448.

Gutstein, E., & Peterson, B. (Eds.) (2005). *Rethinking mathematics: Teaching social justice by the numbers.* Milwaukee: Rethinking Schools.

Hadlock, C. R. (Ed.) (2005). *Mathematics in service to the community: Concepts and models for service-learning in the mathematical sciences.* Washington: The Mathematical Association of America.

Hart, D., Donnelly, T. M., Youniss, J., & Atkins, R. (2007). High school community service as a predictor of adult voting and volunteering. *American Educational Research Journal, 44*(1), 197–219.

Hart, S. M., & King, J. R. (2007). Service learning and literacy tutoring: Academic impact on pre-service teachers. *Teaching and Teacher Education, 23*(4), 323–338.

Jacobs, J. E. (1994). Feminist pedagogy and mathematics. *ZDM, 26*(1), 12–17.

Jacobs, J. E., & Becker, J. R. (1997). Creating a gender-equitable multicultural classroom using feminist pedagogy. In J. Trentacosta & M. J. Kenney (Eds.), *Multicultural and gender equity in the mathematics classroom: The gift of diversity, 1997 yearbook* (pp. 107–114). Reston: National Council of Teachers of Mathematics.

Kaiser, G., & Rogers, P. (1995). Introduction: Equity in mathematics education. In P. Rogers & G. Kaiser (Eds.), *Equity in mathematics education: Influences of feminism and culture* (pp. 1–10). New York: Routledge Falmer.

Keitel, C. (Ed.). (1998). *Social justice and mathematics education: Gender, class, ethnicity and the politics of schooling.* Berlin: Freie Universität.

Knight, K. M., Elfenbein, M. H., & Messina, J. A. (1995). A preliminary scale to measure connected and separate knowing: The knowing styles inventory. *Sex Roles, 33*(7/8), 499–513.

Ladson-Billings, G. (1997). It doesn't add up: African American students' mathematics achievement. *Journal for Research in Mathematics Education, 28*, 697–708.

Ladson-Billings, G. (1995). Toward a theory of culturally relevant pedagogy. *American Educational Research Journal, 32*(3), 465–491.

Lamb, S. (1996). Gender differences in mathematics participation in Australian schools: Some relationships with social class and school policy. *British Educational Research Journal, 22*(2), 223–240.

Lamb, S. (1997). Gender differences in mathematics participation. *Educational Studies, 23*(1), 105–126.

Lave, J. (1988). *Cognition in practice: Mind, mathematics, and culture in everyday life.* New York: Cambridge University Press.

Leder, G. C. (1992). Mathematics and gender: Changing perspectives. In D. A. Grouws (Ed.), *Handbook of research on mathematics teaching and learning* (pp. 597–622). New York: Macmillan.

Leder, G. C., Forgasz, H. J., & Solar, C. (1996). Research and intervention programs in mathematics education: A gendered issue. In A. J. Bishop, K. Clements, C. Keitel, J. Kilpatrick & C. Laborde (Eds.), *International handbook of mathematics education* (pp. 945–985). Dordrecht: Kluwer.

Lim, J. H. (2004). Girls' experiences in learning school mathematics. *Focus on Learning Problems in Mathematics, 26*(1), 43–60.

Lingard, B., Martino, W., Mills, M., & Bahr, M. (2002). *Addressing the educational needs of boys.* Research report submitted to the Australian Government Commonwealth Department of Education, Science, and Training, Canberra City, Australia. Retrieved November 2, 2007, from http://www.dest.gov.au/sectors/school_education/publications_resources/profiles/addressing_educational_needs_of_boys.htm.

Lopez-Claros, A., & Zahidi, S. (2006). *Women's empowerment: Measuring the global gender gap.* Geneva, Switzerland: World Economic Forum.

McGraw, R., Lubienski, S. T., & Strutchens, M. E. (2006). A closer look at gender in NAEP mathematics achievement and affect data: Intersections with achievement, race/ethnicity, and socioeconomic status. *Journal for Research in Mathematics Education, 37*(2), 129–150.

Moll, L. C. (1992). Bilingual classrooms and community analysis: Some recent trends. *Educational Researcher, 21*(2), 20–24.

Moll, L., Amanti, C., Neff, D., & González, N. (1992). Funds of knowledge for teaching: A qualitative approach to developing strategic connections between homes and classrooms. *Theory into Practice, 31*(2), 132–141.

Moll, L. C., & González, N. (2004). A funds-of-knowledge approach to multicultural education. In J. A. Banks & C. A. Banks (Eds.), *Handbook of research on multicultural education* (2nd ed., pp. 699–715). San Francisco: Jossey Bass.

Morrow, C. (1996). Women and mathematics: Avenues of connection. *Focus on Learning Problems in Mathematics, 18*(1, 2, & 3), 4–18.

Moses, R. P., & Cobb, C. E. (2001). *Radical equations: Civil rights from Mississippi to the algebra project.* Boston: Beacon Press.

National Center for Education Statistics (2007). *Status and trends in the education of racial and ethnic minorities.* NCES 2007–039: U.S. Department of Education. Retrieved November 2, 2007, from http://nces.ed.gov/pubsearch/pubsinfo.asp?pubid=2007039.

National Commission on Service-Learning (2002). *Learning in deed: The power of service-learning for American schools.* Battle Creek: W. K. Kellogg Foundation.

National Council of Teachers of Mathematics (1989). *Curriculum and evaluation standards for school mathematics.* Reston: Author.

National Council of Teachers of Mathematics (2000) *Principles and standards for school mathematics.* Reston: Author.

National Science Foundation (2006). *Science and engineering indicators.* Arlington: Author. Retrieved October 12, 2007, from http://www.nsf.gov/statistics/seind06/.

Nunes, T., Schliemann, A., & Carraher, D. (1993). *Street mathematics and school mathematics.* New York: Cambridge University Press.

Oakes, J., Joseph, R., & Muir, K. (2004). Access and achievement in mathematics and science. In J. A. Banks & C. A. M. Banks (Eds.), *Handbook of research on multicultural education* (2nd ed., pp. 69–90). New York: Jossey-Bass.

Riordan, J. E., & Noyce, P. E. (2001). The impact of two standards-based mathematics curricula on student achievement in Massachusetts. *Journal for Research in Mathematics Education, 32*(4), 368–398.

Rogers, P., & Kaiser, G. (Eds.) (1995). *Equity in mathematics education: Influences of feminism and culture*. New York: Routledge Falmer.

Schoenfeld, A. H. (2002). Making mathematics work for all children: Issues of standards, testing, and equity. *Educational Researcher, 31*(1), 13–25.

Secada, W. G. (1992). Race, ethnicity, social class, language, and achievement in mathematics. In D. A. Grouws (Ed.), *Handbook of research on mathematics teaching and learning* (pp. 623–660). New York: Macmillan.

Senk, S. L., & Thompson, D. R. (Eds.) (2003). *Standards-based school mathematics curricula: What are they? What do students learn?* Mahwah: Erlbaum.

Shor, I. (1999). What is critical literacy? *Journal of Pedagogy, Pluralism, and Practice, 4*(1). Retrieved November 12, 2007, from http://www.lesley.edu/journals/jppp/4/shor.html.

Sirin, S. R. (2005). Socioeconomic status and academic achievement: A meta-analytic review of research. *Review of Educational Research, 75*(3), 417–453.

Skovsmose, O. (2004). Critical mathematics education for the future. Unpublished manuscript, Aalborg University. Retrieved April 14, 2007, from http://www.learning.aau.dk/en/department/staff/ole_skovsmose.htm.

Smith, D. E. (1974). Women's perspective as a radical critique of sociology. *Sociological Inquiry, 44*(1), 7–13.

Spielman, L. J. (2006). *Preservice teachers' characterizations of the relationships between teacher education program components: Program meanings and relevance and socio-political school geographies*. Blacksburg: Virginia Polytechnic Institute and State University, Unpublished doctoral dissertation.

Sriraman, B. (Ed.). (2007). *International perspectives on social justice in mathematics education*. Charlotte: Information Age Publishing.

Steen, L. A. (Ed.). (1997). *Why numbers count: Quantitative literacy for tomorrow's America*. New York: College Entrance Examination Board.

Stinson, D. W. (2004). Mathematics as 'gate keeper'(?): Three theoretical perspectives that aim toward empowering all children with a key to the gate. *The Mathematics Educator, 14*(1), 8–18.

Turner, E. E., & Strawhun, B. T. F. (2005). With math, it's like you have more defense. In E. Gutstein & B. Peterson (Eds.), *Rethinking mathematics: Teaching social justice by the numbers* (pp. 81–87). Milwaukee: Rethinking Schools.

Valero, P. (2001). Review of the book social justice and mathematics education: Gender, class, ethnicity and the politics of schooling. *ZDM, 33*(6), 187–191.

Weiner, G., Arnot, M., & David, M. (1997). Is the future female? Female success, male disadvantage, and changing gender patterns in education. In A. H. Halsey, P. Brown, H. Lauder & A. Stuart Wells (Eds.), *Education, economy, culture, and society*. Oxford: Oxford University Press.

Wyner, J. S., Bridgeland, J. M., & Diiulio, J. J. (2007). *Achievement trap: How America is failing millions of high-achieving students from lower-income families*. Lansdowne: Jack Kent Cooke Foundation.

Preface to "Adolescent Girls' Construction of Moral Discourses and Appropriation of Primary Identity in a Mathematics Classroom"

Jae Hoon Lim

My research interest in gender and equity issues in mathematics education has a long history stemming from my dissertation research in which I examined the intersection of race, gender, and class in a middle school mathematics classroom. Though I completed my dissertation as an ethnographic study in 2002, I was deeply puzzled by the contrasting perceptions and experiences manifested in each participating girl's profile. I wondered if further analysis would provide more powerful means to unravel the complexity and subtle dynamics in the girls' emerging social and academic identities.

My 2008 article "Adolescent girls' construction of moral discourses and appropriation of primary identity in a mathematics classroom" published in *ZDM—the International Journal on Mathematics Education, Special issue—"Mathematics Education: New Perspectives on Gender"* was borne of this lingering question. I continued the search for a more robust and effective theoretical and analytical approach to seemingly inconsistent and even contradictory voices of my young adolescent participants. Bakhtin's (1981) theory of language and identity and Gee's (1999) discourse analysis provided me with the means to systematically dissect the multilayered voices of the girls and their identities. To my pleasant surprise, the results from my new theoretical and analytic endeavor largely supported my previous ethnographic findings regarding the sociocultural context of the mathematics classroom and each girl's location in the classroom. As a result, I gained confidence in my arguments in previous work—that there could be a qualitatively different set of challenges faced by a smaller subset of girls in school mathematics—and, therefore, the gender equity discourse should be restructured considering two other powerful sociocultural factors, race/ethnicity and class, in American schooling contexts.

Gender issues in mathematics education have been a controversial topic worldwide during the last four decades. Concerned with girls' lower mathematics per-

J.H. Lim (✉)
University of North Carolina at Charlotte, Charlotte, USA
e-mail: jhlim@uncc.edu

H. Forgasz, F. Rivera (eds.), *Towards Equity in Mathematics Education,*
Advances in Mathematics Education,
DOI 10.1007/978-3-642-27702-3_6, © Springer-Verlag Berlin Heidelberg 2012

formance reported in early studies in the 1970s, feminist scholars raised public awareness facilitating various instructional strategies that supported more positive mathematics learning experiences among girls (Leder 1992; National Center for Education Statistics [NCES] 2005). As a result, many publications in the 1990s and early 2000s reported that traditional gender differences in mathematics achievement favoring males had gradually decreased in the United States and Great Britain (e.g., Department for Education and Skills [DfES] 2005; Fennema 1996; NCES 2005). Witnessing a significant improvement in girls' status in schooling processes, some researchers have argued that gender inequity in mathematics education is no longer an urgent topic for scholarly discussion. Some even contest that it is boys who experience significant disadvantages in the current schooling system (e.g., Sommers 2000; Weaver-Hightower 2003). However, the most recent PISA results (OECD 2010) and NAEP data (NCES 2010) provide a puzzling picture of this long-standing issue. The majority of Western countries, such as United Kingdom, United States, Germany, Netherlands, Canada, and Australia showed a large and consistent gender gap in mathematics favoring boys. In some countries (e.g., Australia), the PISA results testified to the resurge of the traditional gender gap, facilitated by the conservative backlash during the last ten years (Forgasz 2008). The latest NAEP data confirmed a similar pattern in American education. The traditional gender difference favoring boys seems to have resurged in the modified data of 2004, and more evidently in the 2008 data. A statistically significant gender difference between boys' and girls' mathematics performance was found for students aged 13 and 17, while the data for 9 year olds did not show a significant gender difference.

Educational researchers have long argued that the American schooling process can never be properly understood without simultaneously considering the three powerful sociocultural factors: gender, race/ethnicity and class (Campbell 1989). Research on gender and mathematics is unlikely to be an exception to this thesis. Not surprisingly, during the last 30 years, numerous American policy researchers have reported a great disparity in students' mathematics performance across racial/ethnic lines as well as by different socioeconomic spectra (Lee 2004; Kohr et al. 1989; NCES 2010). African Americans and Hispanics consistently perform lower than their White and Asian counterparts. More interesting is that the gender difference that emerged during the last ten years varies across different racial subgroups; 13 and 17 year old Whites and Hispanics exhibit a statistically significant gender difference favoring males while African Americans and Asians do not show any significant gender difference for any of the three age groups. The gender gap observed among Whites is consistent for both middle and working class subgroups. As a result, White working class girls who perform 5–6 points below their male counterparts are the lowest achieving students among all White students. Hispanic boys perform significantly higher than Hispanic girls at age 13 and 17. However, this gender difference favoring boys is relatively small and not statistically significant among Hispanic middle class students. In contrast, Hispanic working class boys significantly outperform Hispanic working class girls by a large margin. The off-set effect of gender is also found among African America students. Even though they show no significant gender difference as a whole, 13 and 17 year old middle class African–American girls score about the same as or slightly higher than their male counterparts, while working class African–American girls consistently score lower than their male counterparts at age 17. Therefore, it is fair to say that both Hispanic and African–American girls from middle SES tend do as well as their male counterparts. On the contrary, both Hispanic and African–American working class girls tend to fall behind their male counterparts.

What can be discerned from these data? Working class girls are the most vulnerable group of students in school mathematics in the United States in almost all racial/ethnic subgroups. As a whole, girls' mathematics performance, though varying in degrees across racial/ethnic subgroups, seems to be more influenced by contextual factors such as family SES than is boys' performance. Girls from middle or upper class families seem to perform at the same level as their male counterparts, while girls from working class backgrounds seem to fall far behind than their male counterparts with the same SES background. Therefore, I argue that the recent NAEP data suggest great challenge and hardship faced by working class girls in their mathematics learning, especially those from Hispanic and African–American backgrounds. The off-set effect of gender observed in the two minority groups, and a more consistent gender gap appearing among the White samples, indicate that working class girls' struggle in school mathematics may be qualitatively different from that of middle class girls.

This emerging pattern of gender gap by race and SES is, in fact, a diversion from previous studies that reported the lack of significant interaction effect by gender by race by SES in student mathematics achievement (e.g., Gilleece et al. 2010; Kohr et al. 1989). The vulnerability of working class girls, as compared to their male counterparts, is an interesting phenomenon that differs from other studies conducted in England and other Western countries (e.g., Machin and McNally 2005; Mensah and Kiernan 2010). However, this is not a phenomenon unique to American schooling either. A similar observation was documented in Australian schools, that is, that girls' participation in higher mathematics is largely influenced by their class status (Lamb 1996). Lamb (1996) reported that the gender gap favoring boys was much weaker among girls from middle class families, and suggested that the higher SES status of their families offset the negative impact of gender. As a result, it was argued that the traditional gender difference is more evident among Australian students from socially disadvantaged backgrounds (Teese et al. 1995).

What, then, could be the major challenges faced by working class Hispanic and African–American girls in their mathematics learning? Though there might be multiple challenges, I would like to highlight one critical element, a profound social disconnection from their (mathematics) teachers and the dearth of social/academic support provided to these girls. In my two current studies, one with White and Hispanic high school students comparing their classroom experiences with a computer-based tutorial program, and the other, a longitudinal study with three racial/ethnic groups of high-achieving middle school girls, the importance of social support provided to poor minority girls in their mathematics learning appeared evident. Not surprisingly, based on some social network and cultural knowledge afforded through their family backgrounds, middle class minority girls seemed to have a higher chance to develop a positive relationship with their teachers. Yet, poor Hispanic and African–American girls, even the high-achieving girls, in the two studies tended to experience a deeper cultural and social disconnection from their teachers and school contexts as a whole. The poor minority girls participating in these two studies were primarily concerned with having a psychologically safe space and developing a social and emotional bond with others in their school contexts. In fact, these minority girls' strong desire for a safe space and social support is not a totally new finding. Several researchers have already reported that girls are more likely than boys to form a more closed and tight social network, and that such a close social network significantly influences academic pursuits (Riegle-Crumb et al. 2006). As a result, with strong instructional and social support from their teachers and peers, coupled with their own desire to make a difference in their lives, some minority girls seem to excel in their academic work (Hubbard 2005). It should be noted that working class minority girls' social disconnection from their teacher(s) poses a critical dilemma in their mathematics learning. Because of the scarcity of instructional support available to these girls through their families or communities, their mathematics teachers are almost their only reliable source of mathematical knowledge. As a result, while feeling disconnected and uncomfortable, these girls have no option but to depend heavily upon their teachers who may not understand their psychological and academic fragility.

The cultural and social dis/connection has been contested in many recent publications (e.g., Tyler et al. 2010) and could be a double-edge sword. Although the social and cultural disconnection between working class minority girls and their teachers has contributed to existing gender, racial/ethnic, and class inequities in mathematics education, the same element could become a powerful means to support these girls' enthusiastic learning in mathematics (Ladson-Billings 2009). As I argued in the 2008 article, what seems most urgent is raising a troupe of in-school practitioners—teachers and school administrators—who willingly proclaim themselves advocates of these most vulnerable groups of students and re-conceptualize their teaching and service in the light of social activism. Ironically, this argument seems to be far removed from the dominant research discourse in mathematics education which has focused on the cognitive process of individual learners to explain how authentic mathematics learning may occur. However, the confluence of social context and mathematics learning outcomes, as reflected in the varied gender inequity phenomena across racial/ethnic and class backgrounds, might be a new, though not unique, dilemma in American schooling contexts where three powerful sociocultural factors, gender, race/ethnicity, and class, have created a complex dynamic shaping the entire schooling experience of individual students, including their mathematics learning. Furthermore, this confluence of three sociocultural factors in mathematics education is clearly one of the most compelling research agendas in the international research community as researchers become increasingly aware of the intricate relationships of gender, race, and class in school mathematics.

References

Bakhtin, M. M. (1981). *The dialogic imagination: Four essays by M. M. Bakhtin*. C. Emerson & M. Holquist (Trans.). Austin: University of Texas Press.

Campbell, P. B. (1989). So what do we know with the poor, non-White females? Issues of gender, race, and social class in mathematics and equity. *Peabody Journal of Education, 66*(2), 95–112.

Department for Education and Skills (2005). *National statistics first release, GCE/VCE A/AS examination results for young people in England, 2003/4 (final)*. Department of Education and Skills. Retrieved June 19, 2011 from http://www.education.gov.uk/rsgateway/DB/SFR/s000586/sfr26-2005.pdf.

Fennema, E. (1996). Mathematics, gender, and research. In G. Hanna (Ed.), *Towards gender equity in mathematics education* (pp. 9–26). Dordrecht, The Netherlands: Kluwer Academic Publishers.

Forgasz, H. (2008). Gender, socio-economic status, and mathematics: Performance among high achievers. In O. Figueras, J. L. Cortina, S. Alatorre, T. Rojano, & A. Sepúlveda (Eds.), *Mathematical ideas: History, education and cognition. Proceedings of the Joint Conference PME 32 – PME-NA XXX* (Vol. 3, pp. 25–32). Morelia, México: Cinvestav-UMSNH [also available on CD-ROM].

Gee, J. P. (1999). *An introduction to discourse analysis: Theory and method*. New York: Routledge.

Gilleece, L., Cosgrove, J., & Sofroniou, N. (2010). Equity in mathematics and science outcomes: Characteristics associated with high and low achievement on PISA 2006 in Ireland. *International Journal of Science and Mathematics Education, 8*, 475–496.

Hubbard, L. (2005). The role of gender in academic achievement. *International Journal of Qualitative Studies in Education, 18*(5), 605–623.

Ladson-Billings, G. (2009). *The dreamkeepers: Successful teachers of African American children*. San Francisco: Jossey-Bass.

Kohr, R. L., Masters, J. R., Coldiron, J. R., Blust, R. S., & Skiffington, E. W. (1989). The relationship of race, class, and gender with mathematics achievement for fifth-, eighth-, and eleventh-grade students in Pennsylvania schools. *Peabody Journal of Education, 66*(2), 147–171.

Lamb, S. (1996). Gender differences in mathematics participation in Australian schools: Some relationships with social class and school policy. *British Educational Research Journal, 22*(2), 223–240.

Leder, G. C. (1992). Mathematics and gender: Changing perspectives. In D. A. Grouws (Ed.), *Handbook of research on mathematics teaching and learning* (pp. 597–622). New York: Simon & Schuster Macmillan.

Lee, J. (2004). Multiple facets of inequity in racial and ethnic achievement gaps. *Peabody Journal of Education, 79*(2), 51–73.

Machin, S. & McNally, S. (2005) Gender and student achievement in English schools. *Oxford Review of Economic Policy, 21*(3), 357–372.

Mensah, F. K., & Kiernan, K. E. (2010). Gender differences in educational attainment: Influences of the family environment. *British Educational Research Journal, 36*(2), 239–260.

National Center for Education Statistics (2005). *Gender differences in participation and completion of undergraduate education and how they have changed over time: Postsecondary education descriptive analysis report*. Washington: U. S. Government Printing Office.

National Center for Education Statistics [NCES] (2010). *The data explorer for long-term trend. 2008 [data file]*. Retrieved June 19, 2011 from http://nces.ed.gov/nationsreportcard/naepdata/.

OECD (2010). *PISA 2009 Results: What students know and can do – Student performance in reading, mathematics and science (Volume I)*. Retrieved July 2, 2011 from http://dx.doi.org/10.1787//9789264091450-en.

Riegle-Crumb, C., Farkas, G., & Muller, C. (2006). The role of gender and friendship in advanced course taking. *Sociology of Education, 79*, 206–228.

Sommers, C. H. (2000). *The war against boys: How misguided feminism is harming our young men*. New York: Simon & Schuster.

Teese, R., Davies, M., Charlton, M., & Polesel, J. (1995). *Who wins at school?* Melbourne: Department of Education Policy and Management, The University of Melbourne.

Tyler, K. M., Uqdah, A. L., Dillihunt, M. L., Beatty-Hazelbaker, R., Conner, T., Gadson, N., et al. (2010). Cultural discontinuity: Toward a quantitative investigation of a major hypothesis in education. *Educational Researcher, 37*(5), 280–297.

Weaver-Hightower, M. (2003). The "boy turn" in research on gender and education. *Review of Educational Research, 73*(4), 471–498.

Adolescent Girls' Construction of Moral Discourses and Appropriation of Primary Identity in a Mathematics Classroom

Jae Hoon Lim

Abstract This qualitative study examines the way three American young adolescent girls who come from different class and racial backgrounds construct their social and academic identities in the context of their traditional mathematics classroom. The overall analysis shows an interesting dynamic among each participant's class and racial background, their social/academic identity and its collective foundation, the types of ideologies they repudiate and subscribe to, the implicit and explicit strategies they adopt in order to support the legitimacy of their own position, and the ways they manifest their position and identity in their use of language referring to their mathematics classroom. Detailed analysis of their use of particular terms, such as "I," "we," "they," and "should/shouldn't" elucidates that each participant has a unique view of her mathematics classroom, developing a different type of collective identity associated with a particular group of students. Most importantly, this study reveals that the girls actively construct a social and ideological web that helps them articulate their ethical and moral standpoint to support their positions. Throughout the complicated appropriation process of their own identity and ideological standpoint, the three girls made different choices of actions in mathematics learning, which in turn led them to a different math track the following year largely constraining their possibility of access to higher level mathematical knowledge in the subsequent schooling process.

1 Introduction

During the last few decades, gender issues in mathematics education have drawn a great amount of attention from educational researchers and school practitioners worldwide. Numerous reports confirm that gender inequity in mathematics education is a world-wide phenomenon found in many countries including Australia,

J.H. Lim (✉)
Department of Educational Leadership, College of Education, The University of North Carolina at Charlotte, 9201 University City Boulevard, Charlotte, NC 28223-0001, USA
e-mail: jhlim@uncc.edu

This chapter is a reprint of an article published in ZDM—The International Journal on Mathematics Education (2008) 40(4), 617–631. DOI 10.1007/s11858-008-0119-7.

Britain, Germany, and USA (e.g., Leder 1992; National Center for Education Statistics [NCES] 2005). Researchers found that female students reported lower self-confidence in mathematics (Hargreaves et al. 2008; Preckel et al. 2008) and, as a result, were more likely to avoid taking advanced math courses in high school (Jones et al. 1996; van Langen et al. 2006). Recently, gender differences in mathematics achievement, however, are gradually decreasing in USA (NCES 2004, 2005) and Great Britain [Department for Education and Skills (DfES) 2005], but gender inequity is still considered a problem in the fields of mathematics and many occupational spheres that are related to mathematics, science, and technology (Chipman 2005; NCES 2005; van Langen and Dekkers 2005). Since mathematical knowledge is an important prerequisite in many professions, unbalanced mathematics achievement in favor of males implies limited access for females resulting in inequity in acquiring high socio-economic status.

Recently, American researchers reported interesting dynamics across gender and other sociocultural factors, such as race/ethnicity and class. Several studies have shown that mathematics achievement of American female students is significantly influenced by age, ethnic group, and socio-economic status (SES) (Ansell and Doerr 2000; Mcgraw et al. 2006; Tate 1997). For example, Tate's study (1997) showed that the achievement disparity significantly differs across various racial/ethnic groups as well as across different class spectra within each ethnic/racial group. Secada (1992) identified that some minority girls—Black and Hispanic—and working class students are the most vulnerable group in mathematics learning and called for a more in-depth examination on the intersection of gender, race/ethnicity, and class.

Researchers have long suggested that various social and cultural factors (e.g., class affiliation, ethnic membership, and gender) may influence female students' experiences in learning mathematics and affect both their motivation to acquire advanced mathematical knowledge and their actual performance in the domain (Reyes and Stanic 1988). However, research examining this critical intersection of gender and other socio-cultural factors (e.g., race/ethnicity and class) has not flourished in mathematics education. The field of mathematics education has been dominated by individual psychological approaches (Atweh et al. 2001, p. ix) emphasizing the cognitive process rather than analyzing the socio-cultural aspect of social inequity (Tate 1997). Concepts of gender, class, and race, which are critical to understanding sociocultural issues in mathematics education, may still be overlooked by the majority of researchers (Lubienski 2002). As a result, few educational researchers have investigated the socio-cultural aspects of mathematics classrooms or examined how class, race, and gender, three powerful systems of social stratification in America, play a role in the learning environment shaping young girls' experience with and identity in school mathematics. Little is known about how female students who come from different class and racial backgrounds respond to the traditional learning environment—known to be so pervasive across the nation—developing positive or negative attitudes toward mathematics learning (Campbell 1989).

Research suggests that the development of an individual's positive or negative identity in the domain of mathematics is a subtle and complex process. Feminist scholars, such as Boaler (1997) and Jungwirth (1993), argue that traditional math instructional methods—prevalent in American classrooms—ignore women's unique way of learning, depriving them of meaningful experience in their mathematical endeavors. Most recently, researchers pointed out that mathematical identity development among boys and girls heavily relies on a gendered binary opposition system (Mendick 2005), and such gendered relationship with mathematics is observed even among primary school children (Bhana 2005). Rodd and Bartholomew (2006) also presented a similar dilemma and struggle experienced by female undergraduate math major students as they constantly engaged in the process of re-defining the relationship between their everyday view of femininity and mathematical endeavors. This reflects, Mendick argues, the gendered discourse of rationality stemming from Western Enlightenment thought that still persists as a powerful source of stereotyped images regarding mathematics. As new literature adds further insight, it becomes clear that we have to admit that we live in a society where competing images of gender and mathematics intermingle making it difficult for many young girls to construct meaning out of such contradictions.

This study examines the way three American young adolescent girls with different class and racial/ethnic backgrounds construct their social and academic identities in their school contexts, particularly within their traditional mathematics classroom. Through the analysis of their use of particular terms, "I," "we," and "they," this study investigates the way how the three girls develop and construct their academic and social identities, and how they appropriate different ideological stances in order to support their emergent identities in the discipline.

1.1 Theoretical Framework: Bakhtin's Theory of Language and Identity

This study is indebt to the sociocultural approaches to mind that emphasizes the importance of the sociocultural, as well as the historical, environment in the formation of an individual's mind (Wertsch 1991, p. 6)—one's identity (Lacasa et al. 2005). One interesting aspect of Bakhtin's theory is the concept of "self:" He rejects the conventional notion of "self" as an entity constituted through a unified, monadic relation to the external world. Bakhtin's entire work, including the conception of self, is based on criticism of individualism and dualism that were prevalent in western society during his era (Morris 1994, pp. 25–26). He stresses that all social and cultural phenomena are "profoundly intersubjective or dialogic in nature (Voloshinov 1973, p. 34)" and that the formation of self is not an exception (Voloshinov). For Bakhtin, and the members of his circle, an individual consciousness, his or her subjectivity, is not a self-sufficient, pre-constituted entity, but is formed through the dialogic struggle between contending voices or discourses. The phenomenon of "self-ness" is constructed through the operation of a dense and conflicting network of discourses, cultural and social practices and institutional structures, which are bound up with the intricate and complex interplay of the self-other relation.

Based on this concept of "self-ness" many researchers have explored issues of identity development through an analysis of discourse—how people construct or content their identities through the use of language, or how the (particular) use of (certain) language contributes to the construction of people's identities in various social contexts (e.g., Lacasa et al. 2005). Numerous studies have confirmed that the process of identity development at the individual level is a social process (Buxton et al. 2005; Jenson et al. 2003), which features language as a key element (Bucholtz and Hall 2005).

Another important contribution made by Bakhtin's circle is that their theory enables educational researchers to see the multiple sociopolitical layers embedded in a person's speech and identity: there is more than one voice and one identity in an individual's spoken word. Bakhtin (1981) states:

> As a living, socio-ideological concrete thing, language, for the individual consciousness, lies on the borderline between oneself and the other. The word in language is half someone else's. It becomes "one's own" only when the speaker populates it with his own intention, his own accent, when he appropriates the word, adapting it to his own semantic and expressive intention. Prior to this moment of appropriation, the word does not exist in a neutral and impersonal language, but rather it exists in other people's mouths, in other people's contexts, serving other people's intentions: it is from there that one must take the word, and make it one's own. And not all words for just everyone submit equally easily to this appropriation, to this seizure and transformation into private property: Many words stubbornly resist, others remain alien, sound foreign in the mouth of the one who appropriates them and who now speaks them; they cannot be assimilated into his context and fall out of it; it is as if they put themselves in quotation marks against the will of the speaker. Language is not a neutral medium that passes freely and easily into the private property of the speaker's intentions; it is populated—overpopulated—with the intentions of others. Expropriating it, forcing it to submit to one's own intentions and accents, is a difficult and complicated process. (p. 114)

Acknowledging the sociopolitical nature of language, the incessant tensions and dynamics among different intentions and powers deep-seated in it, Bakhtin believes that subordinate groups can generate a differentiated not yet complete set of knowledge, which is embedded in traditions and practices, and which is, at least, partially resistant to dominant discourses and ideologies (Morris 1994). Influenced by Bakhtin's idea of language as a discursive practice of social life, some feminist theorists and critical theorists have developed the concept of "language (voice) of possibility" (Giroux 1991, p. 53) and "language (voice) of resistance" (Bauer and McKinstry 1991, p. 4). They view discursive practices through language as the way of dismantling the oppressive social reality in which people are situated.

2 Methodology

This study is primarily based on in-depth interviews with three young adolescent girls who were part of a larger ethnographic study. The ethnographic research was conducted in a middle school located in the Southeast United States and included a sixth grade mathematics teacher and two groups of her students enrolled either in an advanced or regular mathematics class. The three girls—Jessica, Stella, and Amanda—reported in this paper were all enrolled in the advanced math class. The middle school, though located in a rural area, had an ethnically and economically diverse student population consisting of 70% White and 26% Black pupils with a very limited number of Hispanic students. The proportion of students living below the poverty level was 16% and almost half of students (44%) were eligible to receive free or reduced lunches. As a whole, the middle school had a student body consisting of two groups of students, those from White middle class families and a significant number of "rural poor" students including, yet not limited to, many Black students.

I collected three types of data throughout one semester: participant observation data from the advanced and regular math classes taught by the participating teacher; repeated interviews with the teacher, the eight selected participants, and some school personnel including other sixth grade teachers; and I reviewed a collection of school records and other school-based documents. Additional observation data were acquired from other classes, such as reading and remedial mathematics, as well as from various sites at the school, such as the library and playground. Even though the analysis presented in this paper focuses on three girls in the advanced math class,[1] I explained some of the findings from my overall ethnographic research as the background information about the girls and their mathematics classroom contexts.[2] In particular, explaining the cultural characteristics of their mathematics classroom— an authoritative, individuated, and procedural learning environment—is critical to understanding the three girls' identity and ideological appropriation process in the domain of mathematics.

My data analysis is deeply influenced by Bakhtin's theory of identity and language, presenting language as the fundamental source for analysis examining the subtle, yet critical emergence of self-and-other relationship. I conceptualize that my young adolescent participants—with their growing sense of self-ness—were constantly exploring their possible position in the world in general, and in their math classroom in particular. Their use of language manifested their struggles to locate their right position in the given space, and to provide an ideological support for their emerging identity. They exploited and appropriated the existing language—overpopulated by others' voices—to make it serve their purpose so that the language, at least, partially expressed their own voice.

At a more practical level, my analysis is indebted to Gee's method of discourse analysis (1999). Gee explains that people construct six areas of "reality" as they engage in any type of communication (e.g., speaking or writing).[3] Among those six possible areas for discourse analysis, I specifically chose two areas of reality building—socially situated identity and relationship building and political building

---

[1] I interviewed the total of eight girls, four from the advanced and regular class. This paper presents only three girls enrolled in the advanced class. April, my fourth participant in the class, showed results somewhat similar to Amanda's case. Yet, April did not develop as strong a voice as Amanda.

[2] I believe that reporting findings derived from my larger ethnographic research is beyond the scope of this paper. However, I presented a few major findings that are essential to understand the results from my discourse analysis presented in this paper. Those findings are explained in Sect. 3 (pp. 7–8).

[3] The six areas of "reality" are: (1) the meaning/value of aspects of the material world; (2) activities; (3) identities and relationships; (4) politics; (5) connections, and (6) semiotics (Gee 1999, p. 12).

since those two were most crucial to understand the girls' emerging identity and its collective ground as well as their ideological standpoint. Rymes' (2001) example of discourse analysis that presented the intricate, yet consistent pattern in drop-out students' narratives portraying their world divided by antagonistic "they" and victimized "I" (or "we") also inspired my analysis. There have been several studies analyzing (adult) author's use of "I" in various texts (e.g., Hyland 2002; Tang and John 1999) or speech (Fairclough 2003) as a way to examine the identity constructed in those texts or speech. However, the analysis of "we" as opposed to "they" which often "represents a deeper division" (Fairclough 2003, p. 150) between the two groups of people has been rarely found even in linguistic literature.

After an initial analysis of my entire ethnographic data, I noticed that my student participants exhibited contrasting relationships with various groups of students in their mathematics class—and beyond. Expecting that a more detailed, thorough analysis of interview transcripts would reveal critical insights into their emerging identities and their collective ground, I initially decided to examine the participants' use of "I," "we," and "they" in the contexts of school, more specifically in their mathematics classroom. I read through all nine interview transcripts (each participant generated three interview transcripts conducted at different points of time) and identified the total of 55 excerpts (16 from Jessica, 21 from Stella, 18 from Amanda) that met the following three criteria: (a) Were those words used to describe their peers in their school or math classroom? (b) Is it clear whom they refer to? (c) Does the statement or story include (imply) at least two parties either contrasting or associated to each other?[4] Then, I used open codes (e.g., smart, high-achieving, intimidated, scared, nice/kind, pretty) to clarify the characteristics of each "I," "we," and "they" in those identified interview excerpts. This analysis created a data analysis table showing different set of characteristics observed in each participant's "I"s, "we"s, and "they"s. At the same time, I closely examined the relationship/dynamics among those three terms (e.g., if the student's "I" in the story easily transited to or connected to "we"—the groups of students with whom she identified). In my analysis of the three pronouns I noticed Jessica and Amanda articulating a strong ideological standpoint with a clear division between "we" and "they." As a result, I also analyzed the aspect of political building by identifying the type of ideology advocated or repudiated by each student in the given data set.

Gee recommends that discourse analysis should pay attention to "situated meanings" that are inherent in all discourse. He argues that situated meanings are not just individual phenomenon; instead they "are *negotiated* between people in and through communicative social interaction" (Gee 1999, p. 51). Therefore, all the discourses presented in this paper including the three girls' use of "we" and "they" should be

---

[4]Based on the nature and common usage of these three words (I, we, and they) as generic first person or third person pronouns often used as a proxy for a larger group, I limited my analysis of these three terms to a smaller number of interview excerpts that meet the three criteria. Therefore, the participants' use of "we" or "they" that did not meet the criteria were not included in my analysis. For example, the participants, of course, used a lot of "they" as they describe their parents, school teachers, and others. Those "they"s were not the target of my analysis.

understood within the contexts of their interview *with me*—an international graduate student and the only Asian at the school. My student participants viewed me as someone similar to a student teacher who could be a teacher later, but not yet. My role at the school as an observer and instructional assistant when necessary located me somewhere between the teacher and students. Equally important, my racial/ethnic category—being Asian and foreigner—made me land somewhere between, or possibly outside, the two racial groups of students at the school—Whites and Blacks.

3 The Teacher and Classroom Contexts

The three participants reported in this article, Jessica, Stella, and Amanda, attended Mrs. Oliver's advanced mathematics class—one of two advanced sixth grade math classes at the school. Their mathematics teacher, Mrs. Oliver, a White teacher with 25 years teaching experience, was known as one of the most effective and respected mathematics teachers in the county. She was an extremely hard-working, enthusiastic teacher who believed in the value of school work and benefits of education in her students' lives: very few students or parents I interviewed, either White or Black, doubted that she was a devoted and serious teacher.

However, Mrs. Oliver's teaching mathematics also reflected some of the major challenges in mathematics education in USA. She started to teach sixth grade mathematics not based on her professional training or preference but because of a chronic shortage of math teachers in the county. Her professional development as a middle school mathematics teacher was rather isolated from the professional communities of mathematics teachers and researchers. As a result, she perceived mathematics as static knowledge and adhered to a somewhat traditional mode of instruction: authoritative, procedural, and individual work-based (Ball 1990; Boaler 2000; Brown et al. 1999). Mrs. Oliver often limited the number of possible answers for a given problem. She divided the entire process of problem solving into several steps and constantly emphasized those steps. She rarely used cooperative learning strategies in her classroom.

Mrs. Oliver identified a self-motivated, hard-working attitude as one of the most important qualities that students must develop in order to succeed in mathematics—a tough and difficult subject to learn. She, as one of the most "strict teachers" at the school, held a clear set of rules, and exhibited a very firm demeanor with the students in her math class. She consciously created and maintained an authoritative and restrictive classroom culture that intended to foster—even demand—hardworking attitudes among students. Mrs. Oliver strongly believed in the effectiveness of her strict disciplinarian approach to mathematics learning: Without having such firm structure of learning and clear expectation, most students—Mrs. Oliver conjectured—would not be able to master the level of mathematics content knowledge that the state curriculum requires. These specific cultural characteristics of Mrs. Oliver's mathematics classroom—an authoritative, individuated, and procedural learning environment—is critical to understanding the three girls' identity and ideological appropriation process in the classroom contexts.

Many of her students, mostly from white, middle-class backgrounds, interpreted her strict and firm attitude as an expression of high expectations for their academic success. In contrast, those from working-class backgrounds, including many Black students, who deeply doubted the fairness of school practices, experienced her strict disciplinarian approach differently. These students viewed the teacher's firm attitude and strict rules as something that could hurt them at any time—because school was *not* fair to them and teachers treated them differently. These two groups of students brought their prior knowledge and experience with school and teachers into the classroom. As a result, despite the teacher's good intention and hard work, Mrs. Oliver's students tended to experience the math class differently based on their class and racial membership.

4 Three Students

The overall analysis of three young adolescent girls—Jessica, Stella, and Amanda reported in this paper—shows an interesting dynamic among each participant's ethnicity and class background, their social/academic identity and its collective foundation, the types of ideologies they repudiate and subscribe to, the implicit and explicit strategies they adopt in order to support the legitimacy of their own position, and the ways they manifest their position and identity in their use of language referring to their mathematics classroom. Detailed analysis of their use of particular words, such as "we," "they," "should," and "shouldn't" elucidates that each participant has a unique view of her mathematics classroom, developing a different type of collective identity associated with a particular group of students. Most importantly, participants actively construct a social and ideological web that helps them articulate their ethical and moral standpoint to support their positions. Throughout the complicated appropriation process of their own identity and ideological standpoint, the three girls made different choices of actions in mathematics learning, which in turn may lead them to a different social context in which their possible identity and actions in the future are partly determined and largely constrained.

4.1 Jessica: They Call Us "Smarties"

Jessica, a high-achieving student in the advanced mathematics class, represented a typical portrait of an accomplished girl in the mathematics domain. Her motivation, as well as her academic identity, in the domain was firm and strong. Throughout her three interviews, Jessica exhibited a strong sense of "we"ness, frequently referring to a group of her close friends. Her "I" and "We" appeared interchangeably without any visible sign of hesitation or awareness of a subtle difference or possible conflict between the two. "They call us smarties. I take that as a compliment," declared Jessica. Jessica's use of "we" revealed that she had already constructed a firm collective identity with her friends coming from similar cultural and class backgrounds.

Jessica's use of "we" is characterized by four aspects; homogeneity, selectivity, academic accomplishment, and conforming school values and practices—accepting meritocracy ideology. Jessica's "we" is homogeneous in every aspect that she could think of except how they "look." Jessica's "we" share similar personalities, dispositions, hobbies, and learning goals. Jessica explained, "We look nothing alike, but we act just alike. We... we like to have good grades and make sure our work is done. And then after our work is done, that's when we can act silly and tell jokes and everything. But we have to make sure that our work is done so that we can make good grades and make sure that our average is staying the same or going up."

Jessica frequently used "we" in order to indicate a group of selective students which she belongs to. For example, in the very beginning of her first interview, Jessica emphasized the fact that she is in an advanced math class, not a regular one, and used "we" as she explicated differences between her own class and other regular classes. This is one of rare uses of Jessica's "we" referring to other than her social clique, yet she accepted this "we"ness of the entire class as she emphasized the selective process through which the class had been formed. She highlighted that students in her math class were different from those in "regular" classes. Jessica's dominant "we" referring to her social clique—a much smaller subset of high achieving middle class White girls—also implies the high selectivity of the group to which she belongs.

Jessica's dominant "we" was characterized by the fact that they were successful accomplishers—"smarties" paying attention to their academic status at school. This "we" of Jessica's also showed a strong conformist attitude toward school values and practices. Jessica confidently testified, "All my friends love school" and "like mathematics." "All of us like to do well in school. And all of us, we don't interrupt class and we do pay attention," she added. Jessica's "we" stood firmly as strong supporters for current school practices, and school-related ideology and value system including meritocratic ideology.

The construction of Jessica's positive and collective identity in the mathematics domain was often based on an implicit marginalization or accusation of others whom she believed were different from herself and her friends. Separating herself from others seemed to be essential in the construction of her positive identity. Her use of "they," referring to students other than her friends and her, tends to have a negative connotation. From Jessica's point of view, "they" were passive learners and trouble makers. Responding to my question if she knew anybody who was struggling in math and did not like the subject, Jessica's first reaction was to distance herself from those students. She clearly drew a clear line between those students and her.

> Jessica: I don't really have friends that are like that. But I know that there are a few people like that... When they come to school, they're like, oh man, I hate school, I don't want to be here. And they don't really make good grades because they're usually sleeping or not paying attention or drawing. Something that does not have to do with the class. So, I really don't understand why people do that because I think school is very important.
> J.H.L.: And you said you don't have such a friend, right?
> Jessica: All my friends love school.

Believing the benefits of school work wholeheartedly, Jessica could not understand those students who came to school unprepared or did not show eagerness to learn. When some students kept asking the same question to the teacher, Jessica felt "kind of irritated because I think they should know that they need to be paying attention to understand it." Jessica attributed some students' lack of understanding or low grades to their bad choice—"sleeping or not paying attention" to the teacher. Her description of such students often reflected an ethical or moral judgment.

To Jessica, students who seemed not to understand the importance of schoolwork did not deserve a good grade, additional support, or special care. Jessica wanted to give a high B to her mathematics class because there were "six or seven people in the class" who "misbehaved a lot." In the same vein, she was rather selective about giving help to her classmates. She tended to separate or distance herself from those whom she believed did not deserve her care or help. She said, "I usually don't help them if they don't pay attention because I think that it's not right that they didn't pay attention." Jessica's decision not to help her peers who "didn't pay attention" reflected her math teacher's classroom policy—a firm and consistent use of reward and punishment system based on student behaviors in class. Jessica internalized much of the justice principle that Mrs. Oliver practiced in the classroom and used it as her own.

Jessica expressed a strong voice—including several "shoulds"—arguing for students' ethical obligations at the school including her math classroom. Jessica deeply appreciated Mrs. Oliver's dedicated teaching and declared that "they should be thanking Ms. Oliver because she could be doing any other job. But she chose to teach and to show them, to help them understand what they're going to need." Jessica easily identified herself with the teacher and volunteered to be her advocate. She developed a strong sense of justice differentiating what is right and wrong to do in the mathematics classroom: all students should pay due respect to the teacher—follow the classroom rules and pay attention to her lecture. Therefore, it was not right to give her help to those who did not pay their due respect. Distancing herself from those students was the right thing to do from Jessica's ethical point of view. Armed with this strong sense of justice, Jessica effectively separated herself and her close friends from other students who exhibited negative attitudes towards the school and mathematics and, therefore, did "not make good grades."

Mrs. Oliver supported Jessica' sense of "we"ness. She knew that Jessica's social clique consisted of high math achievers and conceived that Jessica was naturally "part of that group." Mrs. Oliver highly evaluated Jessica's eagerness to learn and succeed as signs of academic potential. The teacher felt that it was right for Jessica to "get into that pre-algebra class" along with her close friends.

There was a clear cultural consistency and social alliance between Jessica's family and teachers providing a collective ground for Jessica's development of "we." Both parties shared White middle class cultural values, such as strong self-discipline, hard work ethic, benefits of education, and meritocracy. They initially provided a cultural and social network that made Jessica's "we" develop; they welcomed the formation of the group and supported its stability and cohesiveness through various out-of-school activities. Jessica eagerly accepted a meritocratic ideology, internalized it, and used it to construct her own social world with her close friends—the "smarties." Furthermore, she developed an ethical standpoint that reflected the justice principle adopted by Mrs. Oliver in her mathematics classroom. This emergent ethical standpoint conferred on Jessica a power and logic to distance and marginalize other students who seemed not to share the same attitudes as hers toward school and mathematics learning.

4.2 Stella: Between Two Worlds

Stella, an African–American girl, was one of the highest achieving students in the advanced math class, who would possibly enter the pre-algebra class the following year. Her family was not a middle class family, yet, provided her with a stable and relatively comfortable home environment with two parents. Her social clique consisted of other African–American girls with varying achievement levels: most of them had little positive experience in the school, particularly in Mrs. Oliver's math class.

As a whole, Stella's experience with Mrs. Oliver's mathematics class was much less positive than Jessica's. She suffered from a high level of anxiety and fear in her mathematics classroom. All three of her interviews revealed her uncomfortable feelings about her mathematics class and learning mathematics. The analysis of her use of "I," "we," and "they" poignantly revealed the dislocation of her two different social worlds—the world of her family and friends, and the world of school and mathematics—and her conflicting identities in each of those two worlds.

As a whole, Stella rarely used "we" as she explained any successful academic performance. It was always "I" who accomplished something at school while her close friends slipped into "they." Stella's academic "I" was very lonely without a collective ground of "we." The following excerpt explains the reason for her lonely "I" without a collective sense of "we." During her second interview, Stella gave her math class a B or C and explained why—which revealed her difficult position in the space:

> Well, I probably won't give (the math class) an A because sometimes it's boring. When I ask for help sometimes, she'll (Mrs. Oliver) be like 'why don't you understand it?' or something like that. I'll be like, "Well, we just learned it" or something. And if we do play (a game), if we do group stuff, all my group will say, 'you've got to say this though because you're the smartest' or something like that. I'll be like, 'Why can't ya'll say it?' They'll be like, 'Because you're the smartest.' So, I have to say all the answers. I don't really want to, though.

Second, when Stella connected her "I" and "we," the "we"—her social clique—tended to have many negative experiences at school. Stella's "we" did not feel comfortable being in a strict teacher's classroom, and automatically "got into trouble" if a teacher looked like a "mean" one. Other students, sometimes, appeared as the one treating Stella's "we" unfairly: behind the scene was a "mean teacher" approving the injustice practiced by the students. Stella described one incident when her "we" was unfairly accused of being "so loud" by another student.

> There was this one girl in our class that she (a teacher) really liked and she made her do everything. She would help her out all the time and she won't be so mean to her. And she'd

let her pass out papers, even though we raised our hands to pass out papers, she'd call on her even though she don't raise her hand. When she'd tell us we could have free time and we could do whatever we want, we'll like talk. And she'd be like 'don't talk so loud. And all these folks be over there talking real, real loud... So, I'd be like we ain't talking that loud.' I'd say 'what about them over there?' and she'd say, 'don't worry about them, they're my business.' And so I'd be like, OK.

Interestingly enough, Stella uses "they" as she explained high-achievers at school. Even though Stella is one of most high performing students, she tended to separate herself from other high-achievers in the class. "They make As in math all the time... Sometimes I am watching the teachers having the students to collect something and the... matching their answers, taking the other students answers. It's just the people that got all their answers right and she'll tell them to just go check other people's. Sometimes I'll just miss one or two problems, I didn't get that right. So I'll have to do them over." It was consistent in her interview data that Stella contrasted herself with other high-achieving girls. Her "I" statements often implied that she was not as good as other high-achieving Caucasian girls, yet, the teacher's record showed Stella herself was, in fact, one of top students in the class.

The second group of people that Stella called "they" was actually her own social clique. She tended to jump from "I" to "they" when she described some activities or behaviors of her close friends that seemed undesirable from the school's perspective. Sometimes, such use of Stella's "they" included herself implicitly. Other times, she completely distanced herself from such "they." The following excerpt shows an unnatural transition from "I" to "they," and absence of "we" when "we" seems a more appropriate term to be used in the contexts.

> J.H.L.: Do you act differently when you are at home and when you are in school?
> Stella: Yes.
> J.H.L.: Tell me a little bit about that. How do you act differently in two places?
> Stella: At school everybody thinks I'm quiet.
> J.H.L.: Hmhm (smiling), I thought like that.
> Stella: I'm real, real loud.
> J.H.L.: Hmhm.
> Stella: And they just be talking loud about stuff on the phone.
> J.H.L.: But is it OK to talk a little bit loud at home?
> Stella: Hmhm. To me, it is. Because we can't talk that loud here at school. And well, I ain't saying it about me but when I'm on the phone with them, they don't talk the way they talk here. In here they talk all like they shy and stuff and they don't cuss in there. But they do when they're on the phone.

In the above excerpt, Stella still admits that there was a connection between herself and, "they", her close friends. "We" appears intermittently in order to indicate Stella's "they" actually means "we,"—her social clique. However, Stella often cut off the connection between herself and her close friends as she described her school experience. When the field of her or her close friends' activity was moved from home to school contexts, Stella quickly removed herself by saying "I don't do anything like that" and referred to "they" rather than "I" or "we" in the rest of the story. The following excerpt shows Stella's strategy to position herself in Mrs. Oliver's classroom as she was asked to describe what was happening in the space. To my question asking her to describe the mathematics teacher in detail, Stella replied:

I don't really know her height, but, she's like not short, but like tall, in the middle kind of. That's all I can say about her looks. Well, she's a good teacher. Some folks say she's mean. Because when they ask for help she'll probably... (*pause*) probably don't want them... (*long pause*). Well, if they ask for help and she'll tell them, but they probably won't get it that good, so they'll probably say she's mean like that. But I think she's nice. I get, whenever she be talking about a subject, I get it, because she explains it good. To me, she do. But I don't know to other people, how other people think and (*whether*) she explains it... I don't know.

Stella's position between the teacher and her close friends posed a problem as she tried to understand the meanings of her academic endeavor in the classroom. Most of her close friends did not have a positive relationship with Mrs. Oliver. They were troubled with Mrs. Oliver's heavy-handed disciplinary rules in the mathematics class and did not accept the idea that those rules were the tool to help them learn the subject. As a result, it was very hard for Stella to draw a coherent picture of both the teacher and her friends in order to support her positive identity in mathematics learning. In her interview data, Stella revealed her confusion and inner conflict in understanding her mathematics teacher and friends because her own experience with the teacher seemed a bit different from her friends'; or at least she wanted to believe that the teacher treated her differently from her peers. Stella's desire to maintain a good relationship with both the teacher and her friends forced her to call her close friends "some folks–they," and to emphasize her subjective knowing, "to me, she do," followed by an intentional unlearning—ignorance "I don't know." She first tried to close her eyes to the experience of her friends and turned away from understanding them, yet her final resort was one of her 48 "I don't know"s constantly repeated throughout her three interviews. This response poignantly revealed her inner conflict between two different social worlds and her unconscious desire not to see the conflict residing deep in her heart. Stella's subsequent answers about the teacher and her friends were inconsistent, revealing her insecure feelings and contradictory knowledge about two hostile worlds existing simultaneously within the school.

However, Mrs. Oliver witnessed a negative impact of Stella's "we–they," her social clique consisting of African–American girls. She saw it was gradually eroding Stella's potential for learning and academic success; yet the teacher did not see the psychological conflict deeply residing in Stella's mind. As a result, Mrs. Oliver predicted a rather negative future for Stella's academic endeavor despite her high evaluation of Stella's current intellectual capacity.

She is just an excellent student and I'm a little bit concerned right now. She's sort of falling down a little bit. She doesn't seem to be as interested in her academics as a whole... I don't know if it's from outside this class, but I don't think she wants the attention of being a goody–goody or doing too well. I think she's happy just doing what everybody else does. And she could, I think, do a lot better. Simply because she likes where she is. She's comfortable. She likes that area of comfort. She's making good, she's doing well. And so she doesn't want to get out of the box.

Without significant supporting structure from either side, Stella confessed a deep self-alienation in learning mathematics. It is not surprising that Stella portrayed herself in the mathematics class as "a rat in a maze... trying to get out of" Stella wanted to be a good student and hoped to be different from her brother, who was a "trouble maker" both in school and at home. She tried hard to endorse and accept the legitimacy of school practices and to identify herself with her mathematics teacher. To accomplish these goals, she consciously changed her way of talking and acting at school, conformed school practices, and distanced herself from her close friends. However, her teacher did not see Stella's struggle. It was only the lonely "I" of Stella's vacillating between the two worlds without a strong ground of "we," collective identity.

4.3 Amanda: Political Dissent

Amanda, 1 of 13 Caucasian girls in the advanced mathematics class, used to be a high-performing student in elementary school. However, her grade had plunged this year dramatically. In Mrs. Oliver's classroom, Amanda always sat in the back of the classroom strictly limiting her communications to her close friends sitting nearby. Her social clique consisted of White girls with varying achievement levels; most of their grades had plunged this year. Amanda's view of school and her mathematics class was heavily influenced by her prior experience at home; she witnessed ongoing family violence that victimized her mother, the only person she could rely on in her family. As a result, she had a more critical view of the world of adults, including many school practices and supporting ideologies for those practices.

Amanda had a high level of anxiety in mathematics class. Like other participants in this study, she was worried about grades and test scores: yet, her anxiety was more deeply related to the social and emotional aspects of the class, such as whether she could enjoy learning the subject and working with the teacher. Amanda perceived the mathematics classroom as a lonely and impersonal environment in which she might be embarrassed in public. She refrained herself from asking questions in public. She hardly expected any support from Mrs. Oliver. Rather she viewed the teacher as a cold and distant authority who would "embarrass" students who were already struggling in the situation.

Amanda's "we" in the mathematics class shared some similar characteristics with Stella's "we." For example, as a whole, Amanda's "we" were vulnerable people in the class. Amanda's "we"s were unfairly treated and ignored by the teacher and other "smarter children." This Amanda's "we" frequently emerged in conjunction with "they"—"smarter children" who seemed to enjoy some privileges at school.

J.H.L.: Then tell me a little bit more about how she is treating her students differently.
Amanda: She likes, um... older children, I mean smarter children, alright, Stella, Whitney, and Janet. She don't let us go to the restroom at all. And if we ask to go to the restroom, she'll say no, you should have went [gone] earlier. But if they go ask her, she'll say yeah.
J.H.L.: Can you remember the time when that kind of thing really happened?
Amanda: Stella. She went to ask to go to the restroom and then Mrs. Oliver said, you should have went [gone] earlier, and then um, Janet went to ask her if she could get to the restroom and Mrs. Oliver let her. I don't think that's fair. Because Stella, she's really good in the class.

Getting "embarrassed" was another big issue to Amanda's "we." Amanda viewed the classroom as a risky space in which she and her "we" were getting embarrassed because they could not understand the concept taught by the teacher.

> Amanda: She embarrasses us sometimes, embarrasses us when we can't do a problem. She's like well, I've already taught you that, and... ugh. Like, it took ten minutes to answer a question and I got really embarrassed. And she told me that I needed to know that and I couldn't help it that I didn't understand it.

Associated with Amanda's "we", constantly being embarrassed in the classroom, was another intriguing concept of "we"—nice, pretty and attractive girls. She explained this "we" with pride, often in relation to her important "they"—boys in the class: "They don't treat us that bad. They kind of treat us good. But if they think we're ugly, they say, Oh God, that girl is so ugly. Like somebody (an ugly girl) asked him for a phone number because they (ugly girls) liked them (boys). Then he was like, I don't want to give you that. I don't want to give you my phone number because you're stump ugly." This "we" of Amanda's was depicted as an object of the constant gaze from boys—another important "they" in her data. Amanda perceived that this "they" constantly talked about her and her friends (her "we"): Her "we" defended their action against those boys' senseless judgment. "I can, sometimes I can hear them talking about me and they talk about my friends and stuff. Like if I'm walking with my friends, they'll say, oh, those two are gay. We're like, I don't think we're gay just because we're walking and talking to each other."

However, there are some aspects that were more unique in Amanda's "we." Unlike Stella's "I" and "we" disassociated from each other in school contexts, Amanda's "I" and "We" are firmly aligned to each other. Amanda frequently used both terms interchangeably as she described her own and her "we's" experience in school/classroom contexts. Amanda's "we" is active, even rebellious to what was then given to them, creating a different set of ideas of mathematics learning. Amanda's "we" consciously challenged the authority and rules of the game in the classroom. Naturally, there was a strong presence of "I," subjective voice, questioning the legitimacy of the existing practice, and challenging the given authority in the space.

> Amanda: Well, sometimes she (Mrs. Oliver) just says, "OK, this is how you're going to do this." And *we tell her a different way*, she's like... April, her aunt is a teacher and she told her that her aunt taught her differently. And Ms. Oliver said well, "That's just my way. This is the way I'm going to teach it."
> J.H.L.: So, um, tell me about that time when April and you explained another way to solve the problem. The teacher looked like she didn't like it?
> Amanda: Hmmm (yes). She was mean. She just said, "This is the way I'm doing it and this is the way I'm going to do it and this is the way you're going to learn it."..... (but) *I think it's better if you can learn in a different way that you can learn and understand better.*

Amanda's "we" seemed to grow in terms of scope. Amanda's "we," in her later interviews, tended to include not only herself and her close friends, but also others who were vulnerable and hurt in the space. Amanda's effort to make sense of her own experiences led her to a sense of community with others who had similar experiences. Based on this expanded "we"ness Amanda felt distressed even when the teacher's disciplinary action was directed toward someone else. In the following excerpt, Amanda used "people," "I," and "we" interchangeably reflecting her resistance was not only individual, but based on the collective experience of "we-people."

> J.H.L.: Tell me about what you don't like about math.
> Amanda: I don't like when she yells at *people* and I really don't like that. And I don't like when she embarrasses *people*. Like say, *I* got a problem wrong and she'd get mad and say that she's already taught *us* that (and) *we* should know it. And I really don't like when she does that.
> J.H.L.: How do you feel when she says that kind of thing to students?
> Amanda: I feel mad. I'm like well, you shouldn't be treating *people* like that.

Amanda's understanding of the importance of mathematics and schoolwork was not much different from that of high-achieving middle class girls, such as Jessica. She emphasized that she would be different from others in her family who "want[ed] to go to college, but once they took a job, forgot about it." She did not "want to be like them." She firmly stated, "I want to be the one in my family that goes to college." However, Amanda's social world—"we"—was different from Jessica's: Amanda's "we" consisted of White or Black working class girls who found Mrs. Oliver's mathematics class uncomfortable or even frightening. Amanda's "we" was the one who suffered from high anxiety experiencing a dearth of caring relationship in the mathematics classroom. Her "We" had little support for her academic endeavor.

More important, Amanda possessed a strong subjective voice that she had developed from her own lived experience (Belenky et al. 1986). She did not blindly accept the manifested school ideology, but carefully examined its value and coherence as she experienced it. Having experienced family violence, Amanda held a firm belief in the necessity of caring relationship in the shared world of many people, including her classroom, and desire for such element in the learning environment. To Amanda, her mathematics classroom was not a mere learning space but a place where such ethical principles should be observed.

Apparently, Amanda and her teacher, Mrs. Oliver, had very different views on mathematics learning. Mrs. Oliver believed that Amanda was "not doing as much as she could do." The teacher deplored, "She (Amanda)'s capable, [but] she just doesn't do it. …She is not concerned about it." To Mrs. Oliver, who strongly believed in hard work, Amanda's lack of motivation and attitude of resistance were frustrating. Mrs. Oliver, however, continued to provide Amanda with her best support—close monitoring and strict supervision, which she believed Amanda really needed. The teacher strictly applied her classroom rules to Amanda as she did to many other students who came from lower socioeconomic backgrounds. Mrs. Oliver never doubted the best policy for Amanda was forcing her to follow, and hopefully embody, the strict rules in her class so that she would discipline her for a very slight chance of academic success in the following grade.

However, Amanda saw the situation differently. To her, effective and comfortable learning could not take place unless she had a warm and mutually respectful relationship with the teacher. Being forced to participate in learning without such an interpersonal base made Amanda hate mathematics and the teacher. Amanda who originally hoped her teacher "like me [her]" and "I [she] don't [doesn't] do bad in her class" changed her mind when the teacher "started yelling at kids" who "couldn't understand something." Her mathematics teacher, from Amanda's perspective, had disregarded the ethic of caring, which was central to her relational world especially when students were vulnerable, helpless, and most in need. In her last interview, Amanda declared that "Math is not important." She refused to learn mathematics and attached no value to mastering the subject. Instead, she expressed her fondness in Language Arts taught by a teacher who was often referred to as the most accessible, caring teacher in the sixth grade hall.

5 Discussion

The findings of this study reconfirm the thesis in the literature of mathematics education suggesting the complexity of gender issues in this discipline. As a whole, this study demonstrates that the three participating girls' experiences with school mathematics were grounded in the cultural characteristics of their mathematics classroom, which emphasized authoritative, procedural, individual, and competitive work. However, it also reveals that the class backgrounds and ethnicities, and the characteristics of the peer groups they belonged to, were other important factors that created a significant variation across the three students' profiles.

Jessica, a girl from a white middle class family, had less difficulty in constructing a positive identity in mathematics. Her social world, including friends and family, provided her with a cultural and social context that helped her readily adopt and internalize the various school-related ideologies, such as the legitimacy and goodness of school practices in general, and an ethic of hard work and justice principle as practiced by Mrs. Oliver's classroom in particular. Jessica's use of "we" showed that she clearly separated herself from "others who don't like mathematics and don't care about school," successfully constructing her positive and confident identity in the domain of mathematics. It is noticeable that Jessica' positive identity development in mathematics was accompanied with a path of moral development closely associated with the Justice principle (Strike 1997).

In contrast, Stella, an African–American girl from a working class background, was experiencing a much deeper confusion between her social world and the world of school mathematics. No matter how hard she tried, she could not seem to make these two worlds fit together. Even though she was one of the top-performing students in the advanced mathematics class, Stella had no "we" who could support her academic endeavor: Her social "we" slipped into a rather antagonistic "they" when she tried to develop a more positive academic identity in mathematics. Exposed to conflicting voices and identities within herself, Stella was confused feeling a deep alienation in the class environment. It was difficult for her to develop a firm and stable academic identity when she felt either of those two words was not exactly hers. Caught in a crisis of understanding the gap between her two worlds, she was left to deal with the fracture in her experiences with significant people, her teacher and her friends, and the implications of learning mathematics.

Several Black scholars (e.g., Stiff and Harvey 1988) have argued that the culture of traditional mathematics classrooms poses a challenge to Black students because it lacks cultural relevance. Stiff (1990) suggests that the African–American cultural frame of reference entails a particular set of dispositions, such as working in support groups and using a "conversational style" of discourse in an instructional situation. Yet, traditional mathematics classrooms remain largely based on individual and procedural work and employ an elaborate syntactical discourse as the acceptable form of communication for mathematics learning. The African–American cultural frame of reference or "Afrocultural ethos (Boykin et al. 2005, p. 521)" could be found across different spectra of social class within a black community (Timm 1999), yet such cultural differences/conflicts seem most visible and frequently documented in schools serving primarily working class Black students. For example, Boykin et al. (2005) confirmed that there is a consistent cultural mismatch between working class Black students and their teachers in poor urban schools. Experiencing a significant mismatch between their own cultural dispositions and what is expected of them in a traditional mathematics classroom, Black students—particularly working class Black students—may feel a high level of anxiety and even inadequacy in their endeavor to succeed in mathematics (Reyes and Stanic 1988; Ryan and Ryan 2005).

Amanda's case raises another interesting issue. Even though Amanda possessed the cultural knowledge of how school worked for or against each student, she found that the authoritative and procedural instructional approach that prevailed in her current classroom uncomfortable and even unacceptable. Drawing on her traumatic experiences at home, Amanda paid close attention to the relational aspects of her mathematics classroom. She viewed the mathematics classroom as a space where people should observe some basic ethical principles (e.g., caring) in order for authentic learning to take place. Based on this ethical framework, Amanda constructed her "we" and "they" with a strong present of "I" voices in each of her stories. By positioning herself in this way, Amanda actually lost a connection with the teacher, and other high-achieving peers in the classroom. It seemed rather natural that Amanda voluntarily chose to desert mathematics learning: She directed herself to other subject areas which seemed to readily meet her ethical claim. Working class students' resistance against the culture of school and its negative consequence have been well documented in literature (e.g., Willis 1981; Archer et al. 2007). On one hand, Amanda's decision not to pursue a higher level of mathematics learning was a strategy helping to preserve her own identity and cultural values: on the other hand, such decision limited her academic advancement for subsequent years, and therefore fulfilled the cycle of class reproduction. Researchers observed that effects of class/SES on girls' choice of math and science related fields were stronger than boys' choice of those disciplinary fields (Trusty et al. 2000). As a result, women in mathematics and related fields tend to come from middle/upper class families—often in significantly higher numbers than those of their male counterparts in the same disciplines (Oakes 1990). Furthermore, the two reasons for Amanda's disenchantment with mathematics are hardly new ones. Several studies in the past confirmed that girls and women experience the field of mathematics as an uncaring field (Herzig 2004) filled with authoritative male voices. Even those who have been successful and remained in the field constantly negotiate their identities facing the conflict images of femininity and mathematics as male's domain (Rodd and Bartholomew 2006).

Given the small sample size and the conflation of gender, race/ethnicity, and class in the three young adolescent girls reported here, findings from this study are limited in explaining the concrete relationship between the three socio-cultural factors and the mathematics learning experience of girls. However, this study suggests that there could be a subtle, yet powerful (dis)connection and dynamic existing between the cultural structure of their math classroom and each student' class/race-based disposition, which determines their "right" or "desirable" place in the school's math streaming system.

Despite many limitations of the findings described above this study still poses at least two important discussions concerning the gender and class inequity problem in mathematics education in American schools and beyond. First of all, this comparison suggests a critical outlook toward the prevalent culture of mathematics classroom that can be generalized to middle and high schools across USA. Despite 20 years of criticism on such practice (Ball 1990) and continuous reform efforts (NCTM 1991, 2000), many American middle and high school math classes still repeat the same authoritative, procedural, and individuated mode of instruction. More important, this study illuminates such mathematics classrooms are not merely a neutral, peaceful learning space, but a political arena (Noddings 1993) in which different ideological and moral discourses compete against one another, and individual (Zevenbergen 2001) and collective identities (Atweh et al. 1998; Zevenbergen 2005) are constantly being negotiated and reconstructed. Even though the three students were in the same classroom, each student's interpretation of and experience in the space was dramatically different from each other's. To Jessica, the space was a supportive environment for her mathematics learning; yet, the same environment posed a serious problem to Stella and Amanda. They felt self-alienation or anger being in the space.

It should be noted that the traditional and procedural mode of mathematics instruction has been widely criticized for its lack of relevance to the life-experience of minority and/or working class students who present a different cognitive/learning style—field-dependent (Bond 1981), and collective and mutual (Foley 2005). Therefore, it is not surprising that several mathematics education researchers confirmed that providing a cooperative learning environment in which teachers engage students in group projects and connect their academic work to everyday life experiences/examples is crucial to successful mathematic learning experiences of minority and/or working class students (Boaler 2006; Balfanz et al. 2004). Mrs. Oliver's traditional math classroom failed to foster such connection among different groups of students as well as between her and those coming from other than white middle class families. This may be one of the possible reasons contributing to the dearth of minority and/or working class girls in higher tracks of math class as well as the field of mathematics in later years.

From a gender perspective, this phenomenon of dis/connected identity may pose a critical challenge to minority or working class girls considering women's different ways of knowing: girls, as connected knower, tend to rely on the commonality of experience, rather than authority, as they try to access another person's idea or knowledge (Belenky et al. 1986). Therefore, Jessica who was able to develop a strong psychological, emotional connection with both her high-achieving white peers and teacher appeared to be more comfortable with the authoritative and procedural learning environment than others. In contrast, when the possibility of such connectedness is slim, almost invisible, the task of understanding an abstract mathematical concept or procedure through a distant teacher may become a challenging task to Stella, and even unthinkable from Amanda's perspective. From this perspective, it seems rather inevitable that in the case of Stella, the path to a higher level of mathematics learning did not appear as a pleasant, desirable, and promising one; it looked more like a creepy and unsafe route to her. Amanda's loss of connection from high-achieving peers and the teacher, as well as her emerging connection with "pretty, nice girls" and other students distressed in the class cost her initial desire to build a relationship with the teacher and disregard the importance of mathematics to her ultimate goal-going to college. It should be noted that class and racial membership creates a whole new set of social and psychological dynamic in the three girls' experiences with school mathematics due to such (dis)connected experience and identity with different groups of peers and ultimately with their math teacher.

Findings from this study suggest that the most common rhetoric for learning mathematics based on the instrumental value of school mathematics (either to be a good student or to get into college) or advocating and inculcating the legitimacy of school practices in Stella's or Amanda's heart is not an appropriate or effective method to support their higher level of mathematics learning. Stella's life, her consciousness, cannot be separated from the rich, collective cultural identity of the African–American community and its collective experiences of the fundamental injustice of many school practices. In the same vein, Amanda will not easily give up her critical voice, which sprang from her subjective knowledge on the illegitimacy of an authority and strong commitment to the ethic of care over any other principles.

What appears to be a more relevant—and urgent—approach is to raise mathematics educators' sensitivity and responsiveness toward their students who come from other than White middle class backgrounds. In this study, Mrs. Oliver, although she was one of the most dedicated, hardworking, and competent mathematics teachers at the school, failed to understand the identity dilemmas Stella and Amanda faced. Her strong conviction about the value of schooling, the ethic of hard work/self-discipline, and the effectiveness of traditional math instruction method kept her from reaching a deeper, more reflective understanding of the cultural and social challenges some of her students experienced in her classroom. Teachers with more cultural awareness and sensitivity will develop a sense of advocacy for marginalized students (Parsons 2005) and critically examine their own instructional practices. This process of self-reflection and taking their position as an advocate—rather than a neutral instructor—are critical to shaping their everyday "teaching as social activism." (Gutstein et al. 1997, p. 732).

When the community of educators as a whole, inspired by their reflective and critical examinations of their own practices, many more Stellas and Amandas in the near future will see the possibility of building a meaningful bridge between their seemingly irreconcilable worlds and competing voices—their identities as reflected in their construction of "I," "we," and "they". These students need to find teachers who will eagerly provide culturally sensitive, caring support as they attempt to develop a sense of "we" supporting their academic endeavors.

Equally important, however, is that it involves more than an individual student's or teacher's agency to create the possible strategies that would lead to a more equitable intellectual pursuit and joyful learning in school mathematics across diverse groups of students: it demands that multiple agencies work together critically examining the current structure and practice of school mathematics in light of gender, racial, and class implications, and constantly soliciting strategies that effectively disturb or counteract the reproduction process of the existing status quo across the various pipelines in the discipline. I, therefore, argue that there should be systematic examination and intervention built into the entire US public school system directed toward various groups of underserved students. Without such realization and commitment at a structural level—claiming schools as a space for social justice (Giroux 1997) and "prepare[ing] people for critical citizenship" (Skovsmose 2007, p. 215)—it would be hard to make a significant difference, a positive long-term lasting effect, in the most marginalized groups of students—White and Black working class girls who face multiple challenges in their pursuit of higher level mathematics learning.

References

Ansell, E., & Doerr, H. M. (2000). NAEP finding regarding gender: Achievement, affect, and instructional experiences. In E. A. Silver & P. A. Kenny (Eds.), *Results from the seventh mathematics assessment of the national assessment of educational progress* (pp. 73–106). Reston: National Council of Teachers of Mathematics.

Archer, L., Halsall, A., & Hollingworth, S. (2007). Class, gender, (hetero)sexuality and schooling: Working class girls' engagement with schooling and post-16 aspirations. *British Journal of Sociology of Education, 28*(2), 165–180.

Atweh, B., Bleicher, R. E., & Cooper, T. J. (1998). The construction of the social context of mathematics classrooms: A sociolinguistic analysis. *Journal for Research in Mathematics Education, 29*, 63–82.

Atweh, B. H., Forgasz, H., & Nebres, B. (Eds.). (2001). *Sociocultural research on mathematics education: An international perspective*. Mahwah: Lawrence Erlbaum.

Bakhtin, M. M. (1981). *The dialogic imagination: Four essays by M. M. Bakhtin*. C. Emerson & M. Holquist (Trans.). Austin: University of Texas Press.

Balfanz, R., Legters, N., & Jordan, W. (2004). Catching up: Effect of the talent development ninth-grade instructional interventions in reading and mathematics in high-poverty high schools. *NASSP Bulletin, 88*(641), 3–30.

Ball, D. (1990). The mathematical understanding that prospective teachers bring to teacher education. *Elementary School Journal, 90*, 449–466.

Bauer, D. M., & McKinstry, S. J. (1991). Introduction. In D. N. Bauer & S. J. McKinstry (Eds.), *Feminism, Bakhtin, and the dialogic* (pp. 1–16). New York: SUNY Press.

Belenky, M. F., Clinchy, B. M., Goldberger, N. R., & Tarule, J. M. (1986). *Women's way of knowing: The development of self, voice, and mind*. New York: Basic Books.

Bhana, D. (2005). "I'm the best in maths. Boys rule, girls drool". Masculinities, mathematics and primary schooling. *Perspectives in Education, 23*(3), 1–10.

Boaler, J. (1997). Reclaiming school mathematics: The girls fight back. *Gender and Education, 9*(3), 285–305.

Boaler, J. (2000). Identity, agency, and knowing in mathematics worlds. In J. Boaler (Ed.), *Multiple perspectives on mathematics teaching and learning* (pp. 171–200). Westport: Ablex.

Boaler, J. (2006). Promoting respectful learning. *Educational Leadership, 63*(5), 74–78.

Bond, G. C. (1981). Social economic status and educational achievement: A review article. *Anthropology and Education, 12*(4), 227–257.

Boykin, A. W., Tyler, K. M., & Miller, O. (2005). In search of cultural themes and their expressions in the dynamics of classroom life. *Urban Education, 40*(2), 521–549.

Brown, T., McNamara, O., Hanley, U., & Jones, L. (1999). Primary student teachers' understanding of mathematics and its teaching. *British Educational Research Journal, 25*, 299–322.

Bucholtz, M., & Hall, K. (2005). Identity and interaction: Sociocultural linguistic approach. *Discourse Studies, 7*(4–5), 585–614.

Buxton, C., Caroline, H. B., & Caroline, D. (2005). Boundary spanners as bridges of student and school discourses in an urban science and mathematics high school. *School Science & Mathematics, 105*(6), 302–333.

Campbell, P. (1989). So what do we know with the poor, non-white females?: Issues of gender, race, and social class in mathematics and equity. *Peabody Journal of Education, 66*(2), 95–112.

Chipman, S. F. (2005). Research on the women and mathematics issues: A personal case history. In A. M. Gallagher & J. C. Kaufman (Eds.), *Gender differences in mathematics: An integrative psychological approach* (pp. 1–24). Cambridge: Cambridge University Press.

Department for Education and Skills (DfES) (2005). *National statistics first release, GCE/VCE A/AS examination results for young people in England, 2003–4 (final)*. Department of Education and Skills. http://www.dfes.gov.uk/rsgateway/DB/SFR/s000586/SFR262005.pdf.

Fairclough, N. (2003). *Analyzing discourse: Textual analysis for social research*. London: Routledge.

Foley, G. (2005). Educational institutions: Supporting working-class learning. *New Directions for Adult and Continuing Education, 106*, 37–44.

Gee, J. P. (1999). *An introduction to discourse analysis: Theory and method*. New York: Routledge.

Giroux, H. A. (1991). Modernism, postmodernism, and feminism: Rethinking the boundaries of educational discourse. In H. A. Giroux (Ed.), *Postmodernism, feminism, and cultural politics: Redrawing educational boundaries* (pp. 1–59). Albany: SUNY Press.

Giroux, H. (1997). *Pedagogy and politics of hope: Theory, culture and schooling*. Boulder: Westview Press.

Gutstein, E., Lipman, P., Hernandez, P., & Reyes, R. (1997). Culturally relevant mathematics teaching in a Mexican American Context. *Journal for Research in Mathematics Education, 28*(6), 709–737.

Hargreaves, M., Homer, M., & Swinnerton, B. (2008). A comparison of performance and attitudes in mathematics amongst the 'gifted'. Are boys better at mathematics or do they just think they are? *Assessment in Education: Principles, Policy & Practice, 15*(1), 19–38.

Herzig, A. (2004). Slaughtering this beautiful math': Graduate women choosing and leaving mathematics. *Gender and Education, 16*(3), 379–395.

Hyland, K. (2002). Options of identity in academic writing. *ELT Journal, 56*(4), 351–358.

Jenson, J., De Castell, S., & Bryson, M. (2003). "Girl talk": Gender, equity, and identity discourses in a school-based computer culture. *Women's Studies International, 26*(6), 561–573.

Jones, J., Porter, A., & Young, D. (1996). Perceptions of the relevance of mathematics and science: Further analysis of an Australian longitudinal study. *Research in Science Education, 26*(4), 481–494.

Jungwirth, H. (1993). Reflections on the foundations of research on women and mathematics. In S. Restivo, J. P. V. Bendegem & R. Fischer (Eds.), *Math worlds: Philosophical and social studies of mathematics and mathematics education* (pp. 134–149). Albany: SUNY Press.

Lacasa, P., del Castillo, V., & García-Varela, A. (2005). A Bakhtinian approach to identity in the context of institutional practices. *Culture Psychology, 11*, 287–307.

Leder, G. C. (1992). Mathematics and gender: Changing perspectives. In D. A. Grouws (Ed.), *Handbook of research on mathematics teaching and learning* (pp. 597–622). New York: Simon & Schuster Macmillan.

Lubienski, S. T. (2002). Research, reform and equity in U.S. mathematics education. *Mathematical Thinking and Learning*, *4*(2, 3), 103–125.

Mcgraw, R., Lubienski, S. T., & Strutchens, M. E. (2006). A closer look at gender in NAEP mathematics achievement and affect data: Intersections with achievement, race/ethnicity, and socioeconomic status. *Journal for Research in Mathematics Education*, *37*(2), 129–150.

Mendick, H. (2005). A beautiful myth?: The gendering of being/doing 'good at maths'. *Gender and Education*, *17*(2), 203–219.

Morris, P. (1994). *The Bakhtin reader*. London: Edward Arnold.

NCES: National Center for Education Statistics (2004). *Trends in educational equity of girls and women: 2004*. NCES 2005-016. http://nces.ed.gov/nationsreportcard/ltt/results2004/sub-mathgender.asp.

NCES: National Center for Education Statistics (2005). *Gender differences in participation and completion of undergraduate education and how they have changed over time: Postsecondary education descriptive analysis report*. Washington: U. S. Government Printing Office.

NCTM: National Council of Teachers of Mathematics (1991). *Professional standards for school mathematics*. Reston: National Council of Teachers of Mathematics.

NCTM: National Council of Teachers of Mathematics (2000). *New principles and standards for school mathematics*. Reston: National Council of Teachers of Mathematics.

Noddings, N. (1993). Politicizing the mathematics classroom. In S. Restivo, J. P. V. Bendegem & R. Fischer (Eds.), *Math worlds: Philosophical and social studies of mathematics and mathematics education* (pp. 151–161). Albany: State University of New York Press.

Oakes, J. (1990). Opportunities, achievement, and choice: Women and minority students in science and mathematics. *Review of Research in Education*, *16*, 153–222.

Parsons, E. C. (2005). From caring as a relation to culturally relevant caring: A White teacher's bridge to black students. *Equity & Excellence in Education*, *38*, 25–34.

Preckel, F., Goetz, T., Pekrun, R., & Kleine, M. (2008). Gender differences in gifted and average-ability students comparing girls' and boys' achievement, self-concept, interest, and motivation in mathematics. *Gifted Child Quarterly*, *52*(2), 146–159.

Reyes, L. H., & Stanic, G. M. A. (1988). Race, sex, socioeconomic status, and mathematics. *Journal for Research in Mathematics Education*, *19*, 26–43.

Rodd, M., & Bartholomew, H. (2006). Invisible and special: Young women's experiences as undergraduate mathematics students. *Gender and Education*, *18*(1), 35–50.

Roychoudhury, A., Tippins, D., & Nicols, S. (1993). An exploratory attempt toward a feminist pedagogy for science education. *Action in Teacher Education*, *15*(4), 36–46.

Ryan, K. E., & Ryan, A. M. (2005). Psychological processes underlying stereotype threat and standardized math test performance. *Educational Psychologist*, *40*(1), 53–63.

Rymes, B. (2001). *Conversational boarderlands*. New York: Teachers College Press.

Secada, W. G. (1992). Race, ethnicity, social class, language, and achievement in mathematics. In D. A. Grouws (Ed.), *Handbook of research on mathematics teaching and learning* (pp. 623–660). New York: Macmillan.

Skovsmose, O. (2007). Doubtful rationality. *ZDM: The International Journal on Mathematics Education*, *39*(3), 215–224.

Stiff, L. V. (1990). African American students and the promise of the curriculum and evaluation standards. In T. J. Cooney & C. R. Hirsch (Eds.), *Teaching and learning mathematics in the 1990s* (pp. 152–158). Reston: NCTM.

Stiff, L. V., & Harvey, W. B. (1988). On the education of black children in mathematics. *Journal of Black Studies*, *19*, 190–203.

Strike, K. A. (1997). Justice, caring and universality: In defense of moral pluralism. In M. S. Katz, N. Noddings & K. A. Strike (Eds.), *Justice and caring: In search for common ground in education* (pp. 21–36). New York: Teachers College Press.

Tang, R., & John, S. (1999). The 'I' in identity: Exploring writer identity in student academic writing through the first person pronoun. *English for Specific Purposes*, *18*(1), 23–39.

Tate, W. (1997). Race-ethnicity, SES, gender, and language proficiency trends in mathematics achievement: An update. *Journal for Research in Mathematics Education*, *28*(6), 652–679.

Timm, J. T. (1999). The relationship between culture and cognitive style: A review of the evidence and some reflections for the classroom. *Mid-Western Educational Researcher, 12*(2), 36–44.

Trusty, J., Robinson, C., Plata, M., & Ng, K. (2000). Effects of gender, SES, and early academic performance on postsecondary educational choice. *Journal of Counseling & Development, 78,* 463–472.

van Langen, A., & Dekkers, H. (2005). Cross-national differences in participating in tertiary science, technology, engineering and mathematics education. *Comparative Education, 41*(3), 329–350.

van Langen, A., Rekers-Mombarg, L., & Dekkers, H. (2006). Group-related differences in the choice of mathematics and science subjects. *Educational Research and Evaluation, 12*(1), 27–51.

Voloshinov, V. N. (1973). *Marxism and the philosophy of language.* L. Matejka & I. R. Titunik (Trans.). Cambridge: Harvard University Press.

Wertsch, J. V. (1991). *Voices of the mind: A sociocultural approach to mediated action.* Cambridge: Harvard University Press.

Willis, P. (1981). *Learning to labor: How working class kids get working class jobs.* New York: Columbia University Press.

Zevenbergen, R. (2001). Mathematics, social class, and linguistic capital: An analysis of mathematics classroom interactions. In B. Atweh, H. Forgasz & B. Nebres (Eds.), *Sociocultural research on mathematics education* (pp. 201–216). Mahwah: Erlbaum.

Zevenbergen, R. (2005). The construction of a mathematical "Habitus": Implications of ability grouping in the middle years. *Journal of Curriculum Studies, 37*(5), 607–619.

Shall We Do Politics or Learn Some Maths Today? Representing and Interrogating Social Inequality

Paul Dowling and Jeremy Burke

Abstract In this chapter we shall first introduce a schema that describes strategies of representation in terms of whether representation is explicit or tacit and whether it is oriented in consonance or dissonance with dominant or expected patterns, in this case of social inequality. We shall then use this schema to describe the construction of gender and social class in school textbooks, giving some attention to the contexts of their use. We shall argue that addressing social inequalities demands explicit, dissonant strategies, referred to here as interrogation. However, by reflecting on a particular critical mathematics lesson apparently interrogating racial inequality, we conclude that interrogation itself is likely to lead to misrepresentation where the mathematical activity is foregrounded and mathematics is likely to lose out where it is not. Ultimately, we may be left with the choice of whether to do politics or to teach mathematics.

1 Strategies of Representation

On 23rd March 2010 the online version of *The Guardian* published an article about a photograph shot by Jimmy Sime some seventy three years earlier (Jack 2010).[1] The photograph foregrounds two boys from Harrow School, dressed in their 'Sunday best'—top hats, jackets with buttonhole flowers, wide collars and ties, waistcoats, and carrying canes. They were in town to watch the annual Eton and Harrow cricket match at Lords cricket ground. To the right of the two Harrovians—as we look

---

[1]The original *Guardian* story has been removed, the copyright having expired. A version of the article appeared on p. 8 of the G2 section of *The Guardian* on 24th March 2010. The longer, original story is still available on *The Economist* website at http://moreintelligentlife.com/content/ian-jack/5-boys (last accessed 12th April 2011).

P. Dowling (✉)
Institute of Education, University of London, London, UK
e-mail: p.dowling@ioe.ac.uk

J. Burke
Kings College London, London, UK
e-mail: jeremy.burke@kcl.ac.uk

H. Forgasz, F. Rivera (eds.), *Towards Equity in Mathematics Education,*
Advances in Mathematics Education,
DOI 10.1007/978-3-642-27702-3_8, © Springer-Verlag Berlin Heidelberg 2012

at the picture—and slightly backgrounded are three local boys looking at them, apparently with some amusement. The Harrow boy in centre frame stands erect, one hand resting on his cane, the other on a stone bollard; he looks somewhat detached. The three locals seem scruffy by comparison in rather shabby jackets and trousers and their hands in their pockets. They are also there for the cricket match, but hoping to earn some money by doing minor errands for the presumably wealthy spectators. There are many ironies associated with the photograph that Ian Jack describes in his article. Perhaps most tellingly, the fates of the two groups seem to have contrasted radically with their respective starts in life. All three of the locals—from working class backgrounds—went on to have long and successful working and family lives. The Harrow boy in the foreground —whose father was a career soldier—succumbed to diphtheria whilst visiting his family in India a little more than a year after the photograph was taken; the other, the son of a stockbroker suffered years of mental illness and died in an asylum in 1984.

Sime's photograph was originally published the day after it was taken in *The News Chronicle* under the headline 'Every picture tells a story'. Since then, the image has been used regularly in the media, generally accompanying articles calling for educational reform. Jack reports that the Eton and Harrow cricket match is nowadays a far more low key affair and the dress code is 'smart casual'. As he points out, "if a photographer wanted to re-create Sime's picture, he might be faced with five boys dressed much the same, in jeans and brand names. Giving a superficial impression of equality, the picture would be even more of a lie than before" (Jack 2010). We might need a little more information than the imaginary photo alone would provide in order to accuse it of lying—the caption 'Harrow and local comprehensive boys outside the Lords cricket ground', perhaps—something would need to draw attention to the fact that there is some kind of differentiation going on. The original image achieves this via dress and posture, rendering explicit the representation of the familiar pattern of social class hierarchy, albeit that the story behind the photo—of course, unknown to most viewers—challenges this representation as exaggeration and irony. The imagined image and caption connotes social class, perhaps, via the differentiation of a famously named school and an anonymous school described only by a common category. This imagined image/caption, however, does not tell us who is who and so hints at, as Jack suggests, a representation of equality—they're all dressed the same—a representation that is dissonant with prior expectations, perhaps, but is hardly explicit.

Of course, both images are silent on other dimensions of social inequality; they include only white and apparently able-bodied boys who we are likely to assume (if we think about it at all) are heterosexual. These other dimensions may be connoted in our interpretations of the images, but, certainly, familiar patterns of inequality in these respects will not be challenged. An image from the *Edexcel GCSE Mathematics Foundation Book 1* (Pledger et al. 2006) addresses some of these dimensions visually and seems to share aspects of each of the others—the original Jimmy Sime photograph and its imagined update. The image this time is a cartoon. It shows five individuals who, according to bubblespeech, have just won a lottery: a female of colour; a white male in a wheelchair; a white female wearing glasses; a male of

Table 1 Strategies of representation

Expression	Orientation to pattern	
	Consonance	Dissonance
Connotative (tacit)	*invisibility*	*tokenism*
Denotative (explicit)	*stereotype*	*interrogation*

colour; and a second white male; all but the wheelchair-bound individual are standing.[2] This image resembles the first one discussed above in that social categories are visibly explicit and it resembles the second in that it is representing—albeit tacitly—an equality that perhaps challenges our expectations. Oddly, this presentation of equality is emphasised by the mathematical topic of the verbal text, 'equal shares': 'We are having equal shares', bubbles the wheelchair-bound individual; the text beneath the image reads, "The key words here are **equal shares**. They tell you to divide" (Pledger et al. 2006, p. 18).

In our descriptions of these images (real and imagined) we have introduced two variables. The first distinguishes between tacit or connotative representation of social inequality on the one hand, and explicit or denotative representation on the other. The imaginary photograph and the cartoon fall into the first category, the original photograph into the second insofar as it relates to social class and into the first in respect of other dimensions of inequality. The second variable distinguishes between representations that are consonant or dissonant with expected patterns; Sime's photograph is in the first category, the other two texts are in the second. This organisation gives rise to the strategic space in Table 1.

Now we are referring to the categories, *invisibility*, *tokenism*, *stereotype* and *interrogation*, as strategies, which is consistent with Dowling's *Social Activity Method* or SAM (Dowling 1998, 2009). Thus the Sime photograph is a denotative representation—which is to say that it is clearly marked out in the text—of social class hierarchy that is consonant with the familiar pattern of professional middle class/working class. On the other hand, it can only connote (by absence) other dimensions of social inequality, which thereby remain unchallenged. This image deploys a strategy of social class stereotype and of invisibility with respect to other dimensions. The imaginary update of this photograph, together with its caption, deploys social class tokenism and, again, invisibility with respect to other dimensions. The mathematics text cartoon deploys tokenism with respect to the dimensions of social equality that it represents.

We are claiming that both of the variables here—expression and orientation to pattern—may be helpful in talking about the representation of social inequalities in all contexts, but, in particular, in school mathematics and associated practices as constituted in textbooks, in the media, in the classroom, and in research. Before moving to a consideration of agency in these contexts, we shall offer a discussion of the dimension of gender and social class in terms of the categories in Table 1.

---

[2]It might be noted, in passing, that the wheelchair looks rather like a hospital machine, suggesting, perhaps, that the disability might be only temporary.

2 Texts, Contexts and Patriarchy

In many areas media representations of gender have changed significantly so that, for example, we no longer expect that financial reporters and experts will always be male and the most celebrated media 'mathematician' in the UK is probably Carole Vorderman, who also has an online mathematics school at www.themathsfactor. com/.[3] On the other hand, there are plenty of areas in which stereotypical, patriarchal images persist. In sports presentation ceremonies, for example, the medal carriers are generally glamorous women—irrespective of the gender of the recipients— and the medal presenters are generally men (usually not at all glamorous) and it is rather sickening to see that the main image on the *Car Middle East Online* report on the Lebanon Motor Show shows a female model in a sleek, little black dress posing seductively with a sleek, red and black sports car.[4] It remains the case that most of the business leaders appearing in the media are male—an invisibility strategy in terms of gender, so that women industrial leaders, when they do appear, are tokens. If this kind of difference generalises (which, of course, it may not), then the patterning of gender *may* be shifting on social class lines, which is to say (optimistically) greater equality in (certain) intellectual areas and (less so) traditional inequality in others. To the extent that this is the case, then our alliance of enlightened intellectuals will generally produce an allergic reaction to expressions of gender inequality. Dowling, for example, reported this observation in 1991:

> The Mathematical Experience by Davis and Hersh (1981) includes sixty-six pictures of mathematicians (including one of each of the authors, on the dust jacket) precisely none of which are images of women. Furthermore, as far as I can see, there is no reference to a female anywhere in the book and the expected gender of a mathematician (and that of a mathematician's student) is quite apparent in the authors' choice of pronoun... (Dowling 1991a, p. 3)

...which was very consistent—invisibility indeed! We are surprised that Davis and Hersh (1981) felt able to produce such a text only thirty years ago—perhaps the de-gendering of the intelligentsia proceeds at a slower pace in some areas than in others. Certainly, in 2010, *The Observer* (online edition) appeared embarrassed that 'Women are under-represented in mathematics' (Bellos 2010), but managed to come up with one, Hypatia. The article (which gives two versions of her date of birth for some reason) claims that "her most valuable scientific legacy was her edited version of Euclid's *The Elements*" (Bellos 2010). Euclid himself did not get into *The Observer* top 10, having apparently made way for his editor as the token female, who could—as far as we can see from the article—reproduce but not inseminate. The website Math(blog) might have helped the author of *The Observer* article, providing a list of ten remarkable mathematicians—all women (Cangiano 2008). This list includes Hypatia and the text informs us that she was the inventor

---

[3]Retrieved April 12, 2011.

[4]See www.carmiddleeast.com/article-1-1425-lebanon_motor_show_2010. Last accessed 12th April 2011.

of the hydrometer—an achievement possibly overlooked by *The Observer* author in his haste to insert a token female into his own list. The Math(blog) text moves beyond tokenism by bringing the dissonant orientation to dominant patterns of inequality into denotative expression; Math(blog) here recruits an interrogative strategy. The crucial question, to which we shall return later, is what is being transmitted here?

Research on gender issues in school texts (there is not a great deal of this looking specifically at mathematics textbooks) has focused attention on revealing what we are calling here invisibility and stereotype strategies. A good deal of this research notes either the absence of female figures or pronouns. McKimmie (2002), for example, observes that one mathematics textbook makes reference to thirty-one famous mathematicians, only one of whom is a woman—again, Hypatia of Alexandria; Selzer (2007) notes the absence of women in the public sphere in Australian history textbooks; Stott (1990) found predominantly male pronouns in geology textbooks; Witt (1997) noticed significantly more male than female representations in reading texts and Commeyras and Alvermann (1996), in a study of then recent world history textbooks in the US noted that "language functions in these textbooks to position women in stereotypical ways or to obfuscate the patriarchal system that accounts for women's demeaning experiences and differential treatment throughout history" (p. 47). Some of this work reveals the relative absence of the feminine. This kind of partial visibility can be regarded as a stereotype strategy, which is made more explicit in many texts analysed in the research. Here, attention is on the differentiation of masculinity and femininity in terms of attribution binaries—strong/weak, active/passive, and so forth—or occupations—paid/domestic work, senior/junior. Thus Mikk (2002), looking at Estonian textbooks, finds women portrayed as caring housewives and men with successful paid occupations. Similar differences are reported by Esen (2007) in Turkish textbooks; Sansom (2000), in respect of Italian language textbooks, says "The consistent linguistic, semantic and contextual representation of women in a stereotypical, sexist and limited manner perpetuates negative and restrictive perceptions in society" (Sansom 2000, p. 7). Some of the research indicates that the picture has been changing. Lee and Collins (2009), for example, point out that whilst there had been a relative absence of feminine pronouns in earlier Australian English-language textbooks, there is a recent trend towards using gender-neutral terms, such as they, or paired pronouns, he or she. Tamarra McKimmie (2002) notes that female characters in three Australian mathematics textbooks have moved from stereotypical roles in the home to others including company director, craftspersons, and windsurfing instructors. Philip Clarkson (1993), in a study of 18 texts frequently used in Victoria, Australia, considers the ratio of representations of male to female characters to be "nowhere near the two to four times of earlier years" (p. 16). Helen Forgasz (1997) found "small lingering vestiges of inequity" (p. 136) in cartoon images in one mathematics textbook; nevertheless, when asked to identify the gender of the images, school students often applied stereotypical categories, such as the association of skateboarding with boys rather than girls, where there was ambiguity. Dowling (1991a) described the move from stereotype to tokenism strategies in the development of the major UK textbook scheme produced by the School Mathematics Project (SMP).[5] We shall return to this work later in giving some consideration to agency.

Research concerned with school texts often seems to presume that texts that are consonant with dominant patterns of gender inequality—texts that, in the terms presented here, deploy invisibility and stereotype strategies—are responsible for transmitting or reproducing these patterns. Thus, for example, Zevenbergen (1993) asserts that the absence of feminine images or the limitation of the feminine to the passive powerfully influence girls' construction of their 'mathematical identities'. Arikan (2005) suggests that representations of gender (also of age and social class) in English language teaching coursebooks have an effect on the ways in which teachers and students view the world. Kereszty (2009) suggests that textbooks can present a 'hidden curriculum' that is identity forming in terms of gender. Esen's (2007) analysis of new textbooks in Turkey argues that they perform a function of gender segregation that runs counter to an apparently 'gender-blind' reform of the curriculum in that country.

These transmission models, however, may be accused of unduly simplifying the situation insofar as they remove the textbook from the conditions of its use. After all, we have described Sime's photograph as consonant with dominant patterns of inequality, yet it is used, for the most part, in the context of challenging these patterns, presenting alternatives that are dissonant with them, thus shifting the text, now in context, to interrogative mode. Sunderland et al. (2001) suggest:

> ...rather than looking 'in the text' for bias, or even looking diachronically for improvements in textual representations of gender, a more relevant and fruitful focus in terms of both language learning and gender identity may be the mediation of gender representation in textbook texts by teachers, through their discourse on those texts. (Sunderland et al. 2001, p. 283)

Interrogation is one possibility, of course, though Naureen Durrani (2008) found that the construction of the 'ideal' Pakistani woman in the elementary student school textbooks that she analysed were consistent with stereotyping practices more generally in the school and that student perceptions were also consistent with this stereotype condemning, for example, women who violated *purdah*. Durrani describes the curriculum as

> ...a set of discursive practices that constitute gendered subjectivities which serve the interest of the dominant groups in Pakistan—the military, religious leaders, and men—and marginalise and silence the interests of women, non-Muslims and civil society. The findings suggest the success of the curriculum in interpellating the students as subjects through which they become agents of national, religious and gender ideologies that sustain the existing social relations in Pakistan. (Durrani 2008, p. 608)

Of course, the curriculum is not the only set of discursive practices at play:

> Although students' voices were in line with the curriculum, this does not rule out the influence of other cultural resources such as family, social class, ethnicity, peers and the larger social milieu in the NWFP. For this group of students, it seems that ethnic identity, class and

---

[5]The textbooks were from the SMP numbered series and from the later *SMP 11-16*, both schemes were published by Cambridge University Press.

the influence of the Soviet-Afghan war as well as the current war on terror in Afghanistan may have reinforced the curricular impact. (Durrani 2008, p. 608)

The processes of transmission that are necessarily invisible when looking at closed texts may be revealed by research, such as Durrani's (2008), that also deals with people. Sian Beilock et al. (2010) reveal what was perhaps a less deliberate mechanism in the gendering of 'math anxiety'. They measured the mathematical achievement and gender ability beliefs (whether it is thought that ability in mathematics and reading is differentiated on gender lines) of 117 elementary school students at the beginning and end of a school year; they also measured the levels of 'math anxiety' and mathematical knowledge of their female teachers. The researchers found that:

> By the school year's end, female teachers' math anxiety negatively relates to girls' math achievement, and this relation is mediated by girls' gender ability beliefs. We speculate that having a highly math-anxious female teacher pushes girls to confirm the stereotype that they are not as good as boys at math, which, in turn, affects girls' math achievement. If so, it follows that girls who confirmed traditional gender ability beliefs at the end of the school year (i.e., draw boys as good at math and girls as good at reading) should have lower math achievement than girls who do not and than boys more generally. This is exactly what we found. (Beilock et al. 2010, p. 2)

Teacher 'math anxiety', it is proposed, is realised in girls, but not boys, in such a way as to display stereotype strategies in relation to gender ability and this, in turn, is realised in the form of relatively weak mathematical performances. Now, here we might speculate that the 'math-anxious' teachers are exhibiting invisibility strategies to the extent that they do not actually make explicit (perhaps not even internally) that their anxiety is associated with their gender. Yet the proposition by the authors suggests that girls manage to translate the invisibility strategies into stereotype strategies, so themselves rendering the representation of traditional gender ability patterns explicit (at least, when they are provoked to do so by researchers). Invoking the social learning theory of Perry and Bussey (1979), they argue that "children do not blindly imitate adults of the same gender. Instead they model behaviours they believe to be gender-typical and appropriate" (Perry and Bussey 1979, p. 3). Insofar as children will attribute to textbooks the same kind of authority in terms of gendered identities that they attribute to their teachers, then similar kinds of argument might be made in terms of the transmission of these identities.[6]

This transmission, as described by Beilock et al. (2010) would appear to have installed in the girls a contradiction between a feminine identity and an identity as a successful mathematician, resulting in the attenuation of the latter. Other research has found evidence of similar tensions. Work by Boaler (1997a, 1997b) and Skelton et al. (2010), for example, found that girls in 'top sets' have to find a balance between

---

[6]Testing such propositions experimentally would be likely to generate ethical problems to the extent that one group of children would be working with materials that would be regarded as potentially harmful. A quasi-experimental design would be methodologically problematic in respect of controlling for teacher, student, school etc variation, though this would not render such a study valueless.

the maintenance of a feminine identity—'doing girl'—and academic success. Girls, it is claimed, lack confidence with the pedagogic strategies—"the survival of the quickest" (Boaler 1997b, p. 575)—adopted in these classes. To be a " 'proper school girl' demands that girls present as cooperative, diligent, conscientious, with a care and concern for relationships with teachers and friends and a (heterosexual) interest in boys" (Skelton et al. 2010; p. 187).

All of this might be taken to suggest that girls' mathematics performances in general are likely to be lower than those of boys. At the time of publication— or at least the authoring—of *The Sociology of Mathematics Education* (Dowling 1998) this was just about true at age 16 in the UK. However, even in 1998 girls in the UK were outperforming boys in primary mathematics—they had been for a considerable time—and were overtaking boys in performances at 16, outperforming boys in terms of the number of General Certificate of Secondary Education (GCSE) passes at grades A* thru C in every year from 1999 to 2008.[7] For some time, girls had also been outperforming boys in most other subjects at this age in the UK and in other countries, provoking what Katherine Hodgetts (2008) reports as a "globalised moral panic" (p. 466). The responses to this panic included the setting up, in Australia, of the Australian Parliamentary Inquiry into Boys' Education. Hodgetts conducted an analysis of the accounts presented by witnesses to this inquiry in which she discovered evidence of a prevalent stereotype discourse of gender and ability that constructed girls as passive, compliant, oriented to meaningless learning—focusing attention on unnecessary details of presentation and so forth— and as products of pedagogic manipulation by, for example, rote learning. Boys, on the other hand, were believed to be active learners, resistant to what they perceive as meaningless tasks and as products of integrity in learning, achieving understanding. Ironically, what achieves academic success for girls is constituted as undermining the value of this success, whereas boys' failure seems to be taken to be an indicator of their preference for the more challenging route, which choice was taken as evidence their greater potential. This certainly resonates with earlier work by Valerie Walkerdine (Walkerdine 1989, also Walden and Walkerdine 1985) that showed that whilst boys could be interpreted as having 'flair' in primary mathematics, even though they might make lots of mistakes, girls' successes were often interpreted as the result of hard work, which is to say, not proper learning and unlikely to result in sustained success in the future. Just to add another twist, in the 2009 GCSE examinations, boys performed better than girls for the first time in a decade. This reversal coincided with a change in the form of the examination (see JCQ 2009), though the girls' advantage had been reducing since 2005.

In any event, mathematics education research must address mathematical performances of both boys and girls, and must also address the pervasion of strategies

---

[7]GCSE figures for girls and boys available at http://www.bstubbs.co.uk/gender/fem-g.htm and http://www.bstubbs.co.uk/gender/male-g.htm (last accessed 29 January 2012). Despite girls achieving proportionately better grades at GCSE and at A level (until 2009) fewer girls take maths beyond GCSE. (Around 40:60 girls to boys take A level Maths and around 30:70 girls to boys take A level Further Maths.)

within mathematics pedagogy that are consonant with patriarchy. These strategies are clearly instantiated in some (many?) learning resources, but also seem to be present in the practices of teachers. Can textbook design assist with this? We have mentioned Dowling's (1991a) reference to the tokenism strategies in the scheme *SMP 11-16*. These strategies do constitute a dissonance with patriarchal expectations, but do so tacitly. Dowling's article refers to instances of characters who are out of stereotypical position. These included a task in three parts, in which the first two parts referred to raids on a castle by a Viking chief and his sons; the third part:

> Yet another chief and his daughters stole 6941 gold pieces. The chief took 2417 and his daughters shared the rest equally. If each daughter got 348 pieces, how many daughters were there? (SMP, quoted by Dowling 1991a; p. 3)

"An unusual way, perhaps, to count your daughters" (Dowling 1991a; p. 3). At that time, Dowling drew on the activity theory of Aleksei Leont'ev (1978, 1979) to describe the way in which the text, overall, constituted the solution of the mathematical tasks and the elaboration of mathematical skills as the goal of the pedagogic action, so that the more or less arbitrary construction of the tasks themselves provided the operational means. The connotative dissonance of some of the 'public domain' settings (Dowling 1998, 2010) of these tasks with patriarchal expectations was, he argued, thereby suppressed. This, indeed, did seem to be the case with these texts in use. Another *SMP 11-16* task showed a map of a park with the positions of statues of an unlikely collection of men (Christopher Columbus, William Shakespeare, Isaac Newton, Karl Marx, Albert Einstein and Elvis Presley) and two women (Florence Nightingale and George Eliot). A drawing showed the statue of the Victorian English novelist—still known by her *nom de plume*, George Eliot, though her in real life name was Mary Ann (or Marian) Evans—dressed as one would expect and with the name, George Eliot, clearly visible on the pedestal. In visits by Dowling to classrooms in which this text was used, no one had ever remarked on any aspect of the task. He reported the following conversation with a girl in one of the classes:

> "Who's this?"
> "It says it's George Eliot."
> "Do you know who George Eliot is?"
> "Is he a scientist?"
> "So who's on top of the pedestal?"
> "Florence Nightingale."
> "She's moved over to George's pedestal, has she?'
> "Yes." (Dowling 1991a, p. 4)

In present day England, George might be more readily interpreted as a diminutive of Georgina, of course rendering this instance of tokenism completely pointless. Dowling's intervention had been intended to provide a context for what we are now calling an interrogation strategy, rendering the dissonance between text and patriarchal discourse explicit. However, the text and the more general context of the mathematics lesson privileged mathematical action over the operational public domain content, and so the student, possibly quite rightly, saw no need to pursue the apparent inconsistency. The potential anti-patriarchal petard, without it even being recognised as such, was nevertheless successfully defused.

We—researchers, textbook authors, teachers—seem to be ensnared in a discursive web that offers us a choice between teaching mathematics and interrogating pedagogy by de-privileging mathematics. Girls seem to be offered a choice between femininity and real mathematical success. Is there a way out? Before addressing this question, we shall return to a consideration of social class and related patterns of inequality.

3 Putting the Class into Texts

As well as considering the gendering of mathematics texts, Dowling (1991a, 1991b) also describes a preliminary analysis of the *SMP 11-16* scheme, then widely in use, in social class terms. This preliminary analysis revealed a resonance between certain properties of the textbooks and what used to be referred to as the 'quality' and 'popular' newspapers. Essentially, the books directed at 'high ability' students were weighty, both physically and verbally, whilst those directed at 'low ability' students were lightweight—being stapled booklets rather than bound books—and included far more pictorial text and far less verbal text than the 'high ability' texts. The use of humour in the two series was different as well: 'high ability' humour demanded an understanding of the mathematics; 'low ability' humour was generally disconnected from the mathematics and often made fun at the expense of it. These and other differences corresponded with differences in the form and content of the 'quality' and 'popular' newspapers. Furthermore, it was clear from an analysis of the content of the newspapers—for example, comparing the positions advertised in the jobs sections—and from data relating to their readerships that the 'quality' press would tend to connote professional middle class, whilst the 'popular' papers would tend to connote working class. The textbook differentiation thus connoted a social class hierarchy via its consistency with the media. It was also noted that the 'low ability' books had a marked tendency to emphasise manual activities, which were very much backgrounded in the 'high ability' books, a manual/intellectual distinction that, again, connoted social class. When looked at separately—which would be the students' view as they would generally see (at least, see closely) only one series of books—the two series of books can be described as deploying invisibility strategies in respect of social class. When taken together, the strategy overall is stereotype. In neither view is there any dissonance with traditional class hierarchy. Differentiation in terms of resonances between textbook and newspaper format—still linked with socioeconomic class—remains in mathematics textbooks currently in use in the UK. The foundation level textbooks in the Heinemann/Edexcel series, for example, employs larger font size and more illustrations than the higher tier books.

The classification and listing of differences and similarities in this work does bear some similarity with some of the research on the gendering of mathematics textbooks that is referred to above and is certainly worth doing. However, in and of itself, it lacks any theoretical development. This was undertaken subsequently and

Table 2 Domains of action

	Expression (signifiers)	Content (signifieds)	
		I^+	I^-
$I^{+/-}$ represents strong/weak mathematical institutionalisation.	I^+	*esoteric domain*	*descriptive domain*
	I^-	*expressive domain*	*public domain*

the result formed the initial components of Social Activity Method (SAM) (Dowling 1998, 2009).[8] One move in this direction was to make a distinction between what appeared to be, from a mathematical point of view, arbitrary and what appeared to be non-arbitrary text. Thus some text concentrated on mathematical forms of expression and on mathematical objects, whereas other text recruited non-mathematical expressions and/or content, such as tasks involving domestic settings. Settings that are strictly mathematical are referred to as esoteric domain; settings that are apparently non-mathematical are public domain. An analytic distinction can be made between the form of expression or signifiers of a textual segment and its content or signifieds. Expression and content can then be considered separately in terms of the extent to which they are strongly or weakly institutionalised mathematically. There is always a need to take context into account. Thus, $2x + 5 = 7$ is clearly strongly mathematically institutionalised in terms of expression. In respect of content, however, it would be weakly institutionalised if we are told that x stands for an amount of money in a public domain setting. If there is no public domain setting, then the content as well as the expression would be strongly mathematically institutionalised. The product of the two variables, expression and content, generates the possibility space in Table 2.

This analysis generates four domains of action. The esoteric and public domains are the pure forms in terms of the mathematical institutionalisation of expression and content. The 'multicultural' task about equal shares of a lottery win, the Viking and his daughters, and the George Eliot tasks are all constructed in the public domain. The descriptive domain is the domain of mathematical modelling, where mathematical expression is used to describe a public domain setting. The expressive domain is the domain of pedagogic metaphor —an equation is a balance, for example, or a fraction is a piece of cake—where non-mathematical expression signifies a mathematical content.

Another move (Dowling 1994, 1996) was the introduction of the category, discursive saturation. Texts that exhibit high discursive saturation (DS^+) will tend to render the principles of a practice—in this case, mathematics—explicit within linguistic discourse; texts that exhibit low discursive saturation (DS^-) do not. Clearly, mathematical discourse is only fully available within DS^+ esoteric domain text.

---

[8]In the 1998 book, the term 'social activity theory' was used; this was changed to SAM in the later book to avoid confusion with some Vygotskian work. The analysis that follows has been radically simplified in both theoretical and empirical terms in order to accommodate to the limitation on space. We feel that it nevertheless has some value in respect of understanding the relationship between mathematics education and social injustice.

This discourse, however, is the regulating discourse for text in the other domains. A key implication of this is that public domain text is not simply text in an activity other than mathematics, but is the product of a recontextualising action, a gaze cast from the esoteric domain that privileges mathematics over the regulating principles of the activity from which the recontextualised setting has been taken. Put simply, shopping, for example, in a public domain school mathematics text is very different from domestic shopping, it is mathematised shopping. The public domain, then, is not real mathematics, nor are its actions structured in the same way as those of the activities that appear to be being signified; the Viking/daughters text is, in this sense, not unrepresentative. The public domain constitutes a world of mythical practice.

The domains of action schema and the category, discursive saturation, arose—along with a number of other elements of SAM—out of the engagement with the *SMP 11-16* texts. Analysed in these terms, it was demonstrated that the 'high ability' books—and we can now associate these with relatively high socioeconomic status (SES)—involved movement between all four domains of action. The public domain often served as a portal into mathematics or as, shall we say, a playground for mathematising. Entry into the esoteric domain—often via the descriptive and (less frequently) the expressive domains—was almost always facilitated[9] and DS$^+$ text was prevalent. The 'low ability'/low SES status books, on the other hand, remained almost exclusively within the public domain. This was interpreted by Dowling (1998) as providing a real mathematical career for the 'high ability'/high SES students, whilst 'low ability'/low SES students were confined to a domain that had no value either in mathematics or anywhere else other than in the classroom. There is a sense in which the mathematics curriculum, as represented in the *SMP 11-16* texts, can be understood as a mechanism for translating SES into 'ability'. This is consistent with a good deal of the sociology of education of the 1970s and 1980s, where social class was a key focus.[10] On the basis of the discussion above, it also seems appropriate to consider schooling as a device that translates gender into ability, though the mechanisms are different.

Further reflection on the esoteric domain of school mathematics suggests that the mathematical practices that constitute the esoteric domain are very substantially dissociated from any other practice (Dowling 2010). School mathematics is very different, by and large, from university mathematics and from other school subjects. It is not that there is no relationship between these different activities—clearly there is—but that this relationship is more appropriately thought of as one of recontextualisation, rather than of progression or of direct use-value. The esoteric domain of school mathematics thus emerges as a region of essentially self-referential practice. To this extent, it may be appropriate to regard school mathematics as apprenticing both low SES and high SES students to mythical practices, the difference being that high SES students emerge with the symbolic value of school certification. The question that we now need to address is, can we do anything about this?

---

[9] An interesting exception being chapters on probability, where entry into the esoteric was delayed very considerably (Dowling 1998).

[10] See, for example, Bernstein (1971), Bourdieu and Passeron (1977), Rist (1970), Sharp and Green (1975), Willis (1977), Young (1971).

4 Critical Mathematics Education?

Here is Judith Butler on signification:

> As a process, signification harbours within itself what the epistemological discourse refers to as 'agency'. The rules that govern intelligible identity, i.e., that enable and restrict the intelligible assertion of an 'I', rules that are partially structured along matrices of gender hierarchy and compulsory heterosexuality, operate through repetition. Indeed, when the subject is said to be constituted, that means simply that the subject is a consequence of certain rule-governed discourses that govern the intelligible invocation of identity. The subject is not determined by the rules through which it is generated because signification is not a founding act, but rather a regulated process of repetition that both conceals itself and enforces its rules precisely through the production of substantialising effects. In a sense, all signification takes place within the orbit of the compulsion to repeat; 'agency' then, is to be located within the possibility of a variation of that repetition. (Butler 1990, p. 145)

We can, perhaps, see the 'compulsion to repeat' in the invisibility and stereotyping strategies illustrated by the Sime photograph and by the *SMP 11-16* texts that signify social class—we're now using SES—in a way that is consonant with the dominant pattern. Do we see agency in the use of the photograph in newspaper and magazine articles that seek to change the patterns of privilege in schooling, or in the tokenism of the 'equal shares' text or the Viking and his daughters? Again, we see the 'compulsion to repeat' in Beilock's math anxious teachers and their female students and also in the witness reports analysed by Hodgetts. Do we see agency in Dowling's attempted provoking of interrogation in the 'George Eliot' classroom? Indeed, do we see agency in analysis that renders explicit significations consonant with patterns of dominance such as that reported and presented in this chapter? Perhaps we do, but Butler has taken us only part of the way, leaving open the matter of principles of realisation of repetition and agency; there is no sociology in her account of signification. We can try to activate the social as follows:

> ...borrowing from cybernetics... the sociocultural [is] that which is defined by strategic, autopoietic action directed at the formation, maintenance and destabilising of alliances and oppositions, the visibility of which is emergent upon the totality of such action, rendering them available as resources for recruitment in further action... (Dowling 2010, p. 1)

In this conception, repetition and agency are combined in autopoietic (self-making) action, and 'rule-governed discourses' are socially grounded as alliances and oppositions; action relates to a field of social relations not just to discourses. Thus, invisibility and stereotype strategies constitute action that relates to the maintenance of, for example, patriarchal alliances; tokenism and interrogation, then relate to the maintenance of anti-patriarchal alliances that stand in opposition to patriarchy.

What makes an alliance visible is the regularity of practice that emerges from the totality of action in the field. Thus the field of mathematics education may be understood as an alliance of teachers, textbook authors, guardians of curricula and official assessments, mathematics education researchers, and so forth. This is not a tight alliance. Nevertheless, a practice emerges that articulates mathematics with what we might refer to as pedagogic theory and it is the latter that facilitates the gaze of the esoteric domain and the construction of the other domains of action

(Dowling 2010), generally privileging the mathematical discourse of the esoteric domain.[11] This privileging is sustained by the strong institutionalisation of curricula and the official assessment and certification schemes through which school mathematical performances are held to account and also by the visibility of learning resources primarily textbooks—to institutionalised interests beyond the classroom (for example, the family, media, and so forth).

Clearly, participants in the mathematics alliance are also participants in other alliances including those that are concerned to contest social injustice. We can see the impact of this juxtaposition in the deployment of strategies that are dissonant with dominant patterns of social inequality. We would expect these strategies to have the most impact where the dissonance becomes explicit, which is to say in interrogation, in critical mathematics. An example of such a move can be seen in a mathematics lesson conducted by Eric Gutstein (2002) that is discussed in Dowling (2010). Gutstein was concerned with the teaching of the statistical concept, expected value, and had discovered that the random number generator that is available on graphing calculators was a useful resource. He was also concerned to develop mathematics as a critical discourse in respect of social inequality. In his lesson, he recruited statistical information on the number of traffic stops made by police in Illinois by ethnic group. Gutstein and his class discovered that the number of stops of Latino drivers was very substantially more than the expected value predicted by the calculator and concluded that the Illinois police were racist. However, not only might one suppose that the ethnicity of the driver is often unknown to police officers in advance of the stop, but random traffic stops are actually illegal in the US, being a breech of Fourth Amendment rights; the police have to have a reasonable suspicion that an offence has been committed before proceeding to make the stop. Clearly, any association found between ethnicity and SES that made it more likely that Latinos, in comparison with white professionals, would be driving old and poorly maintained vehicles with visible defects may contribute to the explanation of the disproportionate number of stops of Latinos. This is an indicator of social inequality, but it is not evidence of police racism. On the contrary, accusations of police racism on the basis of such data are explicit representations that are consonant with what appear to be predetermined assumptions about patterns of behaviour in society—just alternative stereotypes. Equally, of course, our interrogation of Gutstein's lesson does not in any sense constitute an assertion that Illinois traffic police are not racist.

The public domain setting of Gutstein's lesson has been achieved by a *fetch* strategy that casts a mathematical gaze onto an area of potential political significance and retrieves mathematically resonant features. This is not a problem, insofar as the students are also to be 'fetched'—perhaps 'pulled' would be more apt—into mathematical discourse. A problem does arise, however, when the results of mathematical manipulation are then *pushed* back into the political arena, providing a simplistic and misleading response to a complex sociological problem. Any serious attempt to address this problem would have involved qualitative research, but that would have entailed losing mathematical control and that, we suggest, is not possible within the

---

[11] Though there are exceptions, see the examples from TIMSS in Dowling 2007.

context of an institutionalised curriculum. Oddly, the Wilmette Police Department [WPD] used the same statistical strategy as Gutstein (though, one imagines, without the use of the graphing calculator) to demonstrate that 'Wilmette police officers are engaging in bias free traffic enforcement' (Carpenter 2004, p. 66). Ironically, their results raise exactly the same sociological questions as Gutstein's lesson and for similar reasons these questions are not going to be followed up by the WPD. Despite his earnest intentions, Gutstein is recruiting politics for mathematical purposes; the WPD is recruiting mathematics for political purposes.

So, what is our answer to the question posed at the end of the previous section of this chapter, and in this chapter more generally: Can we do anything about social inequalities within the context of mathematics teaching? Our first answer is, yes, we can continue to strive to enable all students to acquire erudite, esoteric domain mathematics. The second answer is that, of course, we must eliminate representations that are consonant with prevalent patterns of social inequality. However, attempts to move from tokenism to interrogation, whilst politically praiseworthy in terms of motive, are unlikely to generate appropriate understanding of social injustice, because the public domain settings that we construct will always be mathematically motivated distortions of the alliances that we want to destabilise, resulting only in alternative stereotypes. The way out of this is to privilege our political motivations over our mathematical ones. Of necessity, this involves switching to another activity. It seems to us that this is probably the best strategy: we can be both mathematics educators and political activists, just not at the same time.

References

Arikan, A. (2005). Age, gender and social class in ELT coursebooks: A critical study. *Hacettepe Üniversitesi Eğitim Fakültesi Dergisi, 28*, 29–38. Retrieved January 29, 2012 from www.efdergi.hacettepe.edu.tr/200528ARDA%20ARIKAN.pdf.

Bellos, A. (2010). The 10 best mathematicians. *The Observer (online edition)*. Retrieved January 29, 2012 from www.guardian.co.uk/culture/2010/apr/11/the-10-best-mathematicians.

Beilock, S. L., Gunderson, E. A., Ramirez, G., & Levine, S. C. (2010). Female teachers' math anxiety affects girls' math achievement. *Proceedings of the National Academy of Sciences of the United States of America, 107*(14), 1–4. Retrieved January 29, 2012 from www.pnas.org/content/early/2010/01/14/0910967107.full.pdf+html.

Bernstein, B. (1971). *Class, codes and control: Theoretical studies towards a sociology of language*. London: RKP.

Boaler, J. (1997a). Reclaiming school mathematics: The girls fight back. *Gender and Education, 9*(3), 285–305.

Boaler, J. (1997b). Setting, social class and survival of the quickest. *British Educational Research Journal, 23*(5), 575–595.

Bourdieu, P., & Passeron, J.-C. (1977). *Reproduction in education, society and culture*. London: Sage.

Butler, J. (1990). *Gender trouble*. London: Routledge.

Cangiano, J. (2008). 10 remarkable female mathematicians. *Math(blog)*. Retrieved January 29, 2012 from math-blog.com/2008/09/28/10-remarkable-female-mathematicians/.

Carpenter, G. (2004). *Wilmette Police Department 2004 annual report: Traffic stop data collection and analysis*. Wilmette: Wilmette Police Department.

Clarkson, P. (1993). Gender, ethnicity and textbooks. *Australian Mathematics Teacher*, *49*(2), 14–16.

Commeyras, M., & Alvermann, D. E. (1996). Reading about women in world history textbooks from one feminist perspective. *Gender and Education*, *8*(1), 31–48.

Davis, P. J., & Hersh, R. (1981). *The mathematical experience*. Brighton: Harvester.

Dowling, P. C. (1991a). Gender, class and subjectivity in mathematics: A critique of Humpty Dumpty, *For the Learning of Mathematics*, *11*(1), 2–8.

Dowling, P. C. (1991b). A touch of class: Ability, social class and intertext in SMP 11–16. In D. Pimm and E. Love (Eds.), *Teaching and Learning Mathematics*. London: Hodder & Stoughton.

Dowling, P. C. (1994). Discursive saturation and school mathematics texts: A strand from a language of description, In P. Ernest (Ed.), *Mathematics, Education and Philosophy: An international perspective*, London: Falmer Press.

Dowling, P. C. (1996). A sociological analysis of school mathematics texts. *Educational Studies in Mathematics*, *31*(4), 389–415.

Dowling, P. C. (1998). *The sociology of mathematics education: Mathematical myths/pedagogic texts*. London: Falmer Press.

Dowling, P. C. (2007). Quixote's science: Public heresy/private apostasy. In B. Atweh, A. C. Barton, M. C. Borba, N. Gough, C. Keitel, C. Vistro-Yu, & R. Vithal (Eds.), *Internationalisation and globalisation in mathematics and science education* (pp. 173–198). Dordrecht: Springer.

Dowling, P. C. (2009). *Sociology as method: Departures from the forensics of culture, text and knowledge*. Rotterdam, The Netherlands: Sense Publishers.

Dowling, P. C. (2010). Abandoning mathematics and hard labour in schools: A new sociology of education and curriculum reform. In C. Bergsten, E. Jablonka, & T. Wedege (Eds.), *Mathematics and mathematics education: Cultural and social dimensions. Proceedings of Madif 7* (pp. 1–30). Linköping, Sweden: SMDF. Plenary presentation at Madif 7, Stockholm University, Stockholm, 27th January 2010. Retrieved January 29, 2012 from http://www.pauldowling.me/stockholm2010/Abandoning%20Mathematics%20and%20Hard%20Labour%20in%20Schools.pdf.

Durrani, N. (2008). Schooling the "other": The representation of gender and national Identities in Pakistani Curriculum Texts. *Compare. A Journal of Comparative Education*, *38*(5), 595–610.

Esen, Y. (2007). Sexism in school textbooks prepared under education reform in Turkey. *Journal for Critical Education Policy Studies*, *5*(2), 15. Retrieved January 29, 2012 from www.jceps.com/index.php?pageID=article&articleID=109.

Forgasz, H. J. (1997). Choosing mathematics textbooks: Criteria for equity. In H. Hollinsworth & N. Scott (Eds.), *Mathematics—creating the future. Proceedings of the 16th biennial conference of the Australian Association of Mathematics Teachers* (pp. 132–136). Brunswick, Victoria: Australian Association of Mathematics Teachers Inc.

Gutstein, E. (2002). Math, SATs, and racial profiling. *Rethinking Schools Online*, *16*(4), 18–19 www.rethinkingschools.org.

Hodgetts, K. (2008). Underperformance or "getting it right"? Constructions of gender and achievement in the Australian inquiry into boys' education. *British Journal of Sociology of Education*, *29*(5), 465–477.

Jack, I. (2010). The photograph that defined the class divide. guardian.co.uk. (This article is no longer available on the website, but see note 1.)

JCQ (2009). *Rules for new GCSE specifications – terminal requirements, re-sits and cashing-in*. Retrieved January 29, 2012 from http://www.jcq.org.uk/attachments/published/1209/GCSE%20Entry%20&%20Aggregation%20Rules%20March%202011.pdf.

Kereszty, O. (2009). Gender in textbooks. *Practice and Theory in Systems of Education*, *4*(2), 1–7. Retrieved January 29, 2012 from www.freeweb.hu/eduscience/0901Kereszty.pdf.

Lee, J. F. K., & Collins, P. C. (2009). Australian English-language textbooks: The gender issues. *Gender and Education*, *21*(4), 353–370.

Leont'ev, A. N. (1978). *Activity, consciousness, and personality*. Englewood Cliffs: Prentice-Hall.

Leont'ev, A. N. (1979). The problem of activity in psychology. In J. V. Wetsch (Ed.), *The concept of activity in Soviet psychology*. New York: M. E. Sharpe.

McKimmie, T. (2002). Gender in textbooks. *Vinculum, 39*(4), 18–23.

Mikk, J. (2002). *The roles of women and men in Estonian readers*. 5th ATEE Spring Conference, Latvia University, Riga, 6.

Perry, D. G., & Bussey, K. (1979). The social learning theory of sex differences: Imitation is alive and well. *Journal of Personality Social Psychology, 37*(10), 1699–1712.

Pledger, K., Cole, G., Jolly, P., Newman, G., Petran, J., & Bright, S. (2006). *Edexcel GCSE Mathematics Foundation Book 1*. Oxford: Heinemann.

Rist, R. (1970). Student social class and teacher expectations: The self-fulfilling prophecy in ghetto education. *Harvard Educational Review, 40*(3), 411–451.

Sansom, A. (2000). Dove sono le ragazze? Gender and Italian textbooks, *Tuttitalia, 2*(8).

Selzer, A. (2007). The dis/location of women in secondary school Australian history. Unpublished doctoral thesis, Monash University, Clayton, Victoria, Australia.

Sharp, R., & Green, A. (1975). *Education and social control*. London: RKP.

Skelton, C., Francis, B., & Read, B. (2010). "Brains before 'beauty'?" High achieving girls, school and gender identities. *Educational Studies, 36*(2), 185–194.

Stott, T. (1990). Do geology textbooks present a sex-biased image? *Teaching Earth Sciences, 15*(3), 77–83.

Sunderland, J., Cowley, M., Abdul Rahim, F., Leontzakou, C., & Shattuck, J. (2001). From bias 'in the text' to 'teacher talk around the text': An exploration of teacher discourse and gendered foreign language textbook texts. *Linguistics and Education, 11*(3), 251–286.

Walden, R., & Walkerdine, V. (1985). *Girls and mathematics: From primary to secondary schools*. London: Institute of Education, University of London.

Walkerdine, V. (1989). *Counting girls out*. London: Virago.

Willis, P. E. (1977). *Learning to Labour: How working class kids get working class jobs*. Aldershot: Gower.

Witt, S. D. (1997). Boys will be boys, and girls will be… hard to find: Gender representation in third grade basal readers. *Education and Society, 15*(1), 47–57.

Young, M. F. D. (Ed.). (1971). *Knowledge and control: New directions for the sociology of education*. London: Collier-Macmillan.

Zevenbergen, R. (1993). The production of gendered positions in mathematics. In J. Mousley & M. Rice (Eds.), *Mathematics: Of primary importance. Proceedings of the thirtieth annual conference of The Mathematical Association of Victoria* (pp. 24–29). Brunswick, Australia: The Mathematical Association of Victoria.

Commentary on the Chapter by Paul Dowling and Jeremy Burke, "Shall We Do Politics or Learn Some Maths Today? Representing and Interrogating Social Inequality"

Mathematics Educators as Political Activists: Dissonance, Resonance, or Critical

Bill Atweh

To start on a personal note, I consider the opportunity to respond to the chapter by Dowling and Burke to be an honour and privilege indeed. In the past I have always found Dowling and his colleagues' writings to be a challenge to many of our widely accepted views and practices in mathematics education—even for those mathematics educators, may I add, writing from sociopolitical or sociocultural perspectives. Reading this chapter at this particular time was opportune for me. It spoke directly to findings of a recent research project I have participated in as well as some writing I am currently undertaking on the Australian national curriculum. Reading this chapter, I found myself agreeing with the majority of arguments developed, with the exception of one (minor?) point I will address in the last paragraph below. Here I will restrict my comments to two points that the chapter raised for me: the construction of the relationship between mathematics and the world (in particular the social world), and the corresponding tasks for mathematics education (in particular with respect to the agenda of social justice).

1 Mathematics and the World

The chapter by Downing and Burke has identified four domains of action in mathematics education that are useful to describe the relationship of mathematics to the social world which they call "esoteric" (when mathematical language is used to refer to mathematical content; e.g. mathematical proofs); "descriptive" (when mathematical language is used to refer to aspects of the world; e.g. modelling); "expressive" (when everyday language is used to refer to mathematics; e.g. use of the metaphor

B. Atweh (✉)
Curtin University, Perth, Australia
e-mail: b.atweh@curtin.edu.au

I borrow the subtitle from a discussion by Skovsmose and Valero (2001) on the relationship of mathematics education and democracy.

H. Forgasz, F. Rivera (eds.), *Towards Equity in Mathematics Education*,
Advances in Mathematics Education,
DOI 10.1007/978-3-642-27702-3_9, © Springer-Verlag Berlin Heidelberg 2012

of balance to represent an equation); and finally, "public domain" (when everyday language is used to describe the world; e.g., media reporting on some research findings).

It seems to me that these are very useful analytic tools to describe the different practices of relating institutionalised mathematics in schools as well as its use in society. Here I like to argue that their implication for practice does depend on the view one takes of what we take as "mathematics" in mathematics education. In this context, I can identify two alternative positions one can take (perhaps there are others). On one hand, the mathematics in mathematics education can refer to refer to a *body of knowledge* (without neglecting is cultural and historical roots) that is (re)presented in formal curricula, textbooks and examinations. From this view, the task of mathematics education is ultimately to develop this body of knowledge among its students. This is what Downing and Burke call the domain of the "esoteric" practices. From this particular perspective, the other three domains of action are legitimated as far as their contribution to the esoteric domain. This view is consistent with "generally privileging the mathematical discourse of the esoteric domain" (p. 9).

On the other hand, an alternative understanding of the mathematics in mathematics education may be constructed as an activity of a reading of the world. Mathematics in this view is a *process* rather than a body of knowledge. However, as Skovsmose reminds us, this way of reading the world has also the function of formatting the world (Skovsmose 1994). Here, the task of mathematics education is seen as a development of a capacity to read the world and, perhaps now or in the future, of writing the world. According to this understanding of mathematics, the "esoteric" practices lose their privilege (albeit, unfortunately, not necessarily in formal curricula and regimes of testing!). One can go further and posit the facilitation of participation in "public domain" action as the ultimate value of mathematics education.

Arguably, analytic tools are useful to describe a phenomenon and understand its complexity. However, they may not be sufficient to inform practice that essentially involves questions of values. The challenge that this presents to mathematics education (which I take in this context in its widest possible meaning) is to engage in a continued discussion from different perspectives (including sociology, the discipline of mathematics, politics, general education and philosophy) towards normative decision making at all levels of action. This discussion would lead to the very purpose(s) of mathematics education itself, which Biesta (2010) reminds us is noticeably absent from current educational discourse.

2 Mathematics Education and Social Justice

The above alternative understandings of mathematics and of mathematics education also have implications for the conceptualisation of the role of social justice in mathematics education. According to the first view of mathematics as a body of knowledge, social justice concerns are arbitrary and optional extras in mathematics education. At worst, they may be seen as contrary to the purposes of mathematics education as we will see below; or they may have heuristic value to motivate students to see a value for, and meaning of, mathematical knowledge; or at best, as important in their own right but for non-mathematical reasons. However, from the alternative perspective that constructs mathematics as a particular activity of a reading the world, social justice is intrinsically related to mathematics education since social injustice is a significant feature of the social world.

Certainly Dowling and Burke are right in the analysis they provide about the problem of using social justice in mathematics education. Such a reading of the world necessarily is subjected to the simplification of the phenomenon being read and the simplification of the mathematics used to read it (based on the level of knowledge available to the student). Further, any reading of the world—or writing the world for that matter—is subject to certain assumptions about the lens through which we look at the world, and about which aspects of the world it makes visible. Most crucially, these limitations are subject to questions of values, both personal and communal. Hence, a prerequisite for a productive reading of the world through mathematics is the need to be aware of the limitations and assumptions one makes about the mathematics, the world, and the viewer. Perhaps this makes reading the world through mathematics both necessary yet impossible. The challenge is to find a way to chart a course between the two traps of inaction that leads to failure to achieve what we must, and zealous uncritical action that leads to frustration of what we are proposing to do in the first place.

Finally, the chapter by Dowling and Burke has spoken to me with respect to two teachers who were involved in a recent project we called Socially Responseable Mathematics Education (Atweh and Brady 2009), designed in line with the literature on critical mathematics education (Skovsmose 1994) and social justice pedagogy (Gutstein 2006). We based the conceptualisation of this project on ethics and, in particular, the concept of responsibility. The project aimed to assist teachers to develop and trial middle school activities to teach mathematics through real world activities that also aimed at developing understanding of the social world. We, as project designers, had a strong commitment to social justice; but this commitment varied with the various participating teachers. Reading the chapter by Dowling and Burke reminded me of the experiences of two teachers in the project who we will call T_1 and T_2 (the reason behind this particular reference will be clear below[1]). Both teachers were very experienced in teaching mathematics, and both enjoyed great rapport with their students. They both were acknowledged as leaders in mathematics education in their respective schools.

T_1 was from a primary school teaching background and had transferred to middle school teaching when he moved into a small town. One of the passions of his life was a dedication to industrial matters related to the conditions of work of teachers, and a strong commitment to social justice. He was elated at the commencement

---

[1] It so happened that the two teachers shared the first name. In the common, apparently irreverent but affectionate, Australian humour, the two teachers were referred to in the project as Name 1 and Name 2.

of the project which he saw as an opportunity to put into practice what he always wanted to do. T_2 was a geologist by training and had a strong commitment to the development mathematical knowledge and the achievement of his students that might open doors for their post school aspirations. All through the project T_2 expressed great concerns about introducing questions of values into his teaching, in fear of leading into student indoctrination. He felt comfortable developing activities that incorporated the physical world with some implication for the social world, as long as he did not feel he had to deal with questions of values.

T_2 developed a well conceived activity for dealing with the concept of trigonometric functions using data from cyclones that had hit his town a few years previously. Through these data, students were able to use the spreadsheets to draw and add and subtract trigonometric functions in ways that made sense to them. As a result of the project, the students were able to make a presentation to the local council chief engineer about the dangers to their town from future cyclones. In spite of the great learning that had undoubtedly arisen in this project for both the teacher and his students, social justice was not an issue that was raised. In contrast, T_1 developed activities with very strong social justice links, such as comparing the food that different countries around the world consumed, and raised significant questions about fairness and the relationship between amount of consumption and happiness. Two notable features of the three activities that he developed were the use of very traditional teaching methods of completing worksheets and what can be described as low order thinking demands. Using the categories developed by Dowling and Burke, while the activities developed by T_2 privileged the "esoteric" action, those developed by T_1 privileged the "descriptive" domain.

The questions that this chapter has left me with stem from the very last sentence in which the authors conclude that "we can be both mathematics educators and political activists, just not at the same time". The two teachers T_1 and T_2 have dealt with the potential tension between esoteric and descriptive actions in different ways; each has sacrificed one for the sake of the other based on their own personal beliefs about useful mathematics and their commitment to social justice. It is not clear, however, that it is inherently impossible to be involved in action that develops both esoteric and descriptive mathematics. It is true that teachers face different demands on their time and action. T_1 may have seen his primary role as a mathematics teacher, albeit in the traditional understanding. T_2 may have constructed his primary role as a political activist. Perhaps T_3 may see his/her primary role as nurturer of attitude towards mathematics. T_4 may focus on the understanding of mathematics using everyday language. T_5 may be concerned with maintaining the peace in his/her disruptive and violent class. T_6 may aim for students to achieve the highest scores on national examinations because school funding depends on it. Can a teacher's identity be seen as the sum of a series, while from time to time focusing on one aspect more than others, rather than serial identities to meet different demands? The challenge to mathematics education is to investigate what conditions can support teachers to be productive agents in mathematics education to enable them to meet the demands of their students. Under the right conditions, I would still like to hope that one can be a good mathematics teacher and a political activist at the same time.

References

Atweh, B., & Brady, K. (2009). Socially response-able mathematics education: Implications of an ethical approach. *Eurasia Journal of Mathematics, Science & Technology, 5*(3), 267–276.

Biesta, G. (2010). *Good education in an age of measurement: Ethics, politics and democracy.* Boulder: Paradigm Publishers.

Gutstein, E. (2006). *Reading and writing the world with mathematics: Toward a pedagogy for social justice.* New York: Routledge.

Skovsmose, O. (1994). *Towards a philosophy of critical mathematics education.* Dordrecht, The Netherlands: Kluwer Academic Publishers.

Skovsmose, O. & Valero, P. (2001). Breaking political neutrality: The critical engagement of mathematics education with democracy. In B. Atweh, H. Forgasz, & B. Nebres (Eds.), *Sociocultural research on mathematics education* (pp. 37–55). Mahwah: Lawrence Erlbaum & Associates.

Commentary on the Chapter by Dowling and Burke, "Shall We Do Politics or Learn Some Maths Today? Representing and Interrogating Social Inequality"

Joanne Rossi Becker

This chapter posits the following conundrum: Is it possible to structure a mathematics lesson that integrates questioning of social inequality with worthwhile mathematical material? Dowling and Burke answer this question negatively. They conclude that a lesson must either privilege mathematics, providing all students with appropriate and rigorous mathematical content, or a lesson must privilege political motivations, providing for a full and extensive discussion of critical issues; according to the authors, it is not possible to accomplish both goals at the same time.

The authors present a theoretical model for considering representations of social (in)equality; examples come from photographs and mathematics textbooks. The images the authors analyze are considered on two variables of representation: expression of inequality (tacit or explicit); and consonance (or dissonance) with expected patterns of behavior. The concepts of invisibility and stereotype relate to tacit and explicit expressions of inequality, respectively, representations that are consonant with prevailing views of the target group in society. Tokenism and interrogation, on the other hand, are categories that are dissonant with the prevailing patterns.

This model affords a theoretical framework that can be useful to help analyze mathematics textbooks, but the model is also useful for analyzing representations in other mathematics education contexts such as the classroom. Various studies in Western countries have analyzed textbooks for gender inclusion, but most lack a framework such as this one that allows a qualitative consideration of representations, not just a quantitative one (see, for example, Schniedewind and Davidson 2006, pp. 220–222).

Dowling and Burke provide interesting examples analyzing various representations of gender and social class through the lens of this expression/orientation framework, examples that help the reader construct meaning for their categories. They rightly point out that materials such as textbooks cannot be evaluated fully

J.R. Becker (✉)
San Jose State University, San Jose, USA
e-mail: joanne.rossibecker@sjsu.edu

H. Forgasz, F. Rivera (eds.), *Towards Equity in Mathematics Education*,
Advances in Mathematics Education,
DOI 10.1007/978-3-642-27702-3_10, © Springer-Verlag Berlin Heidelberg 2012

without considering the context and conditions of their use. Thus a text that is absent in significant female representation might be used by a teacher as a means of interrogating prevailing patterns of gender (in)equality in the culture, reducing the potential harmful impact of the text material itself.

Ultimately the conclusion reached by Dowling and Burke that it is not possible to privilege both mathematics and political goals, is a bit discouraging for those committed to social justice issues. One weakness of the chapter is a lack of approaches other than integration of politics and mathematics that might be examined. Perhaps a multicultural perspective on issues of social inequality can provide one alternative lens through which to examine the dilemma posed by Dowling and Burke. This perspective may also provide some guidance for how teachers and teacher educators can meet the challenge posed to teach mathematics for social justice.

Multicultural education in the U.S. grew out of the civil rights movement, and thus offers a vision, yet to be reached, of democracy, social justice and equality. Multicultural mathematics education proposes that schools and teachers should help students develop skills they need to analyze critical social justice issues in their own environment and learn to work collectively to address these issues (see Sleeter 1997). As Banks and Banks (2003) have stated, teachers and students need to be critical consumers of mathematics, while at the same time recognizing how mathematics can be useful (or limited) for questions they themselves might pose about their own environment or cultural context.

Banks (1993) identified four approaches to multicultural education: the contribution approach; the additive approach; the transformation approach; and the social action approach. These approaches to multicultural education would be characterized as dissonant from patterns in society relative to specific cultural groups, to use Dowling's and Burke's framework. The contributions approach utilizes cultural components such as holidays in the curriculum without changing its structure, an approach Dowling and Burke might characterize as tokenism; an example might be including some activity during Black History Month (February in the U.S.) that highlights contributions of African Americans to mathematics and science. In the additive approach, content, concepts, and perspectives are added to the curriculum while leaving its structure unchanged and unquestioned, another example of tokenism, albeit a more thoughtful and extensive approach (see, for example, Krause 1983; Zaslavsky 1991). The transformation and the social action approaches better fit the example from Gutstein that Dowling and Burke critique. [While I was unable to locate the Gutstein article the authors refer to, the reader may consult Gutstein (2003) for other examples.] In the transformation approach, issues and problems from diverse groups are raised and result in a change of curriculum, while the social action approach additionally helps students take action. The latter two approaches clearly better match Gutstein's work, with all the limitations and problems inherent in it that Dowling and Burke enumerate. They also fit the interrogation strategy of the Dowling and Burke framework.

Thus multicultural mathematics education advocates teaching mathematics in a way that cultivates the mathematical capabilities of students from a wide variety of socio-cultural groups traditionally marginalized in mathematics: students of color;

students whose first language is not English; females; the disabled; and, students from low socioeconomic backgrounds, to mention a few. Multicultural mathematics educators endeavor to teach mathematics *for* social justice, rather than attempting to integrate mathematics and social justice issues. Adapting Sleeter's (1997) definition, a good multicultural mathematics teacher:

- knows how to teach students from historically low-achieving groups to achieve well in mathematics by using their cultural backgrounds as a fertile pedagogical resource;
- challenges lowered expectations for such students in mathematics and works to institutionalize higher expectations;
- creates a culturally pluralistic mathematics curriculum that helps all students see mathematics as a human creation; and,
- connects mathematics concepts with students' lives, helping them use complex mathematical reasoning to analyze social issues of concern to them.

As Sleeter characterizes this stance, it is no more political than a traditional stance that ignores cultural differences. Being mindful of this characterization may be helpful to teachers who face administrative criticism for "politicizing" mathematics, criticism that is no less likely today in the U.S. than 40 years ago, when I had a parent complain about my high school bulletin board display of jobs using mathematics (all male figures) with the header "Women can do these jobs too!" Interrogation of prevailing views on gender (or social class, language minority status, gender identity, disability, or ethnicity) can be dangerous, especially for new teachers.

Of course none of this is easy to accomplish. Murtadha-Watts and D'Ambrosio (1997) detail the difficulties inherent in attempting to transform a K-6 mathematics curriculum using a multicultural approach. These educators aimed not just to improve the achievement of minority groups in mathematics, but also to develop a curriculum that posed societal problems usually within the purview of the humanities or social sciences and use mathematics as a tool to solve them. Thus they were attempting to implement all four of Sleeter's (1997) characteristics of multicultural education. Difficulties arose from the shared understanding of multicultural education, the deep-seated understanding of what mathematics is, the complexities of group deliberation, and the demands involved in the teacher-research process. Gutstein (2003) also identifies the complexities inherent in teaching mathematics for social justice. In his two-year study, he worked with a group of 7th and 8th graders in an urban Latino/a school. As Gutstein phrases it, students began to "*read the world* with mathematics, develop mathematical power, and change their dispositions toward mathematics" (Gutstein 2003, pp. 66; emphasis added). These seem like worthwhile goals, albeit difficult to achieve.

I acknowledge that implementing effective multicultural education is not easy, with the potential to fall prey to the criticisms Dowling and Burke delineate in their chapter. However, it seems that the first two or three points that Sleeter (1997) enumerated are the minimum that should be expected from an exemplary mathematics teacher. While these are not simple to achieve in our preparation or professional development of mathematics teachers, they provide goals that those of us committed to social justice can seek to reach, without diminishing the extent and rigor of the mathematics we are teaching.

References

Banks, J. A. (1993). The canon debate, knowledge construction, and multicultural education. *Educational Researcher, 22*(5), 4–14.

Banks, J. A. & Banks, C. A. M. (Eds.) (2003). *Handbook of research on multicultural education* (2nd ed.). Hoboken: Jossey-Bass.

Gutstein, E. (2003). Teaching and learning mathematics for social justice in an urban, Latino school. *Journal for Research in Mathematics Education, 34*(1), 37–73.

Krause, M. C. (1983). *Multicultural mathematics materials*. Reston: National Council of Teachers of Mathematics.

Murtadha-Watts, K., & D'Ambrosio, B. S. (1997). A convergence of transformative multicultural and mathematics instruction? Dilemmas of group deliberations for curriculum change. *Journal for Research in Mathematics Education, 28*(6), 161–182.

Schniedewind, N., & Davidson, E. (2006). *Open minds to equality* (3rd ed.). Milwaukee: Rethinking Schools.

Sleeter, C. (1997). Mathematics, multicultural education, and professional development. *Journal for Research in Mathematics Education, 28*(6), 680–696.

Zaslavsky, C. (1991). World cultures in the mathematics classroom. *For the Learning of Mathematics, 11*(2), 32–36.

Gender Role Stereotypes in the Perception of Mathematics: An Empirical Study with Secondary Students in Germany

Gabriele Kaiser, Maren Hoffstall, and Anna B. Orschulik

Abstract Findings from international educational studies such as TIMSS or PISA show that gender differences in mathematical achievements are now only marginal, or have completely vanished in many countries. Nevertheless, beliefs about the higher achievements in mathematics by boys are still widespread. In this chapter the results of a study of students' views of gender-role stereotyping in mathematics are presented. For this study, 1244 German students from lower and upper secondary level (11–17 year olds) were tested using instruments developed by Forgasz and Leder concerned with gendered views of mathematics and mathematics education. The study clearly shows that as they grow up, boys as well as girls in contemporary Germany still have doubts concerning the equal mathematical performances of girls and boys, and internalise gendered stereotypes of girls being less talented and interested in mathematics, influenced by gender-role stereotypes within society, where mathematics is still described as a male domain.

1 State of the Art

In the last few years, research results have been presented which point out that there are almost no differences in mathematical achievements between boys and girls in primary and secondary schools. For example, the often quoted study by Hyde et al. (2008), in which 7 million students from age 7 to 17 were tested, found that generally there is no evidence for gender differences in mathematics favouring males today. On the other hand, in a re-analysis of the PISA studies from 2003 and 2006 by the OECD (2009) a general gender pattern of males outperforming females in the combined mathematics scale and every subscale was reported, with only a few countries as exceptions; similar results were reported in the European report on gender

G. Kaiser (✉) · M. Hoffstall · A.B. Orschulik
University of Hamburg, Hamburg, Germany
e-mail: gabriele.kaiser@uni-hamburg.de

M. Hoffstall
e-mail: marenhoffstall@web.de

A.B. Orschulik
e-mail: annaborschulik@web.de

H. Forgasz, F. Rivera (eds.), *Towards Equity in Mathematics Education,*
Advances in Mathematics Education,
DOI 10.1007/978-3-642-27702-3_11, © Springer-Verlag Berlin Heidelberg 2012

differences in educational outcomes (Eurydice 2010). However, while performance differences in mathematics tend to be modest, affective data from PISA 2003 and 2006 reveal remarkable gender differences in students' attitudes and approaches towards mathematics, in their levels of interest in and enjoyment of mathematics as well as in their self-related beliefs, emotions and learning strategies. For example, in most countries, participating in PISA 2006, females tended to report lower mathematics-related self-efficacy than males. While males tended to have a more positive view of their abilities than females, females described significantly more feelings of anxiety, helplessness and stress in mathematics classes.

Similar results have just been reported from a re-analysis of the PISA and TIMSS studies. It was concluded that internationally, gender differences in mathematics achievements are no longer pronounced. However, there still exist remarkable differences between females and males concerning their self-confidence in mathematics and their instrumental motivation in mathematics, that is, males seem to be more motivated to learn mathematics because they believe that mathematics will help them in their later careers (Else-Quest et al. 2010).

The role of teachers' and parents' stereotypes was analysed in extensive research by Tiedemann (2000, 2002). He reports various stereotypical teacher beliefs, for example, beliefs that girls think less logically than boys, and that mathematics is more difficult for girls. Teachers' gender stereotypes influence their gender-differentiated attributions of students' mathematical developments. For example, with regard to girls, teachers attributed unexpected failure more often to low abilities and less often to a lack of effort than with boys. Similar biasing effects of parents' gender stereotypes on their children's mathematical abilities relate to the children's self-perceptions of their mathematical abilities.

In their overview of 30 years of international research on gender issues in mathematics education, Leder et al. (1996) emphasised the high impact of affect, which does not seem to be limited to particular countries. They describe, amongst others, the following research results:

- Gender differences are more prevalent among older students and seem to increase as students progress through school;
- Mathematics continues to be viewed as a male domain, more by males than by females;
- Compared to males, females are more likely to attribute mathematical failure to a lack of ability rather than to a lack of effort;
- External sources such as parents, peer groups, socialisation patterns, and the media are strong influential factors on students' beliefs, and quite often emphasise gender-role stereotypes concerning mathematics.

These research results coincide with claims dating back to the 1970s and 1980s when, in the context of the Women's movement, much research was being conducted to identify factors contributing to gender differences favouring males in participation patterns in mathematics and in mathematical achievements. Stereotyping of mathematics as a male domain, that is, mathematics being more suitable for males than for females, was identified as one explanatory factor (e.g., Fennema 1974). In this context, Fennema and Sherman published their *Mathematics Attitude Scales* (Fennema and Sherman 1976) which consisted of nine subscales, including one on *Mathematics as a male domain*; this subscale has been widely used to assess the extent to which respondents regard mathematics as a male domain. Based on critiques concerning the anachronistic wording of some items and the underlying assumption of the subscale (Forgasz et al. 1999), Leder and Forgasz (2002) developed two new instruments—*Mathematics as a gendered domain* and *Who and mathematics*, in an attempt to overcome the limitations of the original scale. These new scales have been widely used in many studies around the world and have shown remarkable, even unexpected, cultural differences in the gendered beliefs about mathematics; but they have also produced some contradictory results.

In their study of Australian lower secondary students using the two instruments, Leder and Forgasz (2002) report that the majority regard mathematics as being a gender-neutral domain, and that boys find mathematics difficult and boring and need much more help than girls. Forgasz and Mittelberg (2008) found that Australian students challenge the traditional stereotype of mathematics as a male domain to a greater extent than their samples of Israeli Jewish and Israeli Arab students, and that the Israeli Jewish students' views were more consistent with the traditional gender stereotype than the Israeli Arabs' views. Brandell and Staberg (2008) report that Swedish secondary students perceive mathematics as a male domain, that older students hold more strongly gendered views than younger students, and that of all participating subgroups, boys in a science programme had the strongest beliefs that mathematics was a male domain. In a comparison of Australian and Swedish data, it was identified that Swedish students were less inclined to view mathematics as a female domain than Australian students of the same age (Brandell et al. 2007).

Against the background of this debate, in this chapter the situation in Germany concerning the acceptance of gendered views of mathematics among German students at different age levels is reported. The study focused on the following: whether mathematics is still seen as a male domain by German students from lower and upper secondary level; whether there are differences between males and females in their gendered views of mathematics; and whether these stereotypes are age-dependent. The chapter is based on the joint master thesis of Hoffstall and Orschulik (for details, see Hoffstall and Orschulik 2009).

2 Design of the Study

2.1 Description of the Instrument and Data Analysis

The study is based on a slightly modified version of the questionnaire *Who and Mathematics* by Leder and Forgasz (2002). The second instrument *Mathematics as a Gendered Domain* by Leder and Forgasz was not used in the study due to time restrictions. The questionnaire *Who and Mathematics* was modified as follows:

First, the statements from Leder and Forgasz (2002) were changed into questions. For example, the original item "Mathematics is their favourite subject" was changed to "Whose favourite subject is mathematics?" The answer options were left unchanged and were as follows:

- Boys definitely more than girls;
- Boys probably more than girls;
- No difference between boys and girls;
- Girls probably more than boys;
- Girls definitely more than boys.

Additionally, the original statements "Tease boys if they are good at mathematics" and "Tease girls if they are good at mathematics" were complemented by the questions "Who likes boys who are good at mathematics" and "Who likes girls who are good at mathematics". Question 23 in the original, "Are not good at mathematics" was changed to positive wording: "Who is good at mathematics?". In addition, the open-ended question, "Who do you think achieves more in mathematics? And why?" was added in order to obtain additional insights into the students' gender stereotypes concerning mathematics.

The questionnaire contained the following questions—for easier comparison with the results of the original studies, the order of the items is retained below; the questionnaire, as it was used in this study, is found in the appendix:

1. Whose favourite subject is mathematics?
2. Who thinks that it is important to understand the work in mathematics?
3. Who is asked more questions by the mathematics teacher?
4. Who gives up, when he/she finds a mathematical problem is too difficult?
5. Who has to work hard in mathematics to do well?
6. Who enjoys mathematics?
7. Who cares about doing well in mathematics?
8. Who thinks he/she did not work hard enough if he/she does not do well in mathematics?
9. Whose parents would be disappointed when he/she does not do well in mathematics?
10. Who needs mathematics to maximise future employment opportunities?
11. Who likes challenging mathematical problems?
12. Who is encouraged to do well by the mathematics teacher?
13. Who do mathematics teachers think will do well in mathematics?
14. Who thinks that mathematics will be important in his/her adult life?
15. Who expects to do well in mathematics?
16. Who distracts other students from their mathematics work?
17. Who gets the wrong answers in mathematics?
18. Who finds mathematics easy?
19. Whose parents think that it is important for him/her to study mathematics?
20. Who needs more help in mathematics?
21. Who teases boys if they are good at mathematics?
22. Who worries if he/she does not do well in mathematics?
23. Who is good at mathematics?
24. Who likes using computers to work on mathematics problems?

25. With whom do mathematics teachers spend more time?
26. Who considers mathematics to be boring?
27. Who finds mathematics difficult?
28. Who gets on with his/her work in class?
29. Who thinks mathematics is interesting?
30. Who teases girls if they are good at mathematics?
31. Who likes boys who are good at mathematics?
32. Who likes girls who are good at mathematics?
33. Who do you think achieves more in mathematics? And why?

Furthermore, as in the original instrument, the students' perception about their proficiency in mathematics was assessed (with the five choices: excellent, good, average, below average, weak).

In contrast to Leder and Forgasz (2002), the questionnaire had no title in order to leave the aim of the questions open. Only one introductory sentence was used at the beginning of the questionnaire to inform the students about the topic. Additionally, the possible answers were enriched by small avatars to help, especially the younger, students understand the possible answers. The complete translation of the questionnaire is found in the appendix.

As Leder and Forgasz (2002) point out in their description of the instrument, the statements in this questionnaire are to stand alone and cannot be combined to form a scale or a subscale. We therefore used the approach developed by Leder and Forgasz (2002) in their introduction of the instrument and used in other studies (e.g., Forgasz and Mittelberg 2008). We arranged the possible answers on a (pseudo) interval scale from 1 (boys definitely more than girls) to 5 (girls definitely more than boys) which allowed the calculation of mean scores on each item and the 'distance' of these average scores from the neutral value of 3, meaning no difference between boys and girls. Mean scores lower than three show that, on average, students believe that 'boys are more likely than girls to exhibit the behaviour described in the item or hold the beliefs', whereas mean scores above three indicate that, on average, students believe that 'girls are more likely than boys to fulfil the described expectation'. Several statistical tests of significance were carried out: one-sample t-tests were conducted on the mean scores to test for statistically significant differences from the neutral value 3, at the $p < .05$ level, the score indicating no gendered view of mathematics and mathematics teaching Independent samples t-tests were conducted to examine the significance of the differences in mean scores by gender and by age group at the .05 level of significance.

The analysis of the open-ended question was done qualitatively, adopting methods of grounded theory. Codes were developed on the basis of the responses provided, and the distributions were then analysed.

The changes made to the original instrument were minor and do not affect the validity of the administered questionnaire. The use of additional items does not create difficulties because the instrument does not function as scale and the statements are treated as stand-alone items. Furthermore, the two additional questions only confirmed aspects already covered by other items in the questionnaire. The change in the direction of the wording of item 23 is considered in the analysis and yielded the anticipated results. The change in the formulation of the items from statements to questions would not challenge any deeply rooted beliefs about mathematics as a gendered domain. However, in the analyses and discussion that follow, it has to be recognised that additional validity tests were not conducted.

2.2 Description of the Sample

In total, 1244 students participated in the study at the end of the school year 2008. These students represented two different age groups, 11–12 year olds and 15–17 year olds. The former consisted of 646 students from years 5–7, of which only 639 responses could be analysed; the latter comprised 598 students from years 9–11.

For the composition of the sample special care was taken regarding the distribution of the students from different school types in order to obtain a sample as representative as possible of the German school system. It is important to know that Germany has a tripartite school system in the lower secondary school, with the lower type secondary school (so-called Hauptschule) as the school for generally lower-achieving students, the intermediate type secondary school (so-called Realschule) for average students, and the higher type secondary school (so-called Gymnasium) for the higher-achieving students. In addition, there are also comprehensive schools in the federal states in Northern Germany. The distribution of the participating students by school type is shown in Table 1.

In this study a total of 48% participating students attended a higher type secondary school, 27% an intermediate type secondary school, 13% a lower type secondary school, and 12% a comprehensive school. This is a good representation of the situation in Germany, where 41.3% attend a higher type secondary school, 23.9% attend an intermediate type secondary school, 16.3% a lower type secondary school, and 18.5% a comprehensive school (Autorengruppe Bildungsberichterstattung 2010).

As well as similarities and differences by age groups, with evidence of changes in beliefs during adolescence, similarities in the beliefs of males and females and persistent gender-specific differences are the focus of the study.

The sample consisted of 581 boys and 663 girls (11–12 year-olds: 310 girls, 336 boys; 15–17 year-olds: 353 girls, 245 boys), a relatively balanced gender distribution (47% male and 53% female). [It should be noted that responses from all boys but only 656 girls could be analysed.]

The participating schools were in the city of Hamburg, as well as in the bordering federal states of Schleswig-Holstein and Niedersachsen (Lower Saxony). As Schleswig-Holstein and Niedersachsen are rural and provincial states, compared to the metropolitan area of Hamburg, the sample comprised a representation of an urban area as well as rural regions. Furthermore, it can be assumed that the sample is an appropriate representation of the socio-economic background of students in Northern Germany because the schools are situated in socially disadvantaged quarters as well as in areas with an average or high social structure. The high proportion

Table 1 Frequency and percentages of participating students by school type

	Lower type secondary school	Intermediate type secondary school	Higher type secondary school	Comprehensive school	Total
11- to 12-year olds	82	200	277	87	646
	12.7%	30.9%	43.0%	13.4%	100.0%
15- to 17-year olds	75	136	323	64	598
	12.6%	22.8%	54.1%	10.6%	100.0%
Total	157	336	601	150	1244
	12.6%	27.0%	48.3%	12.1%	100.0%

of immigrants to Germany is also represented in the sample. Although the sample represented central aspects of the situation in German schools, it still has to be regarded as a convenience and not a fully representative one.

3 Central Results of the Quantitative Study

A central finding of the study is that students still have very stereotypical attitudes towards mathematics and regard it as a male domain. This gender stereotyped perception of mathematics can be observed in both age groups, but was more marked among the older students. Thus, in the answers from 15–17 year-olds, many gender stereotypes were found, while the opinions of the 11–12 year-olds were much less characterised by gender stereotypical understandings with respect to mathematics learning.

Using the evaluation method described in Sect. 2, mean scores were calculated for each item, separately for all boys and girls in the sample, as well as by gender within age group. Independent groups t-tests were conducted on the mean scores to test for statistically significant differences. Table 2 shows the results which are quite striking. The data clearly show that gender stereotyped perceptions of mathematics increase with age. Overall, on 16 of the 32 items, significant differences between males and females can also be seen, with 17 items in the younger age group, decreasing to 11 items in the more mature group.

If the data are examined in more detail, not from the perspective of gender differences in the thinking of the students but in terms of gender stereotyped views, the following results emerge:

- Opinions on girls' and boys' capabilities in mathematics (questions 22 and 23), and the relevance of mathematics to their future careers (questions 10 and 14) show differences by age. In general, all students rated boys' capacity in mathematics higher than girls' (question 23), with an increasing tendency in favour or boys among older students.

Table 2 Means, standard deviations, and *t*-test results for questionnaire items by gender and age group

	11–12 year olds						15–17 year olds						11–17 year olds					
	Males (N = 336)		Females (N = 303)				Males (N = 245)		Females (N = 353)				Males (N = 581)		Females (N = 656)			
	Mean	SD	Mean	SD	t	sig. level	Mean	SD	Mean	SD	t	sig. level	Mean	SD	Mean	SD	t	sig. level
Item 1	2.61	.93	2.75	.87	−1.9		2.18	.84	2.27	.77	−1.3		2.43	.92	2.49	.85	−1.3	
Item 2	3.05[a]	.95	3.43	.86	−5.2[c]		2.85	.99	3.12	.85	−3.4[b]		2.97[a]	.97	3.26	.87	−5.6[c]	
Item 3	2.91[a]	1.15	2.80	.90	1.4		3.02[a]	1.03	2.87	.89	2.0[a]		2.96[a]	1.11	2.84	.90	2.2[a]	
Item 4	3.16	1.15	2.98[a]	1.15	2.0[a]		3.26	1.17	3.42	1.06	−1.7		3.20	1.16	3.22	1.12	−.2	
Item 5	2.97[a]	1.07	3.07[a]	.88	−1.3		3.41	.95	3.35	.83	.8		3.15	1.05	3.22	.87	−1.1	
Item 6	2.80	1.05	2.76	.97	.5		2.42	.87	2.39	.79	.4		2.64	.99	2.56	.90	1.4	
Item 7	2.87	.97	3.39	.96	−6.8[c]		2.75	.95	3.02[a]	.91	−3.4[b]		2.82	.96	3.19	.95	−6.8[c]	
Item 8	3.33	1.01	3.51	1.04	−2.3[a]		3.65	.96	3.69	.94	−.4		3.46	1.00	3.61	.99	−2.5[a]	
Item 9	2.91[a]	.98	3.12	.86	−2.8[a]		2.87	.77	2.97[a]	.76	−1.6		2.90	.90	3.04[a]	.81	−2.5[c]	
Item 10	2.57	1.00	2.82	.81	−3.5[c]		2.42	.79	2.54	.75	−1.9		2.50	.92	2.67	.79	−3.4[c]	
Item 11	2.66	1.05	2.60	.94	.7		2.27	.97	2.28	.86	−.1		2.49	1.03	2.43	.91	1.1	
Item 12	3.09[a]	.97	3.17	.81	−1.0		3.24	.89	3.09	.82	2.0[a]		3.16	.94	3.13	.81	.6	
Item 13	2.77	.96	3.06[a]	.84	−4.1[c]		2.58	.90	2.65	.76	−1.0		2.69	.94	2.84	.82	−3.0[c]	
Item 14	2.61	.96	3.08[a]	.89	−6.5[c]		2.38	.76	2.58	.78	−3.1[a]		2.51	.89	2.82	.87	−6.0[c]	
Item 15	2.76	.98	3.21	.99	−5.7[c]		2.47	.84	2.76	.91	−3.9[c]		2.64	.93	2.96[a]	.97	−6.0[c]	
Item 16	2.43	1.27	2.39	1.21	.4		2.53	1.18	2.45	1.14	.8		2.47	1.23	2.42	1.18	.7	
Item 17	3.02[a]	1.05	2.82	.78	2.7[a]		3.27	.96	3.16	.75	1.5		3.13	1.02	3.00[a]	.78	2.3[a]	
Item 18	2.55	1.11	2.71	.96	−1.9		2.26	.85	2.27	.79	−.2		2.42	1.02	2.47	.90	−.8	

(continued on the next page)

Table 2 (Continued)

| | 11–12 year olds | | | | | | 15–17 year olds | | | | | | 11–17 year olds | | | | | |
| | Males (N = 336) | | Females (N = 303) | | | | Males (N = 245) | | Females (N = 353) | | | | Males (N = 581) | | Females (N = 656) | | | |
	Mean	SD	Mean	SD	t	sig. level	Mean	SD	Mean	SD	t	sig. level	Mean	SD	Mean	SD	t	sig. level
Item 19	2.88	.87	3.17	.68	-4.8[c]		2.97[a]	.79	2.97[a]	.61	.0		2.92	.84	3.07	.65	-3.4[b]	
Item 20	3.21	1.13	3.20	.99	.1		3.59	.87	3.56	.85	.5		3.37	1.04	3.39	.94	-.3	
Item 21	2.89[a]	1.18	2.76	1.12	1.5		2.75	1.05	2.68	.87	.9		2.84	1.13	2.72	1.00	1.9	
Item 22	3.21	1.03	3.53	1.00	-4.0[c]		3.27	.92	3.51	.86	-3.2[b]		3.24	.98	3.52	.93	-5.2[c]	
Item 23	2.54	1.04	2.89[a]	.96	-4.6[c]		2.22	.89	2.51	.83	-4.1[c]		2.40	.99	2.69	.91	-5.3[c]	
Item 24	2.37	1.09	2.56	1.04	-2.2[a]		2.21	1.00	2.27	.94	-.9		2.30	1.06	2.41	.99	-1.8	
Item 25	3.34	.95	3.02[a]	.86	4.3[c]		3.29	1.01	3.06[a]	.92	2.7[a]		3.32	.98	3.04[a]	.89	5.0[c]	
Item 26	2.77	1.19	3.02[a]	1.07	-2.7[a]		3.19	1.11	3.50	.91	-3.6[c]		2.94[a]	1.18	3.28	1.02	-5.2[c]	
Item 27	3.26	1.06	3.25	.89	.2		3.56	.94	3.68	.83	-1.5		3.39	1.02	3.48	.88	-1.6	
Item 28	2.69	.95	3.02[a]	.98	-4.4[c]		2.48	.85	2.75	.80	-3.9[c]		2.60	.91	2.88	.90	-5.4[c]	
Item 29	2.84	1.04	2.76	1.05	.9		2.40	.85	2.46	.80	-1.0		2.65	.99	2.60	.93	.9	
Item 30	2.56	1.01	2.43	.91	1.7		2.70	.93	2.69	.75	.2		2.62	.98	2.57	.84	.9	
Item 31	3.21	.97	3.11	.95	1.3		3.23	1.02	3.20	.79	.4		3.22	.99	3.16	.87	1.0	
Item 32	2.91[a]	1.00	2.90[a]	.95	.1		2.81	.93	2.82	.73	-.2		2.87	.97	2.86	.84	.2	

Shaded regions: items with statistically significant gender differences.
Level of statistical significance of independent groups t-values: [a] $p < 0.5$; [b] $p < 0.1$; [c] $p < 0.001$.
The lettered indices: Mean scores not significantly different from 3 (one sample t-test results).

- The students also indicated that girls find mathematics more difficult than boys and need more help (questions 18 and 20). Moreover, it is interesting that the girls, especially in the younger age group, show a more gender-balanced perception. When asked about working habits, the majority of students assume that girls are especially diligent and ambitious, even though they are regarded as having lower competencies in mathematics (questions 5 and 23). Furthermore, the students are of the opinion that boys show greater interest in mathematics than girls, and that boys are more enthusiastic about mathematics (questions 6, 11 and 29). The students also state that girls perceive mathematics as being boring while boys perceive it as being more interesting (question 26). These results show that gender stereotyped perceptions about who is interested in mathematics are deeply rooted in the students' minds and get stronger during adolescence.
- When the relevance of mathematics for students' future lives was considered, it was found that students stress the importance of mathematics, especially for boys' future careers. They see less relevance of mathematics to girls' futures, although—and this of special concern—the younger girls described mathematics as being of greater importance for girls' than for boys' lives, opinions which change during adolescence (questions 10 and 14).

With respect to gender differences in the responses of the participants, the second important finding of the study is that the females held more flexible and open perceptions of gender roles than their male counterparts. This tendency became apparent in students' answers to various questions, such as about the importance of mathematics in their future life (question 14), or about the distribution of mathematical competencies such as "Who is good at mathematics?" (question 23), as well as questions about working habits in class such as "Who gets the wrong answers in mathematics?" (17). The 11–12 year-old girls generally attributed the same skills to boys and girls and often did not show a tendency towards gender stereotyped perceptions regarding the relevance to future careers or working habits. The 15–17 year-old girls, however, held the same gender stereotyped perceptions as the male students. This means that during adolescence female change their perceptions of gender roles. Indeed the 11–12 year-old girls did not perceive their own gender as being better than boys in mathematics, but regarded boys' and girls' abilities comparable (question 5). Among 15–17 year olds, however, the females believed that girls were less capable than boys in mathematics.

The third important finding is that students perceived gender-differentiated interactions in mathematics teaching settings. Students of both age groups sensed gender-dependent expectations of teachers. This was apparent in their response that teachers expect higher mathematics achievements by boys than by girls (question 13). Nevertheless it is striking that in response to the question of who is asked more questions in mathematics classes (question 3), they indicated that there was no gender difference. In contrast, students believed that teachers spend more with girls (question 25), which might result from the teacher's desire to help the girls and encourage them to participate in class (question 12). The students thus notice lower expectations of girls by the teacher (question 13) due to the gender-differentiated interactions in the classroom. One possible explanation for these results might be

Table 3 Self-perceptions of the mathematics achievement

		Excellent	Good	Average	Below average	Weak
11–12 year-olds	Male	10.4%	35.5%	45.1%	5.4%	3.6%
	Female	2.6%	29.7%	52.3%	6.5%	8.5%
15–17 year-olds	Male	13.9%	31.6%	38.9%	10.7%	4.9%
	Female	6.0%	28.1%	44.0%	9.7%	12.2%

that the students trace the more intensive interactions between teachers and female students back to the teachers' assumptions about girls needing more support than boys.

In order to compare the self-perceptions of mathematics achievements by gender, the students were asked about their own mathematical achievements at the beginning of the questionnaire. The results (see Table 3) show that girls generally perceive their own accomplishments in mathematics as being lower than boys' do of their own. This is in sharp contrast to the general better achievements of German girls in school (WZBrief Bildung 2010), and similar to results in mathematics education (see Eurydice 2010). Remarkably more girls than boys from each age group describe their achievements as average, and significantly fewer girls than boys describe their achievements as excellent or good.

In order to compare the results with the results from the original study (Leder and Forgasz 2002) and further studies, the evaluation method developed by Leder and Forgasz (2002) in their study and used by Forgasz and Mittelberg (2008) and by Brandell et al. (2007) was undertaken. The direction of the thinking of the students is evaluated, that is: mean scores significantly smaller than 3 mean that the students believe, on average, that boys are more likely than girls to behave or hold beliefs described in the item; and mean scores significantly greater than 3 mean that they believe girls are more likely than boys to behave like that. A mean around or not significantly distinct from 3 describes the neutral opinion, that is, there is no difference between girls and boys. Leder and Forgasz (2002) refer to earlier research in this field, providing the basis for the development of the instrument, and developed predictions for the direction of the responses for each of the 30 items. Against these predictions the data were evaluated, "male" meaning that the direction of the response was expected to be that males 'are more like than females' to behave or hold beliefs consistent with the item, "female" meaning that females 'are more likely than males' to behave or hold beliefs consistent with the item wording, male/female means that the research findings related to the predictions were mixed.

The results, displayed in Table 4, show a clear tendency for the German students to hold many beliefs concerning mathematics as a gendered domain as expected by Leder and Forgasz (2002).

For 21 items out of the 30 items for which there were predictions, the expected direction of the beliefs was evident. This result is in contrast to the results of the study by Leder and Forgasz (2002) carried out with 838 Australian 13–16 years old students, for which only 8 items matched predictions. Leder and Forgasz (2002)

Table 4 Predictions for gender differences from Leder and Forgasz (2002), and findings from this study

Item		Prediction	Findings (11–17 year olds)	
			Mean	Direction
1.	Whose favourite subject is mathematics?	male	2.46	male
2.	Who thinks that it is important to understand the work in mathematics?	female	3.13	female
3.	Who is asked more questions by the mathematics teacher?	male	2.90	male
4.	Who gives up, when he/she finds a mathematical problem is too difficult?	female	3.21	female
5.	Who has to work hard in mathematics to do well?	female	3.19	female
6.	Who enjoys mathematics?	male	2.60	male
7.	Who cares about doing well in mathematics?	male/female	3.02	nd
8.	Who thinks he/she did not work hard enough if he/she does not do well in mathematics?	male	3.54	female
9.	Whose parents would be disappointed when he/she does not do well in mathematics?	male	2.97	nd
10.	Who needs mathematics to maximise future employment opportunities?	male	2.59	male
11.	Who likes challenging mathematical problems?	male	2.46	male
12.	Who is encouraged to do well by the mathematics teacher?	male	3.14	female
13.	Who do mathematics teachers think they will do well in mathematics?	male	2.77	male
14.	Who thinks that mathematics will be important in his/her adult life?	male	2.67	male
15.	Who expects to do well in mathematics?	male	2.81	male
16.	Who distracts other students from their mathematics work?	female	2.45	male
17.	Who gets the wrong answers in mathematics?	male	3.06	female
18.	Who finds mathematics easy?	male	2.45	male
19.	Whose parents think that it is important for him/her to study mathematics?	male	3.00	nd
20.	Who needs more help in mathematics?	female	3.38	female
21.	Who teases boys if they are good at mathematics?	male	2.77	male
22.	Who worries if he/she does not do well in mathematics?	male/female	3.39	female
23.	Who is good at mathematics?	male	2.56	male
24.	Who likes using computers to work on mathematics problems?	male	2.36	male

(*continued on the next page*)

Table 4 (Continued)

Item		Prediction	Findings (11–17 year olds)	
			Mean	Direction
25.	With whom do mathematics teachers spend more time?	male	3.17	female
26.	Who considers mathematics to be boring?	female	3.12	female
27.	Who finds mathematics difficult?	female	3.44	female
28.	Who gets on with his/her work in class?	female	2.75	male
29.	Who thinks mathematics is interesting?	male	2.63	male
30.	Who teases girls if they are good at mathematics?	male	2.59	male
31.	Who likes boys who are good at mathematics?	np	3.19	female
32.	Who likes girls who are good at mathematics?	np	2.87	male

Shaded items: study findings consistent with predictions from previous research.
np = no prediction; nd = no direction.

interpreted that finding as an indication of a strong change in the belief system of the Australian students, that is, far more balanced in their thinking about mathematics and gender than expected from previous studies. They wrote:

> The findings described above suggest that students now believe that it is girls rather than boys who are more capable mathematically, enjoy mathematics, find it interesting and challenging, and whom teachers expect to succeed. The students now consider boys more likely than girls to be bored by mathematics, have to work hard to do well, give up when things get difficult, find mathematics difficult, and to need more help. Students believe that parents no longer favor their sons with respect to who they believe need mathematics, and who would disappoint them if they did not do well. These views are in stark contrast to those reported in earlier work in this field. (p. 8)

As was shown in Table 2, independent groups t-tests were conducted on the 32 items used in this study to test for statistically significant gender differences in the students' responses. One sample t-tests were conducted on the mean scores to determine if they were not significantly different from 3 (indicated with an asterisk on Table 2). For only 3 items did the males indicate a neutral position; these items referred mainly to classroom interactions; females were gender neutral on 4 items related to expectations concerning mathematics and classroom interactions. In the Australian data there were 7 and 5 items that males and females respectively took neutral, non-gendered, positions.

Again, following the evaluation method described by Leder and Forgasz (2002), the graph in Fig. 1 illustrates the direction of students' responses, and the strength of the beliefs held. The vertical axis, through the value 3, represents the gender neutral position, that is, no difference between boys and girls. The bars to the right of this axis reflect beliefs expecting the described behaviour to be more likely to be exhibited by girls, bars to the left beliefs expecting the described behaviour to be more likely to be exhibited by boys. One striking result is the uniformity of the

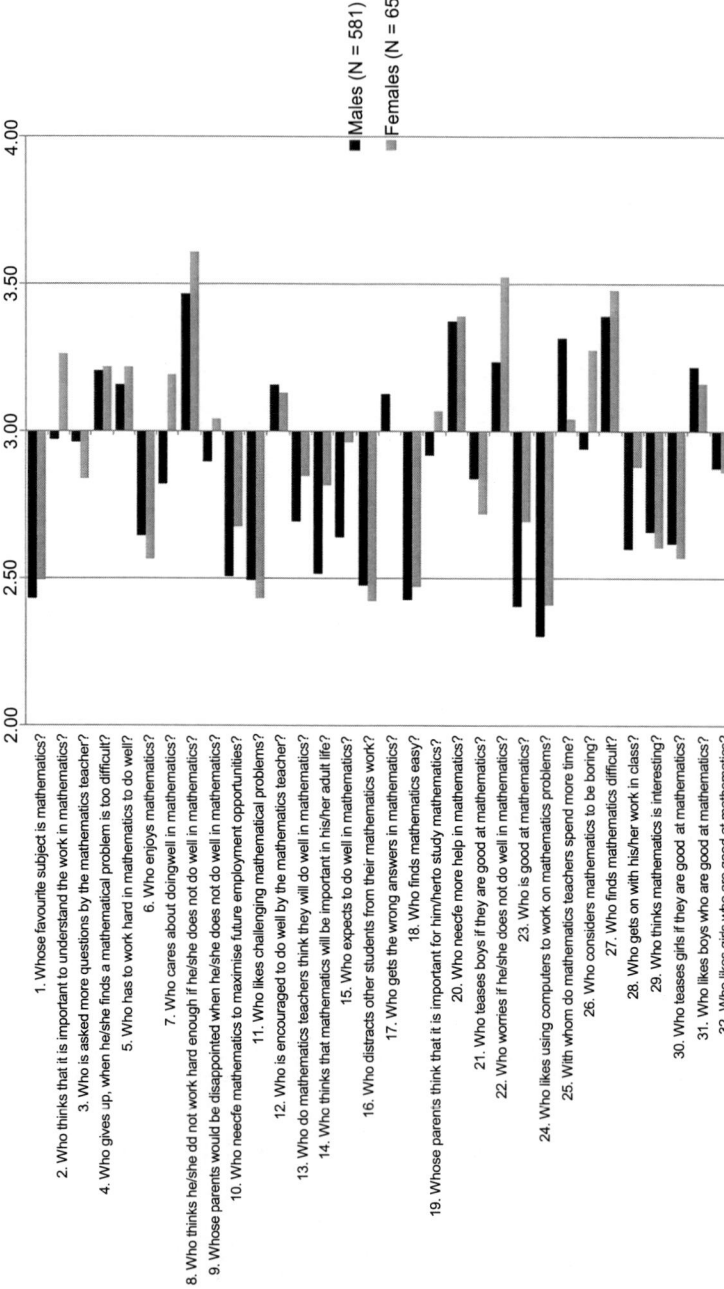

Fig. 1 Means scores for males and females for the 32 items of the test

answers by males and females which are generally in the same direction about which group (boys or girls) was more likely to describe the belief reflected in the items. For only 5 of the 32 items females and males answer in different directions, always in favour of their own gender, namely in questions 2, 7, 9, 19, and 26. Question 17 displays neutral answers by females and a tendency in favour of the girls by the males. Significant differences for both age groups were found on questions about the importance of understanding mathematics (question 2), caring about doing well in maths (question 7), and considering mathematics as boring (26). For two questions on parents' expectations (questions 9 and 19), significant gender differences were only observed for the younger age group; for the older age group both males and females responded similarly, being mainly gender neutral.

If the results of the whole German sample, not disaggregated by age group, are compared to the results of the Australian sample (Leder and Forgasz 2002), only a few similarities can be detected. For only 11 out of the 30 questions students answer in the same direction, namely in the area of teachers' behaviours concerning questioning and encouragement (questions 3 and 12), and for parents' expectations, where in both countries males and females indicate higher parental expectations concerning their own gender (questions 9 and 19). Furthermore, males and females in both countries believe that girls would worry more if they do not well in mathematics, and that girls need to work harder (questions 8 and 22); they describe boys as more distracting in class (question 16) and being more inclined towards computers (question 24). In both countries males and females believe that boys and girls who are good at mathematics are teased more by boys (questions 21 and 30). For all other items, differences in the response patterns in the direction of more traditional gendered views on mathematics and mathematics teaching were found for the German students, as discussed above.

A comparison of the results of the German sample with the Swedish sample, comprising 747 students from year nine, reveals a significantly different picture: for 19 of the 30 items the same response pattern occurs. Apart from question 3 "Who is asked more questions by the mathematics teacher?", where the Swedish males showed a slight tendency to refer to girls being favoured in contrast to the Swedish females, as well as the females and males in Australia and Germany, the similarities in the response patterns for the German and the Australian students—described above—are also found in the comparisons of the Swedish and the German students, together with an additional 8 items. These items confirm the above described similarities such as expectations about doing well in mathematics (questions 7 and 15), importance for future life (question 10), and teachers spending more time with girls than with boys (question 25), a belief especially emphasised by boys. Furthermore, males and females in Germany and Sweden agreed that boys prefer challenging problems (question 11) and find mathematics easy (question 18). Even split reactions are sometimes similar, such as the fact that in both countries males refer to boys as finding mathematics boring, whereas the females refer to girls (question 26). However, the intensity of the beliefs of the Swedish students are quite often less strong and tend towards a neutral position, a pattern which is exemplified by responses to item 29, "Who thinks mathematics is interesting?". For this item, significant differences from the neutral position were found in the answers of the German students, referring to a traditional gendered view on mathematics as boys being more interested in mathematics. Answers with no significant differences between male and female respondents indicate that these traditional gendered views of mathematics are intensified during adolescence. In Sweden the responses go in the same direction, favouring boys, but to a remarkably smaller extent, and in contrast to the Australian students who expected girls to find mathematics more interesting, a view more strongly held by the males than the females.

To summarise, the German students seemed to favour traditional gendered views on mathematics as a male domain in contrast to Australian students who challenged these beliefs and saw girls as more capable and more interested in mathematics. In their responses, the German students were closer to the Swedish students, who also emphasised more traditionally gendered views of mathematics and its teaching; in some areas, the beliefs of the German students seemed to be even more traditional than those of the Swedish students. It is especially interesting that these traditional beliefs about mathematics develop during adolescence, and the females—especially the younger ones—hold less traditional beliefs in some areas.

In the following section describing the qualitative part of the study, more details are provided on the students' thinking about gender differences in mathematics achievements and the development of these beliefs; this helps explain the results from the quantitative study described above.

4 Gender-Specific Differences in the Perception of Mathematics: Qualitative Results

4.1 General Results

A special part of the study consisted of the open-ended questions: "Who do you think achieves more in mathematics? And why?" These questions complemented the multiple-choice questions to allow deeper insight into the reasons for the students' statements. Thereby, not only the gender of those expected to have higher abilities in mathematics would be analysed, but also statements providing reasons for the choice.

Consistent with the results of the 32 questions, it was obvious that mathematics is assessed as a male domain and higher mathematic abilities are attributed to the boys. Regardless of age group, 512 respondents—almost half the sample—declared that boys achieve more in mathematics; 221 decided in favour of the girls, and 378 indicated that there is no gender difference.

The hypothesis that students have stronger stereotypical perceptions of mathematics with increasing age was again confirmed. While the opinions of the 11–12 year-olds were approximately gender-balanced, the 15–17 year-olds strongly tended to favour the boys (see Table 5).

On closer examination, differences between the opinions of the females and the males are clearly noticed for the 11–12 year-olds. However, a relatively balanced

Table 5 Who do you think achieves more in mathematics?

		Boys	Girls	No difference
11–12 year-olds	Male	48.8%	24.1%	27.1%
	Female	30.5%	34.7%	34.7%
	Total	39.8%	29.3%	30.9%
15–17 year-olds	Male	60.2%	8.5%	31.3%
	Female	47.5%	11.6%	40.9%
	Total	52.5%	10.4%	37.1%

opinion distribution can be inferred from the responses of this age group. The whole sample of 11–12 year-olds selected the responses "boys", "girls" and "no difference" with a percentages of 39.8%, 29.3%, and 30.9% respectively. There was only a slight tendency towards the answer "boys". The females' perception of who they believe to achieve more in mathematics appears very well-balanced. Each of the three possible responses was chosen by about a third of them; the response that "boys" achieve more in this subject was chosen by 4% fewer females than those who chose "girls" and "no difference" (34.7% each).

The males' opinions, however, were not so balanced. They seem to be more convinced of themselves and approximated 50% of them identified "boys" as having higher performance in mathematics. Merely a quarter of the male respondents declared that "girls" achieve more; about the same number indicated that there are no differences between boys and girls. Responses to this question again reveal that females possess a more open view and a more flexible perception of gender roles, and that males are more convinced of their own mathematical skills.

The balanced opinions that appear amongst the younger students cannot be observed among the older age group. The overall belief that boys achieve more in mathematics than girls increased by nearly 13% (39.8% for the younger students up to 52.5% for the older group), and the opinion that girls achieve better than boys decreased by 19% (from 29.3% for the younger group, to 10.4% for the older group). Although, the proportion of those students considering girls and boys as equally capable increased, there was a clear overall shift of opinion in favour of boys.

The opinions of the older female students still appear more moderate than those of the male students. An analysis of the changes (see Table 6) shows that, with age, females appear to strongly adopt the gender stereotyped perception of mathematics, and more often decided in favour of boys.

While the male students' perceptions of boys achieving more in mathematics increased by 11.4% (15–17 year-olds compared to 11–12 year olds), the approval of the female students to this statement rose by 17%. A significant difference between the opinion development of the girls and the boys is also reflected in the answer "girls"; a greater variation appears for females (decrease of 23.1 %) compared to the males, whose frequency saying "girls" decreased by only 15.6%.

Table 6 Changes of the answers "boys", "girls" and "no difference" in age comparison

	Male responses			Female responses		
	Age 11–12		Age 15–17	Age 11–12		Age 15–17
"boys"	48.8	→ +11.4 →	60.2	30.5	→ +17.0 →	47.5
"girls"	24.1	→ −15.6 →	8.5	34.7	→ −23.1 →	11.6
"no difference"	27.1	→ +4.2 →	31.3	34.7	→ +6.2 →	40.9

4.2 Analysis of Reasons

The reasons provided by the females and males for their answers are particularly interesting, because they convey attitudes towards mathematics as well as attitudes towards the different capabilities of boys and girls and gender stereotyped perceptions (see Fig. 2).

The explanations for the response that "girls" may achieve more in mathematics than boys were classified into the following nine categories:

- Intelligence;
- Effort;
- Concentration on school in general;
- Ability;
- Concentration/attention;
- Job-related expectations;
- Ambition;
- Interest;
- Logic intellectual power or ability to reason.

It is remarkable that the three most frequently given reasons relate to girls' work habits. Almost two thirds the responses (186 out of 284) included the explanation that girls may achieve more in mathematics because they are more studious, concentrate more during classes, as well as being more ambitious and desiring better grades. The following statements were provided by students in response to the question 'Who do you think achieves more in mathematics?' (translation by the authors):

> Girls because they are more ambitious and more conscientious. (female, 10th grade)
> Girls, because they pay more attention and listen more closely. (male, 6th grade)
> Girls can achieve more. They are more ambitious and keep at a problem until they solve it. (female, 10th grade)

Apart from these three reasons, the frequency of the other categories were lower. For example, only 11% of the students explained that girls' mathematics achievements were due to intelligence (7.4% or 21 entries) or ability (3.9% or 11 entries). The following reasons were found (translation by the authors):

> I think that girls achieve more because they [...] are just clever. (female, 6th grade)
> Yes, I think girls achieve more because mathematics is more in their nature than in boys', and they have greater abilities. (male, 6th grade)

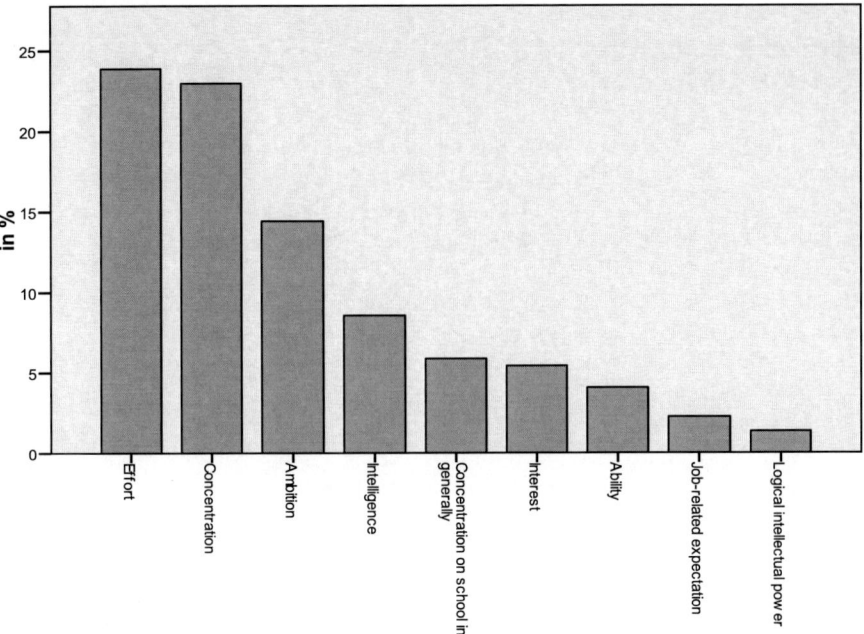

Fig. 2 Distribution of the reasons given to explain why "girls" may achieve more in mathematics

According to what the students wrote, girls' interest in mathematics was not very pronounced. Compared to the other categories, only 4.2% gave this as the reason; this suggests that interest is of little importance in terms of explaining why girls might achieve more than boys.

I think that girls achieve more in mathematics because girls are more interested (in class). (male, 6th grade)

Similar to the results of questions 10 and 14 concerning the relevance of mathematics to future life and career, very few students connected high mathematical capacities to girls' prospects of future careers. This reason was only named five times and thus accounts for only 1.8% of responses. It must be noted here that careers such as cashier were taken into consideration rather than technical occupations.

I believe girls because when they are grown up they will have to apply it [mathematics] more often in their occupation e.g., cashier or salesperson in a shop. (female, 6th grade)

The data showed that the reasons given for girls' higher achievements changed with increasing age, because the explanations provided by the older students often referred to the girls' diligence and effort. In particular, "concentration in school in general", "interest", "job-related expectations" and "ability to reason" are no longer provided by the 15–17 year-old students. Indeed, the 11–12 year-old students also rarely gave these reasons, but some nevertheless connected expected performance of girls in mathematics with ability to reason, future career opportunities, or interest. The missing reason, "interest", reveals that both male and female students of the older age group do not relate girls' high achievement potential with a possible interest in mathematics.

The three most prevalent reasons—"effort", "concentration" and "ambition"— were provided particularly frequently by both the 11–12 year-olds and the 15–17 year-olds. Once again, it became apparent that the 15–17 year-olds offer these explanations more often than the younger students, accounting for 80% of all responses among the older age group. Their proportion in the age group of the 11- to 12-year-olds merely amounts to about 61%. Furthermore, it appears that the percentage growth of the reason "effort" increases significantly from 23.9% up to 40.3%. The 15- to 17-year-old students also often mention the reason "ambition" which increases from 14.4% up to 24.4%.

A contrasting picture was discernible regarding boys. In particular, the categories "interest" and "ability to reason" represented the main reasons provided to explain the better mathematics achievements of boys. The reasons in support of "boys" achieving more could be divided into 18 categories, that is, more reasons featured than the reasons given for "girls". The 18 reasons were (also see Fig. 3):

- Interest;
- Effort;
- Better calculating-skills;
- Females' indifference/boredom;
- Genetic ability;
- Allocation of achievement on sciences and arts;
- Ability to reason;
- Job-related expectations;
- Intelligence;
- Interest/ability in/for computers;
- Concentration;
- Complex thinking of girls;
- Accelerated/clearer understanding;
- Ambition;
- Spatial sense;
- Numerousness of famous male mathematicians;
- Generally better achievements;
- Other.

Of the 605 responses, 110 indicated boys' interest in mathematics as a reason; 75 responses were based on the ability to reason as justifying higher expectations of boys. These two categories accounted for 30% of all reasons given. The following statements are examples:

> I think that the boys achieve more in mathematics because they are just more interested in it. (male, 10th grade)
> I believe that boys achieve more because they have much more pronounced logical thinking and this means that boys achieve correct solutions by thinking, whereas girls use the right formulas. (male, 10th grade)

Career opportunities, accounting for almost 9% of the answers, were more frequently mentioned unlike for "girls". The expectation of "boys" achieving more was

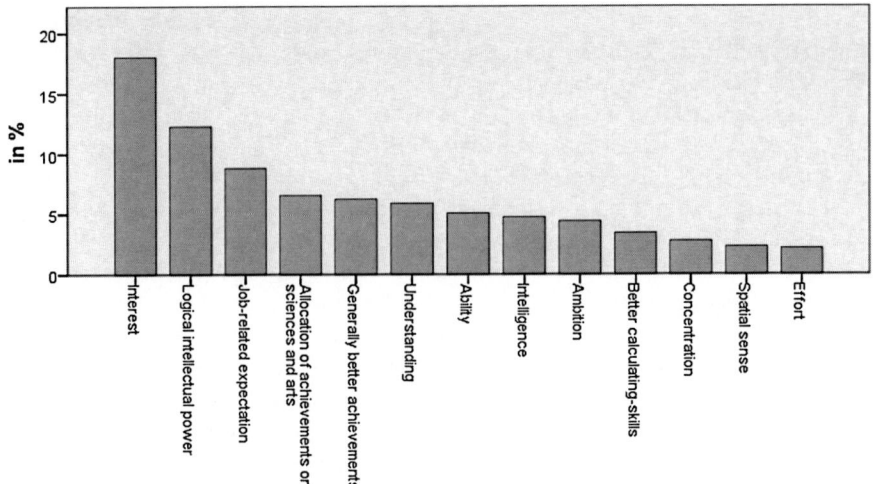

Fig. 3 Distribution of the reasons given to explain why boys achieve more in mathematics

explained in 54 of the 605 reasons as the need for mathematics in boys' occupational careers.

> I think that boys can be expected to achieve more because they will all have occupations which require mathematics. (male, 10th grade)

Less frequent reasons, accounting for 4% to 7% of the responses referred to stereotypical assumptions as well as intelligence. Of the 605 explanations, 40 (6.6%) argued that boys often have better abilities in natural sciences, whereas girls' qualities lie in the humanities and creative areas.

Three further reasons that were mentioned frequently and accounted for 6.3%, 5.9%, and 4.8% respectively, were related to general intelligence, as well as comprehension of mathematics. The reasons, "generally better achievements in mathematics" (38 entries), "better understanding" (36 entries) and "intelligence" (29 entries) were alluded to by a total of 103 students.

> I think boys achieve more in mathematics because they are mostly better. Of course, there are girls who are good at mathematics but boys are still better in the majority of the cases. (female, 6th grade)
> Boys because they have a better understanding in order to solve mathematical and technical problems respectively. (female, 10th grade)
> I do not think that girls achieve more in mathematics because they are not very good at it. Boys often have greater knowledge and are smarter. (female, 6th grade)

The reason "ambition" was identified by only 27 (4.4%) students.

> I think that boys achieve more in mathematics because they do not give up if they do not know a problem. They keep working on the problem until they solve it. (male, 6th grade)

The distribution of reasons for believing that "boys" mathematical achievements were better is displayed in Fig. 3.

These reasons also vary by age group and reveal an increase in stereotypical thinking and that clichés about mathematics are deeply rooted in students' minds. The frequencies of the explanations "ability to reason", "distribution of achievements in sciences and humanities", "better understanding", "genetic ability" and "spatial sense" were noticeably higher for the older age group, whereas the frequencies of the reasons "concentration" and "effort" were lower.

The increase in the reasons related to "ability to reason" are very informative. Males as well as females aged 15–17 named this criterion more often than the younger group, with the frequency increasing from 5.1% to 17.7%. Statements supporting better achievements of "boys" due to superior "spatial sense" increased by only 4%. This is remarkable because the 11–12 year-olds did not take this argument into account at all, it was only provided by the 15–17 year-olds.

Finally, it appears that the reasons for who achieves more in mathematics teaching—boys or girls—are different. The most frequently given reasons for "boys" were: "interest", "ability to reason" and "job-related expectations"; these are barely mentioned in respect to girls. The most frequently given explanations for "girls" were: "effort", "concentration", and "ambition"; these are only occasionally named in regard to "boys". Furthermore, reasons such as "better understanding", "better calculating skills" or "spatial sense" did not occur for "girls" at all.

The varying explanations demonstrate gender specific reasons for high achievements in mathematics. Girls' higher achievements were mainly attributed to effort and diligence, whereas that of the boys were identified as due to intelligence and special mathematical abilities.

Similar to the other items in the questionnaire, it was again noticeable that students' opinions changed for those aged 11–12 and those aged 15–17. The changes were in favour of boys and disadvantaged girls. While attributions of girls' better achievement to effort and ambition in mathematics increased with age, their capabilities and their interest in mathematics receded into the background. The students seem to increasingly internalise gender stereotyped perceptions of mathematics with increasing age. They also assumed different skills and capabilities to be prevalent, but positive appraisal was especially attributed to "boys".

In addition, it was remarkable that there were reoccurring themes within the explanations, clearly indicating that students hold beliefs because they are prevalent and acknowledged by society.

> I think that boys achieve more because girls, from scratch, have the excuse in their minds that 'I am a girl and I don't know and need mathematics'. (male, 10th grade)
> I think boys achieve more because society says so. (male, 6th grade)
> Boys because of stereotypical thinking! (male, 11th grade)

5 Possible Educational Consequences

The results of the study, namely that gender-role-specific stereotypes about mathematics become stronger when students grow up and have effects on the mathematical self-concept, tie in with the results of the various studies described at the beginning of the chapter, such as the OECD study and the newly published meta-analysis by Else-Quest et al. (2010). Furthermore, the study's findings are consistent with the scientific evidence of marked differences between female and male students in their levels of interest, their instrumental motivation in mathematics and the students' mathematics-related self-efficacy beliefs, self-concepts, and anxiety.

In summary, the study shows that the perception of mathematics as a male domain is still prevalent among German students, and that this perception is stronger among older students. This is either reinforced by the peer-group, parents, or teachers. Compared to the results of the studies by Leder and Forgasz (2002) in Australia, the German students are more traditional than the Australians and reflect traditional positions of mathematics as a male domain. They hold views, which are more similar to the Swedish students, as reported by Brandell and Staberg (2008) and Brandell et al. (2007).

These results show that the stereotyping of mathematics as a male domain is still widespread in today's society. This worsens the learning possibilities for girls, and hinders them from freely choosing a career or a course of studies. It thus appears to be necessary to raise parents' and especially teachers' awareness of stereotyping in mathematics because, as the studies by Tiedemann (2000, 2002) have shown, many adults have also internalised these attitudes and appear to hand them on unconsciously. Parents and teachers have to be made aware of the indirect differentiation they make between boys and girls so that they can distance themselves from these stereotypes and acquire a balanced attitude towards boys and girls concerning mathematics.

Finally, it is important for parents as well as for teachers to strengthen girls' mathematical self-concepts and make them believe in their own achievements, because they often doubt their achievements and mathematical potential which can, in turn, have a negative effect on future achievements. When parents and teachers support female students in their beliefs in their own achievements, it is possible for them to be more convinced of their achievements, not tracing their achievements back to external factors like luck or chance. Therefore, the omnipresent view of mathematics as a male domain can be overcome and turned into a view of mathematics belonging to males as well as females.

Appendix

In this questionnaire we would like to find out about your attitudes towards mathematics.
There are no correct or wrong answers.
Please tick your **personal opinion**!

Are you   f ☐   m ☐

How good are you at mathematics?   excellent   ☐   good   ☐
                                    average   below average
                                    weak   ☐

Explanation:
BD = Boys definitely more than girls
JP = Boys probably more than girls
ND = No difference between boys and girls
GP = Girls probably more than boys
GD = Girls definitely more than boys

	BD	BP	ND	GP	GD
1. Whose favourite subject is mathematics?					
2. Who thinks that it is important to understand the work in mathematics?					
3. Who is asked more questions by the mathematics teacher?					
4. Who gives up, when he/she finds a is too difficult?					
5. Who has to work hard in mathematics to do well?					
6. Who enjoys mathematics?					
7. Who cares about doing well in mathematics?					
8. Who thinks he/she did work hard enough if he/she does not do well in mathematics?					
9. Whose parents would be disappointed when he/she does not do well in mathematics?					
10. Who needs mathematics to maximise his/her future employment opportunities?					
11. Who likes challenging mathematical problems?					
12. Who is encouraged to do well by the mathematics teacher?					
13. Who do mathematics teachers expect to do well in mathematics?					

	BD	BP	ND	GP	GD
14. Who thinks that mathematics will be important for his/her adult life?					
15. Who expects to do well in mathematics?					
16. Who distracts other students from their mathematics work?					
17. Who gets the wrong answers in mathematics?					
18. Who finds mathematics easy?					
19. Whose parents think that it is important for him/her to study mathematics?					
20. Who needs more help in mathematics?					
21. Who teases boys who are good at mathematics?					
22. Who likes boys who are good at mathematics?					
23. Who is good at mathematics?					
24. Who likes using computers to solve mathematical problems?					
25. With whom do mathematics teachers spend more time?					
26. Who considers mathematics to be boring?					
27. Who considers mathematics to be difficult?					
28. Who gets on with his/her work in class?					
29. Who thinks mathematics is interesting?					
30. Who worries if he/she is not good at mathematics?					
31. Who teases girls who are good at mathematics?					
32. Who likes girls who are good at mathematics?					

33. Who do you think achieves more in mathematics? And why?

_____

_____

_____

_____

_____

_____

_____

Thank you for your participation!!!

References

Autorengruppe Bildungsberichterstattung (2010). *Bildung in Deutschland – Ein indikatorengestützter Bericht mit einer Analyse zu Perspektiven des Bildungswesens im demografischen Wandel.* Bielefeld: Bertelsmann Verlag.

Brandell, G., & Staberg, E-M. (2008). Mathematics: A female, male or gender-neutral domain? A study of attitudes among students at secondary level. *Gender and Education, 20*(5), 495–509.

Brandell, G., Leder, G., & Nyström, P. (2007). Gender and mathematics: Recent development from a Swedish perspective. *ZDM—The International Journal on Mathematics Education, 39*(3), 235–250.

Else-Quest, N., Hyde, J. S., & Linn, M. C. (2010). Cross-national patterns of gender differences in mathematics: A meta-analysis. *Psychological Bulletin, 136*(1), 103–127.

Eurydice (2010). *Gender differences in educational outcomes: Study on the measures taken and the current situation in Europe.* Bruxelles: EACEA P9 Eurydice.

Fennema, E. (1974). Mathematics learning and the sexes: A review. *Journal for Research in Mathematics Education, 5,* 126–139.

Fennema, E., & Sherman, J. (1976). Fennema–Sherman mathematics attitude scales. *JSAS: Catalog of selected documents in psychology, 6*(1):31 (Ms. No. 1225).

Forgasz, H., Leder, G., & Gardner, P. (1999). The Fennema–Sherman mathematics as a male domain scale reexamined. *Journal of Research in Mathematics Education, 30*(3), 342–348.

Forgasz, H., & Mittelberg, D. (2008). Israeli Jewish and Arab students' gendering of mathematics. *ZDM—The International Journal on Mathematics Education, 40*(4), 545–558.

Hoffstall, M., & Orschulik, A. B. (2009). *Geschlechterrollenstereotype in der Wahrnehmung der Mathematik.* Unpublished masters thesis, University of Hamburg, Hamburg.

Hyde, J. S., Lindberg, S. M., Linn, M. C., Ellis, A. B., & William, C. (2008). Gender similarities characterize math performance. *Science, 321,* 494–495.

Leder, G., & Forgasz, H. (2002). *Two new instruments to probe attitudes about gender and mathematics.* ERIC, Resources in Education (RIE). [ERIC document number: ED463312]

Leder, G., Forgasz, H., & Solar, C. (1996). Research and intervention programs in mathematics education: A gendered issue. In A. Bishop, K. Clements, C. Keitel, J. Kilpatrick, & C. Laborde (Eds.), *Handbook of research in mathematics education* (pp. 597–622). New York: Macmillan.

OECD (2009). *Equally prepared for life? How 15-year-old boys and girls perform in school.* Paris: OECD.

Tiedemann, J. (2000). Parents' gender stereotypes and teachers' beliefs as predictors of children's concept of their mathematical ability in elementary school. *Journal of Educational Psychology, 92,* 308–315.

Tiedemann, J. (2002). Teachers' gender stereotypes as determinants of teacher perceptions in elementary school mathematics. *Educational Studies in Mathematics, 50,* 49–62.

WZBrief Bildung (2010). *Lehrerinnen trifft keine Schuld an der Schulkrise der Jungen.* Berlin: Wissenschaftszentrum Berlin für Sozialforschung.

Commentary on the Chapter by Gabriele Kaiser, Maren Hoffstall and Anna B. Orschulik, "Gender Role Stereotypes in the Perception of Mathematics—Results of an Empirical Study with Secondary Students in Germany"

Sarah Theule Lubienski

Gender patterns in mathematics attitudes and beliefs are remarkably consistent across many countries and cultures, as highlighted in recent international assessments. For example, Else-Quest et al. (2010) examined TIMSS and PISA data and found a strong pattern of gender differences in mathematical affect across most countries, with girls reporting less self-confidence and greater mathematics anxiety than boys.

In order to illuminate the nuances and potential origins of these patterns, it is helpful to have more detailed, country-specific studies such as that conducted by Gabriele Kaiser, Maren Hoffstall and Anna Orschulik. In their study conducted in Germany, they administered an adapted version of Leder and Forgasz' (2002) *Who and Mathematics* questionnaire to roughly 600 11-to-12 year olds and another 600 15-to-17 year olds. They found strong, stereotypical attitudes among the older students, with younger students reporting far less biased attitudes.

Some of the differences they report are striking. For example, there was a strong tendency by the 15–17 year old students to report that girls need to work harder than boys to do well in mathematics. This pattern was not evident among the 11–12 year olds, raising questions about how exactly this belief emerged during the intervening years. Another disturbing pattern they report is that the tendency amongst girls to claim that mathematics is more important for boys' rather than girls' futures increased strongly from the younger girls to the older ones.

Kaiser, Hoffstall and Orschulik also analyzed the reasons students gave for their answers to the question, "Who do you think achieves more in mathematics and why?" When students explained why girls might outperform boys in mathematics, the majority of responses pertained to girls' diligence, whereas the most popular explanations for boys' superiority pertained to their mathematical interest and logic, with effort rarely mentioned. These patterns were, again, stronger among older students than younger students. The authors also report that girls rated their own math-

S.T. Lubienski (✉)
University of Illinois, Urbana-Champaign, USA
e-mail: Stl@illinois.edu

H. Forgasz, F. Rivera (eds.), *Towards Equity in Mathematics Education*,
Advances in Mathematics Education,
DOI 10.1007/978-3-642-27702-3_12, © Springer-Verlag Berlin Heidelberg 2012

ematics achievement lower than that of boys, despite the fact that German girls tend to perform at least as well as boys on mathematics achievement tests.

The authors appropriately include the caveat that their sample of German students was essentially a "convenience sample" and not randomly selected. However, schools were chosen from among four different types—low achieving, mid-achieving, high achieving and comprehensive. One potentially fruitful path the authors could explore further is how gender interacts with school type—both in terms of the composition of students (e.g., whether boys and girls are equally represented at each type of school) and whether stereotypical views are more or less prevalent at the different schools. Given that gender-related disparities in mathematics course taking and outcomes are often more pronounced among higher-achieving students (e.g., Forgasz 2006; Lubienski et al. 2004; Robinson and Lubienski 2011), and given that future mathematicians and scientists are most likely to emerge from high achieving schools, it would be worthwhile to compare gender-related patterns in the lower- and higher-achieving schools.

Overall, the study's results resonate with general findings about gender disparities in attitudes and beliefs from international assessments (e.g., Else-Quest et al. 2010), as well as recent findings from the Early Childhood Longitudinal Study (ECLS), a large-scale, nationally representative dataset in the United States. For example, analyses of the ECLS data have revealed that differences in U.S. girls' and boys' perceptions of their mathematical competence are larger than differences in both actual performance and interest in mathematics. Moreover, students' feelings of competence in mathematics are significant predictors of their later mathematics gains (Lubienski et al. 2011). Additionally, teachers in the ECLS dataset rate girls as better behaved and more diligent than boys (Rathbun et al. 2004; Ready et al. 2005). However, when comparing boys and girls with equal mathematics test scores and similar classroom behavior, teachers rate boys' mathematics proficiency higher, and this "over-rating" appears to relate to the gender gap in mathematics achievement that emerges in the elementary grades (Robinson et al. in preparation). Taken together, the results from both Germany and the U.S. raise concerns about teachers' perceptions and treatment of boys and girls in mathematics classes, and the ways in which that differential treatment might impact boys' and girls' views about themselves and gender in relation to mathematics.

But what are the root causes of these gender-biased perceptions? The conclusion drawn by Kaiser, Hoffstall and Orschulik is that girls and boys internalize teachers' and wider society's faulty stereotypic beliefs about who is good at mathematics. However, this conclusion might be too quick to discount students' observation skills and might not quite capture the full picture.

Perhaps the students in this study, after years of personal experience in the schools, are not simply reporting stereotypes they have internalized from others but are, in fact, reporting what they have observed day after day in their classrooms. So, for example, the male German student who reported that "*boys achieve correct solutions by thinking whereas girls use the right formulas*" might, unfortunately, be accurately reporting the tendencies he has noticed among his fellow students. In fact, the boy's statement is consistent with some prior research, such as one small-scale study of U.S. students' approaches to mathematical problem solving, in which young girls used more standard algorithms and young boys used more invented strategies (Fennema et al. 1998).

Given persistent evidence that girls are more compliant and diligent than boys in school—evidence collected from girls, boys, and teachers alike—it is likely that girls and boys do, in fact, behave differently in the mathematics classroom. This differential behavior could contribute to many of the patterns reported in this study. However, additional questions arise. Why are girls more likely to exhibit "good girl behavior" in the mathematics classroom, making greater efforts to complete tasks as instructed? In what ways might boys be rewarded for acting independently and girls encouraged to conform? In addition to societal stereotypes about mathematics, should we also be concerned about girls' feelings of worth being intertwined with their ability to please others—a notion that seems somehow magnified in the mathematics classroom context?

Another troubling question is, "Why do girls report greater anxiety and insecurity when it comes to mathematics performance, even when they outperform boys?" Comparisons between mathematics and literacy are illuminating in this regard. For example, Correll (2001) found that U.S. males view themselves as better in mathematics relative to females with equal test scores, but the opposite was true for reading. This pattern suggests that societal views about mathematics, literacy, and gender influence students' perceptions of their own abilities.

However, several questions remain. Girls are viewed as more diligent than boys in general, so why does this perception of diligence seem to diminish teachers' and students' assessment of girls' abilities in mathematics, but not in literacy? Is this because mathematics is viewed as something one is either "good at" or not—a measure of fixed, innate intelligence? Hence, are girls' efforts in mathematics signaling that they are not "naturally" good at mathematics in a way that does not carry over to perceptions of their literacy talents? If so, then rectifying the patterns highlighted by Kaiser, Hoffstall and Orschulik will involve overhauling not only gendered stereotypes about who succeeds in mathematics, but also society's beliefs about the nature of mathematics, itself.

As mathematics education scholars attempt to push forward on these issues, they must be prepared to answer the "so what?" question that is increasingly asked in relation to work on gender and mathematics. In the case of this study, its importance could be further emphasized if links could be made, for example, between patterns in German student affect and their later career choices. Economic arguments about the field of mathematics losing potentially strong contributors, and women missing out on lucrative career opportunities, can help highlight the continuing importance of gender in relation to mathematics. These arguments are increasingly necessary to make amid claims that it is boys and not girls who are now "shortchanged" in schools (Sommers 2000; Whitmire 2010).

In more immediate, practical terms, Kaiser, Hoffstall and Orschulik rightly conclude that teachers need to be aware of the patterns exposed in this study and should work to build girls' confidence in mathematics. Still, this admonition likely leaves educators wondering how, exactly, that can occur. Hence, one fruitful line of work would involve identifying the specific aspects that would comprise a successful intervention with teachers and students. What would such an intervention look like?

What faulty beliefs about the nature of mathematics, about who can succeed in mathematics, and about the worth of girls, themselves, should be confronted, and how and when is that best accomplished? Perhaps the most important contribution of this study of German girls and boys is that it helps us get one step closer to identifying those beliefs that are most troubling, and pinpointing adolescence as a key period of development of those beliefs among German students.

References

Correll, S. J. (2001). Gender and the career choice process: The role of biased self-assessments. *American Journal of Sociology*, *101*(6), 1691–1730.

Else-Quest, N. M., Hyde, J. S., & Linn, M. C. (2010). Cross-national patterns of gender differences in mathematics: A meta-analysis. *Psychological Bulletin*, *136*(1), 101–127.

Leder, G. C., & Forgasz, H. J. (2002). *Two new instruments to probe attitudes about gender and mathematics*. ERIC, Resources in Education (RIE). ERIC document number: ED463312.

Fennema, E., Carpenter, T. P., Jacobs, V. R., Franke, M. L., & Levi, L. (1998). A longitudinal study of gender differences in young children's mathematical thinking. *Educational Researcher*, *27*(5), 6–11.

Forgasz, H. J. (2006). *Australian year 12 mathematics enrolments: Patterns and trends—past and present*. International Centre of Excellence for Education in Mathematics. Accessed June 12, 2011 at www.amsi.org.au/index.php/component/content/article/78-education/358-forgasz.

Lubienski, S. T., Crane, C. C., & Robinson J. P. (2011). *A longitudinal study of gender and mathematics using ECLS data*. Final report (grant #R305A080147) submitted to the National Center for Education Research, Institute of Education Sciences, Washington, DC.

Lubienski, S. T., McGraw, R., & Strutchens, M. (2004). NAEP findings regarding gender: Mathematics achievement, student affect, and learning practices. In P. Kloosterman, & F. K. Lester Jr. (Eds.), *Results and interpretations of the 1990 through 2000 mathematics assessments of the National Assessment of Educational Progress* (pp. 305–336). Reston: National Council of Teachers of Mathematics.

Rathbun, A. H., West, J., & Germino-Hausken, E. (2004). *From kindergarten through third grade: Children's beginning school experiences* (NCES 2004-007). Washington: National Center for Education Statistics.

Ready, D. D., LoGerfo, L. F., Lee, V. E., & Burkam, D. T. (2005). Explaining girls' advantage in kindergarten literacy learning: Do classroom behaviors make a difference? *Elementary School Journal*, *106*(1), 21–38.

Robinson, J. P., & Lubienski, S. T. (2011). The development of gender achievement gaps in mathematics and reading during elementary and middle school: Examining direct cognitive assessments and teacher ratings. *American Educational Research Journal*, *48*(2), 268–302.

Robinson, J. P., Lubienski, S. T., Copur, Y. C., & Ganley, C. M. (in preparation). Gender-biased perceptions of performance affect the early mathematics gender gap.

Sommers, C. H. (2000). *The war against boys: How misguided feminism is harming our young men*. New York: Touchstone Books.

Whitmire, R. (2010). *Why boys fail: Saving our sons from an educational system that's leaving them behind*. New York: AMACOM.

Commentary on the Chapter by Gabriele Kaiser, Maren Hoffstall, and Anna B. Orschulik, "Gender Role Stereotypes in the Perception of Mathematics: Results of an Empirical Study with Secondary Students in Germany"

Colleen Vale

1 Affective Factors Matter

Since the mid-1970s gender differences favouring males in mathematics achievement at all levels of schooling and participation at post-compulsory levels of schooling have declined around the world. Yet gender differences in mathematics self-efficacy that favour males have remained intransigent, and the gender stereotyping of mathematics proficiency and interest persist. The study by Kaiser, Hoffstall and Orschulik is the latest in a series of studies commencing in the mid-1970s (Fennema and Sherman 1977) to provide evidence of higher self-concept in mathematics for male students compared to female students, and male stereotyped views of mathematics learning and achievement, especially among senior secondary aged students.

The early gender studies of mathematics set out to confront male hegemony with respect to mathematics. Feminists simply did not accept that males were innately better at and more interested in mathematics than females or that mathematics was a male domain. Along with research on affective factors, researchers also investigated the experiences and work of female mathematicians and the mathematical expertise of ordinary women (e.g., Burton 1995, 1999; Day 1997). They observed, analysed, and exposed mathematics classrooms in schools and tertiary institutions as male-biased environments (e.g., Jungwirth 1991; Leder 1993; Rodd and Bartholomew 2006; Shannon 2004; Vale 2002). Finally feminists exposed gender-biased and stereotyped perceptions of mathematics among the general community including parents (e.g., Forgasz et al. 2000; Walls 2010). Over time this research resulted in changes in curriculum, pedagogies and some methods of assessment to enhance learning, engagement and participation of female students in mathematics at all ages, and especially in secondary school mathematics. And, as noted above, researchers continue to monitor achievement, affect and participation by gender, and

---

C. Vale (✉)
Victoria University, Melbourne, Australia
e-mail: Colleen.vale@vu.edu.au

H. Forgasz, F. Rivera (eds.), *Towards Equity in Mathematics Education*,
Advances in Mathematics Education,
DOI 10.1007/978-3-642-27702-3_13, © Springer-Verlag Berlin Heidelberg 2012

145

to press for developments in mathematics and mathematics teaching and learning that promote gender equity.

The instrument used for this and earlier studies captures the degree of gender bias in student perceptions for a range of factors associated with mathematics learning and use: capability, interest, work ethic, and value for their future. Previous studies have found that male stereo-typing of mathematics varies in intensity according to gender, age, country, and ethnicity. Kaiser et al. (Chap. 11) reported frequencies and means for individual items and compared mean scores by gender and age. For a number of items mean scores did not appear to differ much from neutral and statistical inference tests could have reported effect sizes to establish whether gender stereotyping was statistically evident. The authors note that their findings that male stereotyping of mathematics is more evident among male students and older students, both male and female, than among younger students concur with findings from other studies cited in their chapter.

Gender-stereotyping of mathematics matters, especially if we can establish a relationship with self-efficacy, achievement and/or participation. Kaiser et al. claim that in their study "the gender stereotypical perceptions of mathematics are so deeply rooted in the students' minds that the students... include them into their self-perception." Calculation of correlation coefficients for individual measures of gender role stereotyping and the item on self-perception was needed to establish a relationship that appeared to be present when reviewing the distribution of responses reported. Certainly males rated their achievement more highly than females overall and at the highest level (excellent), but the proportion of males and females rating themselves as excellent at age 15 showed an increase of 3.5% for males on those aged 12, and 3.4% for females, while the proportion of males and females rating themselves below average increased by 6%. Hence gender stereotyping in this study did not result in lower self-perception of females and higher self-perception of males at the older age. However, these data are not paired over time. A longitudinal study is needed to track changes in students' self-perception and perceptions of mathematics as a gendered domain over time. Such a study would provide evidence of the relationship between gendered perceptions of mathematics and self-concept that have been reported in narrative studies such as Walls (2010).

Just as gender differences in mathematics achievement vary from significant to insignificant among countries, the gender stereotyping of mathematics is more evident in Sweden, Germany and Israel than in Australia and the United States where mathematics is a gender neutral domain (Forgasz et al. 2004). Forgasz (2002) previously observed that gender stereotyping of mathematics belief is strongest among students from high SES school communities and weakest or not evident in middle class and low SES school communities. The secondary schooling system in Germany is stratified according to student achievement; this is highly likely to correlate with socio-economic status since, like Australia, German achievement in mathematics is more strongly correlated with socio-economic status than for other European and Asian countries (OECD 2004). It would be interesting to know whether male stereotyping relating to socio-economic status is also evident in the German data. The sample used for this study was dominated by students from "higher type secondary schools" (48.3%). It would be possible to conduct further analyses of these data to determine whether the male stereotyped view pervaded all school types for both age groups. Such analysis would be important as gender stereotyping of mathematics has the potential to impact differently on girls' and boys' participation and achievement according to socio-economic status (Teese 2000).

2 Impact of Affective Factors

So what is the impact of these affective factors, that is, perceptions of mathematics, self-concept, confidence and self-efficacy in mathematics, on participation, engagement and learning, and achievement in mathematics? Intrinsic and extrinsic theories of motivation, along with socio-cultural and feminist theories, have been used when exploring these affective factors (Fennema and Peterson 1983; Forgasz 1995; Leder 1982; Thomson et al. 2004; Watt 2006). These studies reveal the complex gendered relationships between affect and cultural factors including learning behaviours, achievement, and participation.

Kaiser et al. found that students believed that girls achieve well in mathematics when they work hard, are conscientious, and ambitious, whereas they believe that boys are more likely to achieve well in mathematics because they are more interested and have better reasoning capability; findings not dissimilar to those reported in an earlier study by Leder (1982). In this study, and others, there is an assumption that participation and success in mathematics follow for students who don't "get the wrong answers", don't need "help in mathematics," don't have to "work hard to do well at mathematics" and who think "mathematics is interesting" and "will be important in his/her life." Yet the authors of this chapter concede that stereotyping of mathematics as a male domain and girls' lower self perception are "in sharp contrast to the general better achievements of German girls in school." Failure among students to appreciate that constructive learning behaviours such as persistence, diligence, autonomy, and respectful and productive relationships between students are related to success in mathematics may explain these findings and others reporting gender neutral perceptions of mathematics and the closing of the gender gap in mathematics achievement over time around the world (Boaler 2008; Fennema and Peterson 1983; Forgasz et al. 2004).

Thomson et al. (2004) argued that participation of females in tertiary mathematics and related disciplines could be increased if teachers attended to girls' attitudes, while Watt (2006) claimed that we need to find out why girls who perform as well as or better than boys regard themselves as lacking talent. Both of these views imply that girls are the problem. Shannon (2004) on the other hand argued that teachers need to change their practice so that girls come to enjoy mathematics, not feel alienated, and see it as relevant. Kaiser et al. argue that parents and teachers need to be much more aware of how gender stereotyping of mathematics in the community and schools is affecting girls' self concept with regard to mathematics. Their position sits well alongside critical theorists who call for deeper understanding of gendered subjectivity in order to implement strategies and approaches to develop the agency of young women in mathematics (Hirschmann 2003; Walls 2010). Researchers of gender and mathematics need to continue to investigate approaches that promote productive learning behaviours and relationships with peers, teachers, and parents and positive self-concept and interest in mathematics especially for students, both girls and boys, who are most disadvantaged by the stereotyping of mathematics as a male domain.

References

Boaler, J. (2008). Promoting 'relational equity' and high mathematics achievement through an innovative mixed ability approach. *British Educational Research Journal, 34*(2), 167–194.

Burton, L. (1995). Moving towards a feminist epistemology of mathematics. In P. Rogers & G. Kaiser (Eds.), *Equity in mathematics education: Influences of feminism and culture* (pp. 209–226). London: Falmer Press.

Burton, L. (1999). Fables: The tortoise? The hare? The mathematically underachieving male? *Gender and Education, 11*(4), 413–426.

Day, M. (1997). *From the experiences of women mathematicians: A feminist epistemology for mathematics*. Unpublished doctoral thesis, Massey University, New Zealand.

Fennema, E., & Peterson, P. (1983). Autonomous learning behavior: A possible explanation of sex-related differences in mathematics. *Educational Studies in Mathematics, 16*(3), 309–311.

Fennema, E., & Sherman, J. (1977). Sex-related differences in mathematics achievement, spatial visualization and affective factors. *American Educational Research Journal, 14*, 51–71.

Forgasz, H. (1995). Gender and the relationship between affective beliefs and perceptions of grade 7 mathematics classroom learning environments. *Educational Studies in Mathematics, 28*, 219–239.

Forgasz, H. J. (2002). Computers for the learning of mathematics: Equity considerations. In B. Barton, K. Irwin, M. Pfannkuch & M. Thomas (Eds.), *Mathematics education in the South Pacific*. Auckland: MERGA.

Forgasz, H. J., Leder, G. C., & Kloosterman, P. (2004). New perspectives on the gender stereotyping of mathematics. *Mathematical Thinking and Learning, 6*(4), 389–420.

Forgasz, H., Leder, G. C. & Vale, C. (2000). Gender and mathematics: Changing perspectives. In K. Owens & J. Mousley (Eds.), *Mathematics education research in Australasia: 1996–1999* (pp. 305–340). Turramurra: Mathematics Education Research Group of Australasia.

Hirschmann, N. (2003). *The subject of liberty: Toward a feminist theory of freedom*. Princeton: Princeton University Press.

Jungwirth, H. (1991). Interaction and gender—Findings of a microethnographical approach to classroom discourse. *Educational Studies in Mathematics, 22*, 263–284.

Leder, G. C. (1982). Mathematics achievement and fear of success. *Journal for Research in Mathematics Education, 13*(2), 124–135.

Leder, G. C. (1993). Teacher/student interactions in the mathematics classroom: A different perspective. In E. Fennema & G. Leder (Eds.), *Mathematics and gender* (pp. 149–168). Brisbane: Queensland University Press.

OECD (2004). *Learning for tomorrow's world: First results from PISA 2003*. Paris: OECD.

Rodd, M., & Bartholomew, H. (2006). Invisible and special: Young women's experiences as undergraduate mathematics students. *Gender and Education, 18*(1), 35–50.

Shannon, F. (2004). Classics counts over calculus: A case study. In I. Putt, R. Faragher & M. McLean (Eds.), *Mathematics education for the third millennium: Towards 2010* (pp. 509–516). Pymble: MERGA.

Teese, R. (2000). *Academic success and social power: Examinations and inequality*. Melbourne, Australia: Melbourne University Press.

Thomson, S., Cresswell, J., & de Bortoli, L., (2004). *Facing the future: A focus on mathematical literacy among Australian 15-year-old students in PISA 2003*. Camberwell, Australia: ACER.

Vale, C. (2002). Girls back off mathematics again: the views and experiences of girls in computer based mathematics. *Mathematics Education Research Journal, 14*(3), 52–68.

Walls, F. (2010). Freedom to choose? Girls, mathematics and the gendered construction of mathematical identity. In H. Forgasz, J. Rossi-Becker, O. Steinthorsdottir & K.-H. Lee (Eds.), *International perspectives on gender and mathematics education* (pp. 87–110). Charlotte: Information Age Publishing Inc.

Watt, H. M. G. (2006). The role of motivation in gendered educational and occupational trajectories related to maths. *Educational Research and Evaluation, 12*(4), 305–322.

Students' Attitudes, Engagement and Confidence in Mathematics and Statistics Learning: ICT, Gender, and Equity Dimensions

Anastasios N. Barkatsas

Abstract In this chapter the findings of five studies are reported. Two research instruments were used: the *Mathematics and Technology Attitudes Scale* (MTAS), and the *Survey of Attitudes Toward Statistics Scale* (SATS). The aims, methods, data analyses, selected findings and conclusions are presented, as well as implications for the teaching and learning of mathematics and statistics. The studies involved samples from Australia and Greece. Findings from the three MTAS studies revealed that there is a complex nexus of relationships between secondary mathematics students' mathematics confidence, confidence with technology, attitude to learning mathematics with technology, affective engagement and behavioural engagement, achievement, and gender. Findings from the SATS studies indicated that male Greek tertiary students had more positive attitudes toward statistics than female students; there was no gender gap for the Australian tertiary students. Secondary students' attitudes towards ICT use for mathematics learning require further scrutiny in order to bring about gender equity and to facilitate improved outcomes for all students. Gender and cultural sensitivity are paramount in the instructional planning, decision making, and implementation of secondary mathematics and tertiary statistics.

1 Background and Context of the Chapter

Information and Communication Technologies (ICT) or digital technologies, offer exciting opportunities for new approaches to teaching, and for the enhancement of students' mathematical understanding at all levels. Research suggests that the new approaches may enhance learning through cognitive, metacognitive, and affective (emotional) channels. The affective channel for improving the learning of mathematics by using contemporary instructional approaches, including the use of ICT for equitable purposes, is the focus of this chapter. Reed et al. (2010) argued however, that:

> Although the use of computer tools in schools is widespread, actual outcomes of employing such tools have been disappointing. While computer tools are purported to enhance

A.N. Barkatsas (✉)
Monash University, Melbourne, Australia
e-mail: Tasos.Barkatsas@monash.edu

H. Forgasz, F. Rivera (eds.), *Towards Equity in Mathematics Education*,
Advances in Mathematics Education,
DOI 10.1007/978-3-642-27702-3_14, © Springer-Verlag Berlin Heidelberg 2012

the learning experience and to bring learners to higher levels of understanding, motivation, engagement and self-esteem, they are often marginalised within existing classroom practices, or used only for repetitive, delimited activities, rather than to promote complex learning. The potential benefits of employing mathematical computer tools are not always realised.

Their view is shared by Bennison and Goos (2010), who claimed that: "previous research indicates that effective integration of technology into classroom practice remains patchy, with factors such as teacher knowledge, confidence, experience and beliefs, access to resources, and participation in professional development influencing uptake and implementation" (p. 31).

In this chapter the findings of five studies conducted in the past six years are reported. Two research instruments were used: (i) the Mathematics and Technology Attitudes Scale (MTAS) (Barkatsas 2005; Pierce et al. 2007) and (ii) the Survey of Attitudes Toward Statistics Scale (SATS) (Schau et al. 1995). MTAS (Appendix 1) is a relatively new instrument that has been used in a number of research studies in the past five years. SATS is an established instrument that has been used in many research studies for more than a decade. In what follows, the aims, methods, samples, data analyses, and selected findings of the five studies will be presented, and conclusions and implications will be drawn for the teaching and learning of mathematics and statistics.

1.1 The MTAS Hypothesised Model

The development of the MTAS (Appendix 1) scale was based on the assumption that it will measure affective changes which result from technology use and which are likely to have an impact on improving learning. Figure 1 outlines a hypothesised model proposed by the scale developers (Pierce et al. 2007). The authors described it as follows:

> The main row of boxes across the centre of Fig. 1 shows that it was hypothesised that information and communication technologies (ICT) in the classroom can enable more real world problem solving... These effects combine to improve students' behavioural engagement during lessons and hence improve learning. Confidence in using technology (extreme left) is seen as predisposing to full participation in lessons using technology, along with a number of unspecified factors... On the bottom right hand side, a positive attitude to using technology to learn mathematics is seen as an outcome of improved learning. However, a favourable attitude also provides a feedback loop, making further effective use of technology for learning likely. (p. 287)

Since the research of Fennema and Sherman (1976), questionnaires have become standard tools for assessing student attitudes, especially when attitudes alone are not the focus of the study, but rather viewed as one important factor to be monitored when assessing the likely success of a curriculum or teaching innovation. Over the last thirty years, attitudes to mathematics and statistics have been studied extensively. McLeod (1992) claimed that affective issues play a central role in mathematics learning and that "mathematics education research can be

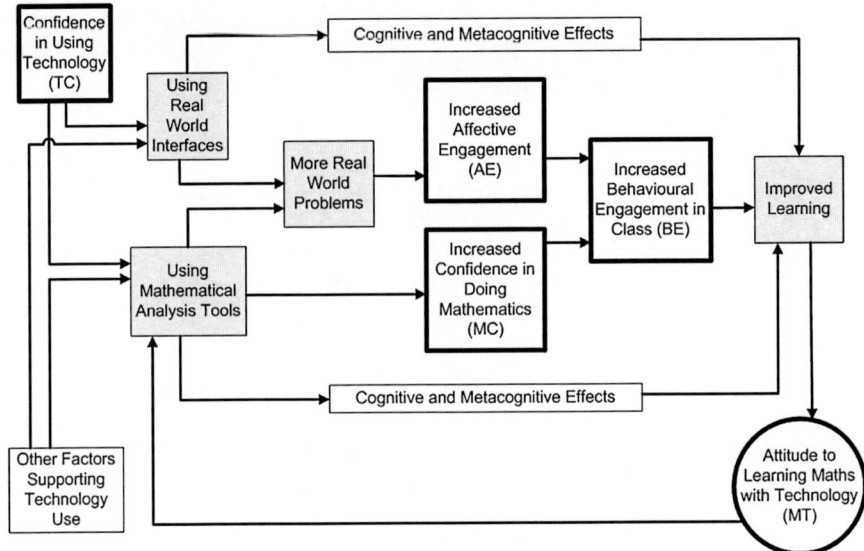

Fig. 1 Illustration of the hypothesised affective channel for technology use to improve mathematics learning. Variables measured in MTAS are in the heavily outlined boxes: MC, TC, AE, BE, MT (Reproduced with permission from Elsevier, License number: 2694491186250)

strengthened if researchers integrate affective issues into studies of cognition and instruction" (p. 575). Attitudes are commonly distinguished from beliefs in that attitudes are moderate in duration, intensity, and stability and have an emotional content, while beliefs become stable and are not easily changed (McLeod 1992; Pajares 1992). McLeod's (1992) definition of beliefs, attitudes and emotions, makes clear the distinction between the cognitive and affective dimensions of beliefs:

> Beliefs are largely cognitive in nature, and are developed over a relatively long period of time. Emotions, on the other hand, may involve little cognitive appraisal and may appear and disappear rather quickly, as when the frustration of trying to solve a hard problem is followed by the joy of finding a solution. Therefore we can think of beliefs, attitudes and emotions as representing increasing levels of affective involvement, decreasing levels of cognitive involvement, increasing levels intensity of response, and decreasing levels of response stability. (p. 579)

1.1.1 Key Concepts in the MTAS

Galbraith and Haines (1998) define a construct similar to attitude towards use of technology which they call 'computer and mathematics interaction'. They claimed that in their context:

> Students indicating high computer and mathematics interaction believe that computers enhance mathematical learning by the provision of many examples, find note making helpful

to augment screen based information, undertake a review soon after each computer session, and find computers helpful in linking algebraic and geometric ideas. (p. 279)

Vale and Leder (2004) view students' attitudes to technology (in their case computers) as being defined by the students' perceptions of their achievement (self-efficacy) and their aspiration to achieve in these disciplines. Galbraith and Haines (1998) take a different view, seeing technology confidence (computer confidence in their case) as evidenced by students who "feel self assured in operating computers, believe they can master computer procedures required of them, are more sure of their answers when supported by a computer, and in cases of mistakes in computer work are confident of resolving the problem themselves" (p. 278).

The use of CAS in mathematics classes and the effects of attitudes and behaviours on learning mathematics with the aid of computer tools have attracted a broad range of attention from researchers in many parts of the world in the past decade (Artigue 2002; Drijvers 2000; Guin and Trouche 1999; Pierce and Stacey 2004; Reed et al. 2010). In a recent article however, Reed et al. (2010) lamented that:

Although the use of computer tools in schools is widespread, actual outcomes of employing such tools have been disappointing. While computer tools are purported to enhance the learning experience and to bring learners to higher levels of understanding, motivation, engagement and self-esteem, they are often marginalised within existing classroom practices, or used only for repetitive, delimited activities, rather than to promote complex learning. In mathematics education also, research has shown that the potential benefits of employing mathematical computer tools are not always realised. (p. 1)

The authors found that promoting learning with mathematical tools must take into account several factors simultaneously. These include the improvement of student attitudes, learning behaviours, and providing ample opportunity for the construction of new mathematical knowledge from mastery of acquired tools. Embedding tool use in meaningful mathematical discourse in which ideas are discussed and reflected upon, was considered the most important aspect in this process. It was also found that students' attitudes and behaviours are influenced by school and classroom factors, in agreement with the outcomes of study three discussed in the selected findings section of this chapter.

It is my conviction that the time is ripe for the mathematics education community to engage in longitudinal studies which will focus on the identification of affective obstacles students may encounter when using CAS in learning mathematics. Heid and Edwards (2001), proposed four possible roles of CAS in mathematics education:

- as the primary producer of symbolic results;
- to create and generate symbolic procedures;
- to assist students in generating many examples from which they can search for symbolic patterns; and
- to generate results for problems posed in abstract form.

Student engagement with the intellectual work of learning is, according to Marks (2000), an important goal for education, leads to achievement, and "contributes to students' social and cognitive development" (p. 154). In this chapter (studies two and three) it is argued that engagement, confidence in mathematics, confidence with technology, and achievement are interrelated, as far as high achievers in secondary school mathematics are concerned. Marks (2000) claimed that "although research examining the effect of engagement on achievement is comparatively sparse, existing studies consistently demonstrate a strong positive relationship between engagement and performance across diverse populations" (p. 155).

Newmann (1989) articulated a theory of student academic engagement. The researcher proposed three dimensions of student engagement: (1) students' need to develop and express competence, (2) students' full participation in school activities, and (3) students being immersed in authentic academic work. It is believed that most students commence their school life inherently motivated, but for many this motivation diminishes or entirely disappears because they are involved in routine and boring activities and they try to get by with as little effort as possible.

Fredricks et al. (2004) considered school engagement to be a malleable concept that is responsive to contextual features, and amenable to environmental change. They claimed that in the research literature engagement is considered a multidimensional concept or even a "meta" construct. They proposed the following three dimensions: *behavioural engagement*, which draws on student participation; *emotional engagement*, encompassing both positive and negative reactions to staff and the school in general; and *cognitive engagement*, which draw on the principle of students making an investment in learning. Two of the dimensions of this framework—behavioural engagement and affective (emotional) engagement—form part of the MTAS instrument. Zyngier (2007) took a different path, proposing that there are three dominant perspectives which could be conceptualised to account for engagement from a social justice point of view: (i) the instrumentalist or rational technical; (ii) the social or individualist; and (iii) critical transformative engagement.

Middleton (1999) put forward a number of reasons that provide a rationale as to why intrinsic motivation for achievement in mathematics is desirable in contemporary mathematics classrooms. He claimed that:

> When students engage in activities in which they are intrinsically motivated they tend to exhibit a number of pedagogically desirable behaviours including time on task, persistence in the face of failure, more elaborative and monitoring of comprehension, selection of more difficult tasks, greater creativity and risk taking, selection of deeper and more efficient performance and learning strategies, and choice of an activity in the absence of an extrinsic reward. (p. 66)

Middleton also argued that intrinsic motivation is more complex than the additive effects of student ability, perceived competence, and achievement desire, even though they significantly contribute to the students' desire to successfully participate in mathematical activities and to do well in mathematics. The importance of intrinsic motivation for achievement and participation in advanced mathematics courses, and the apparent differences between boys' and girls' views, has been demonstrated by Watt's (2006) argument that:

Boys maintained higher intrinsic value for maths and higher maths related self-perceptions than girls throughout adolescence... We need to understand how it is that boys come to be more interested and like maths more than girls; and also why girls perceive themselves as having less talent, even when they perform similarly. (p. 319)

Vale and Bartholomew (2008) cited a finding from the Program of International Student Assessment [PISA] 2003 study relating to female students' confidence in mathematics, that is that compared to males, they appeared to be "less engaged, more anxious, and less confident in mathematics" (p. 279). The authors contended that:

...computer (and technology) confidence is a very different construct to that of mathematical confidence. Mathematical confidence is an affective dimension closely associated with mathematics achievement. Galbraith and Haines (1998) see mathematics confidence as evidenced by students 'who believe they obtain value for effort, do not worry about learning hard topics, expect to get good results, and feel good about mathematics as a subject'. (p. 278)

The role of motivation, intrinsic values, and gendered mathematics related self-perceptions were considered by Watt (2006) as the major influences on "gendered educational participation in senior high school maths, which subsequently predicted maths-related career aspirations over and above prior mathematical achievement" (p. 305).

1.1.2 Key Concepts in the SATS

During the last few decades researchers have been interested in students' anxiety towards statistics. According to Earp (2007):

Attitudes toward statistics and statistics anxiety have been found to be highly correlated, with attitude toward statistics often influencing statistics anxiety. Students with negative experiences from mathematical or statistical courses or instructors are often scared and carry such memories in the form of anxiety. Students with negative attitudes toward statistics are thought to be highly anxious with regard to statistics. (p. 5)

One of the reasons that has been put forward to explain many students' apprehension to attend statistics units or courses, is that statistics activates what Schau et al. (1995) call the "Big M.A. (Math Anxiety)" (p. 868), which could cause students to "break out in hives" (p. 868). The big M.A. summarises many students' anxiety once they realise that they have to undertake university statistics units, a common occurrence nowadays since many contemporary tertiary courses require a certain level of quantitative literacy. When they eventually attend such a course, students express negative attitudes which, in turn, may have negative consequences for their performance.

Several researchers (Araki and Shultz 1995; Elmore and Lewis 1991; Elmore et al. 1993; Elmore and Vasu 1986; Harvey et al. 1985; Onwuegbuzie 1995; Roberts and Bilderback 1980; Schau et al. 1995; Wise 1985; Zimmer and Fuller 1996) have studied a number of factors which have an impact on students' performance in statistics courses, including attitudes toward statistics. A number of instruments have been developed to evaluate these attitudes (see, for example, Bechrakis et al. 2011). The Survey of Attitudes Toward Statistics Scale [SATS] was developed by Schau et al. (1995). The SATS incorporates four subscales which have been designed to measure negative and positive attitudes about statistics. The four subscales are the following: (a) *Affect*—positive and negative feelings about statistics; (b) *Cognitive Competence*—attitudes about intellectual knowledge and skills applied to statistics; (c) *Value*—attitudes about the usefulness, relevance, and worth of statistics in personal and professional life; and (d) *Difficulty*—attitudes about the difficulty of statistics as a subject.

Gender differences regarding students' performance and attitudes toward statistics have been studied extensively. Schram (1996) published a meta-analysis of studies investigating gender differences in achievement in statistics. It was reported that the average effect size ($d = 0.08$) was in favour of female students. Roberts and Saxe (1982) found significant differences when using the Statistics Attitudes Survey [SAS]: the male students had more positive attitudes towards statistics than the female students. Waters et al. (1988) found that the male students showed slightly more positive attitudes than the female students using the Attitudes Toward Statistics [ATS] Course subscale.

2 Selected Research Findings

An overview of selected findings from five of our recent research studies is presented in this section.

In the *first study* the development of the *Mathematics and Technology Attitudes Scale* (MTAS) scale for secondary students is briefly described. This scale aimed to monitor five affective variables relevant to learning mathematics with technology. The subscales measure mathematics confidence, technology confidence, attitude to learning mathematics with technology, and two aspects of engagement in learning mathematics. In the *second study*, the complex relationships between students' mathematics confidence, confidence with technology, attitude to learning mathematics with technology, affective engagement and behavioural engagement, achievement, gender, and year level using MTAS were investigated. In the *third study*, the attitudes towards CAS calculators use, of middle and senior secondary students in Victorian (Australia) independent schools, are being canvassed. In the *fourth study*, the construct validity of the Survey of Attitudes Toward Statistics (SATS) was investigated by examining its factorial structure and determining its gender invariance properties. Finally, in the *fifth study*, postgraduate students' attitudes toward statistics were investigated using the SATS, along with the following supplementary variables: gender, confidence in mathematics, and confidence in statistics.

Each study involved samples from Victoria, Australia and from Athens, Greece. Together they represent a comprehensive picture for cautious conclusions to be drawn beyond the limitations of each individual study.

2.1 (I) MTAS Studies: Aims, Methods, Samples, Data Analyses and Findings

2.1.1 Instrument Used in Studies 1–3

The Mathematics and Technology Attitudes Scale (MTAS) developed by Pierce et al. (2007) was used for the three MTAS studies reported in this chapter. The instrument consists of 20 items. A Likert scoring format is used for each of the subscales: *Mathematics Confidence* [*MC*], *Confidence with Technology* [*TC*], *Attitude to learning Mathematics with Technology* (whether computers, graphics calculators or computer algebra systems in the original scale—computers in this study) [*MT*], *Affective Engagement* [*AE*]. Students are asked to indicate the extent of their agreement with each statement, on a five point scale from strongly agree to strongly disagree (scored from 5 to 1). A different but similar response set is used for the *Behavioural Engagement* [*BE*] subscale. Students are asked to indicate the frequency of occurrence of different behaviours. A five-point system is again used—Nearly Always, Usually, About Half of the Time, Occasionally, Hardly Ever (scored again from 5 to 1). MTAS subscale scores can be calculated by simple addition. With a maximum possible score on any subscale of 20 and a minimum of 4, we consider scores of 17 or above to be high, indicating a very positive attitude, 13–16 to be moderately high and 12 or below to be low reflecting a neutral or negative attitude to the associated factor. The rationale for the selection of the items and the naming of the subscales, as well as the psychometric properties of the scale, have been presented elsewhere (see, for example, Barkatsas 2005; Pierce et al. 2007).

2.2 Study 1. The Development of the MTAS Scale

2.2.1 Aim of the Study

The development of a research instrument for the measurement of MC, AE, BE, TC and MT.

2.2.2 Sample

The scale was trialled with 350 students from 17 classes in grades 8–10 at the six secondary schools which participated in the 2004–2006 Australian Research Council Linkage project, *Enhancing achievement in mathematics achievement by using technology to support real problem solving and lessons of high cognitive demand*, conducted in Victoria, Australia. The technology used was graphing calculators. These schools are typical of the range of secondary schools in Victoria and included two private co-educational schools, two state co-educational schools and one girls' private school and one girls' state school. The schools varied from upper middle to low socio-economic status.

2.2.3 Variation of MTAS Scores Amongst Schools

Scores on three of the subscales AE, BE and TC had similar medians at all schools and, in general, students had high or moderately high scores. In every school, most students agreed rather than disagreed that it was better to learn mathematics with graphing calculators.

2.2.4 Gender Differences

Figure 2 displays the MTAS scores for each of the five subscales by gender. In order to remove between-school influences from the data, only the responses from the four co-educational schools (70 boys and 71 girls) were considered. The breakdown of these scores by gender reveals that boys had statistically significantly higher scores than girls for each subscale except BE ($t = 0.005$, df $= 151$, $p = 0.996$). The differences were greatest for TC ($t = 6.84$, df $= 152$, $p < 0.001$) and MC ($t = 6.13$, df $= 155$, $p < 0.001$). MT ($t = 2.85$, df $= 149$, $p < 0.001$) and AE ($t = 2.56$, df $= 152$, $p < 0.05$) demonstrated less difference. While 50% of boys scored 16+ on MC, this was true for only 25% of girls. MT scores show a similar distribution. TC scores are even more strongly higher for boys, with approximately 75% of boys scoring 16+ and only 25% of the girls. It is important to note that not all the students with negative attitudes to learning mathematics with technology were girls.

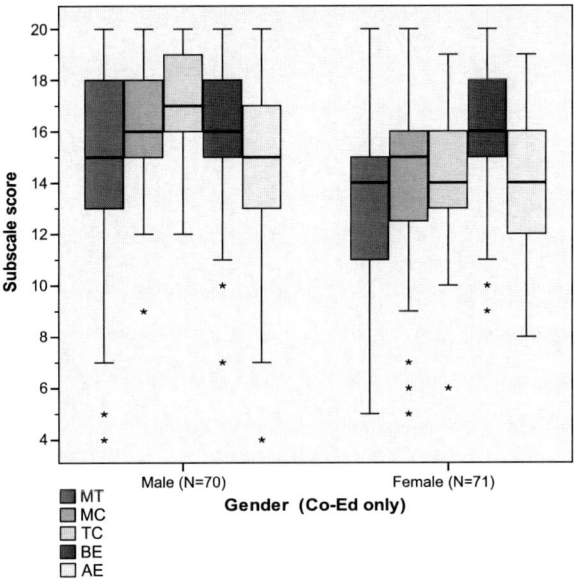

Fig. 2 MTAS scores for subscales by gender (Co-ed schools only)

2.3 Study 2. Investigating the Complex Relationship Between Students' Mathematics Confidence, Confidence with Computers, Attitude to Learning Mathematics with Computers, Affective Engagement, and Behavioural Engagement, Achievement, Gender, and Year Level Using MTAS

2.3.1 Aims of the Study

The aims of the study were:

- to investigate the factorial structure of the following variables: Secondary students' mathematics confidence, confidence with computers, attitude to learning mathematics with computers, affective engagement, and behavioural engagement, and
- to investigate the influence of demographical data and bio-data on students' mathematics confidence, confidence with computers, attitude to learning mathematics with computers, affective engagement, and behavioural engagement
- to investigate the interrelationship between the attitude, engagement, confidence, and achievement variables.

2.3.2 Sample

The participants were 1068 Year 9 students (final year of Junior High school, termed 'Gymnasium' in Greece) and Year 10 students (first year of Senior High school, termed 'Lyceum' in Greece)—see Table 1. The students were from 27 randomly selected state co-educational schools in metropolitan Athens, Greece. In each school, one classroom from each year level was randomly selected. These schools are typical of the range of secondary schools in Greece, and varied from upper middle to low socio-economic status.

Students' mathematics achievements were provided by their teachers, and represented their mathematics grades during the year of the study (2004–05). The grade categories were the following: A (80–100%), B (70–79%), C (60–69%), D (50–59%) and E/F (<50%).

Table 1 Sample by year level and gender	Gender	Year level		Total
		Year 10	Year 9	
	Male	263	286	549
	Female	257	262	519
	Total	520	548	1068

2.3.3 Data Analysis

The results of the factor analysis and the cluster analysis summary will be presented in this section. A more detailed data analysis has been presented elsewhere (see Barkatsas et al. 2009).

Factor Analysis   The questionnaire items were subjected to a Principal Components Analysis (PCA) using SPSS (extraction method: maximum likelihood). Given that the structure could vary, four analyses—one for each of the possible combinations between the gender categories (male and female) and the two year levels (Year 9 and Year 10)—were performed in order to investigate possible differences between year levels and gender. Since no differences were observed in the four initial analyses, a final factor analysis using data from 1068 students' responses to the 20 items forming the MTAS indicated that the data satisfied the underlying assumptions of the factor structure of the MTAS and that five factors (each with eigenvalue greater than 1) explained 67% of the variance, with almost 16% attributed to the first factor—*Mathematics Confidence* (MC) (see Appendix 2).

Reliability analyses yielded satisfactory Cronbach's alpha values for each subscale: MC, 0.92; MT, 0.89; TC, 0.87; BE, 0.77; and AE, 0.68. This indicated a strong or acceptable degree of internal consistency for each subscale. Further, according to Coakes and Steed (1999), if the Kaiser-Meyer-Olkin (KMO) Measure of Sampling Adequacy is greater than 0.6 and the Bartlett's Test of Sphericity (BTS) is significant, then factorability of the correlation matrix can be assumed. For this study, KMO $= 0.679$ and BTS < 0.01, so factorability of the correlation matrix was assumed.

Cluster Analysis Summary   Cluster analysis was used to determine homogeneous and clearly discriminated classes of students. The results of cluster analysis have been used in this study to confirm the results of the factor analysis and to enhance the depth of the analysis by developing more interpretable classes of participating students. The cluster analysis method used in this study was the K-means method. It was hypothesised that there would be a seven cluster partition. The initial seven clusters were formed by using a hierarchical cluster analysis (Ward criterion). The cluster analysis was performed using the SPAD software (Lebart et al. 2001), which uses the students' scores on the first ten factors (by default) as input. A summary of each cluster is provided in Table 2.

Tukey's multiple comparisons tests were used to investigate whether there were statistically significant differences in the mean scores by each of the six dependent variable (mathematics achievement, TC, MT, MC, AE, and BE) for the seven clusters (see Table 3). Statistically significant differences were found by each of the six dependent variables ($p < 0.001$ for each). Chi-square independence tests also revealed statistically significant differences by gender [$\chi^2(3) = 58.86, p < 0.001$] and by year level [$\chi^2(6) = 14.56, p < 0.05$].

The seven clusters resulting from the cluster analysis along with the gender differences are discussed below. It should be noted that the summary is an adapted version of a more extensive summary provided in Barkatsas et al. (2009).

Table 2 Clusters summary (Reproduced with permission from Elsevier, License number: 2694500174584)

Cluster	Mathematics achievement	TC	MT	MC	AE	BE	% sample	% male	% female
1	Low (D)	Very high	Positive	Very low	Strongly negative	Strongly negative	15.82	65.87	34.13
2	Average (C)	Low	Strongly negative	Average	Neutral	Negative	16.47	45.71	54.29
3	Excellent (A)	Very high	Strongly positive	Very high	Strongly positive	Strongly positive	13.04	67.63	32.37
4	Average (C)	Average	Strongly positive	Low	Neutral	Positive	12.12	36.22	63.78
5	High (B)	Very low	Strongly negative	Very high	Strongly positive	Strongly positive	9.90	40.78	59.22
6	High (B)	High	Strongly negative	Very high	Strongly positive	Strongly positive	12.67	60.15	39.85
7	Very low (E/F)	Low	Negative	Very low	Strongly negative	Strongly negative	19.98	41.98	58.02

Legend: Shaded = Mostly Boys, Unshaded = Mostly Girls.
Key: MC: Mathematics Confidence, TC: Confidence with Technology, MT: Attitude to learning Mathematics with Technology, AE: Affective Engagement, BE: Behavioural Engagement.
Note: The achievement grade categories used in the study were: A (80–100%), B (70–79%), C (60–69%), D (50–59%) and E/F (<50%).

- *Cluster 1* (15.82% of participants) consists of students with: low mathematics achievement, low levels of mathematics confidence, and negative levels of affective engagement and behavioural engagement. They are very confident in using computers and have positive attitude to learning mathematics with computers. Cluster 1 type students are statistically significantly more likely to be boys (65.87%) than girls (34.13%).
- *Cluster 2* (12.67% of participants) consists of students with: average mathematics achievement; and average levels of mathematics confidence, affective engagement, and behavioural engagement. They are not very confident in using computers, and have a very negative attitude to learning mathematics with computers. Cluster 2 type students *are statistically significantly more likely to be girls* (54.29%) *than boys* (45.71%).
- *Cluster 3* (13.04% of participants) consists of students with: excellent mathematics achievement, very high levels of mathematics confidence, and strongly positive levels of affective engagement and behavioural engagement. They are very confident in using computers, and have a strongly positive attitude to learning mathematics with computers. Cluster 3 type students are *statistically significantly more likely to be boys* (67.63%) *than girls* (32.37%).
- *Cluster 4* (12.12% of participants) consists of students with: average mathematics achievement, average levels of mathematics confidence, negative levels of affective engagement, positive levels of behavioural engagement, average levels of confidence in using computers, and a strongly positive attitude to learning mathe-

Table 3 Descriptive statistics for the seven clusters[1]

	Cluster number							
	1	2	3	4	5	6	7	Total
% of cases	15.82	16.47	13.04	12.12	9.90	12.67	19.98	
% boys in cluster	65.87	45.71	67.63	36.22	40.78	60.15	41.98	51.23
% girls in cluster	34.13	54.79	32.37	63.78	59.22	39.85	58.02	48.77
% Year 9 students in cluster	44.44	42.13	49.65	42.75	34.58^a	51.82	53.24^a	46.25
% Year 10 students in cluster	55.56	57.07	50.35	57.25	65.42	48.18	46.76	53.75
Means								
Mathematics achievement	2.34^d	3.10^b	4.06^a	2.81^c	3.79^a	3.83^a	1.90^e	2.99
Confidence with technology (TC)	4.47^a	2.12^d	4.18^{ab}	2.52^c	1.48^e	3.93^b	1.97^d	2.95
Attitude to learning mathematics with technology (MT)	3.99^b	1.67^d	4.65^a	4.12^b	1.58^d	2.02^c	2.24^c	2.87
Mathematics confidence (MC)	2.33^c	2.90^b	4.33^a	2.56^c	4.11^a	4.27^a	1.22^d	2.91
Affective engagement (AE)	2.30^d	2.64^c	4.38^a	3.00^b	4.39^a	4.37^a	1.67^e	3.06
Behavioural Engagement (BE)	1.87^d	2.60^c	4.06^a	3.20^b	4.09^a	3.97^a	1.48^e	2.84

Note: The lettered indices represent statistically significant differences: $p < 0.001$.
[1](Reproduced with permission from Elsevier, License number: 2694500174584.)

matics with computers. Cluster 4 type students are *statistically significantly more likely to be girls* (63.78%) *than boys* (36.22%).

- *Cluster 5* (9.90% of participants) consists of students with: high mathematics achievement, high levels of mathematics confidence, and strongly positive levels of affective engagement and behavioural engagement. They are not confident in using computers, and have positive attitudes to learning mathematics with computers. Cluster 5 type students *are statistically significantly more likely to be girls* (59.22%) *than boys* (40.78%).
- *Cluster 6* (12.67% of participants) consists of students with: high mathematics achievement, very high levels of mathematics confidence, and strongly positive levels of affective engagement and behavioural engagement. They are very confident in using computers, and have a very negative attitude to learning mathematics with computers. Cluster 6 type students are *statistically significantly more likely to be boys* (60.15%) *than girls* (39.85%).
- *Cluster 7* (19.98% of participants) consists of students with: very low mathematics achievement, very low levels of mathematics confidence, and strongly negative levels of affective engagement and behavioural engagement. They are not

confident in using computers, and have a negative attitude to learning mathematics with computers. Cluster 7 type students *are statistically significantly more likely to be girls* (58.02%) *than boys* (41.98%).

2.4 Study 3. Attitudes to Using CAS Calculators in the Classrooms of Middle and Senior Secondary Mathematics Students

2.4.1 Aims of the Study

The aims of the study were to investigate:

- the factorial structure of the MTAS, and
- the existence of gender, achievement, year level and CAS experience differences in each of the five MTAS subscales.

2.4.2 Sample

The participants were 835 Year 9–12 students, from nine independent schools, in Victoria, Australia. The head of mathematics in each school selected students from Year 9–12 classrooms; students had one, two or three years of CAS calculator experience (3 schools commenced using CAS calculators in Year 9, 2 schools in Year 10, 2 schools in Year 11, and 1 school in Year 12). The breakdown of students' CAS experience was as follows:

- one year CAS experience: 345 students;
- two years CAS experience: 199 students; and
- three years CAS experience: 300 students.

The study took place in term 4 of the 2009 school year.

The achievement grade categories used in the study were the following: A (80–100%), B (70–79%), C (60–69%), D (50–59%), and E/F (<50%). The achievement data were gathered from the students. For this study, the number of students obtaining grades D and E were combined due to the small numbers of students in these two categories.

2.4.3 Data Analysis

Principal Components Analysis   Using the data from the 835 students' responses to the 20 items forming the MTAS, a PCA using SPSS (extraction method: maximum likelihood) was conducted. The data satisfied the underlying assumptions of PCA, and five components (each with eigenvalue greater than 1) explained 69.3% of the variance, with almost 16% attributed to the first factor—*Mathematics Confidence* (MC)—and a further 14.5% to the second factor—*Attitude towards use of*

CAS for learning mathematics (MT). Reliability analyses yielded the following Cronbach's alpha values for each subscale: MC, 0.93; MT, 0.90; TC, 0.88; BE, 0.78; and AE, 0.70, indicating a strong or acceptable degree of internal consistency for each subscale. The Kaiser–Meyer–Olkin (KMO) measure was 0.874 and Bartlett's test of sphericity (BTS) < 0.001, allowing factorability of the correlation matrix to be assumed.

Additional Statistical Analyses   In order to investigate if there were gender, achievement, year level and CAS experience differences for the MTAS subscales, t-tests, one-way ANOVAs and Multiple Analysis of Variance (MANOVA) tests were conducted. Some of these findings will be discussed next.

2.4.4 Achievement

Barkatsas et al. (2009) investigated the complex relationship between students' mathematics confidence, confidence with technology, attitude to learning mathematics with technology, affective engagement and behavioural engagement, achievement, gender, and year level. High achievement in mathematics was found to be associated with high levels of mathematics confidence, strongly positive levels of affective engagement and behavioural engagement, high confidence in using technology, and a strongly positive attitude to learning mathematics with technology.

In this study, statistically significant differences in mean scores were found on MC by achievement level [$F(4, 37) = 13.429$, $p < 0.001$, $\eta^2 = 0.109$], mean scores on BE by achievement level [$F(4, 37) = 3.470$, $p < 0.05$, $\eta^2 = 0.031$] and mean scores on AE by achievement level [$F(4, 37) = 3.296$, $p < 0.05$, $\eta^2 = 0.029$]. No statistically significant differences in mean scores on MT and TC by achievement level were found. The breakdown of these scores by mathematics achievement grades, illustrated in Fig. 3, also reveals that students with the highest grade, A, demonstrated higher levels of MC, BE and AE than other students, and that the higher the students' achievement the more positive their MC, BE, and AE. The inter-quartile ranges for MT—CAS in this study—are identical for students with grades A and B, with a slightly smaller first quartile (25%) value for students with a C. A smaller inter-quartile range and smaller maximum and minimum values were found for students with a D or an E (this could be due to the effect of the relatively small number of students with a D or an E, but it was decided to include them in Fig. 3 for comparison purposes). The median value for MT, 14, was the same for all grades.

The TC medians and inter-quartile ranges are identical for students with grades A, B and C, with smaller values for students with grades D + E. Interestingly, all the TC inter-quartile ranges are greater than their respective MT inter-quartile ranges.

2.4.5 Gender Differences

Gender differences in attitudes to mathematics have long been of interest (e.g., Barkatsas et al. 2001; Barkatsas et al. 2009; Fennema and Sherman 1976; For-

Fig. 3 MTAS subscales by achievement grade

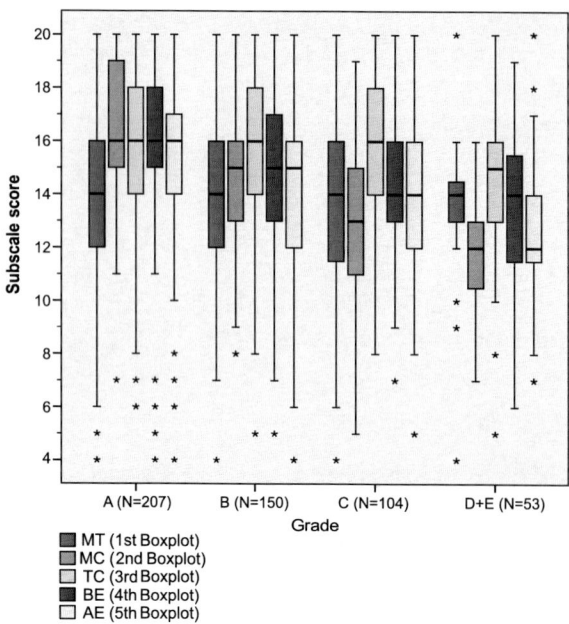

gasz et al. 1998, 1999, 2008; Pierce et al. 2007). A question of current interest is whether using technology to learn mathematics exacerbates differences. In this section, results are reported on the five subscales by gender. In order to remove between-school influences from the data, only responses from students attending the co-educational schools are considered.

The breakdown of the scores on each subscale by gender, illustrated in Fig. 4, reveals that boys have higher median scores than girls for each subscale except MT and BE, and statistically significantly higher mean scores for MC [$t(720) = 6.673$, $p < 0.001$] and TC [$t(727) = 8.438$, $p < 0.001$]. While 50% of boys scored 16+ on MC, this was true for only 25% of girls. TC scores are also higher for boys, with approximately 75% of boys scoring 15+ compared to 50% of the girls. These results reflect the common finding that boys express greater confidence than girls on both mathematics and technology. Note that for BE and MT, the median scores are identical for boys and girls. It can also be seen that the AE median and upper quartile values are higher for boys.

MTAS has been used to investigate the complex nexus of secondary mathematics students' mathematics confidence, confidence with CAS, attitude to learning mathematics with technology, affective engagement, behavioural engagement, achievement, gender, year level, and CAS experience. The results indicate that students with excellent mathematics achievement demonstrate very high levels of mathematics confidence, strongly positive levels of affective and behavioural engagement, are very confident in using computers (Study 2) and CAS calculators (Study 3), and that they have a positive attitude to learning mathematics with computers and with CAS. On the other hand, students with negative attitudes toward mathematics,

Fig. 4 MTAS scores for subscales by gender

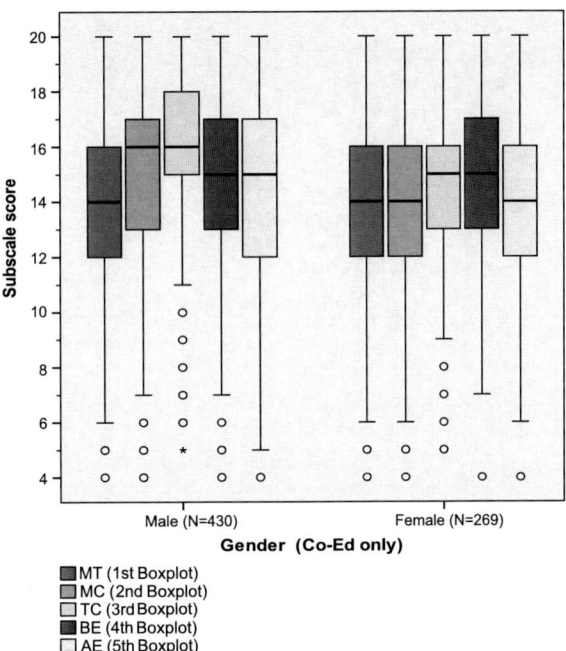

low mathematics achievement, low levels of mathematics confidence, and low levels of affective engagement and behavioural engagement, demonstrated confidence in using computers and positive attitudes to learning mathematics with computers (Study 2).

Overall, high-achieving boys appeared to be more confident than girls about mathematics. Compared to their female peers, boys also demonstrated stronger behavioural and affective engagement, more confidence in using computers and CAS calculators, and had a more positive attitude to learning mathematics with computers and CAS.

2.5 (II) Survey of Attitudes Toward Statistics [SATS] Scale Studies: Aims, Methods, Samples, Data Analyses and Findings

2.5.1 Instrument

The SATS questionnaire (Schau et al. 1995) consists of 28 items (Schau et al. 1995) on a seven point Likert scale (1 = strongly disagree, 4 = neither disagree nor agree, 7 = strongly agree). After re-coding the negatively worded items, the highest values in the scale correspond to positive attitudes toward statistics (for example, when the Difficulty subscale is reverse-scored the higher scores mean that statistics is "perceived as *less* difficult"). The questionnaire comprises four subscales: (a) *Affect*

(6 items); (b) *Cognitive Competence* (6 items); (c) *Value* (9 items); and (d) *Difficulty* (7 items). SATS subscale scores are calculated by addition of responses.

Other variables for which data were collected were gender, year of study, perceived computer competence, mathematics confidence, Year 12 mathematics achievement, confidence in doing well in statistics, and self-predicted achievement in university statistics courses. The participants in Studies 4 and 5 were tertiary students. Some of these findings will be discussed next.

2.6 Study 4. The Construct Validity of the SATS

2.6.1 Aims of the Study

The aims of the study were to investigate the SATS construct validity by:

- investigating the factorial structure of the questionnaire; and
- determining the gender differences.

2.6.2 Participants

SATS was administered to 1033 undergraduate students (25% were enrolled in first year statistics units, 24.5% in second year, 23.5% in third year and 27% in fourth year units) in the Social and Political Studies departments of a large University in Athens, Greece. Only students undertaking studies in courses which incorporated units in statistics were asked to participate in the study. The composition of the final sample was 84% female and 16% male.

2.6.3 Analysis of Data

Initially, a Confirmatory Factor Analysis [CFA] using individual items was conducted. Item-based CFA results showed that none of the items exhibited problems; each item had a higher factor loading on its hypothesised factor. Construct validity evidence for the SATS was assessed using parcels-based CFA correlations, group differences, and multiple regression analysis. Univariate and multivariate distributions of parcel scores were examined for potential outliers. In Fig. 5, the path diagrams for the four-factor, three-factor and the two-factor model of SATS are displayed. (For a detailed analysis of the diagram, see Bechrakis et al. 2011.)

The factor intercorrelations for the four-factor model are shown in Table 4. With one exception, the moderate correlations between the factors suggest adequate discriminant validity between these related aspects of attitudes toward statistics. In other words, the scales measure related but still separate aspects of attitudes. The strong correlation between the Cognitive and the Affect factors may suggest a lack of discriminant validity, although Schau et al. (1995) reported a high correlation between these two scales, as well.

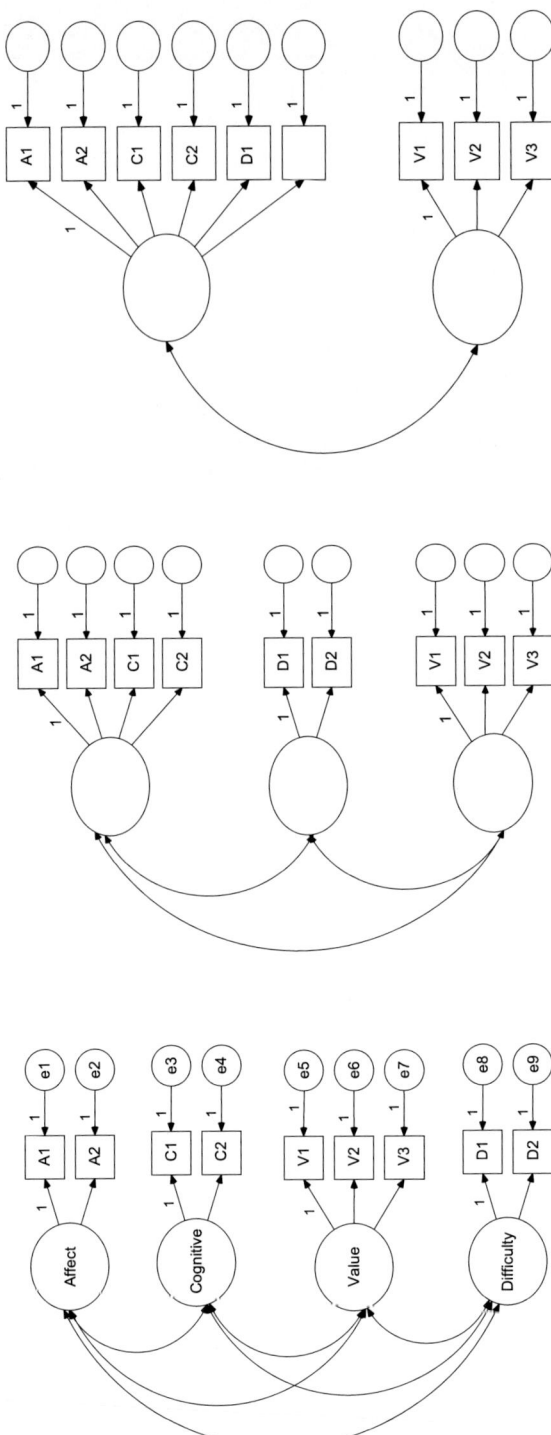

Fig. 5 Path diagrams for the four-factor, three-factor and the two-factor models of SATS

Table 4 Factor
intercorrelations for the
four-factor model

Factor	Affect	Cognitive	Value
Cognitive	0.901[*]		
Value	0.513[*]	0.459[*]	
Difficulty	0.589[*]	0.566[*]	0.151[*]

[*]$p < 0.01$.

2.6.4 Gender Differences

Gender differences for each subscale were investigated by setting the female students' subscale mean scores equal to zero and subsequently estimating the male mean scores (Maximum Likelihood Confirmatory Factor Analysis, SPSS AMOS 6.0). The male mean score for the Affective subscale was 0.45, which represents the difference between the male and the female mean scores (Diff = 0.45–0) and since 0.45 > 0, it was concluded that male students demonstrated more positive attitudes than female students. The mean score differences (Diff) for the other three subscales were the following: Cognitive Competence subscale, Diff = 0.21; Difficulty subscale, Diff = 0.19; Value subscale, Diff = 0.09.

Statistically significant gender differences in the mean scores were found on the Affect subscale $t(1071) = 4.27$, $p < 0.001$, $d = 0.40$, the Cognitive Competence subscale $t(1069) = 2.45$, $p < 0.05$, $d = 0.24$, and on the Difficulty subscale $t(1070) = 2.49$, $p < 0.05$, $d = 0.25$. There were no significant gender differences in the mean scores on the Value subscale. It may be argued that overall, male students demonstrated more positive attitudes toward statistics than female students.

2.7 Study 5. Postgraduate Students' Attitudes Toward Statistics

The aims were to investigate:

- postgraduate (pre-service teacher education) students' affect and cognitive competence about statistics, and perceived value and difficulty of statistics; and
- if there were gender differences in the attitudes toward statistics of the postgraduate students.

2.7.1 Method

Participants   The SATS was administered to 134 pre-service teacher education students enrolled in a Postgraduate Diploma in Education course at an Australian university. The students had previously undertaken one or more statistics courses as part of their first (undergraduate) degree. The composition of the final sample was 77% female and 23% male students.

Analyses Principal Components Analyses (PCA) with a Varimax rotation was conducted and internal consistency reliability estimates were calculated, followed by various box plots and a Multiple Analysis of Variance (MANOVA). The SATS internal consistency reliability estimates in this study were consistent with those reported by Schau et al. (1995). The reliability coefficients ranged from 0.636 to 0.904. The lowest estimate of Cronbach-α was obtained for the Difficulty subscale, a result consistent with the subscale with the lowest estimates reported by Schau et al. (1995) and Cashin and Elmore (2005).

A MANOVA test was also conducted to investigate the differences in mean scores on each of the four subscales by gender and by confidence in doing well in statistics (measured on a seven point Likert scale (1 = not at all confident, 7 = very confident). Assumptions of univariate normality were not violated and no multivariate outliers were detected using the Mahalanobis distance criterion. Linearity among dependent measures was confirmed using scatterplots among pairs of dependent variables across groups. The homogeneity of the variance-covariance matrices was investigated by using Box's M test. Box's M test was not significant at the alpha level of 0.001, and homogeneity of variance was established.

For each subscale, statistically significant differences in the mean scores were found by level of student confidence in doing well in statistics: Affect [$F(1, 128) = 38.328$, $p < 0.001$], Cognitive Competence [$F(1, 128) = 45.751$, $p < 0.001$], Value [$F(1, 128) = 36.829$, $p < 0.001$] and Difficulty [$F(1, 128) = 26.372$, $p < 0.001$]. No statistically significantly gender differences in the mean scores were found on any of the four subscales. This finding indicates that there is no gender gap for the Australian tertiary students regarding their attitudes toward the study of statistics at tertiary level.

3 What do the Findings from the Five Studies Tell Us? Conclusions and Implications

3.1 Secondary Mathematics Students' Attitudes, Engagement and Use of ICT for Learning Mathematics

The findings from the three MTAS studies reveal that there is a complex and interconnected nexus of relationships between secondary mathematics students' mathematics confidence, confidence with technology, attitude to learning mathematics with technology, affective engagement and behavioural engagement, achievement, and gender. The results indicate that:

I. *In a computer environment* (Study 2):

- Students with high and excellent mathematics achievement demonstrated high levels of mathematics confidence, strongly positive levels of affective and behavioural engagement, were very confident in using computers, and had a strongly positive attitude to learning mathematics with computers.

- Students with negative attitudes toward mathematics, low mathematics achievement, low levels of mathematics confidence, and low levels of affective engagement and behavioural engagement, demonstrated confidence in using computers and a positive attitude to learning mathematics with computers.
- The two factors that seem to be associated with the development of a positive attitude to learning mathematics with computers are *mathematics confidence* and *affective engagement*. Overall, high-achieving boys appeared to be more confident in mathematics, demonstrated stronger behavioural and affective engagement, were more confident in using computers, and had a more positive attitude to learning mathematics with computers, than did girls.

II. *In a graphing calculator environment* (Study 1):

- Students having a positive experience using graphing calculators in mathematics demonstrated higher scores in learning mathematics with technology (MT), although there was considerable variability in the students' evaluations.
- Scores on the AE, BE and TC subscales had similar medians at all participating schools, and students had generally high or moderately high scores.

III. *In a CAS environment* (Study 3):

- Statistically significant gender differences were found for Mathematics Confidence (MC) and Confidence with Technology (TC).
- Statistically significant differences were found on Mathematics Confidence (MC), Behavioural Engagement (BE) and Affective Engagement (AE) by achievement level.
- Overall, students demonstrated higher levels of TC compared to MT, irrespective of their years of CAS experience.

Reed et al. (2010) argued that when promoting mathematics learning with ICT the following factors need to be considered simultaneously: "improve students' attitudes, raising levels of goal-oriented learning behaviours, and giving sufficient opportunity for constructing new mathematical knowledge from acquired tool mastery". They also claimed that the use of CAS-related technology may require changes in instructional practices.

As the full spectrum of opportunities offered by new technologies for improving the teaching and learning of mathematics are worked through, it is important that education systems invest in teaching innovations that will boost low and average-achieving students' confidence in learning mathematics. In analysing teacher mediation of ICT and other resources in the classroom, Tanner et al. (2010) argued that the idea of 'orchestration' could be helpful:

> This construct extends the idea of scaffolding' and concerns the planned and responsive manipulation by the teacher of the features of the classroom setting (including students, resources, and less tangible features such as culture and ethos) to support the goal-related actions carried out by students and the development of common or collective knowledge. (p. 548)

Tanner et al. (2010) further claimed that "the impact of interactive whole-class technologies (IWCTs) on learning is dependent on the mediating role of the teacher" and that "effective teaching and learning is often based on serendipity and improvisation" (p. 547).

The origins of gender differences in mathematics motivation, engagement, achievement, and the use of ICT in mathematics, as well as their interrelationships, require further investigation in an era of ICT-driven mathematics curricula. A longitudinal study on secondary students' attitudes towards the use of ICT in mathematics learning and teaching is currently underway by the author.

It has been well-documented that more males than females study the most advanced mathematics courses at senior secondary level, that males choose to enrol in more mathematically-oriented tertiary courses, and that mathematics-related careers are still considered more appropriate for males (Forgasz and Leder 2001). It has proven to be a persistent and difficult area in both research and policy to modify dominant personal and societal attitudes and deeply-rooted beliefs about mathematics and statistics courses and about careers requiring highly technical quantitative skills (Cashin and Elmore 2005; Schau et al. 1995). Contemporary education systems have to find ways to address the gender inequity in highly sought-after careers which require a strong mathematical and/or statistical orientation. Further research is required however, to test the MTAS hypothesised model by using Structural Equations Modelling (SEM) techniques.

3.2 Tertiary Students' Attitudes Toward Statistics

The two factors Affective and Cognitive competence in the four-factor model appear to correlate positively with some of the other factors under consideration. All correlations among subscales were low except for the correlation between the Cognitive and the Affect subscales, a result consistent with the findings of Schau et al. (1995). Furthermore, the study confirmed the structure equivalence for male and female students. It was also found that male and female students' attitudes toward statistics courses differed slightly on the Affective and the Cognitive competence subscales but not on the Value subscale. This last finding is in agreement with the findings of a number of previously conducted studies (Bradley and Wygant 1998; Buck 1985; Harvey et al. 1985; Schram 1996; Ware and Chastain 1991; Waters et al. 1988).

A comparison of the attitudes toward statistics of Australian and Greek tertiary education pre-service students is currently underway by the author, as part of an ongoing study monitoring the attitudes of secondary and tertiary students of mathematics and statistics and their impact on gender and equity factors in education.

Further research is required to investigate relations between secondary and tertiary students' attitudes and performance, and the longitudinal stability or change in students' attitudes in different countries and cultural contexts. Mixed methods studies investigating tertiary students' attitudes toward statistics could be used to identify the factors underlying the reasons why students think highly of the usefulness, relevance, and worth of statistics in personal and professional life, and of the intellectual knowledge and skills applied to statistics, yet consistently report finding tertiary statistics difficult. Finally, measures other than self-reporting could be used to investigate how learning behaviours are affected by cultural, classroom, school and teacher factors and their relationships to students' attitudes and achievement in mathematics and statistics.

The increased confidence in their competence as learners of tertiary statistics demonstrated by the females in the Australian sample (Study 5) challenges the traditional expectations regarding male dominance in tertiary statistics courses and provides fertile grounds for further research. Secondary students' attitudes towards the use of ICT in learning mathematics on the other hand, require further scrutiny in order to bring about gender equity and to facilitate improved outcomes for all students. The issues of gender and cultural sensitivity are, in my view, paramount in secondary mathematics and tertiary statistics instructional planning, decision making and implementation.

Appendix 1: Mathematics and Technology Attitudes Scale

Five Subscales   Mathematics Confidence [MC], Confidence with Technology [TC], Attitude to learning mathematics with technology [MT], Affective Engagement [AE] and Behavioural Engagement [BE].

To tailor to a particular class, change the words "graphics calculators" to the technology used by that class (e.g. computers, graphing calculators, computer algebra systems).

Table 5 MTAS scale showing items and subscale membership (usually invisible)

	Hardly ever	Occasionally	About half the time	Usually	Nearly always
1. I concentrate hard in mathematics [BE]	HE	Oc	Ha	U	NA
2. I try to answer questions the teacher asks [BE]	HE	Oc	Ha	U	NA
3. If I make mistakes, I work until I have corrected them [BE]	HE	Oc	Ha	U	NA
4. If I can't do a problem, I keep trying different ideas [BE]	HE	Oc	Ha	U	NA

(*continued on the next page*)

Table 5 (Continued)

		Strongly disagree	Disagree	Not sure	Agree	Strongly agree
5.	I am good at using computers [TC]	SD	D	NS	A	SA
6.	I am good at using things like VCRs, DVDs, MP3s and mobile phones [TC]	SD	D	NS	A	SA
7.	I can fix a lot of computer problems [TC]	SD	D	NS	A	SA
8.	I am quick to learn new computer software needed for school [TC]	SD	D	NS	A	SA
9.	I have a mathematical mind [MC]	SD	D	NS	A	SA
10.	I can get good results in mathematics [MC]	SD	D	NS	A	SA
11.	I know I can handle difficulties in mathematics [MC]	SD	D	NS	A	SA
12.	I am confident with mathematics [MC]	SD	D	NS	A	SA
13.	I am interested to learn new things in mathematics [AE]	SD	D	NS	A	SA
14.	In mathematics you get rewards for your effort [AE]	SD	D	NS	A	SA
15.	Learning mathematics is enjoyable [AE]	SD	D	NS	A	SA
16.	I get a sense of satisfaction when I solve mathematics problems [AE]	SD	D	NS	A	SA
17.	I like using graphics calculators for mathematics [MTg]	SD	D	NS	A	SA
18.	Using graphics calculators in mathematics is worth the extra effort [MTg]	SD	D	NS	A	SA
19.	Mathematics is more interesting when using graphics calculators [MTg]	SD	D	NS	A	SA
20.	Graphics calculators help me learn mathematics better [MTg]	SD	D	NS	A	SA

Appendix 2: Factor Structure of the MTAS Scale

Subscales   Mathematics Confidence [MC], Confidence with Technology [TC], Attitude to learning mathematics with technology [MT], Affective Engagement [AE] and Behavioural Engagement [BE].

Table 6 Rotated component matrix

	Factor				
	1	2	3	4	5
(MC) I can get good results in mathematics	0.855				
(MC) I know I can handle difficulties in maths	0.840				
(MC) I am confident with mathematics	0.795				
(MC) I have a mathematical mind	0.757				
(MT) Mathematics is more interesting when using CAS calculators		0.862			
(MT) CAS calculators help me learn mathematics better		0.841			
(MT) Using CAS calculators in mathematics is worth the extra effort		0.829			
(MT) I like using Computer Algebra Systems (CAS) calculators for mathematics		0.801			
(TC) I am good at using computers			0.871		
(TC) I can fix a lot of computer problems			0.835		
(TC) I am quick to learn new computer software needed for school			0.820		
(TC) I am good at using things like VCR's, DVD's, MP3's, MP4's, iPods, Wii's and mobile phones.			0.795		
(BE) If I make mistakes, I work until I have corrected them				0.791	
(BE) If I can't do a problem, I keep trying different ideas				0.728	
(BE) I concentrate hard in maths				0.718	
(BE) I try to anwer questions the teacher asks				0.643	
(AE) In mathematics you get rewards for your effort					0.751
(AE) I get a sense of satisfaction when I solve mathematics problems					0.725
(AE) Learning mathematics is enjoyable					0.719
(AE) I am interested to learn new things in mathematics					0.677

Extraction Method: Principal Component Analysis.
Rotation Method: Varimax with Kaiser Normalisation.
Rotation converged in 6 iterations.

References

Araki, L. T., & Shultz, K. S. (1995, April). *Student attitudes toward statistics and their retention of statistical concepts*. Paper presented at the annual meeting of the Western Psychological Association, Los Angeles.

Artigue, M. (2002). Learning mathematics in a CAS environment: The genesis of a reflection about instrumentation and the dialectics between technical and conceptual work. *International Journal of Computers for Mathematical Learning, 7*, 245–274.

Buck, J. L. (1985). A failure to find gender differences in statistics achievement. *Teaching of Psychology, 12*, 100.

Barkatsas, A. N. (2005). A new scale for monitoring students' attitudes to learning mathematics with technology (MTAS). In P. Clarkson, A. Downton, D. Gronn, M. Horne, A. McDonough, R. Pierce & A. Roche (Eds.), *Building connections: Theory, research and practice* (Vol. *1*, pp. 129–137). Proceedings of the 28th Annual Conference of the Mathematics Education Group of Australasia. Melbourne: MERGA.

Barkatsas, A., Forgasz, H., & Leder, G. C. (2001). The gender stereotyping of mathematics: Cultural dimensions. In J. Bobis, B. Perry & M. Mitchelmore (Eds.), *Numeracy and beyond*. Sydney: Mathematics Education Research Group of Australasia.

Barkatsas, A. N., Kasimatis, K., & Gialamas, V. (2009). Learning secondary mathematics with technology: Exploring the complex interrelationship between students' attitudes, engagement, gender and achievement. *Computers & Education, 52*, 562–570.

Bechrakis, T., Gialamas, V., & Barkatsas, A. N. (2011). Survey of attitudes toward statistics (SATS): An investigation of its construct validity and its factor invariance by gender. *International Journal of Theoretical Educational Practice, 1*, 1–15.

Bennison, A., & Goos, M. (2010). Learning to teach mathematics with technology: A survey of professional development needs, experiences and impacts. *Mathematics Education Research Journal, 2*(1), 31–56.

Bradley, D. R., & Wygant, C. R. (1998). Male and female differences in anxiety about statistics are not reflected in performance. *Psychological Reports, 80*, 245–246.

Cashin, S. E., & Elmore, P. B. (2005). The survey of attitudes toward statistics scale: A construct validity study. *Educational and Psychological Measurement, 65*, 509–524.

Coakes, S. J., & Steed, L. G. (1999). *SPSS: Analysis without anguish*. Australia: John Wiley and Sons Ltd.

Drijvers, P. (2000). Students encountering obstacles using CAS. *The International Journal of Computers for Mathematical Learning, 5*(3), 189–209.

Earp, M. S. (2007). *Development and validation of the statistics anxiety measure*. Unpublished Doctoral Dissertation, University of Denver, USA.

Elmore, P. B., & Lewis, E. L. (1991, April). *Statistics and computer attitudes and achievement of students enrolled in applied statistics: Effect of a computer laboratory*. Paper presented at the annual meeting of the American Educational Research Association, Chicago.

Elmore, P. B., Lewis, E. L., & Bay, M. L. G. (1993, April). *Statistics achievement: A function of attitudes and related experiences*. Paper presented at the annual meeting of the American Educational Research Association, Atlanta, GA.

Elmore, P. B., & Vasu, E. S. (1986). A model of statistics achievement using spatial ability, feminist attitudes, and mathematics-related variables as predictors. *Educational and Psychological Measurement, 46*, 215–222.

Fennema, E., & Sherman, J. (1976). Fennema–Sherman mathematics attitudes scales. Instruments designed to measure attitudes toward the learning of mathematics by females and males. *Abstracted in the JSAS Catalog of Selected Documents in Psychology, 6*(1), 31. (Ms No. 1225).

Forgasz, H., Leder, G. C., & Barkatsas, A. (1998). Mathematics—For boys? For girls? *Vinculum, 35*(3), 15–19.

Forgasz, H., Leder, G. C., & Barkatsas, A. N. (1999). Of course I can('t) do mathematics: Ethnicity and the stereotyping of mathematics. In J. M. Truran & K. M. Truran (Eds.). *Making the difference*. Adelaide: Mathematics Education Research Group of Australasia.

Forgasz, H. J., & Leder, G. C. (2001). A+ for girls, B for boys: Changing perspectives on gender equity and mathematics. In B. Atweh, H. Forgasz, & B. Nebres (Eds). *Sociocultural research on mathematics education: An international perspective* (pp. 347–366). Mahwah: Lawrence Erlbaum & Associates.

Fredricks, J., Blumenfeld, P., & Paris, A. (2004). School engagement: Potential of the concept, state of the evidence. *Review of Educational Research, 74*(1), 59–109.

Galbraith, P., & Haines, C. (1998). Disentangling the nexus: Attitudes to mathematics and technology in a computer learning environment. *Educational Studies in Mathematics, 36,* 275–290.

Guin, D., & Trouche, L. (1999). The complex process of converting tools into mathematical instruments: The case of calculators. *International Journal of Computers for Mathematical Learning, 3,* 195–227.

Harvey, A. L., Plake, B. S., & Wise, S. L. (1985, April). *The validity of six beliefs about factors related to statistics achievement.* Paper presented at the annual meeting of the American Educational Research Association, Chicago.

Heid, M. K., & Edwards, M. T. (2001). Computer algebra systems: Revolution or retrofit for today's mathematics classrooms? *Theory and Practice, 40*(2), 128–136.

Lebart, L., Morineau, A., Lambert, T., & Pleuvret, P. (2001). *Manuel de référence de SPAD.* Montreuil: CISIA-CERESTA.

Marks, H. M. (2000). Student engagement in instructional activity: Patterns in the elementary, middle and high school years. *American Educational Research Journal, 39*(1), 153–184.

Middleton, J. A. (1999). Motivation for achievement in mathematics: Findings, generalisations, and criticisms of the research. *Journal for Research in Mathematics Education, 30*(1), 65–88.

McLeod, D. B. (1992). Research on affect in mathematics education: A reconceptualisation. In D. A. Grouws (Ed.), *Handbook of research on mathematics teaching and learning* (pp. 575–596). New York: MacMillan.

Newmann, F. M. (1989). Student engagement in high school mathematics. *Educational Leadership, 46*(5), 34–36.

Onwuegbuzie, A. J. (1995). Statistics test anxiety and female students. *Psychology of Women Quarterly, 19,* 413–418.

Pajares, M. F. (1992). Teachers' beliefs and educational research: cleaning up a messy construct. *Review of Educational Research, 62*(3), 307–322.

Pierce, R., & Stacey, K. (2004). A framework for monitoring progress and planning teaching toward the effective use of computer algebra systems. *International Journal of Computers for Mathematical Learning, 9,* 59–93.

Pierce, R., Stacey, K., & Barkatsas, A. N. (2007). A scale for monitoring students' attitudes to learning mathematics with technology. *Computers and Education, 48*(2), 285–300.

Reed, C. R., Drijvers, P., & Kirschner, P. A. (2010). Effects of attitudes on learning mathematics with computer tools. *Computers and Education, 55*(1), 1–15.

Roberts, D. M., & Bilderback, E. W. (1980). Reliability and validity of a statistics attitude survey. *Educational and Psychological Measurement, 40,* 235–238.

Roberts, D. M., & Saxe, J. E. (1982). Validity of a statistics attitude survey: A follow-up study. *Educational and Psychological Measurement, 42,* 907–912.

Schau, C., Stevens, J., Dauphinee, T. L., & Del Vecchio, A. (1995). The development and validation of the survey of attitudes toward statistics. *Educational and Psychological Measurement, 55,* 868–875.

Schram, C. M. (1996). A meta-analysis of gender differences in applied statistics achievement. *Journal of Educational and Behavioral Statistics, 21*(1), 55–70.

Tanner, H., Jones, S., Beauchamp, G., & Kennewell, S. (2010). Interactive whiteboards and all that Jazz: Analysing classroom activity with interactive technologies. In L. Sparrow, B. Kissane, & C. Hurst (Eds.), *Shaping the future of mathematics education: Proceedings of the 33rd annual conference of the Mathematics Education Research Group of Australasia.* Fremantle: MERGA.

Vale, C., & Bartholomew, H. (2008). Gender and mathematics. In H. Forgasz, A. Barkatsas, A. Bishop, B. Clarke, S. Keast, W.-T. Seah, & P. Sullivan (Eds.), *Research in mathematics education in Australasia 2004–7* (pp. 271–290). Rotterdam, The Netherlands: Sense Publishers.

Vale, C., & Leder, G. (2004). Student views of computer-based mathematics in the middle years: Does gender make a difference? *Educational Studies in Mathematics, 56*, 287–312.

Ware, M. E., & Chastain, J. D. (1991). Developing selection skills in introductory statistics. *Teaching of Psychology, 18*(4), 219–222.

Waters, L. K., Martelli, T. A., Zakrajsek, T., & Popovich, P. M. (1988). Attitudes toward statistics: An evaluation of multiple measures. *Educational and Psychological Measurement, 48*, 513–516.

Watt, H. M. G. (2006). The role of motivation in gendered educational and occupational trajectories related to maths. *Educational Research and Evaluation, 12*(4), 305–322.

Weglinsky, H. (1998). *Does it compute? The relationship between educational technology and student achievement in mathematics*. Princeton: Educational Testing Service Policy Information Center.

Wise, S. L. (1985). The development and validation of a scale measuring attitudes toward statistics. *Educational and Psychological Measurement, 45*, 401–405.

Zimmer, J. C., & Fuller, D. K. (1996). *Factors affecting undergraduate performance in statistics: A review of the literature*. Paper presented at the annual meeting of the Mid-South Educational Research Association, Tuscaloosa, AL. (ERIC Document Reproduction Service No. ED406424.)

Zyngier, D. (2007). Listening to teachers-listening to students: Substantive conversations about resistance, empowerment and engagement. *Teachers and Teaching: Theory and Practice, 13*(4), 327–347.

Commentary on the Chapter by Anastasios Barkatsas, "Students' Attitudes, Engagement and Confidence in Mathematics and Statistics Learning: ICT, Gender, and Equity Dimensions"

Testing the Hypothesis of a Virtuous Cycle

Kenneth Ruthven

The chapter by Barkatsas summarises the findings of five small-scale studies which employ similar questionnaire instruments and statistical methods to assess facets of attitude towards mathematics and statistics. The first three of these studies focus on attitudes to mathematics and to technology use in studying mathematics, as found in student populations in secondary schools in Victoria, Australia or Athens, Greece. The last two focus on attitudes to statistics as found in student populations in tertiary education in one or other of these same locations. Although there is a gender dimension to all of the studies, this is not highly developed or strongly theorised. Nevertheless, the empirical findings merit some comment, particularly where they can be related to the main theoretical hypothesis advanced early in the chapter, a posited mechanism through which use of technology in studying mathematics affects attitudes to the subject and achievement in it. For that reason, I will focus my remarks on the set of three studies that address this.

First, however, I will comment briefly on the validation of the instruments concerned. We are told that the *Survey of Attitudes towards Statistics Scale* [SATS]—used in the last two studies—is a well established one and so perhaps less needful of such validation. My comments, therefore will focus on the less well established *Mathematics and Technology Attitude Scale* [MTAS]—used in the first three studies. None of the studies attempted any form of external validation of the instrument through triangulation with other types of data via other forms of analysis, and so the claim to construct validity rests primarily on the grounding of the proposed facets in prior literature, the face validity of the sets of items for each subscale, and the internal consistency of their structure. The reported statistical analyses establish that the sets of questionnaire items designed to operationalise the five attitudinal constructs posited on theoretical grounds did indeed produce patterns of student response well modelled by the corresponding 5-factor structure; nevertheless provision of complete details of factor loadings in Appendix 2 would have enabled the

K. Ruthven (✉)
University of Cambridge, Cambridge, UK
e-mail: kr18@cam.ac.uk

H. Forgasz, F. Rivera (eds.), *Towards Equity in Mathematics Education,*
Advances in Mathematics Education,
DOI 10.1007/978-3-642-27702-3_15, © Springer-Verlag Berlin Heidelberg 2012

interested reader to scrutinise this more fully. Bearing in mind how further analyses were conducted and findings reported in terms of (potentially associated) subscale totals rather than expressly orthogonal factor scores, a report of the correlations between the different subscales would also have been helpful. Indeed, the proximality of the three mathematical attitude variables in the theoretical model proposed in Barkatsas' Fig. 1 does raise questions about the strength of relationship between the corresponding subscales and the potential viability of more parsimonious models employing a single overarching construct of attitude to mathematics.

Turning now to the substantive findings from the three studies employing MTAS, the main finding from the first study concerns gender differences in scores on the five MTAS subscales from students in Australian co-educational schools. It reports statistically significant trends for boys to respond more positively than girls on two of the subscales relating to mathematics attitude—mathematics confidence [MC] and affective engagement [AE]—but not the third—behavioural engagement [BE]. Later, this same pattern is reported from the third study. On the further two subscales related to technology use—confidence with technology in general [TC] and attitude to learning mathematics with technology [MT]—the first study reports statistically significant trends towards boys responding more positively than girls on both. In the third study, however, no differential trend is found between the MT scores of male and female students, but again a strong differential trend towards higher TC scores amongst male students.

In both of these studies, data from students in girls' schools were excluded from the analysis and no boys' schools were surveyed. This appears to be a missed opportunity to examine a key organisational factor that might mediate the development of gender-stereotyped attitudes, namely the influence of school type (single-sex or co-educational). In particular, it has been argued that pressure of gender-stereotypes—important in relation to mathematics—is stronger in co-educational schools than in single-sex (Sullivan 2009; Sullivan et al. 2010). Although care (and possibly fuller data) would be needed to ensure that this aspect of school type was not confounded with institutional differences in students' social background and academic achievement, such an analysis would strengthen the power and generalisability of findings, and throw light on one type of potentially actionable organisational factor. Indeed, at more readily implementable level, some co-educational schools are now experimenting on such grounds with single-sex classes.

The second study is particularly interesting because it provides an indirect means of testing the theoretical hypothesis that inspired development of the MTAS instrument. Essentially, this study replicates the other two, but in Greek co-educational schools rather than Australian, and then extends the analysis to establish seven prototypical categories of students through statistical clustering methods (with results as shown in Table 1, reorganised from the original presentation by Barkatsas).

The theoretical hypothesis (shown in Fig. 1 of the chapter by Barkatsas) has a mathematics attitude subsystem at its centre; this posits that mathematics confidence [MC] and affective engagement [AE] both influence behavioural engagement [BE], and that these in turn influence mathematical learning (and so presumably achievement). While the available data do not show the effects of *changes* in the

Table 1 Characteristics of student clusters identified in the second Barkatsas study

Cluster	Proportion of sample	Facets of attitude to mathematics			Mathematics achievement	TC	MT	Balance of cluster by gender	
		MC	AE	BE				male	female
3	13%	Very high	Strongly positive	Strongly positive	Excellent	Very high	Strongly positive	68%	32%
6	13%	Very high	Strongly positive	Strongly positive	High	High	Strongly negative	60%	40%
5	10%	Very high	Strongly positive	Strongly positive	High	Very low	Strongly negative	41%	59%
2	16%	Average	Neutral	Negative	Average	Low	Strongly negative	46%	55%
4	12%	Low	Neutral	Positive	Average	Average	Strongly positive	36%	64%
1	16%	Very low	Strongly negative	Strongly negative	Low	Very high	Positive	66%	34%
7	20%	Very low	Strongly negative	Strongly negative	Very low	Low	Negative	42%	58%

first two variables on the third, they do allow us to examine whether *levels* of these variables are broadly compatible with the hypothesis; in particular, whether they display incompatible extremes. It is clear from inspection of the seven prototypes in Table 1 that all the triples of MC, AE and BE are consistent with this part of the model (hence my earlier question about correlation between these mathematics attitude subscales): clusters [3], [5], and [6] combine positive extremes; clusters [1] and [7] combine negative extremes; and clusters [2] and [4] combine more middling values on all three subscales. Moreover, in broad terms, there also appears to be the hypothesised alignment with mathematics achievement. (My reorganisation of the original table is intended to make these patterns readily discernible by bringing out the direct covariation of all four of these constructs.) The same pattern of direct covariation between these measures is also reported from the third study.

It appears, then, that employing a more overarching construct of attitude to mathematics will provide a useful shorthand in pursuing matters further. By radically collapsing this part of the hypothesised model, it becomes easier to examine the larger 'virtuous cycle' in which it is posited that student confidence with (digital) technology in general [TC] influences the quality of student technology-supported mathematical experience [for which the study provides no measure] which, in turn, influences student mathematical attitudes, learning, and achievement (the now collapsed central subsystem); this subsystem, in turn, influences student attitude to learning mathematics with technology [MT] which, finally, feeds back into the quality of student technology-supported mathematical experience. No information is provided about the extent of the surveyed students' technology-supported mathematical experience, but presumably it would have been wholly inappropriate to administer a questionnaire assessing students' attitudes to learning mathematics with technology if they had no experience of this. Again, while it would be preferable to have data capable of showing *changes* in variables attributable to some intervention, any incongruities in *levels* within the available dataset will call the hypothesised model into some question.

Unlike the attitude subscales specific to mathematics alone, there is no form of alignment between the two technology-related subscales (as a quick inspection of the TC and MT columns in Table 1 will confirm) and so no viable overarching construct. Had there been such alignment, of course, the number of emergent clusters from the statistical analysis would very probably have been smaller. It is necessary, then, to examine these clusters in greater detail.

Some of the cluster prototypes are clearly consistent with the virtuous cycle hypothesis: in cluster [3], variables are uniformly high; and in cluster [7], uniformly low. Equally, however, assuming that the students concerned had indeed had appreciable technology-supported mathematical experience, other clusters appear to be incompatible with the hypothesis. In cluster [1], very high confidence with technology and positive attitude to learning mathematics with technology—the posited technology-related drivers of improved mathematical attitudes and attainment—are combined with very negative attitudes to mathematics and low achievement in the subject. Equally, in cluster [6], high technology confidence is accompanied by positive attitudes to mathematics and high achievement in the subject, but associated with strongly negative attitude to learning mathematics with technology. This last pattern may originate from a not uncommon syndrome that the virtuous cycle model seems unable to accommodate: one in which some students (and some teachers) see 'dependence' on technology as indicative of lack of mathematical ability; 'ability' taken as demonstrated through 'independence' from any aid. This syndrome represents an important obstacle to successfully learning mathematics with technology. In conclusion, then, the cluster analysis from the second study appears to challenge the hypothesised model of a virtuous cycle. While it seems that the third study may have attempted to test this model more directly, the reported analysis is persuasive only as regards the central subsystem concerned with mathematics attitudes and achievement. Again it seems that the technology-related variables may not be interacting with this central system in quite the way implied by the 'virtuous cycle' model.

Turning to the gender balance of the clusters emerging in the second study, of those that are strongly positive towards learning mathematics with technology [MT], cluster [3] has a large majority of boys and cluster [4] a similarly large majority of girls. Of those clusters that are strongly negative towards learning mathematics with technology [MT], cluster [6] has a majority of boys, cluster [5] a similar majority of girls, while cluster [2] is more evenly constituted. This relatively balanced overall pattern by gender contrasts with the male predominance in those clusters characterised as very high in confidence with technology [TC], clusters [1] and [3], and the female predominance in that characterised as very low in TC, cluster [5]. Perhaps worthy of note, too, is the male predominance in those two clusters, [1] and [6], which particularly challenge the 'virtuous cycle' hypothesis.

Taking account, then, of findings across the three studies, it seems that attitude to learning mathematics with technology [MT] is less gendered in its expression than confidence with technology in general [TC]. In relation to issues of equity, this is the most important finding to be drawn from these studies. In particular, it challenges any model that posits some overarching attitude towards technology. In terms of gender equity, it suggests that moves towards learning mathematics with technology should be designed with a view to marginalising any influence from differential confidence with technology in general [TC] which would potentially negatively affect more female students than male. More broadly, however, there remains a pressing need to address the strongly negative attitude to learning mathematics with technology [MT] evidenced by a substantial proportion of students, both male and female.

My own view is that progress in this area will depend on more fine-grained analyses of the interaction between particular patterns of technology use and the genesis of specific mathematical competences and attitudes. For example, my own earlier work on the use of graphical calculators in upper-secondary mathematics argued that regular use of the technology mediated the rehearsal by students of specific types of relationships between symbolic and graphic forms, and that availability of the technology facilitated the use of particular cognitive strategies by students (Ruthven 1990). Of particular relevance here, this research showed how these mechanisms could explain not only the broad trend of findings across students, but the observed gender effects. Equally, this same project identified certain attitude profiles (on the part of teachers as well as students) that require attention if students are to learn to make effective use of technology: reluctance to move away from already familiar techniques, particularly where a new technology is not routinely available; and reluctance to use any technology, grounded in a strong sense that this represents a loss of control over the mathematical process (Ruthven 1992).

References

Ruthven, K. (1990). The influence of graphic calculator use on translation from graphic to symbolic forms. *Educational Studies in Mathematics, 21*(5), 431–450.

Ruthven, K. (1992). Personal technology and classroom change: A British perspective on teaching mathematics with advanced calculators. In J. Fey & C. Hirsch (Eds.), *Calculators in mathematics education: Impact and potential* (pp. 91–100). Reston: National Council of Teachers of Mathematics.

Sullivan, A. (2009). Academic self-concept, gender and single-sex schooling. *British Educational Research Journal, 35*(2), 259–288.

Sullivan, A., Joshi, H., & Leonard, D. (2010). Single-sex schooling and academic attainment at school and through the lifecourse. *American Educational Research Journal, 47*(1), 6–36.

Commentary on the Chapter by Anastasios Barkatsas, "Students' Attitudes, Engagement and Confidence in Mathematics and Statistics Learning: ICT, Gender, and Equity Dimensions"

Hazel Tan

Barkatsas' chapter contributes to the discussion on gender issues in the affective aspects of learning statistics (using SATS) and learning mathematics with ICT (using MTAS). My response to the chapter has two parts: contributions of the findings to the original theoretical basis of the instruments, and implications of the studies on equitable policy and practice.

Since the studies provided evidence of the validity and reliability of both the MTAS and the SATS instruments, a next step would be to investigate how the findings of the studies relate back to the theoretical basis of the instruments.

1 Theoretical Basis of MTAS

The development of the MTAS was based on a hypothesised model described by Pierce et al. (2007) in relation to the use of technology to enable more real world problem solving in the mathematics classroom. In the hypothesised model, using a real world problem solving approach driven by technology could lead to increased affective engagement (AE) and mathematics confidence (MC), which in turn would lead to increased behavioural engagement in class (BE) and improved learning. A positive attitude to learning mathematics with technology (MT) could result from the positive learning experience and outcome, which in turn would influence future learning experiences with technology. Other factors supporting technology use such as confidence in technology (TC) were also described in the model. The MTAS was developed to measure five components in the model (AE, MC, BE, MT, and TC).

Although Barkatsas' findings are just a snapshot of the situation, rather than repeated studies or a longitudinal study, I would like to argue that they could be used

H. Tan (✉)
Monash University, Melbourne, Australia
e-mail: Hazel.Tan@monash.edu

H. Forgasz, F. Rivera (eds.), *Towards Equity in Mathematics Education*,
Advances in Mathematics Education,
DOI 10.1007/978-3-642-27702-3_16, © Springer-Verlag Berlin Heidelberg 2012

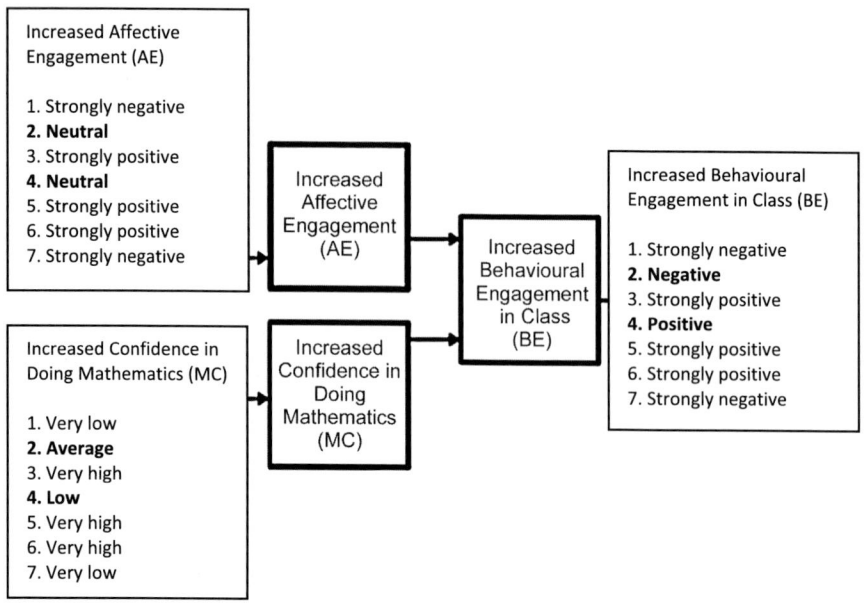

Fig. 1 Part of the hypothesised model on AE, MC and BE, together with the profiles of students in the seven clusters found in Study 2

as a gauge to provide feedback to the hypothesised model to test its applicability. Evidence from Study 3 supports the model, since the three factors AE, MC, and BE were associated with students' attitudes towards using technology to learn mathematics (MT). In Study 1, statistically significant gender differences were found for all subscales except BE from the students in co-educational schools. This suggests that gender could be a factor included in the model, or that perhaps the model might be different for males and females.

In Study 2, seven clusters of types of student characteristics were found among the 1068 secondary school students in Greece. In reading about the study, I assumed that these groups of Years 9 and 10 students had been using technology to learn mathematics, and that they had time to form attitudes about learning mathematics with technology, although it is not certain if teaching approaches such as using real world problems (the basis of the MTAS instrument design) were used. Part of the theoretical model predicts that increases in AE and MC lead to an increase in BE and improved learning outcomes, as shown in Fig. 1. In the corresponding boxes next to the subscales AE, MC and BE are found the characteristics of the students in the seven clusters (numbered 1 to 7), as were reported by Barkatsas. Most of the cluster characteristics seemed to fit the model, except two, clusters 2 and 4 (shown in bold in Fig. 1). Cluster 2 students had neutral AE and average MC, yet negative BE. They also had average achievement and low TC (not shown in Fig. 1). The gender composition of cluster 2 was roughly equal (45.7% males, 54.3% females), although statistically more likely to be female. Cluster 4 students had neutral AE and low MC, yet positive BE. They also had average TC, average achievement, and were statistically more likely to be female than male (36.2% males, 63.8% females).

In addition, cluster 6 students had scored highly on all subscales (AE, MC, BE, TC) except MT (strongly negative), had high mathematics achievement, and were more likely to be male than female (60.2% males, 39.9% females). This group of technologically and mathematically confident students (12.7% of the whole sample) with strongly negative attitudes towards using technology to learn mathematics seemed to contradict the model's prediction that "positive attitude to using technology to learn mathematics is seen as an outcome of improved learning" (Pierce et al. 2007, p. 287).

These discrepancies suggest that the relationships between the factors might need to be revised, and that gender and/or other factors might have a more direct impact on students' behavioural engagement and attitudes towards using technology to learn mathematics.

2 Theoretical Basis of SATS

Barkatsas used the Survey on Attitudes Towards Statistics (SATS), an instrument developed by Schau et al. (1995) to measure students' attitudes towards statistics in four dimensions: affect, cognitive competence, value, and difficulty. The four dimensions were generated empirically, through words and phrases produced by a panel of statistics students and instructors, and from reviews of existing instruments (Schau et al. 1995).

Although Schau et al. (1995) and others (e.g. Chiesi and Primi 2009; Hilton et al. 2004) found four factors using confirmatory factor analysis, there were others with different findings. For example, Cashin and Elmore (2005) compared student pre- and post-course data for three main instruments on attitudes towards statistics (Statistics Attitude Survey, SAS, by Roberts and Bilderback, in 1980; Attitudes Towards Statistics Scale, ATS, by Wise in 1985; and the SATS) and concluded that both ATS and SATS have two domains, contrary to the four-factor solution.

Barkatsas contributes to studies on using SATS, by investigating the construct validity and gender differences for tertiary students in Greece and Australia. Hence I look forward to the detailed findings of Barkatsas' future work (e.g., Bechrakis et al. in press) to see if there is further elucidation of the theoretical framework underpinning the instrument.

3 Implications for Equitable Policy and Practice

The range of applications of technology for learning statistics (Onwuegbuzie and Wilson 2003) as well as for learning mathematics is continuously increasing. An investigation of the attitudes towards ICT use and their relations to student achievement forms only a part of the picture; we should also question if existing teaching approaches, schooling practices, and policies about technology use, situated within a particular educational system and socio-cultural context, contribute to gender differences in students' attitudes towards learning mathematics and statistics with ICT. A contextual perspective could guard against what Gutiérrez and Dixon-Román (2011) termed as "gap gazing", in which systemic and socio-cultural issues perpetuating the gap (achievement gap or, in this instance, gender gap) might fall into a researcher's blind spot. They gave the example of the debate on whether to use calculators or computers in mathematics classes, a debate in which educators and researchers often assumed that "schooling can somehow control what students 'learn' about mathematics, as if they are not already using such technologies outside of mathematics class to make decisions or to educate themselves" (Gutiérrez and Dixon-Román 2011, p. 31). They suggested re-examining the role of mathematics in society (for example the impact of technology on the kinds of mathematics we use) and rethinking schooling practices to match or to further influence that role. Focusing on a broader view of mathematics (including statistics) and education in order to address equity issues is a challenging task. The longitudinal studies proposed by Barkatsas could be a start to a new way of perceiving how gender differences develop and change in relation to different socio-cultural contexts.

The overall argument presented by Barkatsas has a few minor gaps. At the classroom level, Barkatsas recommended some approaches to improve female students' attitude, confidence, and engagement with using ICT to learn mathematics, but did not provide corresponding suggestions for improving attitudes towards statistics. For statistics, I found that Onwuegbuzie and Wilson (2003) provided a comprehensive literature review on studies aimed at reducing students' statistics anxiety. Some strategies that were found to be useful were: use of humour, use of open book and untimed assessments, explicitly addressing anxiety, application of statistics to real-world situations, journal writing, and the interpersonal style of the instructor. Although not specifically targeted at females, these general strategies to reduce statistics anxiety seemed similar to those from a feminist perspective. In her discussion on feminist pedagogy in mathematics classrooms, Jacobs (2010) described some useful teaching approaches which are based on research: use real-world or classroom based experiences; engage learners in inquiry and reflect on their work; encourage alternate methods of solutions; value the finding of alternative methods more than solving similar problems in the same way; emphasise the generation of hypotheses rather than proving stated theorems; use cooperative rather than competitive or individual activities; and make extensive use of writing as a means of learning mathematics. These teaching approaches could be applicable to tackle equity issues in both ICT-enriched mathematics learning and statistics learning settings.

Barkatsas' studies on the attitudes of students are exciting because they add, through the validation and use of the two instruments, to the body of empirical knowledge about how male and female students learn mathematics and statistics in the current information and technological age. The link between the two sets of instruments in the chapter could be made stronger. Since there is an increasing use of ICT for statistics courses, it would be interesting if the two instruments could be used together for a group of statistics students in future studies.

References

Bechrakis, T., Gialamas, V., & Barkatsas, A. N. (in press). Survey of attitudes toward statistics (SATS): An investigation of its construct validity and its factor invariance by gender. *MASAUM Journal of Social Sciences*.

Cashin, S. E., & Elmore, P. B. (2005). The survey of attitudes toward statistics scale: A construct validity study. *Educational and Psychological Measurement, 65*(3), 509–524.

Chiesi, F., & Primi, C. (2009). Assessing statistics attitudes among college students: Psychometric properties of the Italian version of the Survey of Attitudes toward Statistics (SATS). *Learning and Individual Differences, 19*, 309–313.

Gutiérrez, R., & Dixon-Román, E. (2011). Beyond gap gazing: How can thinking about education comprehensively help us (re)envision mathematics education? In B. Atweh, M. Graven, W. Secada, & P. Valero (Eds.), *Mapping equity and quality in mathematics education* (pp. 21–34). Dordrecht, The Netherlands: Springer.

Hilton, S. C., Schau, C., & Olsen, J. A. (2004). Survey of attitudes toward statistics: Factor structure invariance by gender and administration time. *Structural Equation Modeling: A Multidisciplinary Journal, 11*(1), 92–109.

Jacobs, J. E. (2010). Feminist pedagogy and mathematics. In B. Sriraman, & L. English (Eds.), *Theories of mathematics education: Advances in mathematics education* (pp. 435–446). Heidelberg, Germany: Springer-Verlag.

Onwuegbuzie, A. J., & Wilson, V. A. (2003). Statistics anxiety: Nature, etiology, antecedents, effects, and treatments—a comprehensive review of the literature. *Teaching in Higher Education, 8*(2), 195–209.

Pierce, R., Stacey, K., & Barkatsas, A. N. (2007). A scale for monitoring students' attitudes to learning mathematics with technology. *Computers and Education, 48*(2), 285–300.

Roberts, D. M., & Bilderback, E. W. (1980). Reliability and validity of a statistics attitude survey. *Educational and Psychological Measurement, 47*, 759–764.

Schau, C., Stevens, J., Dauphinee, T. L., & Del Vecchio, A. (1995). The development and validation of the survey of attitudes toward statistics. *Educational and Psychological Measurement, 55*, 868–875.

Wise, S. L. (1985). The development and validation of a scale measuring attitudes toward statistics. *Educational and Psychological Measurement, 45*, 401–405.

Part II
Equity and Culture

Ferdinand Rivera and Helen Forgasz

Four reprinted ZDM articles and two new chapters comprise Part II of this book that deals with issues in equity and matters that pertain to culture. By culture, we have in mind a dynamic ensemble of common artifacts and shared relations and understandings that shape the complex identities of groups of individuals (Pellegrino 2007). Hence, so-called diverse mathematics classrooms bring with them cultures that, as a consequence, "challenge our ability to define equity in a straightforward, uncomplicated manner" (Jordan 2010, p. 155). Language as an ideological artifact (see **Radford**'s chapter) constitutes people and their worldviews as a group, including how they see and project themselves to others. Oral mathematical practices exemplify shared relations that operate under the category of rules and, more generally, symbol systems, where such rules and systems pertain to abstracted expressions that have been drawn from familiar, local, and communal ways of representing things over some period of time (Bruner 2001). Across the six studies that comprise this section, we find the pervasive influence of social and cultural factors in the manner students learn and see the value of mathematics. However, from an equity standpoint, it is also important to assess how such learning and value seeing could expand access and enable social mobility among such diverse groups of learners beyond the four walls of the mathematics classroom (Jordan 2010).

Forgasz and **Mittelberg** in their chapter provide large-scale empirical evidence of an interaction between Grade 9 (northern) Israeli students' gendered perceptions of mathematics and the relevant cultural, ethnic, and other social realities that influence their views. While mathematics has consistently and historically been stereotyped as a male discipline, Forgasz and Mittelberg in their interesting study extrapolate reasons for persistent patterns of gender differences in cognitive and affective measures in mathematics. In their study, they compared Jewish and Arab Israelis with their Australian counterparts using two language-translated scaled instruments that have initially been validated in the Australian context. The results indicate similarities in gendered perceptions about the discipline of mathematics between the Australian and Jewish Israeli groups, which could be explained in terms of their shared modern and westernized perceptions. Across the two groups, mathematics is a neutral discipline, that is, it is neither a male nor a female domain. However, more Australians than Jewish Israelis tend to challenge the stereotypical view that mathematics is a male domain. Among the Arab Israeli group, the results are rather ambiguous. While they agree that mathematics is both "a neutral and a female domain, they appear uncertain about it being a male domain." In fact, "the female Arab Israelis agreed, but their male counterparts disagreed, that mathematics is a male domain." Triangulated classroom and school performance data among Arab Israelis drawn from other sources strongly indicate that females tend to be more successful than their male counterparts in those contexts, however, the "strongly gendered expectations of women's roles in the Arab Israel ethnic community and workplace prevent the female group from translating such educational successes into occupational opportunities and participation in the labor force," a situation that markedly differs among the Jewish Israeli group.

In his chapter, **Barton** employs a (Derridean) deconstructivist strategy as he attempts to "talk into existence" a disseminated space where "an alternative relativistic philosophy" could possibly thrive. Marking conceptual constraints in both classical and social constructivist views and practices of mathematics, Barton explores the "possibility of the simultaneous existence of culturally different mathematics" via a "below the surface" extrapolation of QRS systems that are specific to cultural groups who "make sense of quantity, relationships, and space" based on their collective discursive practices or systems that evolve over time. What such systems imply is that "there is no presumed external mathematics or rationality by which one system is judged better than another" as it is "entirely an internal process, a human process, a cultural process." Drawing on evidence from several cultures, Barton illustrates how mathematics conveys its humanly constructed significance, which means to say that the manner in which it is constructed influences outcomes (i.e. the reality that is eventually constructed) and, thus, its validity is not about whether it is "right or wrong" but whether it is "simply useful or not." Hence, for Barton, an ethnomathematical lens into mathematical practices predisposes an (relativist) attitude of seeking for functional meaning and the QRS system that support it.

The chapter by **Knijnik**, **Wanderer**, and **de Oliveira** provides a powerful counterdiscourse to "the school curriculum invisibility of the cultures of nonhegemonic groups" whose peculiar and situated "ways of dealing mathematically with the world" highlight the "centrality of looking at cultural differences in mathematics education." In their chapter, Knijnik, Wanderer, and de Oliveira examine the relationship between cultural processes tied to oral mathematics and those processes that are used with a simple electronic calculator in the context of an adult preservice teacher education that was organized by a Brazilian landless people's social movement. While the authors share the view that there is a relationship between mental calculation skills in different cultural contexts and the corresponding written calculation or numerical procedures, they also foreground how an understanding of such numerical practices is inseparably linked to the cultural setting itself. Oral mathematics, they note, is "part of the cultural practices in which they gain their meaning... [and] as an artifact constituted by and constituting culture." Findings from their two-year ethnographic research with participants who are members of "a peasant culture" indicate a preference toward oral (and not written) mathematical practices that are "connected to labor activities and to the purchase and sale of products for daily consumption," which differ from school mathematical practices that value competence in written numerical processes that are applied oftentimes in the absence of any meaningful contexts. For instance, the participants understand and employ rounding depending on the context in which it is used in everyday activity and in transaction with other people. Also, they add (and subtract) whole numbers and decimals orally by using a decomposition (place-value) strategy that operates within a left-to-right algorithm, which differs from written school-based algorithms that value right-to-left processes. When they learn to use a calculator, they develop an understanding that it should not be used "in a merely mechanical way" but primarily in supporting orally calculated estimations and in providing exact answers especially in situations that involve complex calculations.

Civil, **Planas**, and **Quintos** in their chapter underscore the significance of "recognizing" (low-income) immigrant students' social contexts and, specifically, their parents' perceptions, in explaining how and why they tend to perform and achieve differently in school mathematics. For the authors, "the notion of recognition needs to be interpreted not only as an acceptance of differences but also as an alternative to broaden meanings for the mathematical practices and ideas." Drawing on qualitative data from two countries, Civil, Planas, and Quintos note that the changes in the structures of the students' "learning opportunities are likely to be framed by what their parents think and expect." Their findings show that while the parents in both countries share the view that it is important for their children to acquire the appropriate mathematical behavior at school in order to be successful, they, however, react differently to the "teaching of mathematics in their new country... from accepting them and trying to adapt to the new system to experiencing some form of conflict." For example, the parents in Barcelona (Spain) attribute their children's low mathematics achievement and difficulties to the fact that they are still in the process of adapting to the new environment that has its own systems of language and norms. While some of the parents in Tucson (USA) hold a similar view, there are a few other parents who also express the concern that some arithmetical topics are taught early and with depth in their home country (Mexico, in particular) compared with the approach used in the US system. Proficiency in the use of academic language is also a significant issue, and as the authors note, "to ignore how language can operate as a barrier in the education of immigrant children, including their mathematics education, goes against any attempts to provide an equitable education for all children." Some parents who perceive a separation between learning mathematics and language consequently teach their children to inhibit themselves from participating in classroom discussions until they become language proficient. Other parents who are not provided with (indirect and ongoing) access to bilingual programs feel frustrated at not being able to support their children in their mathematical work.

The chapter by **Barwell** explores "the full richness and complexity of language use" that characterizes linguistically diverse mathematics classrooms and possible implications for equity in the teaching and learning of mathematics. Drawing on his classroom experiences with diverse students in Quebec, northern Pakistan, and England, including his thorough analysis of the available literature base on the topic, he articulates four tensions between: school and home languages; formal and informal language in mathematics; language policy and mathematical classroom practices; and a language for learning mathematics and a language for getting on in the world. For example, heteroglottic epistemologies, especially among second language learners, continue to thrive in a system in which "home languages are often marginalized." Further, while formal or standardized mathematical registers could facilitate collective understanding of institutional mathematical knowledge, tensions arise when there they are not meaningfully linked to students' informal (and everyday) expressions. An unintended consequence of official language policies involves the suppression of minority languages "on the road to monolingual proficiency," which then depicts "diversity... as something to be accommodated rather than an intrinsic and valuable aspect of mathematics classroom life." The political tour-de-force of language, where high-status languages (e.g. English) are loaded with significant sociopolitical value as a way of coping with global demands, oftentimes overshadows the educational value that comes with learning mathematics by drawing on students' aboriginal or immigrant languages. In their commentaries to Barwell's chapter, **Planas**, on the one hand, advances the necessary counteraction of mathematical identity re-construction via the notion of orchestration, while **Radford**, on the other, surfaces the significance of alterity and the ideological dimension of (mathematical) language.

Schlöglmann in his chapter addresses ways in which lifelong learning programs in mathematics for adults in Western industrialized countries can be meaningfully developed in order to redress various social inequities they experience as a result of having had an insufficient level of education. While Schlöglmann's analysis of the general concept (and models) of lifelong learning in the current literature introduces us to its complex, "heterogenous" nature, what is clear at least among adult learners in developed countries, especially the unemployed, is the fact that they find themselves needing to acquire new skills due to the rapid "economic and technological, [and global] changes" that are currently affecting various work environments. Schlöglmann then provides an interesting reconstruction of the history and development of mathematics from a social perspective that enables us to obtain a sense of the different functions in which mathematics has been used in society over time and how it emerged as a central tool in various aspects of human life. Schlöglmann's closing section involves a reflection of a large empirical study on "the state of mathematics education within the adult education system in Austria" that he conducted with his colleagues. In his reflection, which focuses on the participants' motivations for participating in adult education courses and their attitudes toward mathematics and mathematics learning, he points to both the value that the participants place in obtaining "higher education" and the affective (emotional) conditions that influence their relationship to (learning) mathematics. In their commentaries to Schlöglmann's chapter, **Wedege** discusses, among others, the issue of "the so-called justification problem," while **Dindyal** underscores the need to have policies that provide incentives for participating in adult courses in mathematics and the development of an appropriate mathematics curriculum and venue for adults that seriously consider their "approach to learning and interests into account."

References

Bruner, J. (2001). In response. In D. Bakhurst & S. Shanker (Eds.), *Jerome Bruner: Language, culture, self* (pp. 199–215). Thousand Oaks: Sage Press.

Jordan, W. (2010). Defining equity: Multiple perspectives to analyzing the performance of diverse learners. In A. Luke, J. Green, & G. Kelly (Eds.), *Review of research in education: Volume 34 (What counts as evidence in educational settings? Rethinking equity, diversity, and reform in the 21st century)* (pp. 142–176). Washington: American Educational Research Association.

Pellegrino, E. (2007). Culture and bioethics: Where ethics and mores meet. In E. Pellegrino & L. Prograis (Eds.), *African American bioethics: Culture, race, and identity* (pp. ix–xxi). Washington: Georgetown University Press.

Preface to "Israeli Jewish and Arab Students' Gendering of Mathematics"

David Mittelberg and Helen Forgasz

The pursuit of equity in educational outcomes, and gender equity in particular, was a driving force of the research reported in the article that follows. The underlying assumption was that social science can elucidate contributing social and cultural variables that, if acted upon, can change the outcome of educational endeavours. In adopting this perspective the alternative view, that educational advantage is largely inherited and therefore closed to innovative change, was rejected.

In adopting a cross cultural comparison, there was also the potential for insights on how cultures differ and their impacts assessed. Cross cultural gender differences provide evidence of claims that culture matters to the extent that, in certain situations, gender as a variable is dwarfed. The challenge presented is then how to translate these insights into educational strategies.

In Israel the relative disadvantage in educational performance of the national minority of Israeli Arabs is unquestioned and much social and educational research has been dedicated to issues that would narrow the gap between Israeli Arabs and Jews. With this rationale, from the outset our work was predicated on the need for an inter-ethnic comparison. We began by comparing Israeli Jewish and Arab ninth grade high school students' views on the gender stereotyping of mathematics to see if the ethnic divide was evident in that field of education and, if so, what were its characteristics. Consistent with earlier work (Mittelberg and Lev-Ari 1999) on tenth grade Israeli and Arab pupils, counter intuitive findings were revealed that had not been reported in national Israeli documentation despite the continued performance gap favouring Israeli Jews over Israeli Arabs. Israeli Arab girls persisted in defying the male mathematics stereotype that is reported in the west and which was replicated among Israeli Jewish girls.

D. Mittelberg (✉)
Oranim Academic College of Education, Tivon, Israel
e-mail: davidm@oranim.ac.il

H. Forgasz
Monash University, Melbourne, Victoria, Australia
e-mail: helen.forgasz@monash.edu

H. Forgasz, F. Rivera (eds.), *Towards Equity in Mathematics Education*,
Advances in Mathematics Education,
DOI 10.1007/978-3-642-27702-3_17, © Springer-Verlag Berlin Heidelberg 2012

These outcomes led to three important conclusions. First, the classroom need not be an uncritical mirror of conservative forces in the community. Second, members of a disadvantaged minority are capable of generating educational outcomes counter to their relatively inferior position among the majority in their society. Third, gender inequities in mathematics education are not inevitable even in traditional societies where gender norms exclude marriage and career options that are open to young girls in Western societies including for Jews in Israel. The findings described above are detailed in the article that follows. They also challenged the researchers to seek explanations for the somewhat surprising differences.

Interestingly, in the period between 1995 when the research began and the time of writing this preface, further supporting evidence for the trends reported in the article was provided by scholars conducting independent quantitative analyses on a large Israeli national base data (Nasser and Birenbaum 2005). In our quest for explanations for the findings reported in the article, we embarked on a similar analysis of the gendering of mathematics among Israeli Jewish and Arab pre-service teachers at a multi-ethnic Israeli teachers college (Mittelberg and Forgasz 2009). Among the findings, Arab pre-service teacher considered themselves as having higher mathematical competence than did the Jewish pre-service teachers. While both groups generally agreed that mathematics was a neutral domain, the Jews held this view more strongly than the Arab. However, gender stereotyped beliefs with respect to particular behaviours were evident among both groups. Girls, for example, were still thought to have to work hard rather than have the natural ability to succeed in mathematics. Mathematics was considered more important for boys' than for girls' future careers, and the pre-service teachers believed that parents and teachers would think the same way. They also indicated that boys, more than girls, needed to succeed in mathematics and were expected to do well. Interestingly, the Jewish pre-service teachers held more negative view of girls' competence with computers than did the Arab pre-service teachers. With respect to computers for mathematics learning, members of the two groups held some opposing beliefs. The Arab pre-service teachers, for example, believed that boys were more likely than girls to find computers for mathematics learning to be boring, and were less likely than girls to like using computers for mathematics learning; the Jewish pre-service teachers' views were in the opposite directions.

Evidence from a small qualitative study focussing on mathematics teaching that was conducted in one grade 5 Jewish and one grade 5 Druze (Arab) classroom provided evidence to confirm arguments that the Arab community is undergoing a, perhaps belated, academic revolution that accords particular advantages to women (Mittelberg et al. 2011). The Jewish teacher's behaviours towards the male and female students in her classroom clearly reflected gender stereotyped expectations. At interview, this teacher stated that in her view males had more natural flair for mathematics than did girls. The Druze teacher, however, actively challenged gender stereotypes. The extent to which this occurred could have been construed as disadvantaging the males in her class. She was very aware of her behaviours, admitted that encouraging females to be successful in mathematics was one of her teaching goals. She saw education as the way to break out of the gendered expectations of her community.

From a broader Israeli societal perspective, there remain two intriguing questions from the studies we have conducted to date. Will Arab women successfully translate their emerging academic advantage into occupational prestige, income, and autonomy? And, can multicultural teacher education institutions support their Arab pre-service students in these endeavours, while simultaneously challenging the gendered beliefs of their Jewish pre-service teachers?

Future Directions   A large scale survey of high school mathematics teachers in Israeli Jewish and Arab classrooms is planned for 2010; further qualitative explorations in these classrooms is also envisaged.

Methodologically, there is a confounding socio-economic factor at play that also deserves consideration. Each of the studies so far has been conducted in the North of Israel. This is the area in Israel with the major concentration of Israeli Arabs. The comparative Jewish participants in the study were also drawn for this same region. On the face of it, drawing samples from the same geographic area appears to be an appropriate research strategy. Yet, from an educational resource perspective, the samples are not the same. For the Arab participants, the communities under study were representative, while for the Jewish Israelis this was not the case. The best Jewish Israeli teachers, for example, are more likely to be found in and attracted to the major urban centres of Israel. The Arab educational system in the region also includes a strong private educational sector, as does the Jewish system, which belies the popular image of the former as a rural depressed minority educational system. The upcoming planned large scale survey of practising mathematics teachers will be designed to account more comprehensively for the wide spread national and cultural differences within and between the two ethnic groups in an attempt to determine the universality of the trends reported in the current research and to identify more clearly factors that contribute to the gender stereotyping of mathematics in Israel.

References

Mittelberg, D., & Lev-Ari, L. (1999). Confidence in mathematics and its consequences: Gender differences among Israeli Jewish and Arab youth. *Gender and Education, 11*(1), 75–92.

Mittelberg, D., & Forgasz, H. (2009). *Israeli Jewish and Arab pre-service students' gendering of mathematics*. Research Report No. 69. Research Authority, Oranim Academic College of Education, Tivon, Israel.

Mittelberg, D., Rozner, O., & Forgasz, H. (2011). Mathematics and gender stereotypes in one Jewish and one Druze grade 5 classroom in Israel. *Education Research International* (online journal). Retrieved from http://www.hindawi.com/journals/edu/2011/545010/.

Nasser, F., & Birenbaum, M. (2005). Modeling mathematics achievement of Jewish and Arab eighth graders in Israel: The effects of learner-related variables. *Educational Research and Evaluation, 11*(3), 277–302.

Israeli Jewish and Arab Students' Gendering of Mathematics

Helen J. Forgasz and David Mittelberg

Abstract In English-speaking, Western countries, mathematics has traditionally been viewed as a "male domain", a discipline more suited to males than to females. Recent data from Australian and American students who had been administered two instruments (Leder and Forgasz 2002) tapping their beliefs about the gendering of mathematics appeared to challenge this traditional, gender-stereotyped view of the discipline. The two instruments were translated into Hebrew and Arabic and administered to large samples of grade 9 students attending Jewish and Arab schools in northern Israel. The aims of this study were to determine if the views of these two culturally different groups of students differed and whether within group gender differences were apparent. The quantitative data alone could not provide explanations for any differences found. However, in conjunction with other sociological data on the differences between the two groups in Israeli society more generally, possible explanations for any differences found were explored. The findings for the Jewish Israeli students were generally consistent with prevailing Western gendered views on mathematics; the Arab Israeli students held different views that appeared to parallel cultural beliefs and the realities of life for this cultural group.

1 Introduction

Historically, mathematics has been viewed as the domain of white, middle-class, males. Gender inequities in mathematics education are frequently reported favouring males with respect to achievement, participation rates, and in regard to students' attitudes and beliefs about mathematics (e.g., Leder et al. 1996).

H.J. Forgasz (✉)
Monash University, Melbourne, Australia
e-mail: Helen.Forgasz@education.monash.edu.au

D. Mittelberg
Oranim Academic College of Education, Tivon, Israel
e-mail: davidm@macam.ac.il

This chapter is a reprint of an article published in ZDM—The International Journal on Mathematics Education (2008) 40(4), 545–558. DOI 10.1007/s11858-008-0139-3.

H. Forgasz, F. Rivera (eds.), *Towards Equity in Mathematics Education*,
Advances in Mathematics Education,
DOI 10.1007/978-3-642-27702-3_18, © Springer-Verlag Berlin Heidelberg 2012

Some changes have been reported over time with respect to gender-stereotyped attitudes and beliefs (e.g., Forgasz et al. 2004). Yet, regarding achievement, PISA[1] 2003 results (OECD nd) revealed that males generally outperformed females. Gender differences favouring males were found for most of the 46 participating countries in TIMSS[2] 2003; at the grade 8 level, there were nine countries with significant gender differences favouring females (Mullis et al. 2004). Participation data in Australia (Forgasz 2006a) still indicate that males outnumber females in the most challenging mathematics subjects offered at the grade 12 level; with respect to performance levels, higher proportions of males than females achieve the highest grades (e.g., Forgasz 2006b).

Cross-national comparisons of students' mathematics achievements, and/or the factors that underpin them, provide valuable insights into the complexities associated with finding explanations for observed patterns of gender difference in cognitive and affective measures. Cultural, ethnic, and societal factors are found to interact with gender in many such explorations (e.g., Barkatsas et al. 2002). In the study reported in this article, the aim was to examine grade 9 Israeli students' gendered perceptions of mathematics, to compare the findings by ethnicity, Jewish and Arab, and by gender within the two ethnic groups. These findings were then compared with data gathered earlier from Australia, the place where the instruments used in the current study were developed and normed (see Forgasz 2001; Leder and Forgasz 2002), to determine whether there were any clearly discernible similarities or differences in the views expressed by the Israeli and Australian students.

1.1 Contextualising the Study

The study reported in this article was conducted in Israel. In order to contextualise the study, a brief overview of the Israeli education system is described next. Relevant literature is included.

Israeli Arabs are the largest minority in Israel, and comprised 19.7% (Moslems 83.0%; Druze 8.3%; Arab Christians 8.5%) of the overall population in 2006 (Central Bureau of Statistics. Israel 2007). The educational systems for Arabs and Jews are segregated, although both are run by the Israeli Ministry of Education (Ayalon 2002; Birenbaum and Nasser 2006). Nationwide the retention rate from first grade to 12th grade ranges from 74% to 80% (Ministry of Education. Israel 2007).

Each school system is differentiated. The Arab system is comprised of private (mostly religious) and public (secular) schools, while the Jewish system is primarily public, consisting of religious and secular schools. Hebrew and Arabic are the languages of instruction in the pertinent systems, while the intended mathematics

---

[1] PISA: Program for International Student Achievement.

[2] TIMSS: Trends in Mathematics and Science Study.

curricula are the same. Ayalon (2002) reported that the curriculum in both systems is the same in principle, and is:

> Composed of a core of compulsory subjects—civics, Hebrew, Arabic for Arabs, lower level mathematics, English and history in addition to advanced optional subjects (e.g. mathematics, physics, chemistry, biology, literature, history geography) (p. 65).

Students are required to opt for one or more of these advanced courses. This usually results in a cluster of subjects in the sciences, or in the humanities, but rarely a mixture of both. High school is completed by taking the external standardised matriculation examination in 20 units of study: each unit equals two weekly hours of study per year. The Universities further demand at least one middle level (four-point) optional course (equivalent to four units), and offer bonuses for grades in the highest level (five-point) courses.

Ayalon (2002) maintained that Jewish students generally had greater choice of advanced level subjects leading to matriculation than students in Arab schools, where advanced courses were often limited to mathematics, sciences, and history. The restricted curriculum in Arab schools was hypothesised to benefit females with respect to access to valued knowledge such as mathematics (Ayalon 2002). Mittelberg and Lev-Ari (1999) reported that Arab girls' preparedness "to adopt a mathematically based profession in the future is particularly high both when compared to Jewish girls as well as Jewish boys" (p. 88).

Ninth grade concludes middle school, prior to high school placement in optional subjects, streamed by ability level, for the final three years of schooling. Even in middle school (grades 7–9) in Israel, subjects like mathematics and English are ability grouped. While students may ostensibly still share a common school wide curriculum, text books and examinations are often different for each ability grouping. Thus it is generally clear into which streams students in the various ability groups will flow as they advance into the higher grades of schooling.

Based on data from the Israeli Central Bureau of Statistics, Ayalon (2002) noted that "gender inequality among Arab students was relatively moderate" (p. 63), with higher proportions of Israeli Arab than Jewish girls taking advanced courses in mathematics. Ayalon (2002) cited findings indicating that between 1948 and 1980 Arab females' enrolments at all levels of secondary education had equalled Arab males' and that they had higher participation rates in post-secondary education. Mittelberg and Lev-Ari (1999) noted that Arab females also had high levels of perceived achievement and self-confidence in mathematics, and were willing to consider mathematically based studies and professions in the future.

Kashti (2007) reported that recent figures from the Education Ministry in Israel revealed that Jewish females outperformed Jewish males in Hebrew and English but not in mathematics while Arab females outperformed Arab males in all three subjects. Nasser and Birenbaum (2005) indicated that there had been two studies in the United Arab Emirates and one in Lebanon in which females had outperformed males in mathematics. A similar finding was reported in the PISA 2006 results for Qatar, the only country in which there was a significant gender difference favouring females (OECD 2007). For TIMSS 2003 at the grade 8 level, Jordan and Bahrain were among the nine countries with significant gender differences favouring female (Mullis et al. 2004).

Yet, cultural factors appear to work against Arab females being able to capitalise on their mathematics potential. While Jewish families in Israel were considered by Nasser and Birenbaum (2005) to be "characterized as modern and westernized" (p. 294), they claimed that in the "patriarchic Arab family" (p. 294) boys and girls were treated differently: boys were more nurtured and cared for and, when not at school, were freer than girls "to spend more time in various activities outside home" (p. 294). It was claimed that:

> In light of the cultural disparities between the Jewish and Arab communities, particularly in the different standing of boys and girls, it is understandable that the effect of gender on beliefs and motivational variables, and consequently on mathematics achievement through these variables, is more evident in the Arab group. (Nasser and Birenbaum 2005, p. 294).

Nasser and Birenbaum (2005) found that Arab girls had higher levels of self-efficacy, less anxiety, and higher achievement in mathematics than did Arab boys, a pattern described as "contrary to the expected directions" (p. 294). They speculated that these results:

> might also be attributed to the desire of Arab girls to demonstrate their capabilities, which they believe to be underestimated by their family or by their community... in the educational context where they can legitimately compete for status with boys (p. 294).

Compared to the Jewish population in Israel, the Arab population is generally more conservative and "Arab women are not expected to be active outside their homes and labour market participation is still low" (Ayalon 2002, p. 63). Indeed, while females comprise 56% of all Arab matriculants, only 20% of all Arab women are found in the labour force, compared to 51% of all Jewish women and 65% of all Arab men (Fogel-Bizau 2003).

Jewish students in Israel appear to follow more closely the gender-stereotyped patterns of participation reported widely in Western nations such as the USA and Australia. That is, more males than females are enrolled in the most challenging mathematics courses offered at the secondary level (Mittelberg and Lev-Ari 1999). Zohar (2005) (currently the head of the Pedagogical Secretariat of the Israeli Ministry of Education) reported findings from an analysis of gender differences in participation and performance in matriculation mathematics courses. In the highest level (5 point) mathematics course, only 15% of all girls studying for the matriculation are found, whereas 21% of the boys sit for this advanced level of mathematics. In the lowest level of mathematics (three-point), 57% of the girls are found as against 53% of the boys. Zohar (2005) also reported on performance outcomes by gender. In the highest level (five-point) mathematics course, high grades were achieved by 66% of boys and 67% of girls; for the middle level (four-point) mathematics, 60% of girls compared to 50% of boys achieved high grades; and, in the lowest level (three-point) mathematics, 46% of girls and 36% of boys achieved high grades. These data suggest that whilst participation rates in the most advanced mathematics course slightly favour males, females may be slightly outperforming males across all the mathematics offerings.

With respect to the relative performances of students in the Jewish and Arab systems, Lavy (1998) reported on the relative deprivation of economic and educational resources in the Arabic-speaking sector, particularly at the primary level. In 1991 and 1992, failure rates were more than twice as high as in the Jewish sector. Lavy (1998) maintained that there were three variables impacting on students' test scores: "expenditures per student, hours of instruction per student, and the percent of certified teachers" (p. 189). In all three areas, Lavy (1998) claimed that the Arab primary schools in 1991 were seriously disadvantaged.

Based on a secondary analysis of the large Israeli Ministry of Education's nationwide data set, Birenbaum and Nasser (2006) reported a large performance discrepancy at the grade 8 level which was consistent with earlier findings from TIMSS 1999. In addition, their analysis anticipated the wide discrepancy between Jewish and Arab students who earned the Israeli matriculation certificate in 1996, 69% of Jewish students as compared to 49% of Arab students (Birenbaum and Nasser 2006). These findings are not new, however Birenbaum and Nasser (2006) concluded that despite this large performance gap, "Arab students exhibited a more positive attitude to mathematics, and Arab girls exhibited an even more achievement-enhancing pattern than Arab boys" (p. 37).

Based on a comparison between the Israeli data from TIMSS 1999 and TIMSS 2003, Zuzovsky (2006) reported somewhat surprising findings. Despite the ongoing inequality between the Jewish and Arab sectors in Israel noted above, a narrowing of the gap in mathematics achievement in favour of Arab students was found. Specifically, the increase in the mean scores in mathematics over the period was three times higher in Arabic-speaking schools (68 points) than in Hebrew speaking ones (23 points), reducing the achievement gap from one standard deviation in 1999 to a half a standard deviation in 2003. Zuzovsky (2006) maintained that the factors leading to this change stemmed from a change in Government policy in the 1990s, as well as the subsequent increase in enrolment rates and reduction of dropouts in the Arab sector. In addition, Zuzovsky (2006) argued that:

> the higher achievement gains of the Arab sector seem to be a result of adopting mainstream pedagogy in... the conceptual approach in mathematics... [which] did not exclude more traditional, still very effective, modes such as the computational mode and listening to lectures, which provide students in Arabic-speaking schools with suitable instruction (p. 28).

The interaction between ethnicity, culture, participation, and performance is a central theme of this article.

1.2 Gendered Beliefs About Mathematics

During the 1970s and 1980s, the *Women's Movement*, particularly in Western countries, was active in raising awareness of women's social and educational disadvantage. Simultaneously, much research was being conducted to identify factors contributing to the documented gender differences favouring males in enrolment patterns and participation in challenging mathematics studies (e.g., Fennema 1974; Reyes 1984). The extent to which mathematics was stereotyped as a male domain, that is, perceived to be more suited to males than to females, was identified as one of the explanatory variables.

In 1976, Fennema and Sherman published their *Mathematics Attitude Scales* [MAS] (Fennema and Sherman 1976) which comprised nine subscales including *Mathematics as a male domain*. This subscale has been widely used to assess the extent to which respondents stereotype mathematics as a male domain. Forgasz et al. (1999) argued that the wording of some of the items on the sub-scale was anachronistic, and that the assumptions underpinning the scoring of the subscale did not allow respondents to express the view that mathematics was a *female domain*. Leder and Forgasz (2002) described the development of two new instruments— the *Mathematics as a gendered domain* instrument and the *Who and mathematics* instrument—to overcome the identified limitations of the *Mathematics as a male domain* subscale. Both instruments allow respondents to reflect their views of mathematics as a male, female, and/or neutral domain. The two instruments were used in the present study.

Perceptions of mathematics as a male domain persisted well beyond the time Fennema (1974) first drew attention to the issue. Andre et al. (1999) found that primary students in Iowa (USA) viewed mathematics and science related jobs to be more male than female dominated. Greene et al. (1999) found that for students of both sexes, "endorsing the stereotype that mathematics is a male domain was negatively related to reported effort" (p. 421), that is, the extent of stereotyping is linked to students' motivations towards mathematics. In general, males have been found to believe that the mathematical abilities of members of their own sex were superior to those of females (e.g., Andre et al. 1999; Tiedemann 2000); females, too, have generally considered men to be superior in mathematics but not to the same extent as did males (e.g., Meyer and Koehler 1993; Forgasz and Leder 1996).

Although the extent to which gender differences in mathematics achievement has varied depending on the sample studied and the assessment tasks examined (e.g., Cooper and Dunne 2000; Kimball 1989), gender differences in performance are still of concern in many countries. An examination of the 2003 TIMMS results reveals that the extent of gender differences varied greatly among the participating countries. On the whole, males were more likely to outperform females; however, there were many countries with no statistically significant gender differences; and some countries in which females outperformed males. Australian and Israeli grade 8 students participated in TIMMS 2003. Of the 46 participating countries, the Australians (ranked 14) performed slightly better than the Israelis (ranked 19). In both Israel and Australia, there were no statistically significant gender differences in performance, although males did slightly better than females (see Mullis et al. 2004). It must be recognised that Israeli Arabs comprise only 20% of the Israeli population and thus the Israeli TIMMS results reflect more closely the Israeli Jewish students' performance.

In summary, in both Australia and Israel, gendered patterns of participation in the most challenging mathematics subjects offered in the final year of schooling as well as achievements at the highest levels are similar and favour males. Grade 8 TIMMS data also indicated there were similarities in performance among students in the two countries. Forgasz et al. (2004) reported findings from the two instruments developed by Leder and Forgasz (2002) for grades 7–10 students in Australia and the USA. Changes in gendered perceptions of mathematics appeared to have shown some change. They concluded that:

Because the items of the new instruments differ from those on the Fennema–Sherman Mathematics as a male domain subscale, it is not possible to argue definitively about change in attitudes over time. The responses to the new instruments show some change, particularly in situations where females are doing better or have more positive attitudes than males. Taken as a whole the responses indicate that most students see mathematics as relatively gender neutral but there are situations, including distraction of others and completing classwork, where more recent accounts of gender differences have been given more credence (Forgasz et al. 2004, pp. 416–417).

Through the administration of the two instruments developed by Leder and Forgasz (2002), the aim of the present study was to explore the gendered beliefs about mathematics of grade 9 students within Israel's two ethnic sub-cultures and to compare the findings with those already known for Australian students of the same age. Based on the cultural differences between the Israeli Jewish and Arab populations and the statistical data discussed above, we expected to find differences in the two groups' gendered beliefs about mathematics. Since Jews represent 80.3% of the Israeli population and TIMMS 2003 data and previous findings on Israeli students' attitudes towards mathematics appeared to parallel those reported for students in western countries, it was expected that if differences between the Jewish and Arab students were found, it would be the Jewish students' views that would be more similar to those of Australian students than would the views of the Israeli Arab students.

2 Aims and Methods

Answers to the following research questions were sought in the present study:

1. Is there a difference in the gendered beliefs about mathematics of Jewish and Arab Israeli grade students?
2. Do male and female students within each of the two ethnic groups hold different gendered beliefs? How do the males' and females' views across the ethnic groups compare?
3. How do the Israeli students' views compare with those of Australian grade 9 students?

2.1 Instruments

The *Mathematics as a Gendered Domain* instrument and the *Who and Mathematics* instruments (see Leder and Forgasz 2002 for details) were normed on Australian students in grades 7–10, including grade 9 students. The *Mathematics as a Gendered Domain* instrument is comprised of three subscales with 16 Likert-type items scored on five-point scales from SD = 1 to SA = 5 on each. The three subscales are: *Mathematics as a Male Domain* [MD], *Mathematics as a Female* [FD], and *Mathematics*

1 Mathematics is their favourite subject

2 Think it is important to understand the work in mathematics

3 Are asked more questions by the mathematics teacher

4 Give up when they find a mathematics problem too difficult

5 Have to work hard in mathematics to do well

6 Enjoy mathematics

7 Care about doing well in mathematics

8 Think they did not work hard enough if they do not do well in mathematics

9 Parents would be disappointed if they did not do well in mathematics

10 Need mathematics to maximise future employment opportunities

11 Like challenging mathematics problems

12 Are encouraged to do well by the mathematics teacher

13 Mathematics teachers think they will do well

14 Think mathematics will be important in their adult life

15 Expect to do well in mathematics

16 Distract other students fr om their mathematics work

17 Get the wrong answers in mathematics

18 Find mathematics easy

19 Parents think it is important for them to study mathematics

20 Need more help in mathematics

21 Tease boys if they are good at mathematics

22 Worry if do not do well in mathematics

23 Are not good at mathematics

24 Like using computers to work on mathematics problems

25 Mathematics teachers spend more time with them

26 Consider mathematics to be boring

27 Find mathematics difficult

28 Get on with their work in class

29 Think mathematics is interesting

30 Tease girls if they are good at mathematics

Fig. 1 The 30 items on the Who and mathematics instrument

as Neutral Domain [ND]. The full set of items is listed in Leder and Forgasz (2002) and reproduced in the "Appendix". Sample items from each subscale include:

MD: boys understand mathematics better than girls do
FD: girls are more suited than boys to a career in mathematically related area
ND: boys are just as likely as girls to help friends with their mathematics

The *Who and Mathematics* instrument is comprised of 30 statements, presented in the order listed in Fig. 1.

For each of the 30 statements on the *Who and mathematics* instrument, respondents select one of the following responses:

BD: boys definitely more likely than girls
BP: boys probably more likely than girls

ND: no difference between boys and girls
GP: girls probably more likely than boys
GD: girls definitely more likely than boys

The 30 statements stand alone and can not be combined into any sort of total scale or subscale score.

In order to interpret responses to each item, the response categories are scored as follows:

$$BD = 1 \quad BP = 2, \quad ND = 3, \quad GP = 4, \quad \text{and} \quad GD = 5.$$

Mean scores are calculated for each item to determine the direction and the average strength of the beliefs which are represented by the 'distance' from the neutral value of 3 (*no difference between boys and girls*). Thus, mean scores less than three indicate that, on average, respondents believe that *boys are more likely than girls* to match the wording of the item; mean scores greater than three that they believe that *girls are more likely than boys* to do so. One-sample *t* tests are used to determine whether or not mean scores close to 3 are significantly different from 3; critical values are sample size dependent.

The items on both instruments match closely. The two instruments were translated into Hebrew and into Arabic by fluent English-speaking native speakers of Hebrew and Arabic. The reliability of the translations was determined by asking a different pair of Hebrew and Arab speakers fluent in English to back translate into English. Although the two instruments were developed in Australia and normed on Australian students, they have been used outside Australia. They have been administered to students in the USA (Forgasz et al. 2004) and, in translation, to Greek (Barkatsas et al. 2002) and Swedish students (Brandell et al. 2004). The researchers involved in each study did not consider the items culturally insensitive. Similarly, both researchers involved in the present study consulted with Jewish and Arab Israeli colleagues who did not find any of the items to be culturally unacceptable.

The instruments were administered to grade 9 students in four Jewish schools and in three Arab schools in Northern Israel. Northern Israel is the region where a large proportion of Israel's Arab population resides. In fact, as reported by Haidar (2005), data gathered in 2001 showed that 56.6% of Israeli Arabs lived in the Galilee, while an additional 23% live in the Arab triangle North of Hadera. Thus, 89% of all Israeli Arabs lived in the region from which our sample was drawn (Haidar 2005). The Jewish schools invited to participate in the study included two small rural private schools and two urban schools within the public sector. The Arab schools invited to participate included two public regional schools, and one private school in a large city. While recognising that these purposeful samples were not fully representative of the entire Israeli grade 9 population, the samples were large enough to enable robust statistical comparisons to be made between the beliefs of the participating Israeli Jewish and Arab students. The data set was sufficiently large to allow comparisons with the earlier gathered Australian data.[3] Also included on both in-

---

[3]It should be pointed out that the Australian data were gathered in metropolitan Melbourne. Participating students were drawn from the three sectors of education—government, Catholic, and Independent (private)—and represented Australia's extensive ethnic and multicultural profile.

Table 1 Sample sizes of respondents for each instrument by country, ethnic group, and gender

Mathematics as a gendered domain						Who and mathematics					
Australia		Israeli Jews		Israeli Arabs		Australia		Israeli Jews		Israeli Arabs	
M	F	M	F	M	F	M	F	M	F	M	F
122	131	50	55	60	57	125	123	65	68	71	55

struments was an item asking students "How good are you at mathematics" [HGM]. Students responded on a five-point scale ranging from $1 =$ weak to $5 =$ excellent.

2.2 Sample

The sample sizes by gender of the Jewish and Arab Israeli grade 9 students and of the Australian grade 9 students on whom the instruments were normed for each of the two instruments are shown in Table 1.

2.3 Data Analyses

To test for statistically significant differences, independent groups t tests or ANOVAs were used as appropriate (i.e. depending on the number of groups for which comparisons were being made). When ANOVA results indicated that statistically significant differences existed, Scheffe post hoc tests were conducted to identify which pairs had statistically significant differences.

3 Results and Discussion

3.1 Israeli and Australian Grade 9 Students' Perception of Mathematics Achievement

Mean scores for perceived levels of mathematics achievement [HGM] were compared by country using an independent groups t test. The Israelis in this sample were found to have a significantly higher mean level of perceived mathematics achievement than the Australians (Israelis: 3.67, Australians: 3.43, $P < 0.01$). The findings are inconsistent with the TIMSS 2003 findings that revealed that Australians had performed better than Israelis. As noted earlier, Israeli Arabs are a minority of the Israeli population (around 20%) and the difference in the direction of the findings may well be due to the unrepresentative sampling of Israeli students, that is, using approximately equal proportions of Jewish and Arab Israelis.

The Israeli data were disaggregated by ethnic group, and comparisons between the three ethic groups were undertaken.

Table 2 HGM: ANOVA
results by ethnic group

Ethnicity	N	Mean	F, p level
Israeli Jews	238	3.58	
Israeli Arabs	240	3.39	2.47, NS
Australians	499	3.42	

Table 3 HGM: Means and t
test results by gender within
ethnic group

Ethnicity	Male		Female		t, p level
	N	Mean	N	Mean	
Israeli Jews	115	3.57	120	3.59	NS
Israeli Arabs	128	3.45	109	3.31	NS
Australians	246	3.52	251	3.31	−2.66, p < 0.01

NS—not significant.

3.2 Perceptions of Mathematics Achievement: Ethnic Comparisons

A one-way ANOVA was conducted to determine if there were any differences in the mean scores on students' perceptions of their mathematics achievement levels by ethnic grouping: Israeli Jews, Israeli Arabs, and Australians. The results are shown in Table 2.

As can be seen in Table 2, although there was no statistically significant difference between the three groups ($p > 0.05$), the Israeli Jewish students had the highest perceived levels of mathematics achievement (mean = 3.58).

Gender differences within each ethnic group were examined using independent groups t tests. The results are shown in Table 3.

As can be seen in Table 3, the only gender difference found was among Australian students with males (mean = 3.52) believing they were higher achievers than females (mean = 3.31).

3.3 Results from the Two Instruments

The findings from the *Mathematics as a gendered domain* instrument are presented first, followed by the findings from the Who and mathematics instrument.

3.3.1 Mathematics as a Gendered Domain: Differences by Ethnic Group

The mean scores for each of the three ethnic groups on each of the three subscales of the *Mathematics as a gendered domain* scale are illustrated in Fig. 2.

As can be seen in Fig. 2, the order of the belief measures on the three subscales was the same for all three groups: highest score on ND, and lowest score on MD,

Fig. 2 MD, FD, and ND:
Mean scores by ethnic group

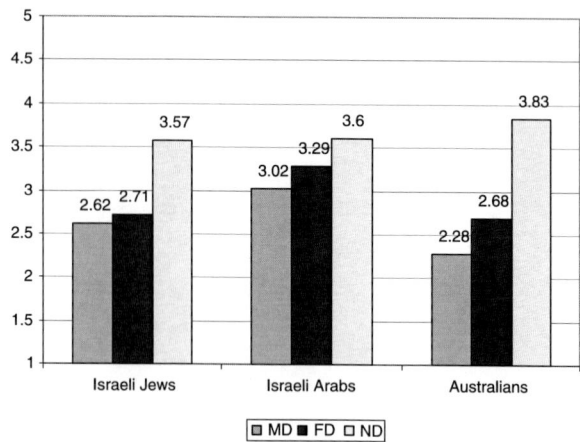

that is, students in each ethnic group agreed most strongly that mathematics was a neutral domain and least strongly that mathematics was a male domain. However, while all three groups strongly agreed that mathematics was a neutral domain (means ≫ 3), for the MD and FD subscales, the directions of the beliefs of the Israeli Arabs differed from those of the Israeli Jews and the Australians. The Israeli Arabs were unsure if mathematics was a male domain (mean approximately 3) whereas the Australians and Israeli Jews disagreed that it was (means < 3), and the Israeli Arabs believed that mathematics was a female domain (mean > 3) whereas the Australians and Israeli Jews disagreed that it was (means < 3).

ANOVAs followed by Scheffe post hoc tests were conducted to determine if there were statistically significant differences by ethnic group in the students' gendered beliefs on the three *Mathematics as a gendered domain* subscales. The results are shown in Table 4.

As can be seen in Table 4, statistically significant differences were found among the mean scores of the three ethnic groups on MD, FD and ND. Scheffe post hoc tests indicated that the only non-significant differences in mean scores were for Israeli Jews and Australians on the FD, and for Israeli Jews and Israeli Arabs on the ND.

The statistically significant findings indicated that:

- MD: Australians (mean = 2.28) disagreed more strongly than Israeli Jews (mean = 2.62) that mathematics was a male domain; Israeli Arabs were not sure if mathematics was a male domain (mean = 3.02).
- FD: Australians (mean = 2.68) and Israeli Jews (mean = 2.71) similarly disagreed that mathematics was a female domain; Israeli Arabs agreed that mathematics was a female domain (mean = 3.29).
- ND: All three groups agreed that mathematics was a neutral domain (means > 3), with the Australians (mean = 3.83) agreeing more strongly than Israeli Jews (mean = 3.57) and Israeli Arabs (mean = 3.60).

Table 4 MD, FD, and ND: Means and ANOVA results

Subscale	Ethnicity	N	Mean	F, p level	Scheffe post hoc
MD	Israeli Jews	103	2.62	42.26, <0.001	Israeli Jews–Israeli Arabs: $p < 0.001$
	Israeli Arabs	107	3.02[a]		Israeli Jews–Australians: $p < 0.001$
	Australians	233	2.28		Israeli Arabs–Australians: $p < 0.001$
FD	Israeli Jews	99	2.71	36.42, <0.001	Israeli Jews–Israeli Arabs: $p < 0.001$
	Israeli Arabs	100	3.29		Israeli Jews–Australians: NS
	Australians	236	2.68		Israeli Arabs–Australians: $p < 0.001$
ND	Israeli Jews	101	3.57	13.89, <0.001	Israeli Jews–Israeli Arabs: NS
	Israeli Arabs	104	3.60		Israeli Jews–Australians: $p < 001$
	Australians	224	3.83		Israeli Arabs–Australians: $p < 0.001$

[a]One-sample t test indicated that this mean was not significantly different from 3.

3.4 Mathematics as a Gendered Domain: Gender Differences Within Ethnic Groups

To determine if there were gender differences in the grade 9 students' gendered beliefs about mathematics within each ethnic group, independent groups t tests were conducted. The results are shown in Table 5.

A few interesting trends are apparent from the data in Table 5.

- There was a similarity in the belief patterns of the Australian males, Australian females, Israeli Jewish males, and Israeli Jewish females. All held that mathematics was a neutral domain (mean > 3) and disagreed that is was either a male or a female domain (mean < 3).
- The pattern was different among the Israeli Arabs. The females believed that mathematics was a neutral domain as well as being both a male domain and a female domain (means > 3), and the males agreed that mathematics was a neutral and a female domain (mean > 3), but disagreed that it was a male domain (mean < 3).

The three statistically significant gender differences, one for each ethnic group, showed that:

- Israeli Jews: males (mean = 2.59) disagreed more strongly than females (mean = 2.86) that mathematics was a female domain [FD].
- Israeli Arabs: females agreed (mean = 3.25) and males disagreed (mean = 2.81) that mathematics was a male domain [MD].
- Australians: females disagreed more strongly (mean = 2.11) than males (mean = 2.47) that mathematics was a male domain [MD].

In summary, the Israeli Jewish students in this sample and the Australians (as well as the males and females in each country) held similar views. They agreed that mathematics was a neutral domain and rejected mathematics as either a male or a female domain. The findings for the sample of Israeli Arabs students were quite

Table 5 MD, FD, and ND: Means and *t* test results by gender within ethnic group

Ethnicity	Subscale	All	Male		Female		*t*, *p* level
			N	Mean	N	Mean	
Israeli Jews	MD	2.62	49	2.51	53	2.73	NS
	FD	2.71	50	2.59	48	2.86	2.346, <0.05
	ND	3.57	49	3.57	51	3.56	NS
Israeli Arabs	MD	3.02	55	2.81	51	3.25	3.241, <0.01
	FD	3.29	49	3.39	51	3.20	NS
	ND	3.60	49	3.58	54	3.62	NS
Australians	MD	2.28	112	2.47	121	2.11	−4.35, <0.001
	FD	2.68	115	2.66	121	2.70	NS
	ND	3.83	108	3.79	116	3.87	NS

different and are ambiguous. They agreed that mathematics was both a neutral and a female domain, and were uncertain if it was also a male domain (mean = 3.02). The gender difference on the MD subscale may well explain the overall uncertainty among the Israeli Arabs; while the females agreed (mean = 3.25), the males disagreed (mean = 2.81) that mathematics was a male domain. An examination of the findings from the second instrument might shed some light on this ambiguity.

3.4.1 Who and Mathematics: Differences by Ethnic Group

In Table 6, the predicted direction [Pred] of responses to each item on the *Who and mathematics* instrument are presented (see Fig. 1 for the 30 items). The predicted response directions have been presented elsewhere and were derived from earlier literature in the field of gender issues in mathematics education. They have been considered consistent with the view that mathematics is a "male domain"—see Leder and Forgasz (2002) for more details. The predicted direction of the response to each item is indicated as follows: M = males more likely than girls to exhibit the behaviour or hold the belief; F = females more likely than males to do so; M/F = mixed results on whether males or females are more likely.

Also shown on Table 6 are the directions of the responses of the Israeli Jews [IJ], Israeli Arabs [IA] participating in this study, and of the Australian [Aus] grade 9 students: M = males are more likely than females to exhibit the behaviour or hold the belief reflected by the wording of the item; F = females are more likely to do so than males; nd = no difference between boys and girls.

As shown in Table 6, responses consistent with the predicated directions of responses were as follows:

Israeli Jews—9 items: 3, 10, 11, 14, 16, 18, 21, 24, 30
Israeli Arabs—10 items: 5, 7, 9, 10, 14, 16, 19, 21, 22, 30
Australians—5 items: 3, 21, 24, 28, 30

Table 6 Predicted directions of responses on the 30 items on the *Who and mathematics* instrument and actual directions of responses for Israeli Jews (IJ), Israeli Arabs (IA) and Australian (Aus) grade 9 students

Item	Pred	IJ	IA	Aus	Item	Pred	IJ	IA	Aus	Item	Pred	IJ	IA	Aus
1	M	nd	nd	F	11	M	M	nd	nd	21	M	M	M	M
2	F	nd	nd	F	12	M	nd	nd	nd	22	M/F	F	nd	F
3	M	M	nd	M	13	M	nd	F	F	23	F	nd	M	nd
4	F	M	nd	M	14	M	M	M	F	24	M	M	nd	M
5	F	nd	F	nd	15	M	nd	nd	F	25	M	F	nd	nd
6	M	nd	nd	F	16	M	M	M	M	26	F	M	M	M
7	M/F	F	nd	F	17	F	nd	M	M	27	F	nd	nd	M
8	M	nd	nd	F	18	M	M	nd	F	28	F	M	nd	F
9	M	nd	M	F	19	M	nd	M	nd	29	M	nd	nd	F
10	M	M	M	nd	20	F	nd	nd	M	30	M	M	M	M

Responses in the opposite direction of prediction (excluding nd responses) include:

Israeli Jews—4 items: 4, 25, 26, 28
Israeli Arabs—4 items: 13, 17, 23, 26
Australians—14 items: 1, 4, 6, 8, 9, 13, 14, 15, 17, 18, 20, 26, 27, 29

In light of the above, it seems that the views of the Australians have shifted more from the gender stereotyped view of mathematics as a male domain than the Israeli Jews and Israeli Arabs participating in this study.

A summary of the findings for the three ethnic groups is presented in a different form in Table 7. Item numbers are listed for which the various ethnic groups held each of the three possible beliefs with respect to the wording of the items: "boys more likely than girls", "no difference between boys and girls", and "girls more likely than boys" (see Fig. 1 for the wording of the items). It should be noted that one-sample t tests were conducted on the mean scores on each item to test if they were significantly different from 3.00—the mean score representing the belief that there is no difference between boys and girls. Hence mean scores for the groups of items representing the belief "no difference between boys and girls" all had mean scores not significantly different from 3.00.

The data in Table 7 indicate that:

1. Students in all three groups agreed that boys were more likely than girls to:

   - Students in all three groups agreed that boys were more likely than girls to
   - Distract other students from their mathematics work (item 16)
   - Find mathematics easy (item 18)
   - Tease boys if they are good at mathematics (item 21)
   - Tease girls if they are good at mathematics (item 30)

Table 7 Summary of findings for the three ethnic groups for the 30 items on the *Who and mathematics* instrument

	Israeli Jews (N = 133)	Israeli Arabs (N = 126)	Australians (N = 248)
"Boys more likely than girls" (mean scores < 3)	Items: 3, 4, 10, 11, 14, *16*[a], 18, *21*, 24, *26*, 28, *30*	Items: 9, 10, 14, *16*, 17, 19, *21*, 23, *26*, *30*	Items: 3, 4, *16*, 17, 20, *21*, 24, *26*, 27, *30*
"No difference between boys and girls" (mean scores not statistically different from 3)	Items: 1, 2, 5, 6, 8, 9, *12*, 13, 15, 17, 19, 20, 23, 27, 29	Items: 1, 2, 3, 4, 6, 7, 8, 11, *12*, 15, 18, 20, 22, 24, 25, 27, 28, 29	Items: 5, 10, 11, *12*, 23, 25,
"Girls more likely than boys" (mean scores > 3)	Items: 7, 22, 25	Items: 5, 13	Items: 1, 2, 6, 7, 8, 9, 13, 14, 15, 18, 22, 28, 29

[a]Italicised item numbers are common across the three ethnic groups.

For three of these items (16, 21, 30), it could be argued that the three groups are likely to share some common classroom experiences with respect to boys' behaviours.

2. Students in all three groups agreed that there was no difference between boys and girls with respect to being "encouraged to do well by the mathematics teacher" (item 12).

3. There were no items on which the students in all three groups agreed that girls were more likely than boys to match the wording of the item.

It can also be seen from Table 7 that, compared to the students in the other two ethnic groups, there were more items on which the Australian students held views that "girls were more likely than boys" (13, compared to 3 for the Israeli Jews and 2 for the Israel Arabs), and fewer items on which they believed there was no difference between boys and girls (2, compared to 10 for the Israeli Jews and 12 for the Israeli Arabs). An examination of the wording of these items and the directions of the responses reveals that the Australian students' views are very different from beliefs traditionally associated with the view that mathematics is a male domain. On the other hand, as shown by the data presented based on Table 6, there were considerably more responses from the groups of Israeli Jews and Israeli Arabs that were consistent with the view that mathematics is a male domain.

In summary, the data in Table 7 indicate that the Australians' views challenge the traditional stereotype of mathematics as a male domain to a much greater extent than would appear to be the case among the Israeli Jews and Israeli Arabs participating in this study. An interesting finding in relation to the Israeli Arabs was noted. There were three closely linked items to which their responses clearly challenge the traditional gender stereotype: items 17, 23, and 13. These responses revealed that this group believed that boys rather than girls "get the wrong answers in mathematics" (17) and "are not good at mathematics" (23) and that girls rather than boys perceive that "their mathematics teachers think they will do well" (13). These responses are likely to reflect their classroom experiences.

Table 8 Results of independent groups t tests by gender for Israeli Jews and Israeli Arabs on the 30 *Who and mathematics* items

Item		Israel Jews	Israeli Arabs
7	Care about doing well in mathematics	NS	M: 3.06[*], F: 2.5; $p < 0.01$
10	Need mathematics to maximise future employment opportunities	NS	M: 2.93[*]; F: 2.43; $p < 0.05$
11	Like challenging mathematics problems	NS	M: 3.16[*], F: 2.71; $p < 0.05$
14	Think mathematics will be important in their adult life	NS	M:2.9[*], F: 2.51; $p < 0.05$
15	Expect to do well in mathematics	NS	M: 3.27[*], F: 2.82[*]; $p < 0.05$
19	Parents think it is important for them to study mathematics	NS	M: 2.93[*], F: 2.51; $p < 0.05$
21	Tease boys if they are good at mathematics	M: 2.75, F: 2.46; $p < 0.05$	NS
22	Worry if do not do well in mathematics	NS	M: 3.46, F: 2.83[*]; $p < 0.01$
24	Like using computers to work on mathematics problems	M: 2.88, F: 2.54; $p < 0.05$	NS
28	Get on with their work in class	NS	M: 3.22[*], F: 2.71[*]; $p < 0.05$
29	Think mathematics is interesting	NS	M: 3.01[*], F: 2.54; $p < 0.05$
30	Tease girls if they are good at mathematics	NS	M: 2.47, F: 3.04[*]; $p < 0.01$

[*]Mean score is not significantly different from 3.00 i.e. the mean score reflects the belief that there is "no difference between boys and girls".

For each ethnic group, independent groups t tests were conducted on the mean scores for each item on the *Who and mathematics* instrument by gender. No statistically significant gender differences were found for the Australian students. In Table 8, only those items for which statistically significant differences for at least one ethnic group, the Israeli Jews or the Israeli Arabs are recorded. Male and female mean scores, and p values for the t tests are also provided.

As shown in Table 8, there were two statistically significant gender differences among the Israeli Jews and ten among the Israeli Arabs. The different patterns of gender differences for the two ethnic groups are discussed below, in turn.

3.5 Gender Differences Among Israeli Jews

The two gender differences found for the sample of Israeli Jewish students indicate that the males and the females believe that boys are more likely than girls to "tease boys if they are good at mathematics" (item 21) and "like using computers to work on mathematics problems" (item 24)—both are traditional gender stereotyped beliefs (see Table 6). Moreover, the females held these beliefs more strongly than did the males. On the remaining 28 items, the males' and females' views concurred, that is, there were no significant gender differences.

3.6 Gender Differences Among Israeli Arabs

On six of the 10 items with statistically significant gender differences (7, 10, 11, 14, 19, and 29), the Israeli Arab males in the sample believed that there was "no difference between boys and girls" (means not different from 3.00), but the Israeli Arab females believed that boys are more likely than girls (means < 3.00) to exhibit the following behaviours: care about doing well (7), need mathematics for future employment (10), like challenging mathematics problems (11), think mathematics will be important in their adult life (14), parents think it is important for them to study mathematics (19), and think mathematics is interesting (29). Here the females' beliefs are more consistent with the gender stereotype of mathematics as a male domain than are the males' views. Consistent with the literature, the responses to three of these items (10, 14, and 19) reflect this group of Israeli Arab students' gendered beliefs about their future roles in post-school society.

Overall all, there were seven items for which the females' views were consistent with the traditional male gender stereotype (7, 10, 11, 14, 19, 28, 29) and three on which the males' views were consistent (15, 22, 30).

4 Conclusions and Implications

The three aims of the present study were:

- to determine if there was a difference in the gendered beliefs about mathematics of Jewish and Arab Israeli grade 9 students
- to examine if there were gender differences among the students in the two groups
- to compare findings from Israelis students with those of Australian grade 9 students.

The major findings from the *Mathematics as a gendered domain* instrument were:

- The sample of Israeli Jewish students and the Australians (as well as the males and females in each country) held similar views. They agreed that mathematics was a neutral domain and rejected mathematics as either a male or a female domain. The Israeli Jews were less convinced than the Australians that this was the case. This finding would appear to be consistent with Nasser and Birenbaum's (2005) claim that Israeli Jewish families are considered "modern and westernised" (p. 294).
- There was an ambiguity in the findings for the sample of Israeli Arabs. They agreed that mathematics was both a neutral and a female domain, and were uncer-

tain if it was also a male domain (mean = 3.02). With respect to the MD subscale, the females agreed and the males disagreed that mathematics was a male domain.

The main findings from the *Who and mathematics* instrument were:

- the Australians' views challenged the traditional stereotype of mathematics as a male domain to a greater extent than did the samples of Israeli Jewish and the Israeli Arabs students.
- The Israeli Jewish students' responses were more consistent with the traditional gender stereotype than were the Israeli Arabs' views. Male and female Israeli Jewish students held similar views, whereas there were many gender differences in the perceptions of the participating Israeli Arabs.

A partial explanation for the ambiguity in the Israeli Arabs' responses on the *Mathematics as a gendered domain* instrument could be found from the *Who and mathematics* results. As identified in the literature review above, Ayalon (2002) and Fogel-Bizau (2003) reported that Arab females fail to translate their educational advantages into occupational opportunities and participation in the labour force compared to Israeli Jewish females. Their claims appear to be echoed in the participating Israeli Arab ninth grade students' responses to two clusters of three items on the *Who and mathematics* instrument (13, 17, 23 and 10, 14, 19). The first three items indicate that girls are perceived to be better mathematics students than are the boys. These responses are consistent with the statistical data cited in the literature review above that Arab females outperform their male counterparts in mathematics and therefore reflect the students' lived experiences in school and the mathematics classroom. The second cluster of three items are also consistent with the findings from the literature related to Arab males and females in the workforce, that is, that it is more than likely that gendered post-school expectations that mathematics will be more useful to males than to females are the reality.

As revealed in the literature, Israeli Arabs have strongly gendered expectations of women's roles in society and the workplace. Nevertheless, the findings of the present study reveal, at least in the traditionally gendered discipline of mathematics, that what happens in the classroom is not necessarily bound to replicate the gendered patterns of the world beyond the walls of the school and are consistent with Nasser and Birenbaum's (2005) view that the classroom is a legitimate environment for Arab girls to demonstrate their abilities which may well have been underestimated, and undervalued, by their families and community.

A limitation of the present study was that students' views of the gendered nature of the adult world were not gathered. To do so would provide a better understanding of the relationship between Arab students' views and behaviours in school and in the outside society in which they will live and work in the future. Gathering such data is a suggested direction for future research. It is also suggested that in future work in Israel, the views of a broader, more representative, sample of students from the two cultures should be tapped.

It would appear that government policy is strongly implicated in affecting a range of learning outcomes, including challenges to traditional patterns of gender differences in mathematics achievement and attitudes. As noted earlier, Kashti (2007) reported that Israel Jewish males outperformed females in mathematics, a trend common around the world, but that Israeli Arab females outperformed males. During the 1990s there was increased investment in the Arab educational sector (Zuzovsky 2006) and it would appear to have had a longer term effect. On the other hand, the recently announced PISA (2006) results revealed that "Australian males performed significantly higher than Australian females. This was in contrast to PISA 2003, where no significant gender differences were found" (Australian Council for Educational Research [ACER] 2007). In recent years, boys have been the focus of government funding attention in Australia, even though advocates have pointed to continued female under-representation in higher level mathematics (e.g. Forgasz 2006b). Thus lack of consistent government effort focusing on girls' disadvantage with respect to participation rates in higher level mathematics in Australia has failed to sustain earlier trends towards closing the gender gap. Lasting change requires continued monitoring of gender patterns of achievement and attitudes, additional research to disentangle factors implicated in the persistence of stereotypes within different cultural groups, and a parallel focus on the classroom and on society at large.

Appendix

The 48 items on the *Mathematics as a gendered domain* scale are presented in Table 9.

Table 9 The 48 items on the Mathematics as a gendered domain scale in the order presented, and subscale for each item

Qn	Item	Subscale
1	Women and men are equally likely to be good mathematics teachers	ND
2	Students who get poor marks on mathematics tests are just as likely to be boys as girls	ND
3	Parents think that getting high grades in mathematics is as important for their daughters as for their sons	ND
4	Being good at mathematics comes as naturally to girls as to boys	ND
5	Mathematics is easier for men than it is for women	MD
6	Girls are more suited than boys to a career in a mathematically-related area	FD
7	Girls have more natural mathematical ability than do boys	FD
8	It is just as difficult for girls as it is for boys to get a job in a mathematically-related profession	ND
9	Boys are just as likely as girls to enjoy mathematics	ND
10	Boys are more determined than girls to do well in mathematics	MD
11	Girls and boys who do well in a mathematics test are just as likely to be congratulated	ND
12	Boys have more use for mathematics than girls do when they leave school	MD
13	Parents believe mathematics is more important for their daughters than for their sons	FD

Table 9 (Continued)

Qn	Item	Subscale
14	Explaining answers in mathematics is harder for boys than for girls	FD
15	Girls and boys are just as likely to be lazy in mathematics classes	ND
16	Boys understand mathematics better than girls do	MD
17	Girls enjoy mathematics more than boys do	FD
18	Boys are distracted from their work in mathematics classes more than are girls	FD
19	Parents are as likely to help their daughters as their sons with mathematics	ND
20	Boys, more than girls, want to do well in mathematics to please their parents	MD
21	Compared to boys, girls do less work in mathematics classes	MD
22	More boys than girls care about doing well at mathematics	MD
23	Mathematics is liked more by boys than by girls	MD
24	The weakest mathematics students are more often boys than girls	FD
25	Students who say mathematics is their favourite subject are equally likely to be girls or boys	ND
26	It is more acceptable for a man than a women to be good at mathematics	MD
27	Career choices make the study of mathematics more important for boys than for girls	MD
28	Compared to girls, boys give up more easily when they have difficulty with a mathematics problem	FD
29	Boys, more than girls, like challenging mathematics problems	MD
30	Men and women are equally suited to careers in the computer industry	ND
31	Girls and boys are equally likely to believe that mathematics is important for their career	ND
32	In a mathematics class with both boys and girls, girls tend to speak up more than boys	FD
33	Men are mathematically more intelligent than women	MD
34	Boys are encouraged more than girls to do well in mathematics	MD
35	Boys, more than girls, say the mathematics test was too hard if they do not do well	FD
36	Girls are encouraged more than boys to do well in mathematics	FD
37	There are more popular boys than popular girls who are good at mathematics	MD
38	Girls are just as likely to work hard in mathematics as boys	ND
39	Girls are more careful than boys when doing mathematics	FD
40	When they leave school, girls will have more use for mathematics than boys will	FD
41	Girls, more than boys, care about doing well at mathematics	FD
42	Boys are just as likely as girls to help friends with their mathematics	ND
43	Girls are more likely than boys to believe they are good at mathematics	FD
44	Girls are more likely than boys to say mathematics is their favourite subject	FD
45	Boys and girls are equally good at using calculators in mathematics	ND
46	The mathematical tasks done in class suit boys more than they suit girls	MD
47	Girls are just as likely as boys to say they want to excel in mathematics	ND
48	Girls are less interested in mathematics than are boys	MD

References

Australian Council for Educational Research [ACER] (2007). *Mathematical Literacy in PISA 2006*. Retrieved on 17 December 2007 from: http://www.acer.edu.au/ozpisa/kf_maths.html.

Andre, T., Whigham, M., Hendrickson, A., & Chambers, S. (1999). Competency beliefs, positive affect, and gender stereotypes of elementary students and their parents about science versus other school subjects. *Journal of Research in Science Teaching, 36*, 719–747. doi:10.1002/(SICI)1098-2736(199908)36:6<719::AID-TEA8>3.0.CO;2-R.

Ayalon, H. (2002). Mathematics and science course taking among Arab students in Israel: A case of unexpected gender equality. *Educational Evaluation and Policy Analysis, 24*(1), 63–80. doi:10.3102/01623737024001063.

Haidar, A. (Ed.) (2005). *Arab society in Israel: Population, society and economics (1)*. Tel-Aviv: Van Leer Institute and Hakibbutz Hameuchad Publishing House (in Hebrew).

Barkatsas, A. N., Forgasz, H. J., & Leder, G. C. (2002). The stereotyping of mathematics: Gender and cultural factors. *Themes in Education, 3*(2), 91–216.

Birenbaum, M., & Nasser, F. (2006). Ethnic and gender differences in mathematics achievement and in dispositions towards the study of mathematics. *Learning and Instruction, 16*, 26–40. doi:10.1016/j.learninstruc.2005.12.004.

Brandell, G., Nyström, P., & Sundqvist, C. (2004). Mathematics—a male domain? Retrieved 9 April 2008 from http://www.mai.liu.se/SMDF/madif5/papers/Brandell.pdf.

Central Bureau of Statistics. Israel (2007). *Israel in figures. 2007*. Retrieved 9 April 2008 from http://www1.cbs.gov.il/www/publications/isr_in_n07e.pdf.

Cooper, B., & Dunne, M. (2000). *Assessing children's mathematical knowledge. Social class, sex and problem solving*. Buckingham: Open University Press.

Fennema, E. (1974). Mathematics learning and the sexes: A review. *Journal for Research in Mathematics Education, 5*, 126–139. doi:10.2307/748949.

Fennema, E., & Sherman, J. (1976). Fennema–Sherman mathematics attitude scales. *JSAS: Catalog of selected documents in psychology, 6*(1), 31 (Ms. No. 1225).

Fogel-Bizau, S. (2003). 'The second earner' in the era of globalization: Women in the Israeli labor market (in Hebrew). *Society, 8*. Accessed 28 December 2006 from http://lib.cet.ac.il/pages/item.asp?item=8060.

Forgasz, H. J. (2001). *Mathematics: Still a male domain? Australian findings*. Paper presented at the annual meeting of American Education Research Association [AERA] as part of the symposium, Mathematics: Still a male domain? Seattle, USA, 10–14 April. [ERIC document: ED452071. http://www.eric.ed.gov/ERICDocs/data/ericdocs2sql/content_storage_01/0000019b/80/16/f7/16.pdf].

Forgasz, H. J. (2006a). *Australian year 12 mathematics enrolments: Patterns and trends—Past and present*. International Centre of Excellence for Education in Mathematics. http://www.ice-em.org.au/pdfs/Mathematics_Enrolments.pdf. Accessed 28 December 2006.

Forgasz, H. J. (2006b). Australian year 12 "Intermediate" level mathematics enrolments 2000–2004: Trends and patterns. In P. Grootenboer, R. Zevenbergen & M. Chinnappan (Eds.), *Identities cultures and learning spaces. Proceedings of the 29th annual conference of the Mathematics Education Research Group of Australasia* (Vol. 1, pp. 211–220). Adelaide: MERGA.

Forgasz, H. J., & Leder, G. C. (1996). Mathematics and English: Stereotyped domains? *Focus on Learning Problems in Mathematics, 18*(1, 2 & 3), 129–137.

Forgasz, H. J., Leder, G. C., & Gardner, P. (1999). The Fennema–Sherman Mathematics as a male domain scale reexamined. *Journal for Research in Mathematics Education, 30*, 342–348. doi:10.2307/749839.

Forgasz, H. J., Leder, G. C., & Kloosterman, P. (2004). New perspectives on the gender stereotyping of mathematics. *Mathematical Thinking and Learning, 6*(4), 389–420. doi:10.1207/s15327833mtl0604_2.

Kashti, O. (2007, 28 November). Wide gaps between test scores in Jewish, Arab schools. Haaretz.

Greene, B. A., DeBacker, T. K., Ravindran, B., & Krows, A. J. (1999). Goals, values, and beliefs as predictors of achievement and effort in high school mathematics classes. *Sex Roles, 40*, 421–458. doi:10.1023/A:1018871610174.

Kimball, M. M. (1989). A new perspective on women's math achievement. *Psychological Bulletin*, *105*, 198–214. doi:10.1037/0033-2909.105.2.198.

Lavy, V. (1998). Disparities between Arabs and Jews in school resources and student achievement in Israel. *Economic Development and Cultural Change*, *47*(1), 174–192. doi:10.1086/452391.

Leder, G. C., & Forgasz, H. J. (2002). *Two new instruments to probe attitudes about gender and mathematics*. ERIC, Resources in Education (RIE). [ERIC document number: ED463312].

Leder, G. C., Forgasz, H. J., & Solar, C. (1996). Research and intervention programs in mathematics education: A gendered issue. In A. Bishop, K. Clements, C. Keitel, J. Kilpatrick, & C. Laborde (Eds.), *International handbook of mathematics education, Part 2* (pp. 945–985). Dordrecht, The Netherlands: Kluwer.

Meyer, M. R., & Koehler, M. S. (1993). Internal influences on gender differences in mathematics. In E. Fennema & G. C. Leder (Eds.), *Mathematics and gender* (pp. 60–95). Lucia: St Queensland University Press.

Ministry of Education. Israel (2007). Matriculation examination data 2006 (translation). Retrieved on 8 November 2007 from http://www.education.gov.il/netuney_bchinot/.

Mittelberg, D., & Lev-Ari, L. (1999). Confidence in mathematics and its consequences: Gender differences among Israeli Jewish and Arab youth. *Gender and Education*, *11*(1), 75–92. doi:10.1080/09540259920771.

Mullis, I. V. S., Martin, M. O., Gonzalez, E. J., & Chrostowski, S. J. (2004). *Findings from IEA's Trends in International Mathematics and Science Study at the fourth and eighth grades*. Chestnut Hill: TIMSS & PIRLS International Study Center, Boston College.

Nasser, F., & Birenbaum, M. (2005). Modeling mathematics achievement of Jewish and Arab eighth graders in Israel: The effects of learner-related variables. *Educational Research and Evaluation*, *11*(3), 277–302. doi:10.1080/13803610500101108.

OECD (nd). First results from PISA 2003. Executive summary. http://www.oecd.org/dataoecd/1/63/34002454.pdf. Accessed 12 December 2006.

OECD (2007). *PISA 2006 science competencies for tomorrow's world. Vol. 1: Analysis*. Paris: OECD.

Reyes, L. H. (1984). Affective variables and mathematics education. *Elementary School Journal*, *84*, 558–581. doi:10.1086/461384.

Tiedemann, J. (2000). Parents' gender stereotypes and teachers' beliefs as predictors of children's concept of their mathematical ability in elementary school. *Journal of Educational Psychology*, *92*, 308–315. doi:10.1037/0022-0663.92.1.144.

Zohar, A. (2005). *Boys' and girls' learning in mathematics and the sciences*. Unpublished internal report, Department of Gender Equality, Ministry of Education, Culture and Sport, Israel.

Zuzovsky, R. (2006). *Capturing the dynamics that led to the narrowing achievement gap between Hebrew-speaking and Arabic-speaking schools in Israel: Findings from TIMSS 1999 and 2003*. Retrieved on 8 November 2007 from http://www.iea.nl/fileadmin/user_upload/IRC2006/IEA_Program/TIMSS/Zuzovsky.pdf.

Preface to "Ethnomathematics and Philosophy"

Bill Barton

We who undertake ethnomathematical studies still have a philosophical problem. I still believe that a version of mathematical relativity is one of our basic assumptions. That is, the study of ethnomathematics rests on the idea that there can be more than one form of mathematics—it is neither absolute nor Platonist. Postmodern writing is exposing more about the contingent nature of mathematics, and the historicity of mathematics is becoming more accepted. Subjectivity and objectivity are becoming blurred (Brown 2011; Radford et al. 2008). But we are still a long way from wide agreement on a philosophical position that "allows" ethnomathematical investigations.

So how do I justify my own continued research activity in this area, and how can I justifiably encourage prospective PhD students to come and study with me? The answer is that, so far, I have not come across an argument that convinces me that a Wittgensteinian perspective is inadequate. Indeed, I feel more strongly that this is a useful way to consider philosophical questions in general. That is, I do not take a classical view. I am not worried about ontology or epistemology so much as I am worried about language and thought.

For example, just as, in my article, I question the nature of mathematical objects by suggesting that they are "objects" merely in a linguistic sense—we use nouns to talk about them—so also I now extend this kind of thinking to mathematics itself. We ask: "What is mathematics?" But this is to create something called "mathematics," which may or may not exist.

Wait a minute. Surely I am not suggesting that mathematics is a chimera. No, I am not. But I am suggesting that we bring mathematics into existence by talking about it, and the way we talk about it changes the questions we can ask. We talk about mathematics as if "it" is something. But perhaps mathematics is, rather, a way of doing things? Or perhaps it is primarily a characteristic? Hence we talk about mathematising, and mathematical, respectively. But usually we regard the verb and

B. Barton (✉)
University of Auckland, Auckland, New Zealand
e-mail: b.barton@auckland.ac.nz

H. Forgasz, F. Rivera (eds.), *Towards Equity in Mathematics Education*,
Advances in Mathematics Education,
DOI 10.1007/978-3-642-27702-3_19, © Springer-Verlag Berlin Heidelberg 2012

the adjective as derivative of the noun. We have this thing called mathematics, and so when we are working with this thing we are mathematising, and when we see connections to this thing then they are mathematical. But this is just English grammar (and the grammar of some other languages). There is nothing to stop us regarding mathematising as the primary expression of our field, and make the noun and adjective the derivative forms.

What difference does this make? Well, it makes us ask different philosophical questions. For example, it would no longer make sense to question the existence of mathematical objects. There would still be mathematical objects, but they would be clearly seen as having been brought into existence by the primary activity of mathematising. There would be no question about whether they exist independently or about how we come to know them. We mathematise, and therefore we create the objects by our thought, and attempt to communicate them to one another. The ontology and epistemology of mathematics simply is not a problem any more.

Philosophically, then, I like to explore the way we talk about mathematics, or mathematising, or mathematicality. But I am not concerned about which of these states is prime. I take a rather contingent view, even of my own philosophical position. Rather than try to find out which is the correct state for mathematics, I would much rather play a "what if" language game.

What if mathematising is the prime state? How would I now talk about ethnomathematics? Suddenly it becomes different (not necessarily easier) to consider cultural relativity. The moment we accept culturally differentiated cognition (or even culturally differentiated propensities in cognition), then, we seem to be able to consider culturally different mathematising.

I do not escape the universalising tendency of mathematicians completely. It would still be possible to talk about universal features of mathematising—but now the universals are not objects (circles, sets, theorems) but are features of thought such as rationality (whatever that may be).

My tendency to play a "what-if" language game has given me another way of justifying ethnomathematical activity. Mathematicians have played along "as if" mathematics was absolute for a long time—so what if we play along with the idea that it is not? Where will that lead us?

And so I am ultimately led into the big challenge for ethnomathematics: Where will it lead us? What good has come of ethnomathematical activity? Do we have new ways of thinking mathematically? Do we have new mathematical objects? Do we have new characteristics to be included in the adjective mathematical?

Yes.

To just take two examples from my own students' work, Alangui's (2010) study of the ethnomathematics of rice terraces in northern Philippines uncovered a model of water flow that included social values. Adam's (2010) ethnomathematical engagement with Malaysian food cover weavers and mathematicians resulted in a 2-dimensional computer weaving template that takes account of the 3-dimensional conical nature of the woven object.

An untapped field of ethnomathematical activity is inside mathematics itself. The article refers to differing perspectives in statistics. What about the emerging algorithmic perspective on old problems made possible by modern computing (Knuth

1985)? Should we describe this as part of the development of mathematics or as a new (or renewed) "culture" within mathematics?

Rather than worry about language–generated philosophical questions, I am more interested in the mathematical creativity of the human mind.

References

Adam, A. (2010). *A study of Tudung Saji weaving: An adaptation of mutual interrogation.* Paper presented to the 4th International conference on ethnomathematics, Towson, South Africa.

Alangui, W. (2010). *Stone walls and water flows: Interrogating cultural practice and mathematics.* Unpublished PhD thesis, The University of Auckland, New Zealand.

Brown, T. (2011). *Mathematics education and subjectivity.* New York: Springer.

Knuth, D. (1985). Algorithmic thinking and mathematical thinking. *American Mathematical Monthly, 92*(3), 170–181.

Radford, L., Schubring, G., & Seeger, F. (2008). *Semiotics in mathematics education: Epistemology, history, classroom, and culture.* Rotterdam: Sense Publishers.

Ethnomathematics and Philosophy

Bill Barton

Abstract Any concept of ethnomathematics must eventually meet philosophical debates about the nature of mathematics. In particular neo-realist positions are anathema to the idea that mathematics is culturally based, but even modern quasi-empiricist philosophies are challenged by the fundamental relativity implied in ethnomathematical writing.

A new way of interpreting mathematical history which may allow for a truly relativist mathematics is described, and some evidence is presented to support this view. The kind of studies which would arise from this perspective on mathematics are outlined.

1 The Problems and the Challenge

We, as ethnomathematicians, have a problem.

For more than two thousand years mathematics has been regarded as the epitome of rational truth, the study of the essential features of quantity, relationships and space. There has been argument about how we come to know these things, and about how we can be sure of them, but few working mathematicians have doubted that they were dealing with essential facts of some kind. Mathematicians seem to say "we know that the mathematics we study tells us truths about numbers and points and lines and circles, and that it can be used to build bridges which don't fall down: it works therefore it must be right. Furthermore it is beautiful, and elegant, and has a long history of great thinkers, and... and..." and so on.

However, perhaps this is the world's greatest circle of self-justification. Describing and justifying mathematics on the grounds of our feelings about it and our per-

B. Barton (✉)
Dept. of Mathematics, The University of Auckland, Auckland, New Zealand
e-mail: barton@math.auckland.ac.nz

A presentation to the First International Conference on Ethnomathematics, University of Granada, Spain, September 2–5, 1998.

This chapter is a reprint of an article published in ZDM—The International Journal on Mathematics Education (1999) 31(2). DOI 10.1007/s11858-999-0009-7.

ZDM-Classification: E20.

H. Forgasz, F. Rivera (eds.), *Towards Equity in Mathematics Education*,
Advances in Mathematics Education,
DOI 10.1007/978-3-642-27702-3_20, © Springer-Verlag Berlin Heidelberg 2012

ception of its usefulness might be a dangerous activity. But, if it is a circle of self-justification, then this circle has convinced a great number of people for many, many years. Who are we to challenge it now? And if we are to challenge it, then we need to be very sure of our ground, and very convincing in our arguments.

But we *are* challenging it—particularly as ethnomathematicians. Where did this challenge come from? On what grounds can we mount this challenge? And how can we explain the lack of previous challenges?

The problems arose when we started to talk about mathematics and culture in the same sentence. Ethnomathematicians were not the first people to do this: Oswald Spengler (1956) in the early part of this century was one of the first to seriously ask whether the mathematics which developed through various cultural eras was the same mathematics. However when these ideas began to move into the educational arena, and were used to challenge the imperialism of education in non-European countries, then much more was at stake.

And this led to a challenge to the classical view of mathematics. D'Ambrosio's writings (1987, 1990) are now well-quoted as the source of the idea that widespread apparent failure in mathematics was actually a social problem being played out through the filtering mechanism of mathematics education. Furthermore this filtering mechanism acted through the cultural nature of the subject: the social terrorism of mathematics.

D'Ambrosio challenges classical views with a social constructivist conception of "mathematics as a system of codification which allows describing, dealing, understanding and managing reality" (1987, p. 37). Hence (p. 74):

> ... we face a need for alternative epistemologies if we want to explain alternative forms of knowledge. Although derived from the same natural reality, these knowledges are structured differently.

D'Ambrosio appeals (among others) to Bachelard, Kitcher, and Lakatos as possible sources for these epistemologies. What is it that these writers offer? What, indeed, is needed as a philosophical basis for ethnomathematics? This paper explains why these sources are not enough, and describes some starting-points for an alternative relativistic philosophy.

Ethnomathematicians need to be able to discuss the possibility of the simultaneous existence of culturally different mathematics. The challenge for anyone attempting a philosophical basis for ethnomathematics is to ensure that there is an account of the way in which mathematics is structured, understood, and communicated which is consistent with sociological and anthropological descriptions of how mathematics is spread and used. In addition, any account must explain how one mathematics culture has come to be dominant, and, apparently, to be so highly developed compared with other mathematics cultures.

So mathematics must be described in a new way: an alternative philosophical position must establish a position which must be argued on conventional terms. Furthermore part of the task is to show why earlier philosophical positions were plausible and firmly held. It is not enough just to say that mathematics is culturally determined and to go on acting as if this is the case simply because we believe it to be so. We must convince others, particularly mathematicians.

How, then, might we conceive of a culturally relativistic picture of mathematics?
It may seem a waste of effort to start by explaining why traditional and non-relativist philosophies do not fulfil the relativist needs of an ethnomathematical position. However it is by examining those aspects of traditional philosophies which preclude relativism that the shape of a relativistic philosophy may emerge. Elaborated arguments explaining the inadequacy of both traditional and recent philosophies for the degree of relativity required by ethnomathematics are given in Barton (1996), or, in a brief version, in the paper prepared for the ICEM1 conference and available on a CD of the Proceedings. The present paper begins by skimming lightly over the major problems, and explaining why even twentieth century positions which seem to embrace some form of relativism fall short of what is required. In the latter half of the paper an alternative way of thinking about mathematical development is described which may lead to the kind of foundation needed for ethnomathematics.

2 Existing Philosophical Positions

The essential feature of realism which makes it unsuitable as a philosophical basis for a cultural understanding of mathematics is the requirement for a universal, *a priori* basis for truth, i.e. a pre-existing world of mathematics. It could be argued that it is not necessary for a cultural conception of mathematics that realism is rejected. It could be that mathematical objects are absolute, but that they are only knowable through human faculties which depend on various factors, including culture. Cultural views of mathematics are therefore an expression of the inadequacy of human understanding of this ideal world.

However, mathematics would then have to be conceived as a cultural approximation to the truth. One consequence is that any such cultural mathematics can be measured in terms of closeness to the ideal—which allows colonial, ethnocentric categorisations of primitive mathematics or sophisticated mathematics, etc. This, indeed, is exactly what some would argue has happened.

Opposed to a Platonist view is one in which mathematics is regarded as a product of human thought. Aristotle's conception of mathematics as the abstractions that the human mind derives from the physical world, and Kantian notions of mathematics in the organising power of the mind, are precursors to logicism, intuitionism, and formalism positions which change the philosophical orientation from "What is mathematics?" to "How can we be sure about mathematical truths?" (Tymoczko 1986, p. xiv). With respect to relativity, the point of all the efforts to establish these positions has been to secure their foundations so well that there is no room for doubt about mathematics, to eliminate the possibility of more than one (competing) conception. Thus logicism, intuitionism, and formalism call upon, respectively, a universal logic, a universal power of intuition, or a universal understanding of form.

Recent philosophical writing appears to have created room for relativistic notions, and thus, it might be assumed, has created an opportunity for establishing a cultural basis for mathematics. The neo-realists, Lakatos and the emerging quasi-empiricists, and the mathematical sociologists are all concerned more with what it is that mathematicians actually do than the status of the mathematics with which they work. Bringing the human (and therefore the social and cultural) element into the philosophy of mathematics has raised relativity as an issue: a spectre for some, a working assumption for others.

D'Ambrosio (1987, p. 30) mentions Bachelard as one possibility. Bachelard, writing in the 1930's, describes a historically relative notion of objectivity which gives rise to changing conceptions of mathematical objects and of rationality itself (Smith 1982; Tiles 1984). Bachelard's key idea is that objectivity is an ideal rather than a reality. At any time we may think we know how to discover truth or that we understand what makes a proof, but these ideas change over time: the sense of objectivity is illusory. However there is a progression towards a better, and then still better, understanding of what objectivity "really" is.

Thus there are many different historical standpoints from which to view mathematics, and each is correct *at that time*, and each explains previous views. Mathematicians are all aware of the demand for rational thought, and it is this which makes it possible to have a changing mathematics which always retains the objectivity required of the discipline.

Thus mathematics may be historically relative. However problems remain for cultural relativism because the changes are evaluated as progressive, i.e. it is directioned to one increasingly objective conception, against which previous conceptions are seen to be inadequate. Ethnomathematics, on the other hand, requires simultaneous progress in different directions under an assumption of equal validity/objectivity.

Neo-realism (e.g. Maddy 1990; Resnik 1993) similarly falls down when two different mathematical worlds meet: it is assumed that they would need to be resolved into a "best" view, although the criteria are not specified. Fallibilists and quasi-empiricists (Tymoczko 1986) at least specify the criteria in such a situation: for fallibilists two mathematical worlds are assumed to contradict each other, so one (or a new mathematics) must emerge as less contradictory; for quasi-empiricists different mathematical worlds would be measured against experience or useful applications. All of these philosophies require a resolution between different mathematical conceptions: simultaneous mathematics in cognisance of each other is impossible in the long term.

Two sociologists of mathematics are also often appealed to by writers in ethnomathematics. They are Bloor (1976) and Restivo (1993). Unfortunately neither of these elaborate a philosophy of mathematics which will support their sociology (admittedly neither is trying to do that). Bloor retreats into metaphysical belief, and Restivo assumes a position close to that of the quasi-empiricists.

Basically, the problem is that they all imply the existence, or the ideal, of some kind of mathematics (or some ideal criteria with which to judge mathematics) "out there", separate from culture. Whenever different conceptions arise, they must be resolved by appeal to this ideal towards which all development is heading. Cultural relativity may happen temporarily, but there is still the presumption that everyone doing mathematics is trying as hard as they can to approach something which is pure and culture-free.

It is as if the mathematics within each culture is a shadow of the "real" mathematics. As cultures interact, that mathematics which is more developed (closer to the "real" one) will subsume the other, and an illusion of one mathematics developing towards a universal perfection is maintained.

However the idea that cultural relativity in mathematics results from imperfections in the culture does not help to explain the problem which gave rise to ethnomathematics: that of apparent culturally-based mathematics failure. It would allow us to say that some cultures are "seeing" mathematics more truly than others, and the hegemony would continue.

3 An Alternative Model

A much more radical version of mathematical relativity is required. In this version it must make sense to talk about Maori mathematics, or English mathematics, or carpenters' mathematics. This writer has always shied away from such phrases because the use of the word "mathematics" presupposes a whole family of preconceptions which are almost impossible to ignore. D'Ambrosio is ambitious enough to force the use of "mathematics" in this wider meaning. This paper introduces the phrase "QRS system".

A QRS system is a system of meanings by which a group of people make sense of Quantity, Relationships, and Space. So the model of mathematics being proposed here is one where each cultural group has its own QRS system.

It is easiest to think about this in an historical way. Imagine two groups who have developed independently of each other. Each has its own way of dealing with quantity, of expressing relationships, and of representing space. As the two cultures begin to interact with each other, their ways of talking and ways of doing things will be mutually translated as far as is possible into each other's systems. Gradually a merging of QRS systems is liable to take place, and, ultimately, it may happen that one will dominate, or that a new system will emerge, probably one which draws more heavily from one system than another. At the end of this process it will seem that both cultures have the same system. The important aspect of this picture is that there is no presumed external "mathematics" or rationality by which one system is judged better than another. This is entirely an internal process, a human process, a cultural process.

One way to think about this picture is to conceive of a range of mountains. Let the QRS system be a range of mountains which circumscribes the landscape of our culture on its horizon. If we are in another culture, then we also see a range of mountains, and they may appear very similar—which is not surprising because they are "about" similar features of our world. It may even seem that it is the same range seen from the other side. If, however, we examine the features of the range, we will see that it is not one range, but two, one behind the other. In silhouette they look the same. If it was possible, we should rise up in the air, and then we would see the two ranges of mountains, and in between we would see a green and fertile valley. As ethnomathematicians we are explorers in that valley, tracing the way the mountains relate to each other in a cross-cultural landscape.

4 Deconstructing the Past

There are huge problems with this picture for mathematicians (and most users of mathematics, and most mathematics teachers). The problems centre around the fixation with "truth" and with "discovery" in relation to mathematics. They will grant that mathematics may be generated by people, but that, for example, once you have a circle and a triangle drawn inside it based on the diameter, then it *must be* a right-angled triangle. Or that $5 + 7$ *must be* 12 no matter what words are used to say it. The modern version is that the beautiful pictures of Chaos theory and Julia Sets were there waiting to be discovered: no-one created them. What is more, much of this mathematics which has been "discovered" is amazingly useful in our physical world—it models that world in fundamental and "true" ways. If one wishes to retain a version of absolute truth, and of a discovery metaphor for doing mathematics, then this culturally relative picture will have to go. But perhaps there is another way of thinking about things.

It is suggested that a philosophy based on Wittgenstein may provide ethnomathematics with the position it needs in order to properly describe the objects of mathematics. The idea being referred to here is the Wittgensteinian idea that we talk mathematics into existence. Shanker's (1987) reading of Wittgenstein proposes that we focus on clarifying what we mean when we talk about mathematics, rather than trying to characterise mathematical knowledge. So, rather than arguing about whether mathematical knowledge is certain or fallible, we should recognise that it is created in our talk. Thus mathematics is neither a description of the world, nor a useful science-like theory. It is a system, the statements of which are "rules" for making sense in that system. (See Barton 1996 for more detail.)

For example, consider a circle. No-one has ever seen, or touched a circle, it is an ideal object. This prompted Plato to hypothesise a world inhabited by such ideal objects, thus, circles exist. Wittgenstein suggests that this is just a "way of talking": circles exist because—and only because—we talk about them. When we do talk about them as if they are real objects then it makes sense to talk as if they had properties, but we should recognise that this is just a convenient figure of speech—literally. When we do not talk about them, they do not exist. In those languages where roundness is embodied as an action, not as an object, circles do not exist. A different QRS system applies. So mathematics is not *about* anything, it *is* a way of talking.

Consider negative numbers. For two hundred years many important mathematicians denied the existence of negative numbers. They were right. For mathematicians who would not talk about them (or write them) negative numbers *did not exist*. It was only as they began to be used, and their properties discussed, that they were talked into existence.

Similarly, during the development of calculus there was huge debate about infinitesimals. Lakatos' (1978) article on the subject makes it very clear that the protagonists were talking past each other. They were talking about different mathematical worlds—one in which infinitesimals *did* exist, and another different mathematical world, in which they *did not* exist.

Now as well as talking things *into* existence, it is possible to talk things *out of* existence. This, it seems to me, is exactly what imperialist mathematics is all about. Mathematical (let us say QRS) concepts are talked out of existence by being ignored, by being superseded, by not acknowledging that that language is mathematics. And Lo! It isn't mathematics (some examples are discussed below).

Another problem to be dealt with is what has been called "the surprising usefulness of mathematics". If mathematics is simply an arbitrary human creation, then how is it that mathematics corresponds to our world so well, and is so useful in it? Furthermore, how is it that there is one world-wide subject called mathematics? The explanation is parallel to that used to explain why many different types of animal have all evolved with an eye, when their common ancestor had nothing remotely resembling an eye. How can evolution, which thrives on divergence, be so convergent? The answer is that things evolve in response to their surroundings. It is true for all types of animals that, if they have a light-sensitive organ, then they are at an advantage. What's more, if that organ is increasingly sharp and specialised, then that animal has an even greater advantage.

So it is for mathematics. Quantification is a powerful tool in social organisation. Thus cultures which develop the notion of quantification are rewarded for doing so, and tend to develop it further, and develop it in ways which correspond to the world in which they live. As cultures increasingly inhabit the same social world, it is not surprising that the QRS systems they evolve will become like each other.

5 What Is the Evidence?

Well, these are interesting ideas, and are the beginnings of a philosophy, but it would be nice to have some evidence for them—particularly in the face of sceptics who point to a single (almost-) universal mathematical world. Fortunately there is some evidence, if only we are willing to see it.

The first piece of evidence is that embedded in the languages we speak. In everyday English number words are describing words, they act like adjectives. "Three glasses" has the same form as "red glasses" or "tall glasses". In mathematics number words become nouns—they refer to objects. "Three" is a thing in itself.

In New Zealand it has just been realised that, in traditional Maori (but not in the modern Maori spoken today), number words are action words, they act like verbs. In order to say "three glasses" you must say "the glasses are three-ing". This is also true of some American Indian languages (Denny 1986). However Denny also describes how, in Inuktutit (the Inuit language), number words are naming words, they are nouns. In order to say "the three glasses" you must say "the glass set-of-three". "Glass" is the adjective.

To me this is evidence that those speaking these other languages think about quantity in a fundamentally different way—and that way has been talked out of existence, or, at the least, it has been talked out of existence as mathematics. For example, consider the traditional questions in the philosophy of mathematics: how do we come to know about mathematical objects? But if the way we talk about number or space uses action words, how are we to make sense of a question about mathematical *objects*? Our QRS system will not be admitted into the realm of things being considered by that philosophy.

The second piece of evidence is one in which we can actually see two opposing mathematical worlds in action. That is, within mathematics there is a contemporary example of simultaneous co-existence. In statistics there are two different conceptions of probability: the Frequentist, and the Bayesian. The Frequentist conception has probability as the result of a long run of similar events, the Bayesian one has probability as a unique function of each event, about which we may have some prior information. Each conception gives rise to its own method of dealing with questions involving probability. In the main, these methods coincide in their results. But it is possible to construct problems which one or other method will give a sensible answer to, and the second method gives no result, or a contradictory one.

This demonstrates clearly that probability is not a "real" thing. It is a human construct, and how we construct it affects the results we obtain. Neither of these worlds is right or wrong. They are simply useful or not.

6 Exciting Horizons

Establishing a firm foundation for conceptions of ethnomathematics leaves us with exciting new directions to follow. It also encourages us to pursue depth in our concepts of ethnomathematics. This can be thought about in several ways.

One way is to see ethnomathematical investigations as going *below the surface*. The above language example encourages us to look, not at the different number words, but at the way they function in the language, i.e. not to see the surface feature of the language, but to examine the QRS system that is implied by those features. Thus rather than examine what base system is used to construct these number words (that is to impose a particular concept of quantity on them), we should examine the concept of quantity which they carry.

Similarly, with weaving patterns it is interesting to analyse the patterns evidenced in the art and crafts of different cultures (and this may have some educational uses although this writer regards that as an open question), but it is also important to examine the concept of symmetry employed by the weavers—a concept which may be different from that with which we are familiar.

Another way of pursuing depth is to ask "What if?".

Much of Gerdes' work (e.g. 1991) is in this vein: what if Lusona are analysed in the abstract, where will that lead me, what new mathematics might emerge. This is not to imply that Lusona drawers understood this mathematics, rather it is to carry out a QRS extension by whatever means is possible to develop new ideas.

A second example in the "what if" category is that generated by the navigation practices of indigenous Pacific navigators. Amongst their many skills was the ability to sit in their boat and, without watching the water, to sense the size and frequency of swells from eight different directions (Kyselka 1987). Mathematically the resolution of waves in one direction is a well-developed field, but 3-D wave analysis is largely an unsolved problem. What would have happened if the scientific effort and resources which went into developing latitude and longitude as a system of navigation, had been turned instead into the 3-D wave analysis problem. Perhaps we would have a shipboard system which would tell the captain when rocks, shoals, or Titanic ice-bergs were in the boat's path.

A final example is the orientation of geometric thinking. Those of us with conventional mathematical backgrounds tend to think in rectilinear grid systems—our graphs are drawn with axes as verticals and horizontals, our talk is full of "ups and downs". Many weavers of indigenous crafts, however, orient themselves to diagonal systems: weaving on the diagonal is an easier technique in many situations. What mathematical functions would interest us if our graphs were drawn using diagonal axes?

Whether "below the surface" or "what if", I am sure of one thing: ethnomathematical research will continue to surprise us, and at the least expected moments, of course. If we conceive of mathematics in a consistently relativist way, then we may be better able to "see" the hidden valleys between our mathematics'. Ethnomathematical research may then be more interesting, more productive, and more useful to both mathematicians and educators.

References

Barton, B. (1996). *Ethnomathematics: Exploring cultural diversity in mathematics*. Unpublished PhD thesis, The University of Auckland, New Zealand.

Bloor, D. (1976). *Knowledge and social imagery*. London: Routledge & Kegan Paul (page references are to the 2nd ed., Chicago: University of Chicago Press (1991)).

D'Ambrosio, U. (1987). *Etnomatematica: Raizes Socio-Culturais da Arte ou Tecnica de Explicar e Conhecer*. São Paulo: Campinas.

D'Ambrosio, U. (1990). The role of mathematics education in building a democratic and just society. *For the Learning of Mathematics*, *10*(3), 20–23.

Denny, P. (1986). Cultural ecology of mathematics: Ojibway and inuit hunters. In: M. P. Closs (Ed.), *Native American mathematics* (pp. 129–180). Austin: University of Texas Press.

Gerdes, P. (1991). Exploration of the mathematical potential of SONA: An example of stimulating cultural awareness in mathematics teacher education. In *Proceedings 8th symposium of the Southern Africa mathematical sciences association,* Instituto Superior Pedagógico Moçambique, Maputo.

Kyselka, W. (1987). *An ocean in mind*. Honolulu: University of Hawai'i Press.

Lakatos, I. (1978). Mathematics, science and epistemology. In J. Worrall & G. Currie (Eds.), *Philosophical papers*, Vol. *2*. Cambridge: Cambridge University Press.

Maddy, P. (1990). *Realism in mathematics*. Oxford: Clarendon Press.

Resnik, M. D. (1993). A naturalized epistemology for a platonist mathematical ontology. In S. Restivo, J. P. Van Bendegem, & R. Fischer (Eds.), *Math worlds: Philosophical and social studies of mathematics and mathematics education*. Albany: State University of New York Press.

Restivo, S. (1993). The social life of mathematics. In S. Restivo, J. P. Van Bendegem, & R. Fischer (Eds.), *Math worlds: Philosophical and social studies of mathematics and mathematics education*. New York: State University of New York Press.

Shanker, S. G. (1987). *Wittgenstein and the turning-point in the philosophy of mathematics*. London: Croom Helm Ltd.

Smith, R. C. (1982). *Gaston Bachelard*. Boston: Twayne Publishers.

Spengler, O. (1956). Meaning of numbers. In *The world of mathematics*, Vol. *IV* (pp. 2315–2347). London: George Allen & Unwin Ltd.

Tiles, M. (1984). *Bachelard: Science and objectivity*. Cambridge: Cambridge University Press.

Tymoczko, T. (Ed.) (1986). *New directions in the philosophy of mathematics: An anthology edited by Thomas Tymoczko*. Boston: Birkhäuser.

Preface to "Cultural Differences, Oral Mathematics and Calculators in a Teacher Training Course of the Brazilian Landless Movement"

A New Theoretical Lens on Cultural Differences, Oral Mathematics, and Calculators in a Teacher Training Course of the Brazilian Landless Movement

Gelsa Knijnik and Fernanda Wanderer

Six years after the publication of the ZDM paper that examined cultural processes involving oral mathematics and its articulations with the use of a calculator, we would like to assign new meanings to what we wrote there using other theoretical tools that now shape our theoretical framework (Knijnik and Wanderer 2008, 2010). In recent papers we showed how we moved from a postmodern perspective to another one, which embraces Foucauldian and later Wittgensteinian ideas (Knijnik 2007; Knijnik and Wanderer 2008; Wanderer 2007). Based on these theorizations we reconceptualized the *Ethnomathematics Perspective*, which is giving support to the studies developed at the Unisinos *Mathematics Education and Society Inter-Institutional Research Group*[1] (Duarte 2009; Giongo 2008; Knijnik and Bocasanta 2010; Knijnik and Wanderer 2010; Wanderer 2007). This *Ethnomathematics Perspective* consists of a toolbox that allows us to analyze the Eurocentric discourses of academic and school mathematics and their effects of truth and to examine the language games that constitute different mathematics and their family resemblances.

This preface aims to highlight what we have gained from a theoretical point of view with the incorporation of both Foucauldian and later Wittgensteinian ideas. First of all it is important to mention that Foucault's notions of *language*, *power-knowledge relations* and *regime of truth* have enabled us to examine the discourses that circulate in school mathematics and outside school contexts like the Landless

---

[1]The Unisinos *Mathematics Education and Society Inter-institutional Research Group* (in Portuguese, *Grupo Inter-institucional de Pesquisa Educação Matemática e Sociedade*) is an inter-institutional research group linked to the Graduate Program on Education at the Universidade do Vale do Rio dos Sinos (Unisinos), Brazil, that developed studies on mathematics education from a socio-political-cultural perspective under Gelsa Knijnik's leadership.

G. Knijnik (✉)
Universidade do Vale do Rio dos Sinos, São Leopoldo, Brazil
e-mail: gelsak@uol.com.br

F. Wanderer
Universidade Federal do Rio Grande do Sul, Porto Alegre, Brazil
e-mail: fwanderer@certelnet.com.br

H. Forgasz, F. Rivera (eds.), *Towards Equity in Mathematics Education*,
Advances in Mathematics Education,
DOI 10.1007/978-3-642-27702-3_21, © Springer-Verlag Berlin Heidelberg 2012

Movement culture in which we identify mathematical practices. The French philosopher's work was also relevant because it led us to question how a single rationality among other rationalities—the rules by which individuals and cultures deal with space, time and quantification processes—all that which Western civilization associates with the notion of mathematics—became a "truth", the only "truth" that could be accepted as mathematics in the school curriculum.

But it was the use of later Wittgensteinian ideas of *forms of life*, *language games*, *grammar* and *family resemblances* (that showed us how we could go deeper in our purpose of problematizing the sovereignty of Modern rationality, which seems to scorn all other rationalities associated with "other" *forms of life*. Briefly, the ideas developed by the Austrian philosopher in what is known as the later Wittgenstein's period were productive in providing us with a philosophical support that made it possible to assume the existence of more than a single mathematics—"the official one"—with its Eurocentric bias and its rules marked by abstraction and formalism.

As one of us has argued elsewhere (Knijnik 2007), Wittgenstein (2004), in *Philosophical Investigations*, criticized not only his earlier work (presented in *Tractatus*) but also "the whole tradition to which it belongs" (Glock 1996, p. 25), "the foundationist schools, and dwells at length upon knowing as a process in mathematics" (Ernest 1991, p. 31). In his remarkable book *The Philosophy of Mathematics Education*, Ernest (1991) examines various philosophical schools and their contributions in the conceptualization of (academic) mathematics. In his analysis of the conventionalist view of mathematics in which "mathematical knowledge and truth are based on linguistic convention" (p. 30) Ernest mentions that Wittgenstein's work "proposes that the logical necessity of mathematical (and logical) knowledge rests on linguistic conventions, embedded in our social linguistic practices" (p. 32).

In shaping a new philosophy of mathematics—social constructivism—Ernest refers to Wittgenstein as one of the philosophers who considers knowledge not only as a product, but gives "great weight to knowing and the development of knowledge" (p. 90). He thinks that "social constructivism employs a conventionalist justification for mathematical knowledge" (p. 64) and assumes that "the basis of mathematical knowledge is linguistic knowledge, conventions and rules, and language is a social construction" (p. 42). Ernest also argues that his philosophical perspective "assumes a unique natural language" and shows that "an alternative (i.e. different) mathematics could result" (p. 64) as a consequence of this position. Mentioning the work of Alan Bishop evidence of a different kind mathematics (p. 67), Ernest claims that "such evidence of cultural relativism strengthens rather than weakens the case in favor of social constructivism" (p. 64).

Viewing mathematics "not as a body of truths about abstract entities, but as part of human practice" (Glock 1996, p. 24), the Austrian philosopher gives us tools for thinking about rationality as forged from social practices of a *form of life*, which implies to consider it as "invention", as "construction" (Condé 2004, p. 29). Moreover, with the support of later Wittgenstein's ideas and using the expressions that he coined we can then admit the existence of distinct mathematics. The basis of this statement can be found in the argument that these different mathematics—in Wittgenstein's words, different networks of *language games*—are produced by different *forms of life*, a term conceived by Wittgenstein as "stress[ing] the intertwining of culture, world-view and language" (Glock 1996, p. 124), as "patterns in the weave of our life" (p. 129).

Analyzing later Wittgenstein's work, especially his new conception of language, Condé (1998, 2004) put forward the crucial role of the notion of *use*:

> In such work, *use* is directly connected to the concept of meaning (...) the meaning is determined by the *use* we make of the words in our ordinary language. (...) The meaning of a word is given based on the *use* we make of it in different situations and contexts. (...) the meaning is determined by the *use*. (Condé 2004, p. 47)

It is within this perspective that the notion of use, according to Wittgenstein, is considered pragmatic and not "essentialist". Meaning is determined by the use of words and such use respects rules that are themselves produced in social practices, which constitute *language games*. As pointed out by Condé (1998), "the notion of *language games* involves not only expressions, but also the activities with which these expressions are linked" (p. 91). *Language games* are produced based on sets of rules (that are rooted in social practices), each of them constituting a specific grammar. So, the grammar that marks each *language game* is itself a social institution. Moreover, authors like Spaniol argue that "the grammar constitutes the logic itself, the grammar is the logic. (...) It is impossible to analyze the logic without considering the language" (Condé 1998, p. 110).

From what was briefly explained here based on the work of the later Wittgenstein and some of his interpreters (like Condé and Glock) it follows that different *forms of life* produce and at the same time, are produced by different *language games* and each form of life is marked by a specific grammar. Such grammar, as a set of rules, constitutes a specific logic. This rationale drives us to ask about a possible relationship between different language games. Here, again, the Austrian philosopher's ideas give support to our answer through the notion of *family resemblances*. The philosopher would say (as shown in aphorisms 66 and 67 of Philosophical Investigations) that *language games* form "a complicated network of similarities overlapping and criss-crossing: sometimes overall similarities, sometimes similarities of detail" (Wittgenstein 2004, p. 320) and adds:

> I can think of no better expression to characterize these similarities than family resemblances; for the various resemblances between member of a family? Build, features, color of eyes, gait, temperament, etc. etc overlap and criss-cross in the same way—and I shall say: 'games' form a family. (Wittgenstein 2004, p. 320)

Looking at our ZDM paper published in this book through later Wittgenstein's theoretical lens led us to understand Landless peasant oral mathematics practices as language games. Such language games are constituted by rules—referred to in the paper as regularities—which shape a specific grammar. It is precisely this grammar that corresponds to the Landless peasant rationality. As easily observed in what we wrote earlier, there are family resemblances between the oral language games practiced by the Landless peasants we studied and the mathematics language games usually transmitted at school. Similarly we can signify the calculator practices analyzed in the paper as a network of language games that have their own specificities but maintain family resemblances to school mathematics language games.

In brief, the theoretical tools that we are using in our current research projects drive us to look further into our own understanding of the analysis we performed in "Cultural differences, oral mathematics and calculators in a teacher training course of the Brazilian Landless Movement".

References

Condé, M. (1998). *Wittgenstein: linguagem e mundo*. São Paulo: Annablume.
Condé, M. (2004). *As teias da Razão: Wittgenstein e a crise da racionalidade moderna*. Belo Horizonte: Argvmentvm.
Duarte, C. G. (2009). *A "realidade" nas tramas discursivas da educação matemática escolar: Tese (Doutorado em Educação)*. São Leopoldo: Universidade do Vale do Rio dos Sinos.
Ernest, P. (1991). *The philosophy of mathematics education*. London: The Falmer Press.
Giongo, I. M. (2008). *Disciplinamento e resistência dos corpos e dos saberes—Um estudo sobre a educação matemática da Escola Estadual Técnica Agrícola Guaporé: Tese (Doutorado em Educação)*. São Leopoldo: Universidade do Vale do Rio dos Sinos.
Glock, H. (1996). *A Wittgenstein dictionary*. Oxford: Blackwell Publishers.
Knijnik, G. (2007). Mathematics education and the Brazilian Landless Movement: Three different mathematics in the context of the struggle for social justice. *Philosophy of Mathematics Education Journal, 21*, 1–18.
Knijnik, G., & Bocasanta, D. (2010). Do ofício da pesquisa: pílulas neoaforísticas—About the craft of research: neoaphoristic pills. In H. Alrø, O. Ravn, & P. Valero (Eds.), *Critical mathematics education: Past, present and future*. Rotterdam: Sense.
Knijnik, G., & Wanderer, F. (2008). *Adult education and ethnomathematics: an analysis of a pedagogical experience with Brazilian Landless Movement leaders*. Paper presented at the (TSG8) 11th International Congress on Mathematics Education, Monterrey, Mexico.
Knijnik, G., & Wanderer, F. (2010). *Mathematics education, differential inclusion, and the Brazilian landless movement*. Paper presented at 6th International Conference in Mathematics Education and Society, Berlin, Germany.
Wanderer, F. (2007). *Escola e matemática escolar—Mecanismos de regulação sobre sujeitos escolares de uma localidade rural de colonização alemã do Rio Grande do Sul: Tese (Doutorado em Educação)*. São Leopoldo: Universidade do Vale do Rio dos Sinos.
Wittgenstein, L. (2004). *Philosophical investigations*. Oxford: Basil Blackwell.

Cultural Differences, Oral Mathematics, and Calculators in a Teacher Training Course of the Brazilian Landless Movement

Gelsa Knijnik, Fernanda Wanderer, and Claudio José de Oliveira

Abstract This paper discusses aspects of a two-year study of a teacher-training course for adult mathematics education organized by a Brazilian landless peoples' social movement. It takes ethnomathematics as a theoretical framework in which cultural differences are central. The paper analyzes some of the oral mathematics practices that mark the landless peoples' culture studied. In particular, it discusses a pedagogical process involving the articulation of oral mathematics practices with the use of the calculator, focusing on how pre-service teachers give meaning to their experience and on how cultural differences operated in this setting.

1 Introduction

This paper presents and discusses some elements of a two-year study of a teacher training course for adult education organized by the Brazilian Landless Movement (Movimento Sem Terra).[1] The main purpose of the paper, is to examine cultural

---

[1]Landless Movement—in Portuguese, Movimento Sem Terra (MST)—is a social peasant movement involving about 250,000 families around the country. Struggling for land reform, it puts education as a central issue in order to achieve their goals. In Knijnik (1998, 1999, 2002a, 2003, 2004) issues about the Landless Movement and its work on education, especially in the field of mathematics are discussed.

G. Knijnik (✉) · F. Wanderer · C.J. de Oliveira
Universidade do Vale do Rio dos Sinos (UNISINOS), São Leopoldo, Brazil
e-mail: gelsak@portoweb.com.br

F. Wanderer
e-mail: fwanderer@uol.com.br

C.J. de Oliveira
Universidade de Santa Cruz do Sul (UNISC), Santa Cruz do Sul, Brazil
e-mail: claudiojo@terra.com.br

A first and condensed version of this paper was presented at ICME-10—Topic Study Group 6, Copenhagen, Denmark, July 2–11, 2004.

This chapter is a reprint of an article published in ZDM—The International Journal on Mathematics Education (2005) 37(2).

H. Forgasz, F. Rivera (eds.), *Towards Equity in Mathematics Education*,
Advances in Mathematics Education,
DOI 10.1007/978-3-642-27702-3_22, © Springer-Verlag Berlin Heidelberg 2012

processes involving oral mathematics and its articulations with the use of a calculator. The paper also examines some repercussions for the peasant pre-service teachers[2] who experienced a pedagogical practice which implemented such articulations between oral mathematics and the use of a calculator. Taking as a theoretical framework the ethnomathematics field with the centrality it gives to cultural differences, the study aims to contribute to the debate produced by diverse research approaches in mathematics education as regards oral mathematics. The paper is organized in six sections. Following this introduction, Sect. 2 presents a summary of the literature on oral mathematics, situating in this landscape the theoretical approach used in this paper. Here the meanings given to notions such as ethnomathematics, culture and cultural differences are discussed. Section 3 discusses methodological procedures of the study. The convergences between the methodological perspective and the theoretical framework are highlighted. Section 4 presents and analyzes some of the oral mathematics practices that mark the landless culture studied. Cultural differences which produce and are produced by peasants' oral mathematics practices are presented. Section 5 analyzes a pedagogical process involving the articulation of oral mathematics practices with the use of the calculator. It focuses on how the students gave meaning to that experience and how cultural differences operated in that setting. The paper ends with remarks on some of the curriculum implications of the study.

2 Cultural Differences and Oral Mathematics[3]

There is a vast amount of literature concerning what has been called "oral mathematics". In such literature, it can be said that oral mathematics is understood as the mathematics practices that are produced and transmitted orally, not including written strategies. It covers research in the field of psychology with repercussions for education. These studies can be thought of as constituted by two main branches. The first investigates mental calculation skills, analyzed independently from their cultural dimensions, as, for instance, the works of Reys et al. (1999), Thompson (2000) and Irons (2001).[4] The second branch analyzes oral mathematics at the interface of cognitive psychology and anthropology. The main initial references of these studies are the works of Gay and Cole (1967), Scribner (1984) and Resnick (1983), which were followed by the studies of Lave (1985) and those of Carraher et al. (1987).

---

[2]Throughout the paper the peasant pre-service teachers who participated in the research will be called "students".

[3]This section presents some of the ideas discussed in Knijnik (2002b).

[4]There is a vast production of studies oriented towards this branch, mainly in the United States and Australia, of which only the most significant were mentioned here. Based on this research there are many studies that, from a more "instrumental" perspective, limit themselves to presenting curricular activities involving mental reckoning.

The theoretical framework developed by these studies influenced much of what has been produced in this area. Analyzing the similarities between them, one could say that they consist of comparative analyses of mental calculation skills in different cultural contexts and their relationships with written calculation procedures, usually taught at school. These studies, inspired by Piagetian clinical interviews and also involving observation of the subjects in 'out of school' settings, mainly emphasize numerical strategies. Moreover, as suggested by Evans:

> [the] subject is asked to solve several sets of problems constructed by the researchers. Typically, the first set of problems are familiar from the work context, but are 'beyond' the familiar tasks; later sets of problems require varying 'levels of transfer'; see below. Transfer is measured in terms of correct performance in solving the different types of problems. (Evans 2000, p. 68)

As will be shown later, this methodological approach differs from the one presented in this paper. What is at stake is the understanding that numerical aspects of a social practice are inseparable from the cultural setting itself. Therefore, it does not make sense to apply tasks that lead to comparisons between 'school mathematics' and 'out of school' practices, since the two contexts are different. As Evans comments about the studies mentioned above:

> When they compared the children's *performances* in street contexts with those in school-like testing contexts, the performances on what appeared to be 'the same task' were superior (in terms of correctness) in the street contexts (Carraher et al. 1987). However, talking about 'the same task in different contexts', and seeking to compare cognition and performance across contexts as different as street markets and testing in school settings, would be seen as highly questionable by researchers accepting the 'situatedness' of cognition This is because different contexts could be expected to differ on a number of aspects, such as the setting, the social relations at play, etc. That is, 'like is *not* being compared with like'. This criticism clearly applies to the production of what I call. . . 'the most celebrated finding of the situated cognition programme' by a number of research teams, and not only by Lave. (Evans 2000, p. 65)

In this paper, a different theoretical and methodological perspective is assumed. It emphasizes that the practices of oral mathematics are studied as part of the cultural practices in which they gain their meaning, i.e., oral mathematics are seen as an artifact constituted by and constituting culture. It is also relevant to stress that culture is not understood as something consolidated, a finished, homogeneous product. On the contrary, culture is seen as a human production, which is not fixed, determined or closed in its meanings once and for all. This way of conceptualizing culture implies it is a conflictive, unstable and tense terrain, undermined by a permanent struggle to impose meanings through power relations.

The study focused on the discussion of how cultural processes involving oral mathematics are produced and their curricular implications for rural adult education. In assuming the centrality of culture to understand oral mathematics, the investigation has as its theoretical framework ethnomathematics. Since it is a broad and heterogeneous field, it is important to clarify the ethnomathematics approach adapted for this study.

First of all, it is worthwhile mentioning that the ethnomathematics approach is aligned with a postmodern perspective, "which rejects a totalized thinking, the illuminist metanarratives, the universal referentials, the transcendencies and essences, that, imploding modern Reason, leave it in the shards of regional rationalities, of particular reasons" (Veiga-Neto 1998, p. 145). It is these "shards of regional rationalities, of particular reasons", such as the peasant mathematics oral practices, that are of concern to ethnomathematics. They constitute its main object of study.

In such a postmodern landscape, ethnomathematics studies the Eurocentric discourses which constitute academic mathematics and school mathematics; it analyzes the effects of truth produced by the discourses of academic mathematics and school mathematics; it discusses issues of difference in mathematics education, considering the centrality of culture and the power relations that institute it; it problematizes the dichotomy between "high" culture and "low" culture in mathematics education (Knijnik 2004).

It also implies considering what has been stressed by authors like Bhabha, who argues:

> If cultural diversity is a category of comparative ethics, aesthetics, or ethnology, cultural difference is a process of signification through which statements *of* culture or *on* culture differentiate, discriminate, and authorize the production of fields of force, reference, applicability, and capacity. (Bhabha 1994, p. 85)

Thus, it is more accurate to use the concept of cultural difference instead of cultural diversity in analyzing oral mathematics practices.[5] These practices will not be considered as a body of "traditional" knowledges that do not re-update their meanings over time, an inert set that is transmitted from generation to generation, as though it were cultural "baggage". The idea that best describes our position is that of "post-tradition":

> The concept "post-tradition" is also meant to inscribe in the master narratives of Tradition and Modernity the contradictions of everyday performances and subaltern life-stories and worldviews. A point will be made to the effect that post-tradition does not summon origin, purity, homogeneity or continuity. Rather, it is a process that is already in the making. Similarly, the task of the post-traditional researcher is not to conceptualize hegemonizing views of history but to document the fractured histories and life-stories of groups and communities as they surface in people's everyday practices. (Graioud 2001, p. 13)

It is from this perspective of "post-tradition" that oral mathematics are examined in this paper, whilst seeking to avoid what Rosaldo (1988) called "imperialist nostalgia". According to Rosaldo (1988), this notion concerns "a particular kind of nostalgia, often found under imperialism, where people mourn the passing of what they themselves have transformed" (p. 69). He states that an imperialist nostalgia "in its attenuated form", refers to a positioning in which a way of life is deliberately changed, and then one regrets that things did not remain as they were before the intervention that produced this change (Rosaldo 1988, p. 70). In the context of this study, avoiding an alignment with this "attenuated position of imperialist nostalgia" means to not regret the disappearance of oral mathematics as a cultural practice that is valued in educational processes, a disappearance in which school is directly involved.

---

[5]In fact, the expression "cultural difference" is redundant, since all differences are culturally constituted. All the same, in this paper it is used, following authors like Bhabha (1994).

Oral mathematics practices, specifically those connected to the processes involving addition, subtraction, multiplication and division, take on a more significant role, precisely because they are even now still part of the knowledge of a significant number of youths and adults. These oral knowledges, when absent in the curriculum of youth and adult education, produce social effects such as failure at school. These effects are further intensified when they are related to youths and adults in the rural areas. As we have observed in our fieldwork with the Landless Movement, oral mathematics practices are present in the ordinary life of this social movement, where mathematics is necessary to meet the challenges of production and the commercialization of what is produced. The low levels of schooling which did not allow them to be aware of written algorithms require a constant use of oral mathematics, which constitute an important cultural artifact. In the sphere of the projects on adult education, however, there is a sort of "forgetfulness" about this world outside school. In curricular terms, it is useful to investigate the meanings produced by this "forgetfulness", by the dichotomization and antagonism of these two logics. It is also useful to examine the implications for the curriculum that can be deduced from empirical studies such as those performed previously (Knijnik 1999, 2002b). The examination of these implications may lead to a localized and partial achievement of a "curricular justice", which Connell defines as curriculum organization that takes as one of its principles consideration of "the interests of those who are at a disadvantage" (Connell 1995, p. 12).

3 The Empirical Part of the Research

The research domain consisted of a group of peasants who were taking part in a regular teacher training course for adult education, belonging to the Landless Movement in southern Brazil. This course is officially recognized and is one of those implemented by this social movement in order to improve the struggle for agrarian reform. The course was organized in six stages. Each of them consisted of "School Time", when the students lived in school, and of "Community Time", in which the students performed specific educational tasks in their own communities.

Data were obtained through the development of the following activities:

1) Fieldwork at two MST settlements, involving
   a) the observation of oral mathematics practices among adults with little or no schooling;
   b) interviews with peasants from those communities (audiotaped and later transcribed); and
   c) making a video with one of the peasants who is an expert on oral mathematics.
2) 100 hours of Mathematics Education classes with the teacher training course students, including:
   a) analysis of the strategies associated with students' oral mathematics practices;
   b) discussion of the videotape mentioned in item 1c);

c) advising the students for their first research experience to be developed in Community time, in which interviews with adults on their oral mathematics practices and a written report on the experience were planned.
3) Advice and participatory observation of a 7-day literacy and numeracy workshop developed by the students in a settlement, in which they gave classes to landless youths and adults.

This paper focuses on the results obtained in items 1 and 2.

The empirical part of the research was supported by direct observation and taped-recorded interviews with the subjects involved in the research. The activities were followed by fieldnotes, which allowed the research team to take notes related to what was observed, and about their own ideas and feelings during the fieldwork.

In choosing such a methodological approach, inspired by ethnography, it was considered important to take into account questions arising in anthropology, strongly marked by its ties to the colonial period and the "description of the 'Other' ". Like many contemporary scholars (e.g. Lather 1992, 2003; Tyler 1992), Lidchi argues that in our decolonized era, anthropologists:

> Have had to question how a discipline which has a growing awareness of its own complicity with colonial forces, whose primary research method—fieldwork—was dependent on colonial support, can ring the changes in the wake of decolonization, globalization and cultural revivalism among indigenous people. (Lidchi 1997, p. 200)

It is precisely in this new social and political reconfiguration of the world, called "Empire" by Hardt and Negri (2000), that new perspectives for anthropology and its related fields are raised. They lead us to understand articles like this one as a discourse about "others", as no more than a representation process, in the sense given by Hall, for whom "representation is the production of meaning through language" (Hall 1997, p. 16). We therefore assume that this paper is our own narrative about what we, with our inevitably partial gaze, observed and heard when performing the fieldwork. Moreover, in narrating oral mathematics practices of landless people, we were aware that we were not "discovering" what "was there". The act of the "writing of one culture by another" implies, in fact, "constructing one culture for another. What is being produced therefore is not a reflection of the 'truth' of other cultures but a representation of them" (Lidchi 1997, p. 200). Questions related to representation—what is represented, who represents it, and how they do it—were therefore taken into account within the research (Knijnik 2004). This position tries to avoid the arrogance of those who claim to speak in the name of others and consider their word as the most valuable.

Taking into account these methodological issues, we found that it would be relevant to analyze the students' voices in the mathematics education classes (mentioned in item 2) considering their experiences in that pedagogical process. Such experiences were understood in the sense giving by Larrosa (2002): "experience could be what happens to us. Not what happens, but what happens *to* us" (p. 137). The pedagogical process developed with the students could be thought of as *knowledge of experience* which, for Larrosa, is, first of all finite knowledge connected to the maturity of a particular subject. Secondly, it is a particular, relative, personal knowledge.

Thus, nobody can learn from someone else's experience, unless this is somehow revived. Thirdly, it is knowledge that is inseparable from the individual in which it is embodied. It is not outside us, as knowledge considered to be scientific. Finally, the *knowledge of experience* has something to do with the *good life*, understood as the unit of meaning of a full human life, which transcends the futility of moral life. Modern science is suspicious of experience and converts it into an element of method, giving rise to the idea of experiment, i.e., a stage of the safe, foreseeable route that leads to science. Knowledge, from this viewpoint, becomes a progressive accumulation of objective truths that will remain external to the subjects (Larrosa 2002). In this study, the meaning given to the act of 'having an experience' goes in another direction: "To have an experience means, therefore, to let us be approached within ourselves by what interpellates us, penetrating us and subjecting us to it" (Larrosa 2002, p. 139).

The empirical part of the research attempted to give voice to the students in order that the experiences they had in the mathematics education classes could be taken into account. Even considering the difficulties involved in the process of hearing the others and writing about them, in making the choices of how to represent the students in this paper we tried to exercise our intellectual humility.

4 The Peasant's Oral Mathematics Practices

A large number of oral mathematics practices were seen in peasant everyday life, mainly connected to labor activities and to the purchase and sale of products for daily consumption. Written mathematics was not very current in the landless peasant culture studied. Pencil and paper seem to be "foreign" artifacts to their culture. To manage them requires a great effort, as one settler said, "It is easier to deal with a hoe than with a pencil".

Three aspects of oral mathematics practices produced by the landless peasants were especially relevant for this paper. The first concerns the close ties between oral calculation strategies and the contingencies in which they are situated. Thus, for instance, a peasant explained that, on estimating the total value of what he would spend to purchase inputs for production, he rounded figures "upwards", ignoring the cents, since he did not want "to be shamed and be short of money when time comes to pay". However, if the situation involved the sale of some product, the strategy used was precisely the opposite. In this case, the rounding was done "downwards", because "I did not want to fool myself and think that I would have more [money] than I really had."

What was observed is that, different from the school mathematics that emphasizes the uses of written processes and the "forgetfulness" of the context, discussed by Walkerdine (1988), the oral mathematics of the peasant culture is strongly contextualized and involves complex reasoning. This finding is convergent with contemporary studies that show the need to challenge the beliefs that throughout history, groups who had no written practices had a less abstract style of thinking.

As authors such as Denny (1998) have argued, what makes the difference between these two diverse cultures is not abstraction, but the decontextualization process, which is the main style fostered by written culture, as opposed to the contextuality of oral culture.

A second aspect refers to the strategy of adding, based on a decomposition of the values to be orally calculated. This is what happened with one of the students in the workshop given by the students, when faced with a situation in which he had to calculate $148 + 239$. He explained that, "first one separates everything [$100 + 40 + 8$ and $200 + 30 + 9$] and then adds up first the numbers that are worth more [$100 + 200$, $40 + 30$, $8 + 9$]. (. . .) This is what really counts". This strategy was found among almost all adults who said that they "were good" at mental calculation. Differently from the addition algorithm taught at school, in oral procedures the peasants considered above all the values of each parcel that was involved and how much difference it would make if it were hundreds, tens or units, i.e., they prioritized the values that contributed more significantly to the final result.

This priority also emerged when the numbers involved in the calculation are decimals. It is observed that recurrently, the peasants use decomposition "to make up integers". This strategy was employed by *Dona Nair*, an already retired settler, who, as a child, attended school for only one year, and did not learn to read or write. On explaining the way she uses mental calculation in her daily activities, she referred to a situation in which two products are purchased, one of them costing R$2.70 and the other R$2.90. She said that to find the amount to be spent, she first of all adds up the integers and then the cents, as follows: "$2 + 2$ makes 4. I complete the 90 [cents] with 10[cents] of the 70[cents] to make another 1 *real*. So $4 + 1$ completes 5 *reais*, plus the 60[cents], and I have 5 and 60." As those before mentioned, also in situations involving decimals, what is prioritized in the calculation process are integer values that, according to the peasants, are "more relevant" to the final sum, a relevance which is marked by their culture.

A third aspect found in this study concerns the duplication strategy present in the oral multiplications, a process similar to that used in ancient Egypt, as indicated by Gillings (1982) and Peet (1970). *Seu* Nerci, an illiterate landless man, whose interview was filmed and later used as pedagogical material in the mathematics education classes with the students, when multiplying $92 \times$ R$0.32 (corresponding to 92 liters of milk produced and sold at 32 cents of real[6] [R$0.32] a liter), first doubled the value of R$0.32, and obtained R$0.64; then he repeated the "doubling" operation twice, finding the amount of R$2.56 (corresponding to 8 liters). He added to this the value of 2 liters calculated previously, and thus found the value of 10 liters of milk: R$3.20. The next procedure was to successively double the values found, i.e., he obtained the result of 20, 40 and 80 liters. Keeping "in his head" all the values reckoned throughout the process, *Seu* Nerci ended the operation adding to the value of the 80 liters, those corresponding to 10 liters and 2 liters (calculated previously), and thus found the result of $92 \times$ R$0.32.

---

[6]The *real* is the Brazilian currency. Its plural is *reais*.

Seu Nerci never went to school. When he was a child, the closest school to his home was 20 miles away and there was no public transportation in the rural zone where his family lived. Since early childhood, boys and girls were introduced into agricultural labor and no children went to school. He did not use pencil and paper to write down the sums as he multiplied them. When the video was made he suddenly withdrew to another room at the back of his house to perform the multiplication, only reappearing after he had come to the final result. Here other regularities found in this study should be presented. The first concerns the need, explicitly mentioned by the adults, "to concentrate to think". Like *Seu* Nerci, most of the adults observed at mental calculation activities became deeply involved in the act of reckoning, in an attitude of isolation and introspection. But, unlike *Seu* Nerci, many of the literate adults observed usually took notes during their mental calculations. The notes were used as "markers" throughout the process, especially in those involving greater complexity. Pencil and paper were used only to take notes. No written algorithms were observed in this study, even among those peasants who had had access to them during their schooling. One of the hypotheses for this absence was that the peasants were aware of our interest in their oral practices, which led them to avoid the written algorithms in the calculations they performed in front of us. Here the nature of the relationship between researcher and researched may be important.

As discussed in another study (Knijnik 2002c), such a relationship is double-faced. On the one hand, during fieldwork, on doing the interviews, it is the researcher who asks, it is he/she who, basically, "is in charge" of the investigative process and later, on writing, it is he/she who will use as he/she considers appropriate, what the interviewees have said. Such a process mobilizes the unequal relationship that is established between researcher and researched, an unequal relationship from which there is no getting away. But such an unequal relationship is not fixed. As can be indicated by the lack of pencil and paper in the peasants' interviews shown in this study, an interchanging power play is exercised: it is not always the researcher who is in a privileged position. When in the field, it is the researched who have the knowledge that one seeks to learn, it is they who know what is unknown by the researcher, it is they who, based on their conveniences and on meanings they assign to the research, select what will be made available to the researcher. This is the other side of the researcher-researched relationship, which, in a way, makes the power slide between the subjects involved in the investigative process. There is no way to escape from this interchanging power play, from those power relations that are permanently exercised in a research study in which ethnographic procedures are used. What remains for the researcher is to include analysis of this issue in their texts, as we have sought to do in this article (Knijnik 2002c, p. 5).

In observing the adult oral practices and hearing the comments of the researched, the importance of analyzing oral mathematics from a cultural perspective has been highlighted. It has been shown that these oral mathematics practices are produced by that peasant culture and that they are its producers. As shown in this section, each strategy used by the interviewees was strongly connected with the situation of which it was part. Moreover, we found that there was not only a diversity of oral mathematics strategies but that the meanings given by the subjects to their oral practices are marked by cultural difference.

5 Oral Mathematics and the Calculator in the Teacher Training Course

The results obtained in the first stage of the research study, which examined the ways landless adults deal with mathematics in everyday life, showed the centrality of oral mathematics for that peasant culture. This conclusion made it possible to challenge the students of the teacher training course to organize processes in adult education that would exclude written algorithms. On excluding them, another pedagogical approach would be experienced, focusing on the articulation of those two artifacts that are part of the peasant culture: oral mathematics and a simple electronic calculator. From the start, it was said to the teachers that the teaching experience component of the course had as its main goal the problematization of this pedagogical approach, with no intention of giving a "final" answer about "what must be done in adult mathematics education". Even considering this important remark, initially there was resistance against the pedagogical experience proposed. Questions were raised by the students: Wouldn't this mean that the opportunity to "develop the capacity of reasoning" was being suppressed, precisely for those who had had less opportunity to study? Wouldn't the students become "addicted" to the calculator? It was necessary to organize and later analyze pedagogical processes focusing on the articulation between oral mathematics and the calculator for a further discussion of these (and other) questions.

The research team judged that it would be important to begin with a pedagogical process involving the students themselves so that they could have an "experience" of such an articulation, understanding it in the sense given by Larrosa (2002). The students' experiences with oral mathematics and its articulation with the calculator were analyzed taking as a central theme the cultural differences of that peasant culture.

The starting point of the experience was an activity in which each of the students interviewed one colleague about their own trajectory as a mathematics learner and their experiences in dealing with oral mathematics and the calculator in everyday life. The data obtained through these interviews were discussed in the classroom. They provided evidence that both artifacts are rooted in the students' everyday life, but not all of them practiced the mental calculations easily. Those who had previous experience of teaching adults expressed their need to learn more about oral mathematics since their students usually bring these cultural practices "naturally" to the class and, in order to follow the Landless Pedagogy[7] in which they are involved, it would be "dissonant" to ignore these oral practices in their teaching. Their worries were also extended to the calculator itself. They argued that this cultural artifact can be found in almost all settlements where they worked but most of the members of those communities (and also they themselves) ignore some of its potentials, such as the use of the memory keys. Apparently, for those students who did not yet have teaching experience in adult education, these arguments did not produce much effect, but they, at least formally, accepted to go further in the learning of oral

---

[7]About Landless Pedagogy, see Knijnik (1999).

mathematics and of calculator uses. This attitude was not an isolated one. During the two-year project, it was observed that the narratives which were most valued in the classroom were those enunciated by the students who had "the" experience, here understood not in the sense given by Larrosa, mentioned above. "The" experience by itself, i.e. "any" teaching practice, validated the narratives and functioned as a "guarantee" for the arguments, even if what was called "the" experience was not marked by what "happens to them". As part of our analysis, therefore, during the mathematics education classes we discussed with the group what was changed in them by their experience of the pedagogical process.

It was precisely such a discussion which drove the next stage of the work with the students: the analysis of their own oral mathematics practices and others obtained in the fieldwork done previously by the research team. To take them as a curriculum subject in the teacher training course could give the students the opportunity to experience the complexity of the oral reasoning strategies and their potential in bringing to adult mathematics education major mathematics notions usually taught in a de-contextualized way.[8]

Here it is worth mentioning an episode[9] that can show the meaning given by the research team to these processes. We were analyzing the dairy production of a peasant community, when it became necessary to calculate 17% of R$240.00. Different oral strategies arose among the group of students to solve the question. The most usual strategy was first to calculate 10% of 240 (performed immediately through dividing 240 by 10), obtaining R$24.00. The remaining 7% were decomposed into $5\% + 2\%$. In order to calculate 5% of R$240.00, they divided into half the R$24,00 found previously (taking 5% as half of 10%). As to the 2%, they decomposed it into $1\% + 1\%$, each of them obtained by dividing 240 by 100.

A second strategy was used in the group: 240 was decomposed into $100 + 100 + 40$, and the student who had done this explained that he reckoned 17% of each one of the parcels. In the first two, he immediately found R$17.00. As to the R$40.00, he repeated the decomposition operation $(40 = 10 + 10 + 10 + 10)$, stating that R$1.70 corresponded to each of these sub-parcels.

A third oral strategy was presented by a member of the group. He explained that, instead of working with R$240.00, it was simpler to think initially about R$250.00, since this could be decomposed into $100 + 100 + 50$. Immediately he stated that, through this decomposition, 17% of R$250.00 corresponded to $R\$17.00 + R\$17 + R\$8.50 = R\$42.50$ (explaining that the latter value is half of one of the previous parcels). At this stage of the process, there were many questions from his classmates, about how he would "discount what he had added". For the student, however, this "was simple". Since he had added R$10.00 to the total amount, now he had to take 17% off R$10.00, which would be the equivalent to R$1.70. Thus, he found the final result of R$40.80.

---

[8]Other activities were implemented during those 100 hours of mathematics education classes.

[9]Even though this episode occurred with another group of landless students, the ideas it brings can show interesting elements for this discussion.

It was only after discussing all these strategies in class that the calculator was used pedagogically in order to check the result. In other situations in which the amounts involved were more complex (with decimals and/or very large amounts), the oral strategies were used to obtain approximations to the precise value. Once these approximations had been obtained, the calculator was used, supported by the previous estimations.

Episodes such as this one have contributed to the problematization of oral mathematics and its articulation with the calculator, as long as the latter is not used in a merely mechanical way. The discussions also point to the constraints of work that does not include preparing and implementing pedagogical activities through which the students may become familiar with the multiple possibilities of dealing with this technology, such as the use of memory keys.

The work with the students also involved the discussion of theoretical issues linked to cultural differences in mathematics education that could qualify the problematization of oral mathematics and its articulations with the calculator. Concepts such as culture and difference, as conceived by the research team's ethnomathematics approach, were emphasized.

Even considering the influence of Freire's thinking in what is called popular education in Latin America (Knijnik et al. 2004), a cultural perspective which also marked the Landless Pedagogy, it was a complex task to go further into the theoretical issues that support the ethnomathematics approach presented in this paper. In the debate, what seemed to be "natural" for other fields of knowledge (such as history, geography, Portuguese) was hard to understand in the case of mathematics. It took a long discussion to deconstruct the idea that what is called mathematics is a particular mathematics ('academic mathematics') which, with its formalism, produces a neutral narrative that aims to be universal—and to see other ways of giving meaning to the world also as mathematics. The research team was aware of the risks of simplifying this rationale, which can lead to the misconception that academic mathematics and popular mathematics are equal from the sociological point of view (Knijnik 2003). This naïve position was carefully avoided in order not to trivialize the ethnomathematics approach. We considered that academic mathematics and its curricular recontextualization, i.e. school mathematics, are socially legitimized. Everyone has the right to acquire such legitimized knowledge. We are not discussing what kind of knowledge (academic or popular) is superior from an epistemological point of view. Rather, what is highlighted here is a sociological perspective.

6 Final Words

The two-year research project constituted an experience (in the sense of Larrosa 2002) for those who participated in it. More than seeing the knowledge built as an "objective" one, we considered that its production produced us as subjects, touching our subjectivities and transforming them. In this process, we identified curricular implications for adult mathematics education in rural areas.

The main implication concerns the challenges involved in incorporating the sophisticated mental calculation strategies, which are part of the peasant culture, to the pedagogical processes developed in rural educational projects. Such an incorporation seeks to problematize the politics of dominant knowledge through what Behdad (1993, p. 43) calls "wild" practices that, to him, are generally "in opposition to the system, contesting and anti-disciplinarian". According to Behdad, "the problematics and the politics of post-Colonial conditions require an anti-disciplinarian way of knowledge that will undermine the social, political and economic reasons that underlie the principle of compartmentalization" (Behdad 1993, p. 43). The ethnomathematics perspective which supports this study is aligned with Behdad's anti-disciplinary approach but, associated with the need to undermine compartmentalization, it also highlights another dimension of these "wild" practices. This is to problematize the school curriculum invisibility of the cultures of non-hegemonic groups, which include their own ways of dealing mathematically with the world, such as the culturally mediated handling of oral mathematics. Such problematization takes into account the centrality of looking at cultural differences in mathematics education.

A second curricular implication refers to the challenges involved in articulating oral mathematics with another peasant cultural artifact, the calculator. In the south of Brazil, where the research was done, the use of this "new" technology has spread in the Landless Movement camps and settlements, but usually its study is not part of their school practices. Therefore, the possibility of the articulation of oral mathematics and the calculator can also be seen as a "wild" practice, insofar as the ways in which the groups deal both with their orality and with the technological and cultural artifact were at the center of the pedagogical process, even if it was not limited only to them. Again, as discussed in Knijnik et al. (2004), cultural differences are a key issue in the mathematics curriculum, not just a "starting point" of the pedagogical work.

The research project produced repercussions in the Landless Movement educational perspective (Knijnik 2003). Similar work was done in other Landless Movement teacher training courses and the results of the research were disseminated in regional and national meetings with in-service landless teachers. Such dissemination, as far as could be controlled by the research team, tried to highlight that what we had in mind was to problematize the oral mathematics and the calculator in adult education. We did not have the intention of giving "prescriptions" for mathematics classes. Nevertheless, sometimes it seemed that we failed to achieve this goal. Maybe the fact of working with a social movement which has, as a peasant said, "historical urgency to give answers also to education" led us to assume a more "positive" and emphatic position about the theme researched. When writing this paper, we realized that possibly, we had forgotten to keep explaining that we considered the study results local, provisional and partial and that our research approach is marked by uncertainty.

References

Behdad, A. (1993). Travelling to teach: Postcolonial critics in the American academy. In C. McArthy & W. Cricholow (Eds.), *Race, identity and representation in education* (pp. 40–49). New York: Routledge.

Bhabha, H. (1994). *The location of culture*. London, New York: Routledge.

Carraher, T. N., Carraher, D. W., & Schliemann, A. D. (1987). Written and oral mathematics. *Journal for Research in Mathematics Education, 18*, 83–97.

Connell, R. W. (1995). Justiça, conhecimento e currículo na educação contemporânea. In L. H. Silva & J. C. Azevedo (Eds.), *Reestruturação Curricular—Teoria e prática no cotidiano da escola* (pp. 11–35). Petrópolis: Vozes.

Denny, J. P. (1998). El pensamiento racional en la cultura oral y la descontextualización escrita. In D. R. Olson & N. Torrence (Eds.), *Cultura escrita e oralidad* (pp. 95–126). Barcelona, Espanha: Gedisa.

Evans, J. (2000). *Adults' mathematical thinking and emotions: A study of numerate practices*. London: Routledge.

Gay, J., & Cole, M. (1967). *The new mathematics and an old culture*. New York: Holt, Rinehart & Winston.

Gillings, R. J. (1982). *Mathematics in the time of the pharaohs*. New York: Dover Publications.

Graioud, S. (2001). *Decolonizing theory: Post-tradition as an everyday practice*. Paper presented at the International Conference PostColonialismS/Political correctnesS, Morocco (Africa), April.

Hall, S. (1997) *Representation: Cultural representation and signifying practice*. London: Sage.

Hardt, M., & Negri, A. (2000). *Empire*. Cambridge: Harvard University Press.

Irons, C. (2001). Mental computation: How can we do more? *Teaching Mathematics, 26*(1), 22–26.

Knijnik, G. (1998). Ethnomathematics and political struggles. *ZDM, Zentralblatt für Didaktik der Mathematik, 30*(6), 119–134.

Knijnik, G. (1999). Ethnomathematics and the Brazilian landless people education. *ZDM, Zentralblatt für Didaktik der Mathematik, 31*(3), 188–194.

Knijnik, G. (2002a). Ethnomathematics, culture and politics of knowledge in mathematics education. *For the Learning of Mathematics, 22*(1), 11–15.

Knijnik, G. (2002b). Curriculum, culture and ethnomathematics: The practices of 'cubagem of wood' in the Brazilian landless movement. *Journal of Intercultural Studies, 23*(2), 149–166.

Knijnik, G. (2002c). A perspectiva teórico-metodológica da pesquisa etnomatemática: Apontamentos sobre o tema. In *VI Encontro Brasileiro de estudantes de Pós-Graduação em Educação Matemática. A Pesquisa em Educação Matemática: Múltiplos Olhares sobre sua produção*. (Vol. 2, pp. 3–6). Campinas, São Paulo: Graf. FE.

Knijnik, G. (2003). *Matemática: A etnomatemática na luta pela terra*. Caderno de Educação. Educação de Jovens e Adultos 11 (pp. 72–83). São Paulo: Setor de Educação do MST.

Knijnik, G. (2004). Lessons from research with a social movement. A voice from the South. In P. Valero & R. Zevenbergen (Eds.), *Researching the socio-political dimensions of mathematics education: Issues of power in theory and methodology* (pp. 125–142). Dordrecht: Kluwer Academic Publishers.

Knijnik, G., Wanderer, F., & Oliveira, C. J. (2004). *Etnomatemática, currículo e formação de professores*. Santa Cruz do Sul, Brasil: Edunisc.

Larrosa, J. (2002). Literatura, experiência e formação. In M. V. Costa (Ed.), *Caminhos inevstigativos: novos olhares na pesquisa em educação* (pp. 133–160). Rio de Janeiro: DP&A (entrevista concedida a Alfredo Veiga-Neto).

Lather, P. (1992). Critical frames in educational research feminist and poststructural perspectives. *Theory into Practice, 31*(2), 87–99.

Lather, P. (2003). Applied Derrida: (mis) reading the work of mourning in educational research. *Journal of Philosophy in Education, 35*(3), 257–270.

Lave, J. (1985). Introduction: Situationally specific practice. *Anthropology and Education Quarterly, 16*(3), 171–176.

Lidchi, H. (1997). The poetics and the politics of exhibiting other cultures. In S. Hall (Ed.), *Representation: Cultural representation and signifying practice* (pp. 151–208). London: Sage.

Peet, T. E. (1970). *The Rhind mathematical papyrus: Introduction, transcription, translation and commentary*. London: Hodder & Stoughton (British Museum 10057 and 10058).

Resnick, L. B. (1983). A developmental theory of number understanding. In H. P. Ginsburg (Ed.), *The development of mathematical thinking* (pp. 109–151). New York: Academic Press.

Reys, R. E., Reys, B. J., McIntosh, A., Emanuelsson, G., Johansson, B., & Yang, D. C. (1999). Assessing number sense of students in Australia, Sweden, Taiwan, and the United States. *School Science and Mathematics, 99*(2), 61–70.

Rosaldo, R. (1988). *Cultura & truth: The remaking of social analysis.* Boston: Beacon Press Books.

Scribner, S. (1984). Studying working intelligence. In B. Rogoff & J. Lave (Eds.), *Everyday cognition: Its development in social context* (pp. 9–40). Cambridge: Harvard University Press.

Thompson, I. (2000). *British research on mental calculation methods for addition and subtraction.* Preliminary reading for the BERA Symposium, University of Exeter, February.

Tyler, S. A. (1992). La etnografía posmoderna: de documento de lo oculto a documento oculto. In C. Geertz & J. Clifford et al. (Eds.), *El surgimento de la antropologia posmoderna* (pp. 297–333). Barcelona, Espanha: Gedisa.

Veiga-Neto, A. (1998). Ciência e pós-modernidade. *Episteme, 3*(5), 43–156.

Walkerdine, V. (1988). *The mastery of reason.* London: Routledge.

Preface to "Immigrant Parents' Perspectives on Their Children's Mathematics Education"

Marta Civil, Núria Planas, and Beatriz Quintos

In the article originally published in 2005 we addressed immigrant parents' perspectives of the teaching of mathematics in two geographic contexts, Barcelona (Spain) and Tucson (USA). The article illuminates two specific aspects of marginalization of those who move to a new country with hopes of a better life, including education, for their children. Since then we have expanded upon our work in these two specific aspects—parents' perceptions about the teaching and learning of mathematics and the role of language in mathematics instruction. We argue that these two issues are central to the marginalization of immigrant parents' and therefore, their children in the area of mathematics education. Therefore it is critical for educators to understand and address these issues to improve the mathematics education of immigrant students.

As part of an ICME survey on the teaching and learning of mathematics with immigrant students, Civil (in press) found several similarities in immigrant parents' perceptions of their children's mathematics education in various countries. A common thread across different research studies with parents from Mexico, Morocco, Pakistan, former Soviet Union, and Turkey was their concern about some aspects of their children's mathematics education. For instance, some parents say that in the receiving country there is a lack of emphasis on the "basics" (e.g., multiplication facts). Responding to what is important in mathematics, a father from a country from the former Soviet Union (and now living in Germany) says:

M. Civil (✉)
School of Education – CB 3500, University of North Carolina at Chapel Hill, Chapel Hill, NC 27599, USA
e-mail: civil@email.unc.edu

N. Planas
Universitat Autònoma de Barcelona, Cerdanyola, Spain
e-mail: nuria.planas@uab.cat

B. Quintos
University of Maryland, College Park, USA
e-mail: bquintos@umd.edu

H. Forgasz, F. Rivera (eds.), *Towards Equity in Mathematics Education*,   261
Advances in Mathematics Education,
DOI 10.1007/978-3-642-27702-3_23, © Springer-Verlag Berlin Heidelberg 2012

Well, the basics. In primary school it's the times' tables. Later, it's formulae and things like that. Maths means being able to think and calculate logically. And here they only talk about the exercises. And once they've finished talking, it's the pocket calculator that does the calculating, not the pupils. They don't do any calculating. And if you don't have a foundation, then there's no basis for you to build a theory upon. (Hawighorst 2005, p. 1173).

Parents also shared the perception that the level of mathematics instruction is less advanced or that schools are less strict (e.g., in terms of discipline; dress code). As a Pakistani parent living in England said:

But I think that the maths in Pakistan is of a higher standard... I think so, because when I go to Pakistan and see my sisters' and brothers' children, their maths is better than the maths of our high school children here. (de Abreu and Cline 2005, p. 706).

And a Mexican mother of a middle school student in Tucson said:

I see with my girl that what they have them do at the school here and the school in Mexico.... I think they should demand a little more of all children. Besides that they are at the age they are at. They are at an age where they don't care, "it doesn't matter," it doesn't matter to them and that's why I think they should demand more of them.... And that the school, besides being demanding, should also be a little tougher on them, because it seems they are giving them a lot of freedom, they let them loose. And because of the age they are at, as I said, they start going to things that are easy. [Focus Group, February 2008]

The prevalence of these themes across different contexts underscores the need for schools to establish meaningful communication with immigrant and other minoritized parents that is specific to the mathematics education of their children. Project MAPPS (described later in this chapter) was a step towards developing this communication. Civil (2009) argues for the need for teacher education programs to address this disconnect. Also, on this topic of communication between parents and schools we point to the research by Schnee and Rodriguez (2009) with Latino immigrant parents in New York City in which parents in their study noted their dissatisfaction with the standard modes of communication (via notes sent home with their children). They viewed these as impersonal and would have preferred face-to-face interactions with teachers.

In our recent work as part of CEMELA (Center for the Mathematics Education of Latinos/as),[1] we have continued to investigate the two topics in the original article (parents' perceptions and language and mathematics) as well as the topics of valorization of knowledge, parents' and children's interactions around mathematics, and parents as adult learners (Civil et al. 2008; Civil and Planas 2010; Díez-Palomar et al. 2011; Menéndez et al. 2009).

In CEMELA we use three kinds of activities to open spaces of communication with parents in mathematics: 1) Math For Parents series in which we engage with the parents in explorations of a topic in mathematics (e.g., geometry) for six to eight sessions (sometimes with their children participating too); 2) "Tertulias Matemáticas" in which we hold focus groups to engage in dialogue about the teaching and learning

---

[1]CEMELA (Center for the Mathematics Education of Latinos/as) is funded by the National Science Foundation—ESI 0424983. The views expressed here are those of the authors and do not necessarily reflect the views of the funding agency.

of mathematics. We use the expression "Tertulias Matemáticas" (mathematical get-togethers), inspired by the concept of literary circles where people gather to discuss literature. In these "Tertulias" we discuss issues related to mathematics education, often from a critical perspective (Civil and Planas 2010; Quintos et al. 2005); and 3) Classroom visits and debriefing in which we take a small group of parents to visit a mathematics classroom and then we discuss it (Civil and Quintos 2009).

In the 2005 article, we used the concept of "Before and Now" to capture the situation of immigrant parents who are often moving back and forth between their own schooling and their children's schooling. As we argue there, while this is true of any group of parents (as a generational element), for immigrant parents it takes an added dimension since the before took place in their country of origin and the now is in their current country of residence, which is marked by their current status as "minorities". This notion of "Before and Now" continues to mark our research with immigrant parents. Their comparisons across systems are closely related to the concept of valorization of knowledge which we allude to in the piece that follows but we have explored it further since then (Civil and Planas 2010; Civil and Quintos 2009). Parents consistently corroborate the three themes highlighted earlier (there is a lack of emphasis on "basics"; level is lower; schools should be more strict). If we take a closer look at their emphasis on "basics" we can see that valorization of knowledge plays a role in that the parents in our study value memorization of the multiplication facts mostly based on their own schooling but also on a comparison with what they think that children in Mexico know. Parents make multiple references to nephews and nieces in Mexico and how they already know their times tables by a certain grade, while their children in the U.S.A do not know them. In the article that follows we refer to the representation of the algorithm for division in the U.S.A and in Mexico as a source of conflict (see Civil and Planas 2010, for more on this and for another example of conflict around subtraction in a father-daughter interaction). Once again, valorization of knowledge is at play. We argue for the need for teachers to become familiar and encourage these different approaches to use them as resources towards their teaching rather than becoming potential sources of conflict between children and parents (and parents and teachers) (Civil 2009; Civil and Planas 2010), though we are aware that it is difficult for teachers to know all the differences. The classroom visits and the Math for Parents sessions that we refer to in the article that follows continue to be important arenas in which to address some of these differences. For example, in the debriefing of a classroom visit, the topic of not knowing the multiplication facts came up once again. Yet one of the mothers said,

> With my son, who is in second grade, maybe it's a technique for learning how to multiply. For us, it was by singing, but we didn't know how to reason, that's the difference. Maybe now they are using a method to make them think, without having to be just singing them, you know, and them not knowing what they are saying. [Marina—March, 2008, debriefing a classroom visit]

This excerpt points to an awareness for different approaches and in particular for one feature of "reform-based" mathematics instruction that puts emphasis on reasoning over memorization. While this mother seems to be appreciative of this emphasis, our research shows that this is not always the case, even when parents experience these approaches themselves (e.g., in the Math for Parents sessions) (Civil 2006). Thus, in our current research we are trying to get a better understanding of perceptions such as "more emphasis on basics", or "the level was higher in our country of origin" by probing further into parents' comments, as a way to tease out the complexity of these apparently "absolute" views (such as, "in Mexico the education is much better").

Another key topic in the 2005 article is the role of language in the teaching and learning of mathematics and in situations where the students' first language is different from the language of instruction. In the article we discuss the case of Barcelona with the "special classes" that immigrant students who are not proficient in the official language of instruction (Catalan) attend. The parents in that study seemed to be in agreement with this arrangement since they viewed it as a way to help their children become familiar with not only the language but also the approaches to teaching and learning in this new setting. In the case of Tucson, the issue with language was more related to parents not being able to help their children with homework. More recently we have looked into the effects of a restrictive language policy in Arizona (approved in the year 2000) on parents-children interactions (Civil and Planas 2010). Civil (in press) gives an overview of the topic of language and the teaching and learning of mathematics for immigrant students and includes perspectives from different parts of the world. We agree with Setati (2005) on the political role of language use and we argue for the need to further examine these different language policies (many of which tend to segregate students who do not speak the official language of instruction). In Acosta-Iriqui et al. (2011), we look at two states (Arizona and New Mexico) with different language policies (in Arizona, bilingual education is extremely restricted, while in New Mexico it is endorsed in their state constitution). This allows us to contrast the effect of such different language policies on parents' participation in their children's mathematics education.

Teachers and parents often focus on the language as being the main obstacle in immigrant children's learning of mathematics (Civil in press). This is an important topic to pursue because we wonder if the issue of language may take precedence over children's learning of mathematics. In particular, we are concerned about placement issues: are students placed in the appropriate mathematics classroom (based on their knowledge and understanding) or are schools basing their placement on their level of proficiency in the language of instruction? (Civil 2008; Civil and Menéndez 2011; Planas and Setati 2009). In Civil and Menéndez we look at the most recent language policy implementation in Arizona that reminds us of the case of Barcelona, since English Language Learners (ELLs) are now being segregated for four hours every day to focus on their learning of English. Although cautious in their comments, the three mothers in that report express an awareness of this segregation. Two of the mothers comment on how their children do not feel comfortable in this setting and are trying their best to move out of "Section A" (which refers to the part of the building where the classrooms for the ELLs are located); the third mother expresses more her concern in terms of whether they are really improving on their English given that they spend much of their time speaking Spanish to their friends.

In the article that follows we concluded highlighting the importance of the notion of recognition, in particular, in terms of acknowledging and viewing the differences as opportunities for learning. This is still the guiding principle in our work. As we argue in our current research, the recognition of parents' knowledge, experiences, and beliefs has implications for the teaching of children:

> Our findings point to the need for respect and positive valorizations of different groups' knowledge. This has implications for schools. Immigrant children interpret their values and meanings according to the mediation of their families' values and meanings. In this process of mediation, both parents and children interact with positive and negative valorizations, suggested by themselves and by other members of the educational context—teachers, non-immigrant students, school administrators, other immigrant students, etc.... Children who experience respect for their home culture at school, and for their school culture at home, may see their learning as a smooth transition process among cultures instead of a place for conflict. (Civil and Planas 2010, p. 147)

References

Acosta-Iriqui, J., Civil, M., Díez-Palomar, J., Marshall, M., & Quintos-Alonso, B. (2011). Conversations around mathematics education with Latino parents in two Borderland communities: The influence of two contrasting language policies. In K. Téllez, J. Moschkovich, & M. Civil (Eds.), *Latinos and mathematics education: Research on learning and teaching in classrooms and communities* (pp. 125–147). Charlotte: Information Age Publishing.

Civil, M. (2006). Working towards equity in mathematics education: A focus on learners, teachers, and parents. In S. Alatorre, J. L. Cortina, M. Sáiz, & A. Méndez (Eds.), *Proceedings of the twenty-eighth annual meeting of the North American chapter of the international group for the psychology of mathematics education* (Vol. *1*, pp. 30–50). Mérida, Mexico: Universidad Pedagógica Nacional.

Civil, M. (2008). Language and mathematics: Immigrant parents' participation in school. In O. Figueras, J. L. Cortina, S. Alatorre, T. Rojano, & A. Sepúlveda (Eds.), *Proceedings of the joint meeting of PME 32 and PME-NA XXX* (Vol. 2, pp. 329–336). México: Cinvestav-UMSNH.

Civil, M. (2009). Inmigración y diversidad: Implicaciones para la formación de profesores de matemáticas (Immigration and diversity: Implications for the preparation of mathematics teachers). In M. J. González, M. T. González, & J. Murillo (Eds.), *Investigación en educación matemática XIII* (pp. 63–87): Santander, Spain: SEIEM.

Civil, M. (in press). Mathematics teaching and learning of immigrant students: An overview of the research field across multiple settings. In B. Greer & O. Skovsmose (Eds.), *Critique and politics of mathematics education*. New York: Routledge.

Civil, M., Díez-Palomar, J., Menéndez-Gómez, J. M., & Acosta-Iriqui, J. (2008). Parents' interactions with their children when doing mathematics. *Adults Learning Mathematics: An International Journal, 3*(2a), 41–58.

Civil, M., & Menéndez, J. M. (2011). Impressions of Mexican immigrant families on their early experiences with school mathematics in Arizona. In R. Kitchen & M. Civil (Eds.) *Transnational and borderland studies in mathematics education* (pp. 47–68). New York: Routledge.

Civil, M., & Planas, N. (2010). Latino/a immigrant parents' voices in mathematics education. In E. L. Grigorenko & R. Takanishi (Eds.), *Immigration, diversity, and education* (pp. 130–150). New York: Routledge.

Civil, M., & Quintos, B. (2009). Latina mothers' perceptions about the teaching and learning of mathematics: Implications for parental participation. In B. Greer, S. Mukhopadhyay, S. Nelson-Barber, & A. Powell (Eds.), *Culturally responsive mathematics education* (pp. 321–343). New York: Routledge.

de Abreu, G., & Cline, T. (2005). Parents' representations of their children's mathematics learning in multiethnic primary schools. *British Educational Research Journal, 31*, 697–722.

Díez-Palomar, J., Menéndez, J. M., & Civil, M. (2011). Learning mathematics with adult learners: Drawing from parents' perspectives. *Revista Latinoamericana de Investigación en Matemática Educativa (RELIME), 14*(1), 71–94.

Hawighorst, B. (2005). How parents view mathematics and the learning of mathematics: An intercultural comparative study. In M. Bosch (Ed.), *Proceedings of the fourth congress of the European society for research in mathematics education* (pp. 1165–1175). Sant Feliu de Guíxols, Spain: FUNDEMI IQS, Universitat Ramon Llull.

Menéndez, J. M., Civil, M., & Mariño, V. ((2009, April). *Latino parents as teachers of mathematics: Examples of interactions outside the classroom.* Paper presented at the annual meeting of the American Educational Research Association (AERA), San Diego, CA.

Planas, N., & Setati, M. (2009). Bilingual students using their languages in the learning of mathematics. *Mathematics Education Research Journal, 21*(3), 36–59.

Quintos, B., Bratton, J., & Civil, M., (2005). Engaging with parents on a critical dialogue about mathematics education. In M. Bosch (Ed.), *Proceedings of the fourth congress of the European society for research in mathematics education* (pp. 1182–1192). Sant Feliu de Guíxols, Spain: FUNDEMI IQS, Universitat Ramon Llull.

Schnee, E., & Rodriguez, M. (2009, April). *Amarrados de pies y manos: Non-English speaking Latino immigrant parents' school engagement.* Paper presented at the annual meeting of the American Educational Research Association (AERA), San Diego, CA.

Setati, M. (2005). Learning and teaching mathematics in a primary multilingual classroom. *Journal for Research in Mathematics Education, 36*(5), 447–466.

Immigrant Parents' Perspectives
on Their Children's Mathematics Education

Marta Civil, Núria Planas, and Beatriz Quintos

Abstract This paper draws on two research studies with similar theoretical backgrounds, in two different settings, Barcelona (Spain) and Tucson (USA). From a sociocultural perspective, the analysis of mathematics education in multilingual and multiethnic classrooms requires us to consider contexts, such as the family context, that have an influence on these classrooms and its participants. We focus on immigrant parents' perspectives on their children's mathematics education and we primarily discuss two topics: (1) their experiences with the teaching of mathematics, and (2) the role of language (native language and second language). The two topics are explored with reference to the immigrant students' or their parents' former educational systems (the "before") and their current educational systems (the "now"). Parents and schools understand educational systems, classroom cultures and students' attainment differently, as influenced by their sociocultural histories and contexts.

1 Introduction

The under-achievement of many low-income immigrant students in school mathematics has become a globalized phenomenon in the modern world. Recently, studies about identity issues are helping to better understand what is happening to

M. Civil (✉) · B. Quintos
Department of Mathematics, The University of Arizona, 617 N. Santa Rita, Tucson, AZ 85721, USA
e-mail: civil@math.arizona.edu

B. Quintos
e-mail: bquintos@email.arizona.edu

N. Planas
Departament de Didàctica de la Matemàtica, Universitat Autònoma de Barcelona, D-140, E-G5, Bellaterra, 08193, Barcelona, Spain
e-mail: nuria.planas@uab.es

This chapter is a reprint of an article published in ZDM—The International Journal on Mathematics Education (2005) 37(2).

H. Forgasz, F. Rivera (eds.), *Towards Equity in Mathematics Education*,
Advances in Mathematics Education,
DOI 10.1007/978-3-642-27702-3_24, © Springer-Verlag Berlin Heidelberg 2012

low-income immigrant students when learning mathematics in mainstream schools
(e.g., Abreu and Cline 2003; Planas and Gorgorió 2004). These studies have shown
that cultural/ethnic identity is an essential construct to consider when interpreting relationships as well as differential treatments in the mathematics classroom.
Immigrant children (and here we include children who were born in the "new"
country but whose parents are immigrants) are often caught between at least two
cultures. As Suárez-Orozco and Suárez-Orozco (2001) write in their study on
immigrant children in the U.S., "immigrants are by definition in the margins of
two cultures. Paradoxically, they can never truly belong either 'here' nor 'there' "
(p. 92). These authors write about the identity issues that immigrant children confront in feeling caught between their parents' culture and the culture in their new
country:

> Children of immigrants become acutely aware of nuances of behaviors that although 'nor-
> mal' at home, will set them apart as 'strange' and 'foreign' in public... Immigrant parents
> walk a tightrope; they encourage their children to develop the competencies necessary to
> function in the new culture, all the while maintaining the traditions and (in many cases)
> language of home (Suárez-Orozco and Suárez-Orozco 2001, pp. 88–89).

This attempt to navigate two cultures extends to all aspects of their life, including
schooling. Our research has shown that immigrant parents often go back and forth
between their own experiences learning mathematics in their home country and what
their children are experiencing in their new country (Civil et al. 2003).

In this paper, we argue that in order to gain a better understanding of the situation
surrounding immigrant students' performance in mathematics, we need to know
more about these students' social contexts and, in particular, about their parents'
perceptions of their children's mathematics education. Although the students' perspectives on their achievement are not necessarily a manifestation of their parents'
perspectives or of any other pre-existing perspective, their learning opportunities are
likely to be framed by what their parents think and expect. As Marisol,[1] a mother in
one of the research projects discussed in this paper reminds us, parents and children
come together:

> This is the first problem, the teachers want to do a better job with the kids that come from
> México or from people in Spanish speaking [places], but they don't start thinking that it is
> not just the kids, it is the parents and they go together[2].

Marisol is a parent leader in a research project[3] that we have had in Tucson, USA,
for five years. Like many other parents in this project, she advocates for the need for
teachers (and schools in general) to pay attention to parents' voices and opinions.

---

[1] All names are pseudonyms.

[2] Most of the quotes in this paper have been translated from Spanish; we have edited the language
(taken out pauses and other characteristics of spoken language) since our main interest is in the
content of what the parents have to say.

[3] Project MAPPS (Math and Parent Partnerships in the Southwest) is funded by the National Science Foundation (NSF) under grant—ESI-99-01275. The views expressed here are those of the
authors and do not necessarily reflect the views of NSF.

In this paper, we draw on data from this project as well as from a research study[4] in Barcelona, Spain to illustrate parents' perceptions about their children's mathematics education. The main focus of the study in Barcelona was on the immigrant students' knowledge of mathematics. In order to gain a deeper understanding of these students' experiences, their parents were invited to participate in a mathematics project, along with their children. The research in Barcelona took place within this mathematics project.

2 Theoretical Framework

Our orientation to research is grounded on a sociocultural approach to education. In this approach, the classroom is both a constructed reality by its participants and a reality in many ways given by external groups of people and facts (Khisty and Chval 2002). Explanations for inequity within the classroom must be focused on a complex set of social relationships, rather than on particular relationships in isolation, and are to be interpreted as a product of larger social problems. From this point of view, the learning conditions and opportunities of immigrant students within the mathematics classroom are seen as being symptomatic of a social structure. In particular, when analyzing the mathematics classroom, the social structure may be reflected in many different ways by looking at what the different participants are doing there, what their interests and intentions are, which strategies help to maintain particular positions and how these strategies contribute to maintain or reduce other positions.

Zevenbergen's (2000, 2003) use of Bourdieu's notion of habitus in her research on mathematics education is particularly relevant to our work. She has interpreted habitus as a structure of practices and beliefs that shapes the expectations on working class students' mathematical attainment. She has documented that the preferences, attitudes, values, behaviors and demands made by the dominant groups do not always fit with those of the marginalized groups. From her perspective, the way in which action and practice are structured according to social and cultural differences highly influences the social and individual construction of identities when learning mathematics.

In our work, the notion of habitus leads us to connect the micro and macro levels of our analysis, moving back and forth between these levels. On one hand, our micro level of social analysis in contexts of mathematical practice consists of the study of relationships among individuals and among individuals and practices. On the other hand, our macro level of social analysis in this type of contexts looks at relationships among different cultural and social groups somehow involved in the mathematical practices. The Bourdieuian notion of habitus permits us to interpret what happens at the micro level by introducing considerations from the macro level. Taking into account the notion of habitus, the difficulties that an individual may encounter when learning mathematics have to do with his/her membership to specific social and

---

[4]This research project was funded by the Ministerio de Ciencia y Tecnología (SEJ2004-02462).

cultural groups, the value given to their mathematical practices by other groups, and the relationship between his/her social and cultural groups and other groups.

Social and cultural differences and how these may affect the construction of identities when learning mathematics are sometimes constructed as obstacles or as resources. Within our socio-cultural lens with which we situate the people, places and contexts in which we work, we move away from descriptions of obstacles and deficiencies toward a description of resources and competencies. Hence, in this orientation we view parents as intellectual resources (Civil and Andrade 2003) in their children's education. We explicitly reject the deficit model (García and Guerra 2004) that is often attached to the education of immigrant or ethnic and language minority working class students. A deficit view often positions the homes and communities at the root of students' academic failure, without taking into account the institutional biases inherent in schools that have contributed to the mismatch between home and school. Instead, we base our work in a growing knowledge of the particular families. We consider it is crucial to learn about the daily-life contexts, resources, values and beliefs that shape their learning experiences at the schools.

As Abrams and Gibbs (2002) mention, creating strong ties between parents and schools is a challenging task since, "these relationships mirror the contexts and inequitable power arrangements of the larger society" (p. 385). Reay's (1998) interviews with immigrant women underscored the difficulties that many of them encountered as they tried to build on their knowledge for their children's benefit. Their experiences with schooling were so different from what their children were experiencing in their new country that their knowledge was of little use in their current situation. As Lareau and Horvat (1999) highlight, "parents' cultural and social resources become forms of capital when they facilitate parents' compliance with dominant standards in school interactions" (p. 42). It is this idea that their knowledge, their experiences, and their forms of language/discourse is not necessarily recognized or valued by schools that puts working-class parents and immigrant parents in a position of disadvantage next to other mainstream groups.

3 Context and Method

The context for our research in Tucson is a school district that is largely Hispanic (82%) (mostly of Mexican origin) and with 81% of children qualifying for free or reduced cost school lunch. Students in this district do not fare well overall in terms of academic achievement (as determined, for example, by results on standardized tests). The research in Tucson took part within a large parental involvement project in mathematics education, MAPPS (Math and Parent Partnerships in the Southwest). A main goal of MAPPS is to develop leadership teams (parents and teachers/administrators) that will help in the mathematics education outreach effort throughout the community. Parents in the leadership teams took part in a series of activities including taking Math for Parents courses and facilitating mathematics workshops for other parents in the community (for a more detailed description of the project see Civil et al. 2003). The data for this paper come from three different sources: Math for Parents sessions; individual interviews with five immigrant mothers (audio- and video-taped); and classroom visits. These visits were developed to include a focused conversation with some of the parents on specific issues related to the teaching and learning of mathematics. We invited parents in the leadership teams to join us in a visit to a mathematics class at one of the project participating schools. After each visit, we held a debriefing conversation (audio- and video-taped). To start the conversation, we usually asked general questions such as: what were your impressions about the class? What was important for you in the class? Was the class what you expected or not, and why? What did you think of the math content? (see Anhalt et al. 2002; Civil and Quintos 2002; Civil et al. 2003, for more details on the classroom visits with parents and the method we use in the debriefing.) The data analysis for the overall research component in MAPPS follows Glaser and Strauss's (1967) constant comparative method. The different pieces of data (e.g., field notes and transcriptions of the tapes) are looked at and codified. This process leads to the development of themes, such as the one we discuss in this paper, "before and now". We follow a phenomenological methodology (Van Manen 1990) that relies heavily on participants' contributions to the experience. The lived experience of each parent is considered significant and thus we try to capture it in our analysis and writing.

The context for our research in Barcelona is a mathematics course taught by the second author to 15 and 16 year-old students in a school that has a large number of immigrant students. The main goal of the larger study was to uncover possible reasons for the difficulties with school mathematics that the thirteen immigrant students in that class (there were a total of 24 students) were experiencing (Planas 2004). Although other teachers had interpreted these difficulties as inherent in the students, in our research we did not find evidence of this. As the study evolved, data pointing to the influence of the parents' perspectives on the children's behavior (e.g., "My parents tell me not to bother you, Miss") emerged. To gain a better understanding of these possible influences, the second author carried out individual interviews (audiotaped and transcribed) with twelve parents (five Moroccan, four Pakistani, two Bangla Deshi, and one Dominican). Before these interviews took place and in order to establish rapport with the parents, the teacher invited the parents together with their children to collaborate in a mathematics project. They worked on the project for two months, one meeting per week. The idea was to organize a one day open-air mathematical fair. At the beginning of the project, most parents did not talk and seemed to feel intimidated by their children's teacher. By the latter stages of the project, they spoke up and made decisions. It would have been very difficult to obtain information from the parental interviews without having created such an atmosphere of mutual confidence. The interviews were based on the discussion of an initial key question: "There are students who get good grades in school mathematics and others who get bad grades. Why do you think your son/daughter gets bad grades [in school mathematics]?" The twelve immigrant students, whose parents were interviewed, were all considered by the school staff to be at risk of school failure and were getting very low grades, particularly in mathematics.

As with the questions used to start the conversation after the classroom observation in the Tucson case, this initial, general question in the Barcelona case served as a prompt to access parents' perceptions about teaching and learning mathematics and about their children's mathematics education experience. We adopted an eth-nomethodological perspective (Garfinkel 1984) to explore all data. In our analysis, language is not a neutral and descriptive tool but a tool that needs to be interpreted within an interactive context where some parents influence other parents within the community context and the researcher-interviewer influences all of them and is also influenced by them.

In this paper, we draw on data from these conversations and the interviews to illustrate our findings, as parents in both cases try to make sense of their children's current experience with school mathematics while seeing such experience through the lens of their own experience in a different country.

4 Results: "Before and Now"

Our goal is not to do a comparison between the two contexts, Barcelona and Tucson. Actually, several key differences would prevent us from doing this. Tucson, as is the case with many other places in the U.S. has a long history of immigration. Tucson was part of Mexico until the Gadsden Purchase in 1854 (Sheridan 1995). Thus, the "Mexican presence" in Tucson is certainly not a recent phenomenon. This is not the case with Barcelona, which has seen an increase in immigration in the last few years. In certain neighborhoods of Barcelona, the immigrant population from Africa, Asia and South America represents more than half of the population and many schools, like the school where the second author used to work, have become highly multiethnic and multilingual.

Furthermore, the parents in the Tucson research were part of a project (MAPPS) that was aimed at developing their leadership in mathematics education and were in most cases already associated with the schools (e.g., as teacher aids, as volunteers, as active members of the Parent Teacher Organizations). Hence, we do not claim that they are representative of all parents in that school district. The researchers (and authors of this paper) were not their children's teachers, nor were they associated with the schools their children attended. We developed a close relationship with the group of parents whose voices will be heard in this paper. Because we were not seen as part of the school system and we developed close ties with several of these parents, we think that they may have felt comfortable sharing information with us that maybe they would not have shared with, for example, their children's teachers.

On the other hand, in the Barcelona case, the researcher was their children's teacher. This may explain the parents' apparent acceptance and even agreement with the school's approach towards the education of immigrant students (e.g., these stu-dents are taught separately from the "local" students for part of the day; see Civil and Planas 2004, for a description of the system). Despite the teacher's effort at es-tablishing rapport (e.g., through the mathematics project described earlier), the fact remains that she was in a position of power. Thus, the information gathered from the interviews has to be seen through this lens.

A final difference is that in the Tucson case we had a wide variety among the parents in terms of how long they had lived in the U.S. (in fact some were born there, but in this paper we focus on the voices of immigrant parents). In the Barcelona case, the parents were all recent immigrants.

What was particularly striking to us, as we met to discuss the data from the two contexts, is that despite the differences and the possible limitations in either case, there were some common themes. It is this commonality that we want to stress in this paper. We argue that gaining a better understanding of the issues that these two very different groups of parents seem to be facing with respect to the mathematics education of their children, can help researchers and educators develop strategies to improve the current educational arena for immigrant children.

As we mentioned earlier, the research in Tucson has recurrently shown that immigrant parents tend to bring in their experiences with education in their country of origin and compare those with their children's schooling in their new country. This was also the case with the parents in Barcelona. Not surprisingly, the key aspect in common has to do with their immigrant condition and thus their moving across two different frames of reference in terms of educational systems—that of their country of origin (the "before") and that of their new county (the "now"). We are aware that there is also a generation difference and that indeed, even among non-immigrant groups, parents and children are likely to have experienced different educational systems. Nevertheless, in the case of immigrants, these differences across systems appear to be particularly value-laden (as we will show). This may create more anxiety, particularly for those immigrants who tend to occupy the lower positions in the social hierarchy in their new country (as is the case of the parents in this paper) and even more so for those who may have actually "moved down" when they left their country of origin (e.g., we have several examples of parents who were teachers and engineers in their country and are now working as custodians and in factories in the new country).

The "Before and Now" distinction must be understood as a result derived from analysis of our data. In the two research projects, parents faced an education system and, in particular, approaches to the teaching and learning of mathematics that were different from what they expected. Parents' reactions to these differences varied from accepting them and trying to adapt to the new system to experiencing some form of conflict. The "Before and Now" distinction permits us to explore processes of reconstruction of immigrant children's habitus. Although some parents maintain high expectations of their children's school achievement, they feel limited in their influence on the host school system because of their relationships with other mainstream groups that do not seem to value their knowledge. These relationships coming from the macro level have an effect on the parents' perception of their children's influence on the school micro level. Changes in the parents' perception of their children's influence can be seen as changes in the structure that shapes their children's opportunities, that is, as changes in the construction of their habitus.

In this paper, the data that we present is focused on two themes within the "Before and Now" frame of reference. The first theme we discuss has to do with parents comparing how mathematics was taught to them and how it is being taught to their children now in their country of adoption (in particular, we address parents' reflections on their children's achievement and parents' reactions to different approaches to arithmetic). The second theme addresses issues of language, as immigrant families are faced with a language different from their native language(s) (in particular, we look at parents' perceptions on the role that language plays in the mathematics class and in the support they can provide at home).

4.1 About the Teaching of Mathematics

Immigrant parents are active participants in their children's learning. They admit talking with their children about the mathematics classroom and they say they often give advice to them concerning proper behaviors (e.g., "Our children know that we care about mathematics and they listen to our advice", "We talk a lot about school and sometimes about school mathematics"). Parents in both settings, Barcelona and Tucson, often commented on the differences in the teaching of mathematics in their country of origin versus their new country. The parents in the Barcelona case seemed to be more accepting of those differences and viewed their children's difficulties with the subject as part of their process of adaptation. When describing differences in achievement, these differences were treated as "natural" consequences of the process of accommodation of their groups to the new reality which also involves "the new mathematics" (e.g., "This is all new to us, new school, new country, new people. Our children did not have such problems with mathematics before"; "It is nobody's fault! They must get used to the new mathematics"). The immigrant children's low grades in school mathematics are seen as something inevitable and explained in terms of group characteristics and long transitional processes (e.g., "You cannot avoid it. It is very hard for all immigrant students", "Our children need time, maybe years till they learn how things work for other people here"). For example, a Pakistani father said:

> It is natural that Pakistani children don't get good marks in mathematics, even those that used to get good marks in Pakistan. They learn quickly the language but their former schools were very different. Pakistani children are not familiar with how the classroom works here.

When asked how Pakistani children can become more familiar with how the mathematics classroom works in Barcelona, he says [to the teacher/second author]:

> When Pakistani children get familiar with how you multiply here, there are still many other things that are very different. Sajid has learned different mathematics, but it's not about what Sajid has learned, but about what he must learn from now on. He must listen very carefully to you and must not insist on what his former teachers taught him, not in the classroom. You cannot waste your time talking and talking.

There seems to be an acceptance of the situation among the parents in the Barcelona case. Although they are aware that their children may have different ways to do mathematics (e.g., use different algorithms for arithmetic operations), they seem willing to have them put those aside and learn the "new" methods. This is different from what we have found with some of the mothers in the Tucson case. One of these mothers, Marisol, mentioned how she thought the education in Mexico was much better (in that children in Mexican schools learned arithmetic operations earlier than those in U.S. schools). She explained that the teacher was teaching her son a certain way to divide, which Marisol thought was a mess and an inefficient approach (i.e., slow because there is more writing involved). She had opted for teaching her son her method for doing division, which is how she learned to divide in Mexico (by doing some of the operations in her head instead of writing it all out). When she shared this with the group of mothers in MAPPS, a lively discussion on how to handle these different approaches took place. One of the mothers, Marissa, viewed the differences as an opportunity for reciprocal learning, children learning from parents and parents learning from their children. Marissa is very confident and knowledgeable in mathematics and had been part of MAPPS for three years when this discussion took place. She understood and believed in the values of different methods in mathematics. This allowed her to easily adapt and take advantage of the situation as a learning opportunity. She said:

> My son did learn a different method, (referring to an alternative algorithm for multiplication) and he taught me how to do it. I had a choice, whether to push my way or to go with what my kid understands. If your child brings something home that you don't know then you should not say "I don't know" right away, don't be so negative, "I'm going to try and we'll learn together" so they don't get discouraged and so they'll come to you later. I waited to see if my kid understood what the teacher taught him, I wouldn't say one way was better than the other but say that there are different ways to do all sorts of things. I'll try and find a way to do it, stay positive and not negative.

Marissa offers an approach to handling differences that not only values the different methods but may actually also help expand the repertoire of approaches that her son will have to do mathematics. Marissa may be an exception, though. Most of the other cases do show a certain level of tension and a preference for one or the other method. In one of our debriefing sessions after having observed a class, two mothers, both recent immigrants from Mexico, seemed to have opposite views on how to handle the differences they viewed in the educational systems. Lucinda, who was a teacher in Mexico (and is now a custodian at a local school), experienced conflict between how she thought her daughter should be learning and how she was being taught in school. The other mother, Gabriela, who is a homemaker, seemed happy to accept the system in the U.S., because she explained that this is where they live now, and this is where her children will live:

> Lucinda: Well, what I say is, for example my daughter tells me "come to learn how they teach here, come see that I am right," when we are upset at each other here around the table, and sometimes she is the one who makes me upset, because I want to explain things to her as I know them, and I tell her "mija, the way I explain it to you, I know it's much better for you," but she sticks to her...
> Gabriela: but for one thing, here we are in the U.S. and here is where they are going to grow up, they are going to study here, and I wanted to do the same thing as you, but then I say, but why, if they are teaching him things from here, and he is going to stay here, and so, one wants to teach them more so that they know more, but what they are teaching them is because they are going to stay here, and they are going to follow what they teach them here.
> Lucinda: When I came from there [Mexico], [my daughter] was in 3rd grade; when we came here, she said that the school looked like play, "why, mijita?" "Because they are making me do 4 + 3, mom, I don't want to go to this school. It's weird." And I would tell her, "but you are going to learn the way from here", well, at that time, that's what I thought, but then I visited my relatives [in Mexico]...

The conversation on the differences between the two systems (U.S. and Mexico) inevitably touches upon the fact that when they compare what their children are learning here with what their relatives' and friends' children who live in Mexico are learning, the latter seem to be learning things earlier than their children. In this paper, it is not our goal to elaborate on the nature of these differences or whether one system teaches topics earlier than the other. We are bringing it up here mostly because this is where our debriefing conversation took us, even though that was not our intention (that is to ask them to compare Mexico vs. U.S.). Yet, this is a topic that comes up in every discussion that we have with immigrant parents. Lucinda felt that in the U.S. they were not teaching her daughter with as much depth as she would have liked ("how they explain it here it's easier and over there [in Mexico] they go in depth for everything, and here no, here they only tell you how and how and that's it"); on the other hand, her daughter rejected her mother's way and was telling her to come to school to see how they do it in the U.S. Gabriela reminds us more of the parents in the Barcelona case. She seemed to have opted for "giving in" to the school culture in which their children are going to grow up in (from the point of view of the teaching of mathematics; we are not addressing here language or larger cultural aspects). What we do not know, however, is whether the parents in the Barcelona case may express a different view if they were to participate in a project such as MAPPS, where we pushed towards the establishment of a critical dialogue. Although the parents in the Barcelona case acknowledge the differences between mathematics content and teaching in their home country and in Catalonia (the north-eastern autonomous region in Spain whose capital is Barcelona), they seem to focus more on language and class norms. Their belief that their children must first learn the new language and how to operate under different classroom norms may account for their apparent agreement with the dual system that exists in Catalonia schools that have immigrant students. In that dual system, immigrant students are separated from the local students for half of the lessons in language and mathematics (see Civil and Planas 2004, for more on these two parallel systems). The approach to the teaching of mathematics in the "special classes" at the school in Barcelona where this research took place was based on the practice of mathematical rules and algorithms, rather than on using problem-solving, which is what the local students in the "regular" class (taught by the second author) were experiencing. Most immigrant parents preferred that their children attend the special classes even though they thought that they would not learn as much mathematics there. They did not interpret the parallel system of regular and special classes as boundaries to be contested, but as a given social fact ("It is as it is") that must be respected. This system is seen as especially convenient for the local students who are supposed to go faster with their learning of mathematics and not especially adverse for the immigrant students. These parents not only predispose their children to accept this situation but also to avoid promotion. Mourad's father (B) clearly exposes this finding to the teacher (T) (second author):

> T: Why do you think Mourad gets bad marks in school mathematics?
> B: All immigrant students get bad marks in school mathematics.
> T: Why do you think this happens?

B: I don't know, but they are placed in different classes. That should help them.

T: Does Mourad like attending special classes?

B: He doesn't, but I always tell him "you cannot leave it, you need to be at school and you need special mathematics."

T: Would he prefer always attending regular classes?

B: It is as it is. Immigrant students attend special classes and they share with local students some regular classes. Our children must learn special mathematics. If they learn special mathematics, they will be left better off than if they learn quick mathematics.

T: What do you mean by quick mathematics?

B: In regular classes local students go faster.

T: Can Mourad be promoted to regular classes?

B: He is OK now.

At the time of the study, Mourad and his family (originally from Morocco) had been in Barcelona for four years. During the interview, Mourad's father started talking about the parallel system of classes but he did not do so in order to complain of discrimination or unequal treatment. On the contrary, he made clear attempts to understand and justify a system that, in our view, locates his son in a disadvantaged and subordinated position. Although the dual system includes the possibility for immigrant students to move out of the special classes and join the local students for all their classes, the reality is that very few immigrant students leave the dual system while in High School.

On one hand, the structure of special and regular classes constrains the immigrant students' learning of mathematics ("They learn special mathematics"). On the other hand, the parents' perspectives help to generate and reproduce the structure that constrains their children's actions ("You need special mathematics"). Mourad's father does not interpret his son's position as transitional but as permanent. This father gives his tacit consent to the arrangement of special and regular classes by encouraging his son not to leave the school and to attend the special classes. Some of the immigrant parents may find the special classes rather attractive because their children are given more individualized attention. For example, a Pakistani mother shared that she did not want her son to transition to only regular classes because then he probably would not receive enough attention. This mother appeared to prefer what we view as the general discrimination given by the local educational system to the expected discrimination that she associates with the regular classrooms.

As we said earlier, for the parents in the Barcelona case, the issue of language seemed to play a prominent role in their views about why their children were not doing well in mathematics and why they may need to be in special classes. In the Tucson case, the theme of language also played a role in the mothers' views about their children's education. The next section addresses this theme.

4.2 About the Language

One of the factors that determines whether a student attending school in Catalonia needs to be in "special classes" is his/her level of proficiency in the official languages (Catalan and Spanish). Thus, many immigrant students from countries such as Pakistan or Morocco end up in the special classes since they are often not comfortable enough in either of those two languages. In Tucson the situation is somewhat different. Although one would perhaps expect that more experience (in terms of a longer history) in the U.S. with immigration from groups who speak languages other than English would be reflected in its educational policies, especially when it comes to the education of English Language Learners (ELLs), the reality is different. The fact is that schools are still faced with uncertainty as to how to best educate students who are ELLs. For example in recent years, several states, including Arizona, have mandates that are dismantling bilingual education and are now following more a model of English immersion. The current tendency in Tucson is to place ELLs in Structured English Immersion (SEI) classrooms. In the communities where our work takes place, it is often the case that every student in the SEI classroom is an English Language Learner. The point we want to make is that the situation is complex and that it is still the case that language issues are often mixed up with cognitive aspects. In fact, at the national level, Hispanic students in the U.S. are frequently over-represented in special education because characteristics of second language acquisition often are mistaken for learning disabilities (Merino and Rumberger 1999). At our local level, the story of one of the mothers makes the point quite clearly. The teacher of Eugenia's daughter suggested testing her child for a learning disability. Her child's teacher was insistent that she needed to be in Special Education or switched to another school with special services. Eugenia tried to explain to the teacher that her daughter did not always understand English yet. Still, the referral was done and Eugenia accepted the assessment to be able to contest the teacher's judgment. Fortunately, she was able to prove that the diagnosis was not a learning disability but rather a language issue.

The problem-solving approach that the teacher in the Barcelona case used, as well as some of the current approaches to teaching mathematics in the U.S., place an emphasis on communication, hence on language. The parents in the Barcelona case, however, tended to see mathematics and language as two separate domains, and advised their children not to use the teacher's time during the mathematics class to ask questions about language. Khati's mother's illustrates this point:

> Khati can hardly participate. She doesn't know the language enough. She always says that fractions are easy but word problems are not easy at all (...). She needs time.

When asked how she can help her daughter at home, she says:

> We cannot help our children because we have difficulties as well.

When the teacher asked how she could help Khati, the mother replied:

> M.: You cannot help her. You are the mathematics teacher. I always say to Khati "don't bother the mathematics teacher with your practical questions about language, you must only ask questions about mathematics, if you don't understand a word, then, before you ask, you must be certain that you will not bother the teacher and your peers". If she is not certain, she must not speak. When she is certain, she will get good marks.
> T.: Why can't she ask me about language?
> M.: You should not waste your time with language questions. The other Bangla Deshi students can help Khati.

At the time this study took place, Khati and her family (originally from the center of Bangla Desh) had been in Barcelona for three years. They were not fluent in the local school language, Catalan, but they were already competent speakers of Spanish. In her mother's opinion, Khati must manage to learn the meaning of Catalan words and become a competent speaker within the mathematics classroom before she can learn mathematics. She could not help her daughter improve her Catalan because she herself had difficulties with it, neither could Khati's out-of-school context help her as she spent most of the time with the Bangla Deshi community. Moreover, the mathematics teacher was not expected to help her daughter because, in this mother's view, the teaching of language was not and should not be a goal of the mathematics classroom. Khati should not ask for help and interrupt the dynamics of the mathematics classroom unless she was certain that her question would not interfere with the lesson. This may have put Khati in a difficult situation—how to determine whether her question was appropriate or not in the mathematics classroom? She needs to be 'certain' and this is not an easy goal to achieve for students experiencing various and simultaneous transitional processes.

In the Tucson case, the mothers did not address so much the issue of separation between language and mathematics, perhaps because as MAPPS participants they had been learners of mathematics themselves and therefore had had a chance to experience the interplay of mathematics and language. But language did play a prominent role in the interviews we conducted with immigrant mothers. Here we only highlight the issues that are more directly related to the teaching and learning of mathematics (see Bratton et al. 2004 for an expanded version of this topic). One key issue relates to parents' frustration at not being able to help their children with homework because of the language barrier. This, in a sense, is implicit in Khati's case above when her mother says that she could not help her because she herself did not know Catalan well. In the Tucson case, we have examples in which the children do speak both languages, English and Spanish, but do not have a command of academic Spanish. Their parents, on the other hand, may be proficient only in Spanish. Some children have to translate the problems to their Spanish-speaking parents in order to receive their help on the homework. This situation requires the children to be able to explain and translate the problems, which is a process that involves proficiency in the mathematical register of two languages (Khisty 1997; Moschkovich 2002; Ron 1999). As we see in the following quote for Verónica, this has resulted in her son's lack of trust in her ability to help and her own frustration about her possibility to help them:

> When I sit with him to see what it is he is doing, it seems that translating the problem so that I can help him is too much trouble for him; so, when it's hard to translate, he kind of prefers to go early or ask someone else and I don't like that.
> He doesn't feel very sure that I am understanding him because the problem is written in English. I don't know how to read it and he doesn't know how to translate well for me because he speaks Spanish and reads Spanish, but we say different things for the same words and questions, I think he thinks I studied differently.

Verónica is particularly frustrated because she knows the content and wants to help and support her son but cannot. She feels that when her son was in the bilingual program, she was able to help him more. She would often volunteer in her son's classroom. But then in second grade, he was placed in an English only class and with a teacher who only spoke English and Verónica stopped going into the classroom. She explained that her limited English proficiency has strongly hindered her possibility to help her children with school work, "I feel as if I don't help enough," she said. She believes she has the knowledge to help both her children who were 7 and 11 years old, at the time of this study. However, even with college studies and with some teaching experience in Mexico, she had to turn to after school tutoring.

In our view, to ignore how language can operate as a barrier in the education of immigrant children, including their mathematics education, goes against any attempts to provide an equitable education for ALL children. The experiences of Eugenia, Khati's mother, and Verónica illustrate only a few of the many obstacles that immigrant parents face when trying to advocate for their children's education. Eugenia succeeded; Verónica has not yet been able to overcome her frustration at the separation that she feels is being established between her son and her. As we saw in the case of Khati, her mother's perception of the need to keep language and mathematics separate, may contribute to Khati's lack of participation in the mathematics classroom, hence limiting her opportunities to learn.

5 Conclusion

Our data, both from Barcelona and Tucson, lead us to the notion of recognition. We have parents, such as Lucinda, who do not totally recognize the host school mathematics. We also have parents, such as Sajid's father, who do not recognize their children's former school mathematical knowledge. And we have children, such as Verónica's son, who do not recognize their parents' mathematical knowledge. There are many differences between the "before" and the "now" in relation with the experience of school mathematics, but these differences are not always recognized for their intrinsic value, neither are they recognized for the value they have for those who sustain them. Most immigrant parents seem to have established a clear distinction between the local mathematical practices, which are taught in the local official language(s) and need to follow particular classroom norms, and their former and current group practices, which often show identities that want to adjust to both the "before" and the "now."

The opinions, values and beliefs of Lucinda, Sajid's father, and Verónica's son show a common interpretation of the notion of recognition. Lucinda is concerned about the effect of the host school mathematics on her daughter's former school knowledge. This mother's recognition of certain practices implies an opposition to other practices. Sajid's father also shows an opposition, not to the host school mathematics but to his son's former educational system ("[Sajid] must not insist on what his former teachers taught him (...)"). In the case of Verónica's son, the idea of recognition is also linked to the idea of opposition and, in particular, to that of conflict between opposing practices. Most parents seemed to experience a form of conflict between the different practices. Some, like Gabriela and Sajid's father seemed to be willing to accept the new practices, since these are the ones in place in their new country.

It is not clear how this notion of recognition-opposition may influence their children's learning of mathematics. However, it is reasonable to think that the recognition-opposition scheme sustained by the parents has an important influence on the children's habitus, that is, on the structure that shapes their opportunities when learning mathematics. To help overcome the recognition-opposition scheme is not an easy task at all. One of the goals of the MAPPS project in Tucson, as well as of the research project in Barcelona, was to consider the content of differences and to explore the perception of differences among members of the educational community. In both research contexts, recognition was interpreted as acknowledging the value of differences, but only in Tucson did the initial research design contemplate enduring actions on and with parents. As we pointed out earlier, in the Barcelona case, the researcher was also the teacher. This may account for some of the differences between the two cases. But further research should also explore whether (and how) the parents' different cultural backgrounds in Barcelona and Tucson (as well as the different immigration experiences in both places) played a role in parents' perceptions of the host school mathematics.

The public recognition of differences in the mathematics classroom is a problematic issue. The individual and group experiences that children have of the school mathematics in their former schools and of their parents' mathematical knowledge may provoke these children to prefer their practices not to be publicly appreciated as different. We have found parents that consider differences as particularities, independently from their content, that must not interfere with their children's mathematical learning (e.g., "He (...) must not insist on what his former teachers taught him"). However, some other parents, most of them in Tucson where the interviewer was not their children's teacher, have learned to develop an inclusive attitude towards differences (e.g., "I wouldn't say one way was better than the other but say that there are different ways to do all sorts of things"). The involvement in the MAPPS project has probably had an important positive effect on the parents' perception of their children's differences. It is reasonable to think that children will hardly recognize their former school practices and their current family mathematical practices as legitimate if they do not perceive recognition in their nearer contexts. The notion of recognition needs to be interpreted not only as an acceptance of differences but also as an alternative to broaden meanings for the mathematical practices and ideas. Further research is needed into the effects of projects such as MAPPS on parents' perceptions and potential action-taking towards their children's schooling.

References

Abrams, L., & Gibbs, J. T. (2002). Disrupting the logic of home-school relations, parent involvement strategies and practices of inclusion and exclusion. *Urban Education, 37*(3), 384–407.

Abreu, G., & Cline, T. (2003). Schooled mathematics and cultural knowledge. *Pedagogy, Culture and Society, 11*, 11–30.

Anhalt, C., Allexsaht-Snider, M., & Civil, M. (2002). Middle school mathematics classrooms: A place for Latino parents' involvement. *Journal of Latinos and Education, 1*(4), 255–262.

Bratton, J., Quintos, B., & Civil, M. (2004). *Collaboration between researchers and parents for the improvement of mathematics education.* Paper presented at the 1st annual binational symposium of education researchers, Mexico City, Mexico, March.

Civil, M., & Andrade, R. (2003). Collaborative practice with parents: The role of researcher as mediator. In A. Peter-Koop, A. Begg, C. Breen, & V. Santos-Wagner (Eds.), *Collaboration in teacher education: Working towards a common goal* (pp. 153–168). Dordrecht: Kluwer.

Civil, M., & Planas, N. (2004). Participation in the mathematics classroom: Does every student have a voice? *For the Learning of Mathematics, 24*(1), 7–12.

Civil, M., & Quintos, B. (2002). *Uncovering mothers' perceptions about the teaching and learning of mathematics.* Paper presented at the annual meeting of AERA, New Orleans, USA, April.

Civil, M., Quintos, B., & Bernier, E. (2003). *Parents as observers in the mathematics classroom: Establishing a dialogue between school and community.* Paper presented at annual conference of NCTM: Research Pre-session, San Antonio, USA, April.

García, S., & Guerra, P. (2004). Deconstructing deficit thinking: Working with educators to create more equitable learning environments. *Education and Urban Society, 36*(2), 150–168.

Garfinkel, H. (1984). *Studies in ethnomethodology.* Cambridge: Polity Press.

Glaser, B., & Strauss, A. (1967). *The discovery of grounded theory: Strategies for qualitative research.* Chicago: Aldine.

Khisty, L. L. (1997). Making mathematics accessible to Latino students: Rethinking instructional practice. In J. Trentacosta & M. Kenney (Eds.), *Multicultural and gender equity in the mathematics classroom—The gift of diversity* (pp. 92–101). Washington: NCTM (97th Yearbook).

Khisty, L. L., & Chval, K. (2002). Pedagogic discourse and equity in mathematics: When teachers' talk matters. *Mathematics Education Research Journal, 14*(3), 154–168.

Lareau, A., & Horvat, E. (1999). Moments of social inclusion and exclusion race, class, and cultural capital in family-school relationships. *Sociology of Education, 72*(1), 37–53.

Merino, B., & Rumberger, R. (1999). Why ELD standards are needed for English learners. *University of California Linguistic Minority Research Institute Newsletter, 8*(3), 1.

Moschkovich, J. (2002). A situated and sociocultural perspective on bilingual mathematics learners. *Mathematical Thinking and Learning, 4*(2, 3), 189–212.

Planas, N. (2004). Análisis discursivo de interacciones sociales en un aula de matemáticas multiétnica. *Revista de Educación, 334*, 59–74.

Planas, N., & Gorgorió, N. (2004). Are different students expected to learn norms differently in the mathematics classroom? *Mathematics Education Research Journal, 16*(1), 19–40.

Reay, D. (1998). Cultural reproduction: Mothers' involvement in their children's primary schooling. In M. Grenfell & D. James (Eds.), *Bourdieu and education: Acts of practical theory* (pp. 55–71). Bristol: Falmer.

Ron, P. (1999). Spanish-English language issues in the mathematics classroom. In L. Ortiz-Franco, N. Hernández, & Y. de la Cruz (Eds.), *Changing the face of mathematics: Perspectives on Latinos* (pp. 23–33). Reston: NCTM.

Sheridan, T. E. (1995). *Arizona: A history.* Tucson: The University of Arizona Press.

Suárez-Orozco, C., & Suárez-Orozco, M. (2001). *Children of immigration.* Cambridge: Harvard University Press.

Van Manen, M. (1990). *Researching lived experience: Human science for an action sensitive pedagogy.* London, Ontario: The University of Western Ontario.

Zevenbergen, R. (2000). "Cracking the code" of mathematics classrooms: School success as a function of linguistic, social and cultural background. In J. Boaler (Ed.), *Multiple perspectives on mathematics teaching and learning* (pp. 201–224). Westport: Ablex.

Zevenbergen, R. (2003). Ability grouping in mathematics classrooms: A Bourdieuian analysis. *For the Learning of Mathematics, 23*(3), 5–10.

Mathematics Education for Adults: Can It Reduce Inequality in Society?

Wolfgang Schlöglmann

Abstract Adult education in mathematics is considered from a number of perspectives. It is a key element in the general concept of lifelong learning. Lifelong learning is essential in a rapidly changing economic and technological world. For example, quickly changing production conditions in companies lead to quickly changing demands on skills possessed by company employees. Skills that are taught in schools and institutions of vocational education also need to be adapted to fulfil the requirements of a changing job environment. Lifelong learning is a process that corrects omissions in basic education. In this sense, lifelong learning is viewed as an opportunity to reduce societal inequalities. However, not only is economic life changing, but conditions within society are changing, too. In this essay it will be argued that lifelong learning is undoubtedly a prerequisite for making democratic participation possible for all members of a society. All adult learners have a learning history that is intimately connected with their experience of learning at school. The school experience has a great influence on learning as adults, particularly if the adult is forced to, acquire new skills because of unemployment. In this chapter we begin with a discussion of the general conditions of lifelong learning. We continue with an account of the role played by mathematics in societies. We then conclude with problems faced by adult learners of mathematics.

1 Introduction

While social inequality—differences in living conditions, income, job opportunities or opportunities to participate in social and political life—is a problem in all societies, the way it is manifested depends on each particular society's characteristics. This paper focuses on inequality in highly industrialised societies. Clearly, social status can impact an individual's education. For this reason, Article 26 of the "Universal Declaration of Human Rights" asserts the "right to education for everyone." However, in industrialised countries, even if every child is able to go to school, the educational system alone is unable to eliminate social inequality. In particular, chil-

W. Schlöglmann (✉)
Universität Linz, Linz, Austria
e-mail: wolfgang.schloeglmann@jku.at

H. Forgasz, F. Rivera (eds.), *Towards Equity in Mathematics Education*,
Advances in Mathematics Education,
DOI 10.1007/978-3-642-27702-3_25, © Springer-Verlag Berlin Heidelberg 2012

dren of parents with a low level of education, as well as children of immigrants, are likely to leave the school system with a low education level. This restricts their chances of getting a job, increases the probability of unemployment, and ultimately limits their opportunity to participate in cultural and social life. In other words, highly industrialised countries, too, have a need to reduce inequality within society. The aim of this chapter is to investigate whether adult education—particularly adult education in mathematics—can help reduce education-related aspects of inequality in the sense of inequality caused by an insufficient level of education.

Adult education is a key element in the general concept of lifelong learning (Aspin et al. 2001; FitzSimons et al. 2003). Lifelong learning is essential in a rapidly changing economic and technological world. For example, quickly changing production conditions in companies lead to quickly changing demands on skills possessed by company employees. Skills that are taught in schools and institutions of vocational education need to be adapted to fulfil the requirements of a changing job environment. Lifelong learning can also be considered a process that corrects omissions in basic education. In this sense, lifelong learning can be viewed as an opportunity to reduce societal inequalities. However, not only is economic life changing, but conditions within society are changing, too. In this chapter it will be argued that lifelong learning is a prerequisite for making more equitable participation in society possible for all members of a society.

All adult learners have learning histories that are intimately connected with their experiences of learning at school. The school experience has a great influence on learning among adults, particularly if they are forced to acquire new skills because of unemployment. In this chapter we begin with a discussion of the general conditions of lifelong learning. We then continue with an account of the role played by mathematics in society. We conclude with problems faced by adult learners of mathematics.

2 Lifelong Learning: A Consequence of Technological Development and Globalisation?

Clearly everybody continues to learn over his or her entire lifespan; nobody finishes learning when he or she has completed his or her formal school education. However, the term "lifelong learning" is usually not used only in the simple sense of "everyday learning" or so-called "situated learning in vocations." In many discourses, lifelong learning also means a specific form of organised learning during adulthood.

For a long time, only philosophers and pedagogues discussed lifelong learning (Aspin et al. 2001). In the final decades of the 20th century, however, lifelong learning became a key concept in proposed solutions to social and economic problems in highly industrialised countries. International bodies such as the United Nations Educational, Scientific, and Cultural Organization, the Organization for Economic Cooperation and Development, and bodies within the European Union (EU) declared that education has to be a lifelong undertaking (Aspin et al. 2001; FitzSimons et al. 2003). It meant that education could no longer be considered as a process that ends at the beginning of adulthood, and adults could no longer be considered experts in their jobs based only on the competencies acquired in their youth. To understand this development we must understand the nature of recent societal change and its driving force. Highly industrialised countries are characterised by the use of technologies because technology determines the structure of their societies (Hülsmann 1985). While Hülsmann emphasises the significance of the natural sciences in the development of technologies, it is important to note that other sciences, particularly the social sciences and economics, play a role in the technological formation of our society (Schlöglmann 1992). Thus scientific research leads to technological development, which in turn leads to a change in the competencies required to fulfil the demands of the technological change.

Lifelong learning was the societal solution to conditions arising from rapid economic and technological change. It has a number of functions, as follows:

> [We] suggest that lifelong learning policies currently introduced across the globe could be classified into four categories: (1) a compensatory education model, which aims at compensating for inequality in access to initial school education, and improving basic literacy and vocational skills; (2) a continuing vocational training model, which aims at coping with changes occurring in the workplace and solving problems arising from unemployment; (3) a social innovation model, or civil society model, which aims at overcoming social estrangement and promoting socio-economic transition and democratisation; (4) a leisure orientated model, which aims at enriching the leisure time of individuals and personal fulfilment." (Aspin et al. 2001, p. XXII)

The concepts embodied in the four categories address lifelong learning and its role within a society in a very general way. Within specific nations and societies, the four categories are realised in a variety of ways. It should also be kept in mind that even in highly industrialised countries, the context of lifelong learning can be very heterogeneous. Although in industrialised countries, too, a need exists that will eliminate inequalities in education and improve basic literacy and vocational skills. Hence, distinctions nevertheless ought to be made, for instance, between people who have passed through the education system with a low level of achievement and people who have migrated to these highly industrialised nations from countries with a weak education system. Moreover, in economically advanced countries, more people want a higher level of education.

When authorities speak of lifelong learning, they usually have in mind the continuing vocational training model (Aspin et al. 2001, p. XX). From the point of view that lifelong learning is a means of coping with the consequences of technological development and globalisation—bearing in mind that governments and supranational bodies often resort to rhetorical set phrases when referring to lifelong learning (FitzSimons et al. 2003)—we must extend the concept of lifelong learning to encompass the needs of people in a democratic society. The concept of lifelong learning ought to be seen as a social innovation model or civil society model that stresses structural and political aims. This concept is often called "critical citizenship" (Evans and Thornstad 1994, p. 65).

While the third category introduced by Aspin et al. (2001) in the quotation above emphasises people as social beings (the social innovation model), the fourth category (the leisure model) emphasises people's individuality. In order to account for all of these aspects of lifelong learning, Aspin and Chapman (2001) suggest three necessary purposes of lifelong learning, as follows: economic progress and development; personal development and fulfilment; and social inclusiveness and democratic understanding and activity (Aspin and Chapman 2001, p. 29). Aspin and Chapman further express an important point. Lifelong learning is often seen only as institutional organised learning, but this view obscures the fact that life and work are always combined with learning. Learning in context, "situated learning" and workplace learning are not usually the focal point when one speaks of lifelong learning. Therefore, in light of the importance of lifelong learning to a well-functioning democracy, democratic societies ought to broaden the scope of institutional learning and thereby create more opportunities for adults to pursue further education.

3 Mathematics and Society: Some Remarks

In the preceding section, we discussed general aspects of the concept of lifelong learning. If we are to focus on a special subject like mathematics, we must consider its status within an industrial society, within the technology used in this society, and within the society's economic and social life. Understanding the influence of mathematics in these three areas is essential in understanding how adult education in mathematics can help reduce inequality in society.

To understand the various functions of mathematics in society, we begin with a few historical remarks. The historical development of mathematics in Western culture is sketched below since the nature of this development sheds light on the role of mathematics in industrialised societies (Schlöglmann 2002). Incidentally, while we speak here of Western societies, it should nonetheless be borne in mind that developments in mathematics existed in other cultures and these developments followed the demands in those cultures (Bishop 1991).

3.1 The Development of Mathematics as a Tool for Economic Growth—The Relationship to Culture

The development of mathematics is always concomitant with the development of a society and activities within it. Bishop (1991) describes six key universal activities (counting, measuring, locating, designing, playing and explaining) as driving forces in the development of mathematics. The importance of these activities in the life and culture of a society leads to the observed strong relationship between mathematics and culture.

> I have presented the case that six key 'universal' activities are the foundations for the development of mathematics in culture. I have also demonstrated that it is the case that all cultures have necessarily developed their own symbolic technology in response to the 'demands' of the environment as experienced through these activities. As a result, however, of certain within-cultural developments, and also of different cultures interacting and conflicting, so a particular and traceable line of development has emerged. (Bishop 1991, p. 59)

The following remarks sketch some important steps in the development of mathematics as well as the demands and activities that are at the root of this development. The first signs of numerical concepts can be found in the early Stone Age, and for thousands of years the signs remained rudimentary (Struik 1967). The development of mathematics took a leap forward following the emergence of settlements in Mesopotamia at about 10,000 BC. Agriculture demanded planning and documentation and trade between villages contributed to this demand. Small artefacts made of clay, so-called "tokens," were used to count and document goods. The system of tokens evolved into a hierarchical system with various subsystems according to local commercial demands. Arithmetic appeared in a primitive form (Schmandt-Besserat 1977).

The next leap occurred after the formation of cities. These bigger societal units created a demand for new and better organisational methods. Division of labour and the demand for a growing economy required new planning methods as well as the standardisation of quantitative measurements for the different kinds of goods. The first systems of writing and numerical symbols emerged in Mesopotamia and Egypt. It is important to note that a single unique number system did not exist at that time. According to the system of tokens, there were different systems with different bases for different objects (people, animals, grain, areas, time and calendar, milk products etc.). We can deduce that neither an abstract conception of numbers existed nor a well developed arithmetic. Arithmetical operations were strongly related to operating with clay tokens (Nissen et al. 1991). The development of a unique number system and arithmetic was the result of a longer process. At the end of this process, mathematics was a tool used in many professions for documentation, planning, measuring, ordering, organisation of trade, astronomical and calendar calculations. As it has comprised an essential component of these important activities, mathematics has always been a part of culture.

Many important developments in mathematics occurred in ancient Greece. While mathematics was developed as a tool for solving practical problems in the Orient, it was the Pythagoreans—the first group of mathematicians—who first used mathematics as a means to understand the laws of the world (Bishop 1991; Van der Waerden 1979). In their philosophy, the laws of the world were written in mathematical language. Just as musical harmony was based on proportions of numbers, the harmony of the universe for them was based on mathematics. To the Pythagoreans, studying mathematics was a way of discovering the eternal laws of the universe. Therefore, mathematical objects like numbers and geometrical figures became objects of interest. The aim of mathematical development amongst them was not to solve practical problems but to discover mathematical laws. New and systematic ways of defining mathematical objects and of justifying were necessary. Real-world applicability was not the main justification for the correctness of a mathematical algorithm, but only proofs that followed the laws of logic were acceptable. Furthermore, only human reason was able to discover eternal truth. Mathematics was a means for applying reason. The result of this new philosophy was a close relationship between mathematics, logic, and truth. Mathematical thinking and rational thinking were closely related (Pichot 1995).

A most important impact of the work of the Pythagoreans in the further development of mathematics was the separation of mathematical terms from their real representations (Van der Waerden 1979). This separation alone led to the development of mathematics as a means for rational thinking and a general means of communication. As a basis for rationality, mathematics became a part of the culture of societies that have rationality as their societal foundation.

3.2 What Kind of Society Needs Mathematics as a Tool?

Demands created by the changing forms of society required new means for documentation, planning, trade and so on. The solution in all cases was the development of some kind of mathematical objects such as a system of tokens or much later, systems of written signs for numbers.

Is mathematics a necessary tool in a special kind of society? To discuss this question, we make use of a concept introduced by the philosopher Heintel (1992), that societies can be classified according to the kinds of communication that occur within them. Small groups can be organised by direct communication. All information is transmitted from person to person. No other mediation channels are necessary. Learning processes are also organised directly. In larger societies, such direct communication between all members of the community is impossible. The society changes from a society organised by direct communication to one organised by indirect communication. Such an organisational form demands common standards, values, and so on. Societies therefore need a more sophisticated language because the standards and values must be formulated in new, more abstract concepts. In larger societies, labour processes are mostly specialised and often organised in a hierarchical way. Specialisation and division of labour demand more planning and more documentation to organise labour in a rational way. To fulfil these demands, a need exists for a means of communicating advice, regulations, as well as various skills in a more abstract way. Written documents are able to reach more people.

In short, mathematics is a tool that fulfils the demands of societies organised by indirect communication. Mathematical terms allow documentation through mathematical procedures, and reliable calculations in the context of labour and trade processes become possible. All processes in such societies evolve from the particular to the general. Mathematics, too, has developed like this. It first appeared in forms whose terms and procedures depended strongly on context, and evolved into a general tool for many problems.

3.3 Mathematics and Democracy

The roots of modern democracy lie in the ideas of the Enlightenment. Rationality is fundamentally important for the structure of society. Mathematics is an important background for rationality. Bishop (1991) argues that rationalism is at the heart of Mathematics. If one had to choose a single value which has guaranteed the power and authority of Mathematics (and the ideal of Mathematicians) it is rationalism. (Bishop 1991, p. 62)

Democratic societies are, as we have seen, usually organised by indirect communication. Therefore, democratic societies have a need for, on the one hand, values and principles, and on the other hand, procedures that implement some of the values and principles that facilitate democratic practice by communicating the principles in an unequivocal and reliable way, and "rationalism, logic and reason have a very well elaborated vocabulary" (Bishop 1991, p. 62). For example, consider the democratic election procedure. Ideally, the election procedure has to implement the principle of same value per vote. This basic principle is often combined with other principles, such as the principle that the result of the election should lead to a stable majority in parliament. Taking into account particular historical features of a society, the democratic system has many instantiations—many different election procedures are possible. But every procedure is formulated as a mathematical algorithm.

In recent political discussions in the EU and elsewhere, a new interpretation of equality has been introduced, according to which equality should not only be a principle that is valid at the individual level, but that it should also be valid for social groups. This means that within a democracy, each social group should be represented as an important subsystem of society with its own appropriate societal share. This principle, too, is based on mathematics, in particular, on statistical arguments. Statistical methods are important for analysing societies (Evans and Rappaport 1998; Frankenstein 1990).

In discussing the relationship of mathematics to democracy, the status of mathematical models that underlie various aspects of democracy ought to be mentioned at least briefly. These models are based on norms, and norms are the consequence of a social process. Norm-based models, which are used in many applications (such as business), are not justifiable by mathematics. Mathematics is only used to guarantee a reliable implementation of a norm; but mathematics is not able to guarantee the correctness of the norm. A norm is a consequence of a social process.

3.4 Mathematics and New Technologies: New Functions of the Tool "Mathematics"

In the preceding sections, it was argued that societal demands led to the development of mathematical methods and procedures. Mathematics affected the structure and condition of society. Trade and administration, in particular, with the extensive use of money, stimulated the search for better methods, and mathematics was applied extensively in these areas. Later, mathematical methods were used in astronomical calculations and more and more professions began to use mathematics. The new status of mathematics in Antiquity led to a relationship between mathematics and rationality and to research in "pure" mathematics as well. In particular, the separation of mathematics and its applications opened the way to the development of mathematics as a "general tool" that could be used in sciences and numerous other fields. The use of mathematics in more and more areas of social, economic and professional life increased. In many cases, mathematics became integrated into a field to such an extent that it became a standard fixture in the background while losing its visibility in the foreground. Great changes have occurred in recent years. With the dawn of new technologies, mathematics began to play a new role within the society. We begin to describe this new role with a quotation from a report from the US National Research Council on the resource Mathematics: "When we entered the era of high technology, we entered the era of mathematical technology" (David 1984, p. 435). This can be interpreted to mean:

- Mathematics is the basis of all new technologies since algorithms are the basis of software and materialised mathematical logic is the basis of computer hardware (microprocessors).
- Mathematical theories and models are becoming increasingly important as the basis of a variety of forward-looking alternatives and in simulating planning in economic and technical fields, for example, in control, automation and construction, or in political and social life.
- Mathematics has long been established as the scientific core of the natural sciences and, to an increasing extent, also of social science (Maaß and Schlöglmann 1988).

An important necessity of technology is that it should facilitate the distinction between the development of knowledge and the application of knowledge—between the "why" and the "how." The user of knowledge needs an understanding not only of how to apply the knowledge, but also of what the knowledge really is in the first place. We all use computer programs, but we do not know what goes on behind the scenes in the computer box. Mathematical black boxes have a long tradition (Maaß and Schlöglmann 1988), however, with the rapid ascent of the computer, their use has increased very strong. The increased use of mathematical black boxes has contributed to the "disappearance of mathematics". Further, these black boxes determine our work processes, economic, and social life. It is an open question what a user of a black box should know about the implemented program.

3.5 Why Are Mathematical Methods so Credible?

The status of mathematics in our society is rather paradoxical. On the one hand, many people see mathematics as abstract, remote from the life, and incomprehensible. On the other hand, the same people have full confidence in mathematical methods—they pay invoices, accept calculation of election results, and accept the use of complex mathematics in technology and the economy. Fischer (1999) notes how "mathematics is the materialisation of the abstract" (p. 89).

Mathematics puts abstract thought into concrete form. Mathematical objects are seen as abstract, but in their concrete form, consisting of numerical symbols, formulas, graphs, they make manipulations and presentations possible. For example, we use graphs to present complex, abstract relationships in a concrete way. This specific characteristic of mathematics—the construction of abstract concepts on the basis of concrete representations—could be the reason why many people have confidence in results obtained through mathematical methods.

3.6 Should Everybody Learn Mathematics?

The foregoing discussion regarding the various functions of mathematics within our society provides a strong argument for mathematics to be a part of all school curricula and also of many courses in adult education. We raise the following two main points: that mathematics is a tool; and that mathematics is an integral part of culture.

1. Mathematics in our society is a tool used to organise our everyday lives. Mathematics is also used as a tool in many occupations. The use of mathematical black boxes in the form of computer programs gives them a new quality and a new challenge for didactics.
2. Mathematics is a part of our culture. Democratic principles such as equality, righteousness and so on need an operational concretisation. On the one hand, democracy demands a means for communicating and discussing principles in a rational way. Mathematics, with its close relationship to rationality, is our concept that will enable us to do this. On the other hand, democracy demands operational procedures for its concrete implementation. Mathematics is again the tool that facilitates this.

The concept of the "responsible citizen" as a citizen who is able to participate in societal processes in a rational way is a part of numerous educational philosophies.

4 Adult Education in Mathematics

The field "adults and mathematics" is characterised by great heterogeneity, particularly in the course system (FitzSimons et al. 1996). To gain a better insight into the situation it is useful to reduce the heterogeneity and complexity by categorising the various concepts into appropriate classes. The guiding goal in the categorisation process is to reproduce the important subject areas that constitute the field. One class is "mathematics as a system," in which the ideas of mathematics are represented as a science. Another is vocational practice, together with the use of mathematics in everyday life situations. The third and newest class concerns new technologies. Mathematics is a key science for computer hardware and software. Software programmes in particular contain mathematical structures behind the scenes, which has led to an immense change in the use of mathematics in many vocational situations.

The focus of educational *intention* ought to be a structuring (i.e. organisational) element. This is because nowadays the educational goal can be any of the following: mathematics; *content* that is not in itself mathematics yet uses mathematics to formulate results (mathematics used as a language); and the use of software that is based on mathematics. Numeracy courses are ubiquitously organised according to a mathematical systematic, and the goal of the students taking them is to learn mathematics (e.g. courses for adults that supplement high school examinations). Some numeracy courses are organised according to mathematical content (e.g. percentage calculation) and their aim is to calculate credits. Today many computer programmes exist that contain mathematical algorithms but the goal of a course that utilises the programme is to teach the student to use the programme—not to teach mathematics. For many occupations, mathematics is an important tool and therefore mathematics is the (often hidden) content of many vocational courses.

Two different conceptions of course organisation are used when dealing with the problem of mathematics learning. In the first conception, the required mathematics is separated from the vocational field and taught as a separate course (for instance, mathematics for building construction) and the application of mathematics to the field takes place at a later stage. In the second conception, the mathematics is not separated and the required mathematical algorithms are taught as part of the vocational content. Here the mathematical algorithms are often very specific and only usable in isolated situations. Many computer programmes used in production processes, for instance, contain mathematical algorithms, but the user is only trained to handle one specific programme and is often unaware that mathematical algorithms in the background control the process. In training courses nowadays one may find learning software for practising mathematical operations and algorithms. This type of software belongs to the category "training programmes" and its goal is to provide mathematical practice. Other software used in adult education courses is classified according to vocational criteria (for instance, bookkeeping) but nonetheless require mathematics for their operation. Finally, some courses aim to teach proficiency in the use of specialised software (for instance, Excel), and these, too, require mathematical concepts (variables and so on) for their use.

5 Vocational Mathematics Versus School Mathematics: Some Remarks

The aforementioned cases demonstrate the complex situation of adult education in mathematics. The aim of education is, on the one hand, to learn mathematics, and on the other hand, to acquire occupational and professional skills in which mathematics is part of the context. The relationship between school or academic mathematics and vocational mathematics has long been discussed (Benn 1997). In the following paragraphs, I discuss the problem from the point of view of lifelong learning.

Generally, two different forms of learning organisation can be distinguished, that is, learning in practice and learning at school. Learning in practice was studied by Lave (1988). In collaboration with Wenger, Lave developed the concept of "situated learning" (Lave and Wenger 1991). The basic idea is that a learner can participate in the working process with the guidance of an experienced person. This participation is possible on the basis of a contract that legitimises the learner for "peripheral participation" in the working process. It is important to analyse the structure of knowledge that is required in such a learning process. An experienced person's vocational knowledge is often a repertoire of actions that can be seen as a unit and which belongs to a specific task. Many of these actions are used in some form of routine. Such "action units" contain a plan to organise the courses of action, the use of different tools, the need for collaboration and so on. The organising elements are vocational conditions and the specific context. If mathematics is part of such an action unit then it is integrated into actions that comprise the action unit—and that means mathematics is part of an action course; and the mathematical tools used in these actions are part of the vocational process. During the learning process the learner does not acquire a system composed of disjoint parts that must be combined to form a whole. Instead, a unit is acquired that is to be used in a specialised situation. The mathematics is often not visible to the learner because it is part of an action system that is realised through the use of specific tools. This action and tool relationship leads to optimal integration into the vocational situation and prevents transfer to other situations (Evans 1999). For our problem we have to bear in mind that vocational mathematics is context-related; requires specific tools to fulfil the necessary actions; and is often hidden within the context and actions. "Mathematical terms" derive their meaning from their position within the vocational process and not from their position within a mathematical subject systematic.

School mathematics is based on the criteria of mathematics and uses a content-related systematic (arithmetic, algebra, geometry, analysis, probability). For our problem it is important that the historical development of mathematics led to a theory system whose terms are described according to their properties, which in many cases have no corresponding reference objects in reality (Dörfler 2002). Operations on such mathematical objects are based on their properties and are rule-guided. The status of mathematical objects within some context is controlled only by their properties. This possibility to use mathematical objects and operate with them inside the mathematical structure has certain consequences:

- Mathematics is a "general language" that is applicable in numerous different contexts. The correspondence between context and mathematics is a problem of the structure of some context.
- A mathematical formulation of a contextual problem is context free. The mathematical terms and signs acquire a contextual meaning only by interpretation within a context.
- Operating with mathematical objects entails operating within the mathematics and not within the context.

Let us summarise these brief reflections. If mathematics is seen as a "general tool," which means that the same mathematical method can be used for many problems in very different contexts, then the method needs only to fulfil the same structural properties in each case. On the other hand, mathematical methods cannot be part of a vocational action unit. The use of mathematics requires reflection on the structure of a problem. Such reflection is not a part of situated learning processes. Situated learning deals with a problem "as it is" and not with reflecting on the situation to discern its structure.

W. Schlöglmann

The concept of lifelong learning emerges naturally in a society where qualification requirements are rapidly changing. The skills that are acquired in situated learning processes are optimally adapted to a specific vocational situation and are therefore very useful in this specific situation. But they are also very strongly connected with this specific context, and a transformation to another context is often very difficult or impossible. Therefore lifelong learning requires both general skills that are transferable and specialised skills for specific vocational requirements.

6 Some Results from an Empirical Study of Adult Education in Austria

In the first section, we argued on theoretical grounds that lifelong learning in general, and lifelong mathematics learning in particular, is important for the functioning of highly industrialised societies. However, it is the "reality" of lifelong learning that is important for research and practice. Hence, between 1993 and 1997, Jungwirth, Maasz, and Schlöglmann explored the state of mathematics education within the adult education system in Austria (Jungwirth et al. 1995, Schlöglmann 2006). Adult education in Austria is characterised by a diversity of agencies offering a variety of programs in community services or vocational education. These agencies include the following: state-run institutions of adult education; trade unions or the federation of industry; industrial enterprises; and the education departments of public or private sector enterprises.

This diversity is also reflected in the numerous courses and offers of study programs, in mathematics as in other subjects. In the above empirical study, we distinguished the courses according to whether mathematics teaching explicitly took place or whether mathematical concepts and methods were used in an implicit way (for instance, in courses in CAD, Excel, or other computer software). Furthermore, courses were distinguished according to the level of mathematics (basic, upper secondary, or university) as well as whether they were part of general education or vocational education. To explore the state of adult education in mathematics we selected 419 participants (80.5% men and 19.5% women) from 19 courses in seven institutions within the Austrian adult education system. In all the courses, mathematics was either explicitly or implicitly included and the courses ranged from basic (10.3% of the participants) to upper secondary level (36.3%) and vocational education (53.4%). The extensive questionnaire elicited each participant's personal data, including sex, age, vocation and education. It also contained the following questions: the course of study; the institution and the participants' attitude toward mathematics and school; their use of computers and mathematics at work; their motivation for participation; their appraisal of the value of further education; the goals and priorities of the course they were studying; their learning problems; and suggestions for improvements in mathematics education. In the second part of the questionnaire, we asked for their individual beliefs and attitudes toward mathematics. Besides the questionnaire for participants, we also asked the teachers of the courses to answer a questionnaire with partly analogous questions. This quantitative part of the study was complemented by interviews with participants and teachers. Even though the data were collected between 1993 and 1997, they are still relevant today because the situation in adult education in Austria has not changed.

In this chapter we only refer to a small portion of the results of the study, namely, those that pertain to their motivation for participating in an adult education course and the individual attitudes toward mathematics and mathematics learning. To investigate these motives, we used a 5-point Likert scale (completely agree, partially agree, undecided, partially reject, reject). The main motives for participation in adult education were: "joy of learning new subjects" (86.8% completely agree or partially agree); "improvement of personal education" (86.3%); "acquisition of latest professional knowledge" (79.8%); "increased vocational demands" (73.3%); "financial gain" (72.4%); "security in economically unstable times" (69.9%); and "higher position in the company" (64.2%). To get a better insight into the structure of participant motives, principal component analysis was used to extract the factors of motivation. The results showed four significant factors at work in motivating the participants to take part in adult education courses. The first factor was "professional and economic advancement" and includes preparation for a better position as well as protection of the position already attained. The second factor was "personal motives" such as the desire for more knowledge as part of the individual's general thirst for knowledge. The third factor was "general professional performance orientation", which means that the participants considered their job to be significant to a greater degree than usual. The fourth factor was a "change in job" with change being either an aspiration or a forced move.

In order to obtain a better understanding of the participants' motives, it was fruitful to consider their opinion of the value of higher education. As a word of explanation about this issue, we should say that the concept of "Bildung" has a long history in the German educational tradition. "Bildung" means more than "education" in a narrow sense. An individual that is "gebildet" not only has knowledge and the ability to acquire new knowledge, but he or she is also able on the basis of this knowledge to develop their inner abilities to the full extent (Klafki 1975). The significance of higher education (höhere Bildung) in German tradition can give a better insight into the participants' motives.

Principal component analysis was also used to extract the independent factors. The results suggested a model with three factors: the value of higher education lies in (1) its effect on vocational advancement, financial security, and personal prestige; (2) its potential for personal development; and (3) its contribution to a better understanding of society. We can see that these factors—explaining the value attached to higher education by the participants—are similar to the factors of the participants' personal motives.

Regarding the mathematics content in the courses, we have found that in many cases, the adult participants did not have any choice in studying the mathematics component. For instance, if a participant desired to acquire a certificate for completing a lower secondary or an upper secondary school, he or she had to study mathematics. Therefore, in order to understand the learning conditions experienced by adults in their further studies in mathematics, it was of interest to know their attitudes toward mathematics education at school. A heterogeneous picture emerges, as follows:

- Interested in mathematics in school (57.7% yes, 37.7% no).
- Had ability in mathematics in school (50.2% yes, 42.5% no).
- Good rapport with mathematics teachers at school (60.3% yes, 30.3% no).

The correlation between interest and claimed ability is high ($R = 0.63$). Those who were interested in mathematics had no problems with mathematics (75%) and had good rapport with teachers (72.9%).

Regarding the adult education system in Austria, two groups of adult learners with very different motives and personal situations could be discerned. Members of the larger group chose to participate in adult education. They viewed learning as an opportunity to increase their chances in the labour market, to raise their prestige, and so on. They were highly motivated to learn. They usually had a positive recollection of their school experiences. Members of the other group, in contrast, had been compelled to participate in adult education (in many cases they had lost their job or had to change their job on health grounds). The existence of this group did not emerge from the data of the questionnaires (we had not asked the right questions to identify this group) but from the interviews with the participants as well as the teachers. From the interviews, we received hints to the problems that the adult learners faced. The number of persons for whom these conditions were relevant depended strongly on the state of the labour market. However, the existence of this group is not specifically an Austrian problem (see, for e.g., Wedege and Evans (2006) that considered the problem of adult learners who had a resistance to learning in childhood schooling).

Mathematics learning is not just a cognitive process since it is also influenced by affective conditions (Evans 2000). From narratives given by adult learners, in particular, we know that many adults had negative experiences with mathematics learning in school (Ingleton and O'Regan 2002; Stroop 1998). Hence, in the next section below, we consider influences on the emotional states of members of the second group above.

7 "Must Learn Mathematics:" Some Remarks on the Emotional Constitution of a Group of Adult Learners

Consider individuals who have just lost their job possibly through an accident that incapacitated them or because of company restructuring. They are now forced into further education in order to prepare for a new job. Learning for a new job is not their decision; it is forced on them by circumstances. To understand the circumstances of such a person, we use the concept of "situated learning" (Lave and Wenger 1991). This concept is applicable to many participants in the Austrian adult education because most participants, after their compulsory schooling, are educated in companies as apprentices. They acquire their qualification by working in companies. Therefore, in most cases, they were members of a "community of practice" (Lave and Wenger 1991) in their previous workplaces. In order to be accepted into such a community, a newcomer or apprentice must go through a long process of practice within a framework of legitimate peripheral participation (Lave and Wenger 1991, p. 53). The goal of all learning processes amounts to becoming an experienced member of the community. Such a position requires numerous skills, experience of responsibility for various tasks, knowing codes within the community, and so on. The result is a place within the community. A person is part of the community and the routine of a company (Schlöglmann 2003). All of these together bestow a sense of prestige and identity (Lave and Wenger 1991, p. 122).

In recent years it has been commonplace for experienced staff to be replaced by inexperienced newcomers. For a significant number of these "old-timers" this has led to a loss of self-confidence (Wenger 1998, p. 148). Furthermore, many of these people have had bad recollections of their time at school. Their narratives give us an insight into these recollections (Ingleton and O'Regan 2002; Stroop 1998) and can give us also hints to their implications for mathematics learning. Taken together, loss of prestige and identity, marginalisation of experience, low self-confidence, and low confidence in one's ability to learn mathematics as a consequence of the school learning experience can lead to emotional barriers during new mathematics learning processes.

8 Is Mathematics Learning Possible for Adults?

Possible reasons for adult students who are not studying by choice have been discussed in the previous section. Successful learning often requires motivation of learners by teachers, particularly learners who must learn mathematics. Teachers in adult further education report anxiety and learning blockages in mathematics learning processes (Lindenskov 1996; Wedege 1998). To understand learning better, we use a model introduced by Hannula (1998) with a continuously changing "landscape of mind". The process of change is stimulated by information arriving from the senses, as well as by mechanisms and structures within the brain. One central principle is the distinction between dynamic and static systems of representation. Dynamic representation includes all systems that are activated at a given moment. Static representation encompasses all the information stored in the memory systems. The structure of each static representation—cognitive and affective schemata—is crucial for all processes. The schemata representing the static representations are changeable via learning processes. In all situations of interest to us, many brain systems are activated; in particular, so is a certain system that generates so-called "background emotions" (Damasio 1999). Background emotions influence cognitive processes in a positive or negative way.

Teachers in adult education classes can help adult students manage their learning situation. In the previous section, we have discussed reasons for negative background emotions. The adult student has lost the position as an experienced professional and this position is likely to have been a crucial part of their identity. Now, like children, they have to learn new things. In many courses, adult participants have had no formal education since their school years. They often feel unable to learn because their learning processes have occurred over a long time, whereas they recognise that mathematics learning requires intense processes of abstraction and generalisation. A frequent consequence of this is an adult student's fear that his or her memory is unable to hold all the abstract concepts required. Formal learning often brings back memories of mathematics learning in school and it is generally accepted that for many adults these memories are bad. So speak Wedege and Evans (2006) from a resistance to learn:

> In adult education, resistance to learning is a well known phenomenon. There is an apparent contradiction between many adults' problematical relationships with mathematics in formal settings and their noteworthy "mathematics-containing" competences in everyday life. However, there is very little research done on the subject, and resistance is often explained purely as a lack of motivation and the symptom described as non-learning. In order to investigate adults' resistance to learning, we must take into account the set of conflicts between the needs and constraints in adults' lives. In this paper people's resistance is seen as interrelated with their motivation and their competence and thus as containing the potential to be a crucial factor in all types of learning. (Wedege and Evans 2006, p. 28)

It is important to impart on adult students the importance of their knowledge in situated learning processes. They also need to feel that they have valuable knowledge at their disposal, and using situated knowledge for mathematics learning processes is possible (Schliemann 1999). But it is important for teachers to help open the eyes of adult learners to the mathematics that is contained in the situated knowledge. Processes of abstraction and generalisation must be done very carefully viz a viz their situated knowledge. Creating a successful learning process ought to include steps to help students better manage their fear and to change their emotional background. However, it is important that students actually understand mathematics by doing real mathematics and not learning a diluted form of mathematics via what is called "replacement strategies" (Schlöglmann 1999). Mathematics teaching in adult education is a very demanding task that requires handling the emotional difficulties of the students and developing mathematical concepts with care.

9 Can Mathematics Education for Adults Reduce Inequality in a Society?

Inequality within a society is, first and foremost, a societal problem with many aspects. Solving social problems require a coordinated strategy involving several activities; silver bullets do not exist. Mathematics education for adults can at most comprise one measure in a coordinated strategy and can only have an effect on inequities that are consequences of the education system, like low education or insufficient qualifications for a new job. As discussed in a previous section, four models of lifelong learning strategies can be discerned, as follows: a compensatory education model; a continuing vocational training model; a social innovation or civil society model; and a leisure-oriented model. Mathematics is woven into, and competencies are necessary in, everyday economic, vocational and social life in our society.

It is important to reduce inequality arising from inadequate formal schooling. This concerns both weak learners who emerge from the local school system and immigrants with a low education level. The number of jobs for persons with a low education level is diminishing. Therefore, improving basic literacy can open the way to a higher qualification and to a better job. Given the importance of mathematics in the modern workforce, this compensatory education needs to succeed in improving basic mathematical literacy. Underskilled adults can then acquire at least some vocational skills necessary for a new and better job even when the learning phase is emotionally difficult, a phenomenon that is felt by some unemployed adults. In this sense, compensating the weakness in basic mathematics literacy opens the way to continuing vocational training. It also helps the learner handle changes that are occurring in the new workplace with the concomitant demands for new qualifications.

Also significant is the reality that mathematics lies at the heart of new technologies and strongly influences the use of these technologies. Being able to deal with computer software is a necessary prerequisite in many professions. Hence, mathematics education can improve computer literacy even if this is not its direct aim.

Mathematics education for adults helps develop and improve skills for further professional and economic advancement or at least provide the qualifications necessary to meet the new demands in a job and can therefore have an effect on personal financial security and prestige. In this sense, mathematics education for adults can help develop the skills necessary for living well. Importantly, mathematics education for adults can also offer the potential for personal development and can contribute to a better understanding of society. Particularly in democratic societies, responsible citizens such as those who participate in political life need mathematics.

References

Aspin, D. & Chapman, J. (2001). Toward a philosophy of lifelong learning. In D. Aspin, J. Chapman, M. Hatton, & Y. Sawano (Eds.), *International handbook of lifelong learning* (pp. 3–33). Dordrecht: Kluwer Academic Publishers.

Aspin, D., Chapman, J., Hatton, M., & Sawano, Y. (2001). Introduction and overview. In D. Aspin, J. Chapman, M. Hatton, & Y. Sawano (Eds.), *International handbook of lifelong learning* (pp. XVIII–XLV). Dordrecht: Kluwer Academic Publishers.

Benn, R. (1997). *Adults count too: Mathematics for empowerment.* Leicester: NIACE.

Bishop, A. J. (1991). *Mathematical enculturation. A cultural perspective on mathematics education.* Dordrecht: Kluwer Academic Publishers.

Damasio, A. R. (1999). *The feeling of what happens.* New York: Harcout Brace & Company.

David, E. E. (1984). Renewing U.S. mathematics: Critical resource for the future. *Notices of the American Mathematical Society, 31*(5), 435–466.

Dörfler, W. (2002). *Instances of diagrammatic reasoning.* Preprint. Austria: University of Klagenfurt.

Evans, J. (1999). Adult maths and everyday life: Building bridges, facilitating transfer. In M. Groenestijn & D. Coben (Eds.), *Mathematics as part of lifelong learning* (pp. 77–84). London: Goldsmiths University of London.

Evans, J. (2000). *Adults' mathematical thinking and emotions: A study of numerate practices.* London: RoutledgeFalmer.

Evans, J. & Thornstad, I. (1994). Mathematics and numeracy in the practice of critical citizenship. In D. Coben (Ed.), *ALM 1: Proceedings of the inaugural conference of Adults Learning Mathematics—A Research Forum* (pp. 64–70). London: University of London.

Evans, J. & Rappaport, I. (1998). Using statistics in everyday life: From barefoot statisticians to critical citizenship. In D. Dorling & S. Simpson (Eds.), *Statistics in society: The arithmetic of politics* (pp. 71–77). London: Arnold.

Fischer, R. (1999). Technologie, mathematik und ewußtsein der gesellschaft. In G. Kadunz et al. (Eds.), *Mathematische bildung und neue technologien* (pp. 85–102). Stuttgart-Leipzig: B.G.Teubner.

FitzSimons, G., Jungwirth, H., Maaß, J., & Schlöglmann, W. (1996). Adults and mathematics. In A. J. Bishop, K. Clements, C. Keitel, J. Kilpatrick & C. Laborde (Eds.), *International handbook of mathematical education* (pp. 753–784). Dordrecht: Kluwer Academic Publishers.

FitzSimons, G., Coben, D., & O'Donoghue, J. (2003). Lifelong mathematics education. In A. J. Bishop, M. A. Clements, C. Keitel, J. Kilpatrick, & F. Leung (Eds.), *Second international handbook of mathematics education* (pp. 103–142). Dordrecht: Kluwer Academic Publishers.

Frankenstein, M. (1990). Incorporating race, gender, and class issue into a critical mathematics literacy curriculum. *Journal of Negro Education, 59*(3), 336–347.

Hannula, M. (1998). Changes of beliefs and attitudes. In E. Pehkonen & G. Törner (Eds.), *The state-of-art in mathematics-related belief research: Results of the MAVI activities. Research Report 184* (pp. 198–222). Finland: University of Helsinki.

Heintel, P. (1992). Skizzen zur "technologischen formation". In W. Blumberger & D. Nemeth (Eds.), *Der technologische Imperativ* (pp. 267–308). München-Wien: Profil.

Hülsmann, H. (1985). *Die technologische formation*. Berlin: Verlag Europäische Perspektiven.

Ingleton, C. & O'Regan, K. (2002). Recounting mathematical experiences: Emotions in mathematics learning. *Literacy & Numeracy Studies, 11*(2), 95–107.

Jungwirth, H., Maasz, J., & Schlöglmann, W. (1995). *Abschlussbericht zum forschungsprojekt mathematik in der weiterbildung*. Austria: Linz.

Klafki, W. (1975). *Studien zur bildungstheorie und didaktik: Durch ein kritisches vorwort ergänzte auflage*. Weinheim/Basel: Beltz Verlag.

Lave, J. (1988). *Cognition in practice*. Cambridge: Cambridge University Press.

Lave, J. & Wenger, E. (1991). *Situated learning. Legitimate peripheral participation*. Cambridge: Cambridge University Press.

Lindenskov, L. (1996). Kursundersogelse pa AMU—centre om alment—faglig kompetence I matematik. In T. Wedege (Ed.), *Faglig profil I matematik* (pp. 73–84). Copenhagen: Arbedsmarkedstyrelse.

Maaß, J. & Schlöglmann, W. (1988). The mathematical world in the black box: Significance of the black box as a medium of mathematizing. *Cybernetics and Systems: An International Journal, 19*, 295–309.

Nissen, H. J. et al. (1991). *Frühe schrift und techniken der wirtschaftsverwaltung im alten vorderen orient*. Bad Salzdetfurth: Franzbecker.

Pichot, A. (1995). *Die geburt der wissenschaft*. Darmstadt: Wissenschaftliche Buchgesellschaft.

Schliemann, A. (1999). Everyday mathematics and adult mathematics education. In M. Groenestijn & D. Coben (Eds.), *Mathematics as part of lifelong learning* (pp. 20–31). London: Goldsmiths University of London.

Schlöglmann, W. (1992). Mathematik als technologie. In W. Blumberger & D. Nemeth (Eds.), *Der technologische Imperativ* (pp. 189–198). München-Wien: Profil. Verlag.

Schlöglmann, W. (1999). On the relationship between cognitive and affective component of learning mathematics. In M. Groenestijn & D. Coben (Eds.), *Mathematics as part of lifelong learning* (pp. 198–203). London: Goldsmiths University of London.

Schlöglmann, W. (2002). Mathematics and society—must all people learn mathematics? In L. Ostergaard & T. Wedege (Eds.), *Numeracy for empowerment and democracy?* Proceedings of the 8th international conference of adults learning mathematics (ALM8) (pp. 139–144). Roskilde: Roskilde University Printing.

Schlöglmann, W. (2003). Education in mathematics for adults today. In J. Evans, P. Healy, V. Seabright, & A. Tomlin (Eds.), *Policies and practices for adult learning mathematics: opportunities and risks*. Proceedings of the 9th international conference of adults learning mathematics (ALM 9)—A Research Forum (pp. 143–150). London: Kings College.

Schlöglmann, W. (2006). Lifelong mathematics learning—a threat or an opportunity? Some remarks on affective conditions in mathematics courses. *Adult Learning Mathematics—an International Journal, 2*(1), 6–17.

Schmandt-Besserat, D. (1977). An archaic recording system and the origin of writing. *Syro-Mesopotamian Studies, 1*, 31–70.

Stroop, D. (1998). *Alltagsverständnis von mathematik bei erwachsenen*. Frankfurt: Peter Lang Verlag.

Struik, D. J. (1967). *Abriss der geschichte der mathematik*. Berlin: VEB.

Van der Waerden, B. (1979). *Die Pythagoreer*. Zürich–München: Artemis.

Wedege, T. (1998). Adults knowing and learning mathematics. In S. Tosse et al. (Eds.), *Corporate and nonconform learning. Adult education research in Nordic countries* (pp. 177–197). Tapir: Trondheim.

Wedege, T. & Evans, J. (2006). Adults' resistance to learning in school versus adults' competencies in work: The case of mathematics. *Adults Learning Mathematics—An International Journal, 1*(2), 28–43.

Wenger, E. (1998). *Communities of practice: Learning, meaning, and identity*. New York: Cambridge University Press.

Commentary on the Chapter by Wolfgang Schlöglmann, "Mathematics Education for Adults: Can It Reduce Inequality in Society?"

Tine Wedege

Adults, education, Bildung, inequality, lifelong learning, and mathematics are the key terms in the chapter written by Schlöglmann. In the 1990s, he was one of the pioneers that cultivated the borderland between mathematics education, adult education, and vocational education as a subfield of mathematics education research (see Wedege 2000). Together with Jungwirth and Maasz at the University of Linz, he conducted a large empirical study exploring "the state of mathematics education within the adult education system in Austria" (Jungwirth et al. 1995, p. 13). In this study, the authors made an important distinction between courses where mathematics is explicitly taught and courses where mathematical concepts and methods are used implicitly. In order to label the latter they constructed the term "Mathematikhaltige Weiterbildung" (translation: "Mathematics-containing continuing education"[1]) presumably to remind people that mathematics in vocational training, as in the workplace itself, is integrated with other subjects and vocational competences. Elsewhere I claimed that within the scientific domain of mathematics education they paved the way for research on vocationally oriented adult education, where mathematics is an integral part (Wedege 2000).

---

[1] For the purpose of a study on workers' mathematical knowledge in the workplace (1996–1998), I translated the German "matematikhaltige" directly into a Danish construction of the word: "matematikholdige" (Eng: mathematics-containing) and defined *mathematics-containing technology* as technology "where mathematics is an integrated but potentially identifiable part" in technic & machinery, work organisation, and workers' competences (Wedege 2004, p. 103)—well knowing that a difficult issue is raised by this definition: Identifiable—by whom?

T. Wedege (✉)
Malmö University, Malmö, Sweden
e-mail: tine.wedege@mah.se

H. Forgasz, F. Rivera (eds.), *Towards Equity in Mathematics Education,*
Advances in Mathematics Education,
DOI 10.1007/978-3-642-27702-3_26, © Springer-Verlag Berlin Heidelberg 2012

1 The Problématique

Schlöglmann in his chapter has chosen a broader perspective on adult mathematics education. The socio-political context is Western highly industrialised countries and the focus is inequality. The notion of Lifelong Learning (LL) is the frame for analysing adults, mathematics, and democracy. From this position, with Bildung as the perspective on adult mathematics education, he formulates the following statement:

• Lifelong learning is prerequisite for making democratic participation possible for all members of a society.

Like numeracy in the international field of adult mathematics education, Bildung is a contested concept in the field of education. However, there is agreement within educational discourses on contrasting Bildung in relation to socialisation and vocational qualification (see Wedege 1993). Schlöglmann has chosen the concept of Bildung as defined by the German didactician, Klafki. This concept, *categorical Bildung*, distinguishes itself by including two dimensions (formal & material Bildung) and the didactical interplay between them (see Wedege 1993).

From the very beginning, the theme mathematics/numeracy for citizenship has been raised in various conferences of the international forum for research on Adults Learning Mathematics (ALM). It is, for example, visible in the debate involving numeracy versus mathematics. Adult numeracy for empowerment and democracy has been problematized in single papers as well as in whole conferences. For example, in Johansen and Wedege (2002), the issue was raised in the call for papers. Also, in Lindberg (2005), Bildung and vocational training were contrasted.

"Som man råber i skoven, får man svar". This is an Old Danish saying that translates into something like "You get what you are asking for". Schlöglmann has given the answer simply by formulating the following question in the title: "Can mathematics education for adults reduce inequality in society?" The same goes for a series of other yes/no questions brought forward in his chapter. For example: "Should everybody learn mathematics?" and "Is mathematics learning possible for adults?" When these questions are formulated from an equity perspective, "'no'" is not a possible answer. A central assumption in Schlöglmann's essay is that:

• lifelong learning is a consequence of technological development in societies.

In such contexts, mathematics is understood as Western, academic/school mathematics, following Bishop (1988). Also, the socio-political context is industrialised democratic countries. Thus, the problématique in Schlöglmann's essay is situated within the OECD discursive hegemony on LL, which adopts a human capital approach (see Rubenson 2008).

Schlöglmann announces the genre of his chapter to be "an essay". Thus, the approach should begin with a presentation and discussion of a problem complex from a subjective or general point of view and end with a personal conclusion. At the same time, the context should be mathematics education research. However, Schlöglmann's approach is between an essay and a scientific article, which makes it possible for him to explore the issue from different angles.

2 Why Teach and Learn Mathematics?

In an international context, Niss (1996) has studied the so-called justification problem in mathematics education. From an educational, societal, and political perspective he reviewed the official reasons for teaching and learning mathematics. The result of his analysis is a grouping of reasons into three general types: (1) Contributing to the technological and socio-economic development of society at large; (2) Contributing to society's political, ideological, and cultural maintenance and development; (3) Providing individuals with prerequisites that may help them cope with life in education or occupation, private life, and as a citizen. In a debate on reasons and motives for mathematics education, following this analysis, I see two main types of purposes at two agentic levels—that is, one that refers to the needs of society (general) and the other to the needs of individuals (subjective). Furthermore, the focus can be global, which concerns mathematics education, or local, which deals with the question of why a certain mathematics education is needed for a particular category of students in some specific educational programme (see Jensen et al. 1998). Schlöglmann's discourse resonates well with Niss's articulation of the three types of reasons for teaching and learning. When Schlöglmann talks about adult mathematics education he uses the following two expressions (my italics):

- Mathematics education *for* adults,
- Adult education *in* mathematics.

The distinction between the two expressions has not been explicitly presented. The first expression has been used in the title and in the concluding section only. The second expression has been used throughout the text. "Education *for* adults" resonates with the idea of "providing" adults with skills, knowledge, competences, etc. Talking about adults, mathematics and education, other discursive possibilities are Adults learning mathematics (ALM), which is related to the international research forum, or Adult mathematics education. The latter I understand is a general neutral expression: The agent is hidden and this makes possible different interpretations. Johansen (2006) has located two different discourses in the argumentation for preparatory adult mathematics education (PAE) in Denmark. The politicians, on the one hand, talked about adults with poor skills that are not prepared for societal and labour market demands. The researchers, on the other hand, were involved as educational planners who talked about competent adults who are not aware of their own mathematics-containing skills in everyday life (Johansen 2006).

Elsewhere (Wedege 2010), I have argued that teacher education has to raise the justification issue and do it from a general as well as a subjective point of view in order to move the discussion from "Why do I have to learn division of two fractions" into a broader context about relevance versus utility (Ernest 2004) and needs versus demands. It is obvious that the answer to the question ('Why mathematics education') is closely related to the other two classical didactical questions 'What' and 'How'. Mathematics education per se does not support democracy as Skovsmose (1990) has argued in a discussion about mathematics education and democracy. He identified a dilemma between high level mathematics for all and democratic teaching methods in the classroom.

My concluding claim is that a fourth aspect ('Who') has to be added into the so-called didactical triangle ('Why—What—How') when it is used for educational planning, teaching, and research in adult mathematics education. Schlöglmann's main question: "Should everybody learn mathematics?" cannot be answered with a simple 'Yes' or 'No' within adult mathematics education without including 'Who'. I propose that the question be reformulated in order to include the needs and constraints experienced by all individuals in their respective social contexts (Wedege and Evans 2006).

Finally, I want to emphasize that we need to address gender issues in mathematics and society in the form of mainstreaming (see Henningsen 2008). This has not been the case in Schlöglmann's chapter and in my short commentary. By mainstreaming, I mean that any problématique explicitly has to consider the extent to which gender is a relevant issue. In the case of mathematics and equity in societies, in particular, where adults are concerned, I find that the gender aspect is a necessary component in the discussion.

References

Bishop, A. J. (1988). *Mathematical enculturation. A cultural perspective on mathematics education*. Dordrecht: Kluwer Academic Publishers.

Ernest, P. (2004). Relevance versus utility: Some ideas on what it means to know mathematics. In B. Clarke et al. (Eds.), *International perspectives on learning and teaching mathematics* (pp. 313–327). Göteborg: National Center for Mathematics Education, NCM.

Jensen, J. H., Niss, M., & Wedege, T. (Eds.) (1998). *Justification and enrolment problems in education involving mathematics or physics*. Frederiksberg: Roskilde University Press.

Henningsen, I. (2008). Gender mainstreaming of adult mathematics education: Opportunities and challenges. *Adults Learning Mathematics—an International Journal, 3*(1), 32–40 (available 25.04.11 at www.alm-online.net ALM journal).

Johansen, L. Ø. & Wedege, T. (Eds.) (2002). *Conference proceedings ALM8* (available 25.04.11 at www.alm-online.net ALM proceedings).

Johansen, L. Ø. (2006). *Hvorfor skal voksne tilbydes undervisning i matematik? En diskursanalytisk tilgang til begrundelsesproblemet*. [Why offer mathematical instruction to adults?] Doctoral thesis, DCN, Aalborg University, Aalborg.

Jungwirth, H., Maasz, J., & Schlöglmann, W. (1995). *Abschlussbericht zum Forschungsprojekt Mathematik in der Weiterbildung*. Linz (Unpublished).

Lindberg, L. (Ed.) (2005). *Conference proceeding from ALM-11* (available 25.04.11 at www.alm-online.net ALM proceedings).

Niss, M. (1996). Goals of mathematics teaching. In A. J. Bishop et al. (Eds.), *International handbook of mathematics education* (pp. 11–47). Dordrecht: Kluwer Academic Publishers.

Rubenson, K. (2008). OECD education policies and world hegemony. In R. Mahon & S. McBride (Eds.), *The OECD and transnational governance* (pp. 242–259). Vancouver: UBC Press.

Skovsmose, O. (1990). Democracy and mathematics education. *Educational Studies in Mathematics, 21*, 109–128.

Wedege, T. (1993). Fra kvalificering til dannelse. 'Fleksibilitet' som en progressiv dannelseskategori. [From qualification to Bildung: Flexibility as a progressiv Bildung category]. *Dansk Pædagogisk Tidsskrift, 3*, 135–143.

Wedege, T. (2000). Technology, competences and mathematics. In D. Coben, J. O'Donoghue, & G. E. FitzSimons (Eds.), *Perspectives on adults learning mathematics: Research and practice* (pp. 191–207). Dordrecht: Kluwer Academic Publishers.

Wedege, T. (2004). Mathematics at work: Researching adults' mathematics-containing competences. *Nordic Studies in Mathematics Education, 9*(2), 101–122.

Wedege, T. (2010). Needs and demands: Some ideas on what it means to know mathematics in society. In B. Sriraman & S. Goodchild (Eds.), *Relatively and philosophically E^{a}rnest: Festschrift
in honor of Paul Ernest's 65th Birthday* (pp. 221–234). Charlotte: Information Age Publishing.

Wedege, T. & Evans, J. (2006). Adults' resistance to learning in school versus adults' competences
in work: The case of mathematics. *Adults Learning Mathematics—An International Journal,
1*(2), 28–43 (available 25.04.11 at www.alm-online.net ALM journal).

Commentary on the Chapter by Wolfgang Schlöglmann, "Mathematics Education for Adults: Can It Reduce Inequality in Society?"

Jaguthsing Dindyal

Education is a social mechanism whereby one generation passes on its knowledge, skills, and values to the next generation. However, for various reasons, access to educational opportunities and successful completion of formal education are not provided to some individuals. As Apple has claimed:

> Education does not exist isolated from the larger society. Its means and ends, the daily events of curriculum, teaching, and evaluation in schools, all of this is connected to patterns of differential economic, political, and cultural power. (Apple 1992, p. 412)

Accordingly, education is deeply rooted in the political domain of a society. In whatever political context, it is seen to be the means, par excellence, for social emancipation, a fact acknowledged by international organizations. In the 2000 World Education Forum in Dakar (see UNESCO 2000), six goals for achieving *Education for All* by the year 2015 were promulgated two of which are as follows:

- Goal 3: Ensuring that the learning needs of all young people and adults are met through equitable access to appropriate learning and life-skills programmes;
- Goal 4: Achieving a 50 per cent improvement in levels of adult literacy by 2015, especially for women, and equitable access to basic and continuing education for all adults.

Although there is no direct reference to mathematics education, these two goals highlight the importance of adult education, which is Schlöglmann's major focus in his chapter.

Mathematics education is situated within the broader notion of education in all societies. In 1989, the National Council of Teachers of Mathematics (NCTM) proposed four new societal goals for mathematics education through the publication of the *Curriculum and Evaluation Standards for School Mathematics*, namely: (1) the development of mathematically literate workers, (2) the pursuit of lifelong learning, (3) the possibility of opportunity for all, and (4) the development of an informed

J. Dindyal (✉)
Nanyang Technological University, Singapore, Singapore
e-mail: jaguthsing.dindyal@nie.edu.sg

H. Forgasz, F. Rivera (eds.), *Towards Equity in Mathematics Education*,
Advances in Mathematics Education,
DOI 10.1007/978-3-642-27702-3_27, © Springer-Verlag Berlin Heidelberg 2012

electorate. These societal goals underscore the importance of mathematics education in contexts beyond the four walls of the classroom to a wider democratic society.

Schlöglmann to some extent addressed the four goals stated above, although not from the perspective of school mathematics. From his chapter, we draw four implications.

First, we need mathematically literate workers everywhere in the world. In the context of highly industrialized countries, knowledge-based industries are expanding rapidly and labor market demands are also changing very fast. Further, highly automated machinery does not reduce the need for workers; it merely changes the required skills and biases them more in the direction of mathematics education (see Walsh 1990).

Second, adult education is a key element in the general concept of lifelong learning. Schlöglmann uses ideas from a group of authors (Aspin, Chapman, Hatton, and Sawano) to introduce the following four models of lifelong learning: (1) a compensatory model, (2) a continuing vocational model, (3) a social innovation model, and (4) a leisure orientated model. These models of lifelong learning clearly describe an adult learner as somebody who is no longer in school or college but somebody who has had some life experiences after leaving school. On the other hand, the World Bank (2003) describes lifelong learning as something more than just adult continuing education: "Lifelong learning encompasses learning over the entire life cycle (from early childhood to retirement) and all learning systems (formal, non-formal, and informal)" (p. 2).

Let us look at the idea of mathematics education for adults. Any discussion about mathematics education for adults has to grapple with the idea of an adult learner of mathematics. Is an adult learner somebody who is legally an adult, say, age 18 years in most countries? Such an age-bound definition may include all students who are still in high school or who are entering college courses. Or, is an adult learner of mathematics somebody who comes to study mathematics after having been away from school or college either because he or she did not have an opportunity to study the subject earlier of if ever he or she did was moderately successful or even unsuccessful in mathematics? Schlöglmann does not dwell on who is an *adult learner* of mathematics but sees adult education as having close connections with the general concept of lifelong learning. However, the term "adult" in any country is a very generic term. Background factors such as gender, race, ethnicity, socioeconomic status, immigrant or local, prior formal education, even physical fitness and age cannot be taken for granted. These background factors strongly influence an individual's choice to take up or not adult education courses.

Third, the idea of *opportunity for all* is encapsulated in the issue of inequality that Schlöglmann highlights in his chapter. The main point he makes is that there is an uneven acquisition of knowledge of mathematics in society that then gives rise to inequalities. He strongly points out that there exists inequalities in highly industrialized societies that have to be reduced. For Schlöglmann, lifelong learning provides an opportunity that will reduce societal inequalities. He raises a valid claim in that "mathematics education for adults can help reduce inequality in societies, but it is not a silver bullet: a coordinated strategy consisting of several measures is required". Schlöglmann does not give details about how the inequality is to be reduced or eliminated altogether, however, he attempts to distinguish between people who have passed through the educational system with a low level of achievement and those who have migrated to highly industrialized nations from countries with weak education systems. Although there are no negative comments, perhaps there is a need to clarify why this distinction is important. The distinction also seems to contrast with his interpretation of equality, that it should not only be a principle that is valid at the individual level but that it should also be valid for all social groups.

Fourth, the notion of an informed electorate is couched on fundamental principles that govern any democratic society. Schlöglmann points to the Universal Declaration of Human Rights in which one of the rights asserts education for everyone. The word *education* is used in a broad sense but is an essential element for all democratic societies. He states that: "We must extend the concept of lifelong learning to encompass the needs of people in a democratic society". To be educated is to be literate. Does general literacy subsume basic knowledge and skills in mathematics? Some questions are pertinent here: What kind of mathematical literacy is appropriate for adult learners? What would be the correct term to use: numeracy, quantitative literacy, financial literacy, mathematical literacy, democratic numeracy or functional mathematics (see Galligan and Taylor 2008, p. 100)? An interesting point to note regarding this matter is from Apple (1992) who claims that mathematical knowledge is often produced, accumulated, and used in ways that may not be totally democratic, which necessitate thinking carefully about the definitions of mathematical literacy with which we now work. Apple also notes that "literacy" is a slippery term and that mathematical literacy is a sliding signifier that can be used to cover a multitude of social goals. How do we, then, reconcile Apple's comments with the notion that mathematics education can reduce inequalities in society?

Mathematics educators relish having more students learn mathematics in schools. Folk knowledge, if nothing else, tells us that mathematics is an important component of the school curriculum. What about mathematics in the context of lifelong learning? Better still, the question should be: Why should mathematics education be considered as a key element of lifelong learning? Schlöglmann claims that democracy "demands a means for communicating and discussing principles in a rational way. Mathematics with its close relationship to rationality, is our concept to do so". Mathematical reasoning is a very powerful tool, and perhaps this is the connection that he wishes to make with mathematics, the harbinger of rationalism. Schlöglmann aptly states that, "The status of mathematics in our society is rather paradoxical. On the one hand, many people see mathematics as abstract, remote from the life, incomprehensible. On the other hand, the same people have full confidence in mathematical methods—they pay invoices, accept calculation of election results, accept use of complex mathematics in technology and the economy". He also describes a responsible citizen as someone who is able to participate in societal processes in a rational way.

Schlöglmann clearly directs attention to the issue of adult education and more so in the area of adult mathematics education. Adult education may not be on the priority list in many countries, which makes *Education for All* a very ambitious goal. This goal cannot be achieved if there is no clear policy about adult education in any given country. Education for all, or more specifically *Mathematics Education for All*, cannot remain at the level of rhetoric. It would be too simplistic to believe that teaching mathematics to all adults will reduce inequality in society. Mathematics education for adults has to be looked at from the same vantage point as school mathematics education. The guiding questions should be along the following Tylerian lines: Why must we teach mathematics to adults? What mathematics should we teach to adults? How should we teach mathematics to adults? And, how do we know that we have been successful in teaching mathematics to adults? Additional questions include the following: Who should teach mathematics to adults? Where should the teaching of mathematics to adults take place? Who should pay for the costs? Accordingly, I raise the five points below for consideration to further the discussion in this area.

1. There should be a clear commitment from policy makers to support mathematics education for all adults. Some kinds of incentives will encourage adults to participate more in adult education courses.
2. A strong mathematics curriculum specific to the perceived needs of adult learners in a democratic society should be developed. There must be a strong emphasis on developing decision making and problem solving skills, which is a necessity for knowledge-based economies (World Bank 2003). The curriculum needs a strong content base. The teaching and assessment has to be adapted to the level of the individual learners. However, the needs should not only be utilitarian or vocational.
3. Countries need to recruit a well-trained teaching force for the specific purpose of teaching adult learners and with specific knowledge of andragogy. More specifically these teachers need to have some type of Mathematical Andragogical Content Knowledge (MACK) similar to Mathematical Pedagogical Content Knowledge (MPCK) that takes adult learners' approach to learning and interests into account.
4. A venue suitable for running adult education courses and at times more suitable for adult learners should be carefully chosen. There is a need to avoid mass adult education practices using technology that do not necessarily respond to the individual needs of adult learners (see FitzSimons 2007).
5. A cost-sharing approach with the state would be more suitable than an individually financed scheme for adult education courses.

Mathematics education is an essential component of any adult education course. Certainly, enhancing the mathematics education of adults will reduce some inequalities. Schlöglmann, who writes from the perspective of highly industrialized countries, highlights this matter in his chapter. What is needed at this time is how mathematics education for adults can be developed and effectively implemented in order to reduce social inequalities.

References

Apple, M. W. (1992). Do the standards go far enough? Power, policy and practice in mathematics education. *Journal for Research in Mathematics Education, 23*(5), 412–431.

FitzSimons, G. E. (2007). Globalisation, technology, and the adult learner of mathematics. In B. Atweh, A. C. Barton, M. Borba, N. Gough, C. Keitel, C. Vistro-Yu, & R. Vithal (Eds.), *Internalisation and globalization in mathematics and science education* (pp. 343–361). Dordrecht, The Netherlands: Springer.

Galligan, L., & Taylor, J. (2008). Adults returning to study mathematics. In H. Forgasz et al. (Eds.), *Research in mathematics education in Australasia 2004–2007* (pp. 99–118). The Netherlands: Sense Publishers.

National Council of Teachers of Mathematics (1989). *Curriculum and evaluation standards for school mathematics*. Reston: National Council of Teachers of Mathematics.

UNESCO (2000). *The Dakar forum for action—Education for ALL: Meeting collective commitments*. Paris, France: UNESCO.

Walsh, J. B. (1990). The NCTM curriculum and evaluation standards for school mathematics: A business perspective. *School Science and Mathematics, 90*(6), 566–570.

World Bank (2003). *The knowledge economy and the changing needs of the labor market*. Washington: World Bank.

Heteroglossia in Multilingual Mathematics Classrooms

Richard Barwell

Abstract What does linguistic or cultural diversity look like in a mathematics classroom? How does such diversity influence the teaching or learning of mathematics? In this chapter, I address these and related questions. Specifically, I draw on Bakhtin's notion of heteroglossia to analyse the literature on teaching and learning mathematics in linguistically diverse classrooms. Based on this analysis, I describe and discuss four tensions that arise in linguistically diverse mathematics classrooms: tensions between school and home languages; between formal and informal language in mathematics; between language policy and mathematics classroom practice; and between a language for learning mathematics and a language for getting on in the world. These tensions can all be traced to an underlying tension between what Bakhtin calls centripetal and centrifugal forces in language. I conclude by considering some of the implications of my analysis for equity in mathematics teaching.

What does linguistic or cultural diversity look like in a mathematics classroom? How does such diversity influence the teaching or learning of mathematics? And what does it mean for equity in mathematics classrooms? My own experience of teaching and researching in mathematics classrooms around the world leads to an awareness that linguistic and cultural diversity is itself diverse. Attempts to categorise different 'types' of multilingual mathematics classroom (e.g. Barwell 2005a; Clarkson 2009), for example, while not entirely unhelpful, tend to vastly oversimplify the full richness and complexity of language use that exists in mathematics classrooms around the world. To at least partially answer my opening question, let me give some examples.

Example 1 The children of immigrants to Québec must generally attend French medium schools. If they do not speak French, they spend their first year in a classe d'accueil, the main purpose of which is for them to learn French. I have recently been visiting a classe d'accueil made up of eighteen 9–11 year-olds from South America, West Africa, the middle East and South Asia. The languages they speak

R. Barwell (✉)
University of Ottawa, Ottawa, Canada
e-mail: richard.barwell@uottawa.ca

H. Forgasz, F. Rivera (eds.), *Towards Equity in Mathematics Education,*
Advances in Mathematics Education,
DOI 10.1007/978-3-642-27702-3_28, © Springer-Verlag Berlin Heidelberg 2012

at home include Spanish, Swahili, Hindi and Arabic. During their mathematics lessons, the requirement that they use French presents challenges. For example, during some work on the properties of different geometric forms, students struggle to explain their thinking e.g. why they think a shape is non-convex. They also make extensive use of deixis (words like 'this' or, in French, 'ça') often combined with gestures e.g. one student explains that a shape is curved by saying comme ça (like this) and sketching a curve in the air with his pencil. In this class, the acquisition of French is the primary objective, even in mathematics. After the year is over, the students will join mainstream classes. Their various other languages are acknowledged and referred to from time to time but they are expected to use French. And they are expected to learn mathematics even as they learn French.

Example 2 In a classroom in northern Pakistan, mathematics is taught in a mixture of English, Urdu and Burushaski. The school in which I taught was established by the local community to provide an English-medium education. As such, parents pay relatively high fees and the children all live close to the school. English is seen as the language of the elite and ruling classes in Pakistan and widely used in higher education, the civil service and the army. Urdu is the national language and Burushaski is the local language. Mathematics textbooks are in English. Mathematics teachers in the school do not necessarily have teaching qualifications and are not necessarily highly proficient in English. Furthermore, if complex topics like the formal arithmetic properties of associativity, distributivity, etc., are taught entirely in English, many students will not understand. Hence teachers use a mixture of three languages. English is at least used for key terms like associativity and sentences from the textbook, Urdu for more surrounding explanation and discussion and Burushaski for more informal discussion (see Halai 2009, for a more detailed discussion of mathematics classrooms in Pakistan).

Example 3 A few years ago, I spent a year visiting a class of 9–10 year-olds in an inner city school in England. The class included middle class and working class White children, middle class and working class Black children, recently arrived children from Hong Kong and a girl from Somalia whose family had reached England via the Netherlands and Wales (and who consequently knew some Dutch and Welsh). There were also several children from Pakistani backgrounds who had spent various periods in school in Pakistan as well as in England and who spoke Urdu and Punjabi and were learning English. My research focused particularly on word problems (see, for example, Barwell 2005b). The students who were learners of English struggled to make sense of word problems, although my research suggested that they were able to relate mathematics to 'real life' contexts if they created these contexts themselves.

It is apparent from these briefly sketched classrooms that mathematics is taught and learned in a wide range of diverse settings, each diverse in their own particular way. In Québec, all the students in the class are learners of French and all are recent arrivals to the province, but they come from widely differing linguistic and cultural backgrounds. In Pakistan, the children were all from close-by and, in broad terms, shared a common cultural background. And yet they studied mathematics in the context of great linguistic diversity, through a language that they were not likely to encounter much outside of school. In the third class I described, the teacher had to take into account a wide range of proficiency in English and a wide range of familiarity with life in the UK (these two not necessarily coinciding). In terms of mathematics teaching and learning, do these situations and others like or unlike them have anything in common? Are there common issues or challenges that arise? How are language, diversity and mathematics related? How can teachers deal equitably with such diversity?

In this chapter, I will address these questions with a specific focus on language-related issues, although the discussion applies by extension to questions of cultural diversity. I begin with a summary of the main substantive findings of research on teaching and learning mathematics in linguistically diverse classrooms. I then set out a theoretical framework for language diversity based on the work of Bakhtin. These ideas are used to discuss four tensions that arise across the existing research and in a range of different contexts. A key point that emerges from this analysis is that these tensions are an inevitable part of linguistic diversity. In the final part of the paper, I return to the above questions and consider some possible ways in which mathematics teachers could respond to linguistic diversity in the light of my analysis.

1 Research on Teaching and Learning Mathematics in Linguistically Diverse Classrooms

There is a growing body of research on teaching and learning mathematics in multilingual, bilingual or second language classrooms i.e. any setting in which participants could draw on a repertoire of different languages. This work has produced a number of useful substantive findings.

First, research has shown that there is a relationship between students' proficiency in the language used for teaching and assessing mathematics and their attainment, although this relationship is not straightforward. In particular, students who have a high level of proficiency in their first language and who develop a similar level of proficiency in the classroom language tend to outperform monolingual students in mathematics. By contrast, students who do not develop a suitable level of proficiency in any language tend to under-perform in mathematics. Students who develop such proficiency in one language, whether at home or at school, tend to perform at a similar level of monolingual students. These findings have been found in a variety of contexts, including ESL settings in Australia (Clarkson 1992, 2007; Clarkson and Galbraith 1992) and in immersion settings in Ireland (Ní Ríordáin and O'Donoghue 2009) and Canada (Bournot-Trites and Reeder 2001; Lapkin et al. 2003; Swain and Lapkin 2005; Turnbull et al. 2001).

Second, research has established that teaching mathematics in multilingual settings presents challenges for teachers. Such challenges relate to choice of language, to enabling students' expression of mathematics and to the socio-political context. Such issues have been identified in a variety of settings including in South Africa

(Adler 2001; Setati 1998, 2005), in the context of immigration in Catalonia (Gorgorió and Planas 2001) and in Spanish-English bilingual education settings in the United States (Khisty 1995; Moschkovich 2002).

Finally, research has examined some of the strategies that bilingual students use to participate in and make sense of school mathematics. Such strategies include the use of two or more languages, sometimes known as code-switching; and connecting mathematics problems with their experience outside of school. Code-switching in mathematics classrooms has been investigated in, among other places, Australia (Parvanehnezhad and Clarkson 2008), South Africa (Setati 1998), Malta (Farrugia 2009a) and Pakistan (Halai 2009). Bilingual students' ways of engaging with mathematical problems have been examined in ESL settings in the UK (Barwell 2003, 2005b, 2005d), in bilingual education classrooms in the United States (Moschkovich 2008) and in indigenous language contexts in Brazil (Mendes 2007).

This growing body of work underlines the tremendously diverse nature of language diversity in mathematics classrooms around the world and the range of challenges this diversity can sometimes present. These challenges are related to the need to develop proficiency in the language of teaching and learning mathematics at the same time as learning mathematics, as well as the use of different languages by teachers and learners in the mathematics classroom. In this chapter, however, I want to go beyond these basic substantive findings to examine some of the underlying issues and tensions that emerge from all this research. These issues and tensions arise from language diversity but have repercussions for teaching and learning mathematics. To take forward this collected work, I propose a deeper theorisation of language diversity in mathematics classrooms based on Bakhtin's notion of heteroglossia. These ideas are set out in the next section.

2 The Diversity of Language

Much of the history of linguistics is based on the idea that languages are unified and describable—that there is, for example, a standard kind of English or French, for which the rules can be determined. This idea has deep roots, drawing on an ideology of language as a unifying force heavily implicated in the politics of national cohesion. This ideology becomes apparent, for example, in calls for immigrants to be compelled to learn the national language (see, e.g., Blackledge 2002). The European origins of this view of language have been highlighted in African (e.g. Makoni and Meinhof 2004) and South Asian (e.g. Canagarajah 2009) critiques of linguistics: a model of unified, discrete languages does not fit well with the kind of continuous variation in languages found in these parts of the world. Indeed, one of the linguistic legacies of colonialism is the struggle to apply a model of unified national languages in the midst of a language diversity for which it is ill suited.

In mathematics education, meanwhile, there is an assumption that mathematical language, often referred to as 'the mathematical register' (Halliday 1978) is as unified and describable as French or Japanese. This kind of assumption is apparent both in research, such as O'Halloran's (2005) extensive analysis, and in mathematics curricula (cf. Barwell 2005c) and often works in concert with a view of mathematics as equally unified and as language and culture free. From this perspective, the different students in the classe d'accueil referred to above do not bring different mathematics, only different languages. Similarly, the students in the mathematics class in Pakistan may use any language to learn mathematics, since the mathematics is always the same.

Bakhtin's work offers a somewhat different perspective from those I have summarised. Working in Russia during the Soviet era, Bakhtin was primarily a literary theorist—he was interested, for example, in what makes a novel a novel and not, say, poetry. His examination of questions relating to literature led him to develop a highly distinctive theory of language. In particular, his thinking about language is relational rather structural (cf. Holquist 1981). For example, he was more interested in the relationship between actual utterances and the contexts that affect how they are interpreted, than in a structural account of meaning or context. His work is also notable for its attention to the great diversity of language as it is used. Bakhtin's ideas have not been widely influential in mathematics education, although traces are apparent in Sfard's (2008) thinking. Van Oers (2001) draws particularly on some of the ideas that I refer to in this chapter, although, like many interpreters of Bakhtin in the field of education (e.g. Wells 1999), his reading seems to me to be more oriented to psychology than to linguistics. In this chapter, then, I take up Bakhtin's distinction between unitary language and heteroglossia.

Bakhtin (1981) used the term 'unitary language' to refer to the sense of language as coherent and unified. Unitary language, he writes, "gives expression to forces working toward concrete verbal and ideological unification and centralisation, which develop in vital connection with the processes of socio-political and cultural centralization" (p. 271). Unitary language is, however, a theoretical construct; language is not inherently unitary, although the idea of unitary language reflects the common patterns that make meaning possible. Language does not exist in this theoretical form; language is always instantiated in use. Language in use, however, is inherently diverse, multiple and fluid:

> At any given moment of its evolution, language is stratified not only into linguistic dialects in the strict sense of the word [...] but also [...] into languages that are socio-ideological: languages of social groups, "professional" and "generic" languages, languages of generations and so forth. (Bakhtin 1981, pp. 271–272)

A key point about this stratification and about language use is that specific utterances can reflect multiple layers simultaneously. Speech necessarily entails 'the social diversity of speech types' (Bakhtin 1981, p. 263), for which Bakhtin's translators introduce the Latinate term heteroglossia. This idea captures well the nature of interaction in the three mathematics classrooms described above. Each classroom swirled with bits of different languages, some more prominent than others, some filtered through others. Each classroom swirled with the 'social languages' of each student's background: the languages of their social class, race, gender and so on. And each classroom swirled with the social languages of school: the languages of mathematics, of curriculum, of textbooks, of students being students and of teachers being teachers.

For Bakhtin, these two ideas of language—unitary language and heteroglossia— are in constant struggle, a struggle characterised in terms of centripetal and centrifugal forces. Duranti (1998) explains these terms as follows:

> The centripetal forces include the political and institutional forces that try to impose one variety of code over others [...] These are centripetal because they try to force speakers toward adopting a unified linguistic identity. The centrifugal forces instead push speakers away from a common center and toward differentiation. These are the forces that tend to be represented by the people (geographically, numerically, economically, and metaphorically) at the periphery of the social system. (Duranti 1998, p. 76)

The struggle between these two sets of forces (for me 'force' is metaphorical) is present each time we speak and, moreover, shapes what we say:

> Every concrete utterance of a speaking subject serves as a point where centrifugal as well as centripetal forces are brought to bear. The processes of centralization and decentralization, of unification and disunification, intersect in the utterance; the utterance not only answers the requirements of its own language as an individualized embodiment of a speech act, but it answers the requirements of heteroglossia as well; it is in fact an active participant in such speech diversity. And this active participation of every utterance in living heteroglossia determines the linguistic profile and style of the utterance to no less a degree than its inclusion in any normative centralizing system of a unitary language. (Bakhtin 1981, p. 272)

On the one hand, then, every utterance must conform to some recognisable pattern of language (or it would be incomprehensible). On the other hand, every utterance contributes to the continuous variation and reinvention of human communication. These forces are apparent in the classrooms described above. The heteroglossia in the mathematics classrooms I have described is resisted by the requirement to speak French or English and the goal of learning to communicate a fixed version of mathematics in a recognisable way. This sense of a struggle between unifying and stratifying forces of language (and of mathematics) makes possible a deeper analysis of some of the issues that have arisen in research in second language or multilingual mathematics classrooms. In the next part of this chapter, I describe and analyse four such issues, which I present as a set of tensions. By *tension*, I am drawing on Bakhtin's sense that each utterance simultaneously represents both diversity and uniformity: there is a kind of struggle between the uniqueness of each utterance and its conformity to the patterns of language. I have synthesised four tensions from the existing literature on teaching and learning mathematics in linguistically diverse classrooms. Following Bakhtin, each tension involves a unitary pole and a heteroglossia pole. They are illustrated with examples from relevant studies. Of course, these tensions are simplifications: they interact and overlap and in many ways and can be seen as different lights cast on the deeper underlying tension inherent in diversity itself.

3 Tension Between School and Home Languages

Perhaps the most explicit manifestation of unitary language in mathematics classrooms is the stipulation of an official language of schooling. In the classe d'accueil, for example, French is the language of teaching and learning. The curriculum, classroom texts, classwork and the teacher are almost always presented exclusively in French. The expectation is that students will become more proficient in the classroom language over time. In practice, Spanish, for example, is heard, sometimes directly in conversations between students, sometimes indirectly in the accents or interpretations of French words. On one occasion, I observed a student ask if he could explain in Spanish how he knew a shape was convex; he was asked to try in French. There is an inevitable tension, then, between the languages that students use in or out of the classroom and the requirements to use a particular language to talk about mathematics.

The research literature provides many examples of the tension between home and school languages, examining the presence and use of students' home languages, or practices such as code-switching in which students switch between two or more different languages. Adler (1995, 2001), for example, observed secondary school mathematics lessons and recorded interviews with teachers in multilingual South Africa, a nation with 11 official languages. Adler (1995) reports the following remarks from interviews with mathematics teachers in two different types of secondary school:

> ...in Std 7, where I asked a question and one answered in Tswana. And I said: "Can you please try answer that in English—I don't understand that?" and he said, crossly "No, mam, but you are Tswana—you are not white!" He was angry.

> Sometimes you find that you get stuck because students cannot communicate—then, though not much, you resort to Tswana. You are careful because if you do that then they want you to do it all the time.

> ...there are Xhosa speakers in the class—so if I am speaking Tswana then they complain I am favouring them.

> The problem (in group work—with discussion in any language, report back in English) is... if all your discussion is in Zulu you get to the concept then you can't report back in English so you can't talk about it in English 'cos you never developed it in English. (All from Adler 1995, pp. 268–269)

Both centripetal and centrifugal forces are much in evidence in these comments. The enforcement of English is in tension with the heteroglossia of the students. The school, the curriculum and the teacher are pushing for the use of a single, unitary language, while the students and at some points the teacher are used to using several different languages. This tension leads to what Adler (e.g. 2001) calls 'teaching dilemmas'—for example, the dilemma of whether it is better to use English, knowing that students may not always understand, or to use students' home languages, knowing that they will not develop proficiency in the English of mathematics. The notion of 'dilemma' corresponds closely to my use of the word 'tension' in this chapter: the dilemmas identified in Adler's research represent the expression of these tensions by teachers. Some teachers may not see these situations as dilemmatic: the tension, however, would still be present, since it is part of human communication.

Adler's work shows how the tension between home and school languages can be experienced by teachers. Clarkson has investigated students' perspectives, seeking to understand when and how multilingual students 'switch' between languages when working on mathematics in Australia (e.g.: Clarkson 2007; Clarkson and

Dawe 1997; Parvanehnezhad and Clarkson 2008). This work has led to recognition that 'home' languages can have a role in understanding and learning mathematics:

> some students appeared to rely in part on what they had learnt from parents, related some of the mathematics items to their lived experiences embedded in L1 contexts in their own community, and used some Persian mathematical words which they habitually use because of their home background. [...] For bilingual students, some of what they know, and indeed are, is embedded in a L1 social context, and some of their ideas are clearly more easily expressed in their L1. (Parvanehnezhad and Clarkson 2008, p. 77)

Essentially, what Parvanehnezhad and Clarkson have uncovered is an often hidden heteroglossia that maintains a level of language diversity in mathematics. In the Australian context, in which there is a strong ideology around the (monolingual) use of English for teaching and learning, Clarkson's work illustrates the tension for students; if multiple language use were accepted in Australian classrooms, there would be little need for Parvanehnezhad and Clarkson to argue for the kind of position set out in the above quote.

One of the challenges of researching the tension between home and school languages in mathematics classrooms is that home languages are often marginalised. I spent a full year visiting the classroom in the UK described at the start of this chapter but rarely heard use or acknowledgement of students' home languages. Their 'other' languages must have been present for these students—heteroglossia does not simply disappear in the face of the centripetal forces of a unitary language ideology. Indeed, in my research in this classroom, I did find evidence of their home languages and experiences informing their work in mathematics (e.g. Barwell 2005d). The danger is that in such classrooms, performance in mathematics is assessed from a unitary language perspective, as Moschkovich (2009) has suggested:

> Too often, descriptions of bilingual students focus on the obstacles they face in understanding text or utterances in English and these misunderstandings are invariably ascribed to their lack of proficiency in their second language. (Moschkovich 2009, p. 79)

There is a danger, then, that our research contributes to the centripetal forces that serve to marginalise the kind of diversity that is, for the most part, ever present. Students are used to using a mixture of languages, regularly incorporating bits of one language into fragments of another in creative and entirely comprehensible ways. School, including in mathematics, requires them to use just one language. My purpose here is not to argue for more heteroglossia, nor for that matter to defend a unitary language approach. At this point, I want merely to highlight the tension that arises—although it is worth noting that it arises more strongly for second language learners of the classroom language, who are by and large, as Duranti (1998) observed, at the periphery of the system.

4 Tension Between Formal and Informal Language in Mathematics

The second recurring tension observable in studies of mathematics learning in multilingual settings concerns students' informal expression of their mathematical thinking and the need for students to learn to use more formal mathematical language.

In classrooms in which multiple languages are actively used, such as the classroom in Pakistan that I described above, it is common for formal mathematical terms to be presented and used in the official language. For example, in the classroom in Pakistan, words like 'algebra', 'divide' or 'axis' would commonly be in English, whether as part of a discussion in English or incorporated into a discussion in Urdu or Burushaski. Furthermore, much of the English used in mathematics lessons quoted the textbook, either through reading it aloud or through subsequent reproduction of the text. Students' informal discussion of mathematical ideas were, however, more likely to be in Urdu or Burushaski. This kind of phenomenon has been more systematically documented by Setati (2005) who in primary school mathematics classrooms in South Africa, found exactly this division. Moreover, she shows how discussion of mathematics in English tended to be more procedural in nature, while discussion of students' thinking or mathematical ideas was more likely to be in their home languages. This work shows how centripetal forces, in this case in the form of a unitary language view of the mathematics register, are in tension with the centrifugal forces of students' 'informal' ways of expressing their mathematical thinking. In Bakhtin's terms, these informal forms of expression reflect the many social languages the students bring with them into the mathematics classroom, whether they are social languages within the official language of the classroom or within the other languages (and combinations of languages) they speak.

The tension for teaching and learning arises from a need for students to develop recognisable ways of communicating mathematics set against a desire on the part of teachers that their students should meaningfully discuss mathematics. This tension is often seen as a trade-off, in which teachers allow a degree of informal expression to ensure that students understand the mathematics, while seeking to gradually enhance students' use of more formal or standard forms of expression (Adler 2001; Khisty 1995). Indeed both Setati and Adler (2000) and Clarkson (2009) develop explicit models to deal with this tension. In each case, the approach consists of recognising and building from students' informal expression of mathematics in any language, towards more formal expression of mathematics in any language, with perhaps the ultimate goal that students are proficient users of formal mathematical English. These models encapsulate the underlying tension between the centripetal forces of standard mathematical language and the centrifugal forces of language diversity, where diversity includes both natural and social languages.

In settings in which students are seen as second language learners, rather than as multilingual, the same kinds of tension are present. Khisty (1995), for example, compared three second-grade and two fifth-grade mathematics classes containing some Spanish-speaking bilingual students. The class teachers were all English-Spanish bilingual to some degree, although Spanish was not used much for mathematical discussion. In one classroom considered to be an effective learning environment, mathematics was negotiated through discussion, challenge and debate. This environment frequently required students to explain their ideas and to draw on previous experience to make sense of new situations. For Khisty, the culture of this classroom led to students making mathematical meaning for themselves through interacting with both the teacher and other students. Teachers in the study that seemed to be more effective paid more attention to the language of mathematics as well as to the mathematics itself. Through this attention, as in the models described above, students' informal expressions of their thinking are shaped into more formal mathematical expression. The same kinds of forces are at work, then, in various settings, with students' social languages seen as a starting point for developing a more recognisably standardised use of mathematical language. Again, I do not see this tension as easily resolvable; the purpose of a fairly standard mathematical register is to facilitate communication of mathematical ideas. Equally, however, there is a danger that this same mathematical register marginalises students at the peripheries.

5 Tension Between Language Policy and Mathematics Classroom Practice

Language policy is a specific manifestation of Bakhtin's centripetal forces in language diverse mathematics classrooms. The purpose of such policy, after all, is, in most cases, to mandate some form of unifying order. By language policy in the context of this chapter, I am referring to any official statement of how language or languages should be used in mathematics classrooms. Such statements range from general language policies set out a national level to school-level policies. Such policy may arise within mathematics curriculum or other policy documents. Government guidance for primary school teachers in the UK, for example, includes statements about how teachers should work with learners of English as an additional language. Such statements have tended to emphasise the precise and the unambiguous nature of mathematical language and the importance of learning to do mathematics in English (cf. Barwell 2005c). More recent statements have highlighted the value of students' home languages in learning mathematics, but only as a useful strategy in developing proficiency in English (e.g. DCSF 2007). Such guidance therefore generally reflects a unifying perspective on language and mathematics. In particular, diversity tends to be presented as something to be accommodated, rather than an intrinsic and valuable aspect of mathematics classroom life (Barwell 2005e, p. 318). Such policies represent an idealised view of language use in mathematics classrooms; what actually transpires is inevitably rather messier. In the UK classroom described at the start of this chapter, for example, while languages other than English were rarely heard, their presence was nevertheless felt in the form of non-standard pronunciations or discussions of linguistic features of English (see, e.g., Barwell 2005d). While in unitary language terms English was the only language used, the nature of this English was diverse and clearly interacted with how students' interpreted mathematical texts such as word problems (Barwell 2005d).

The same kind of tension arises in officially multilingual societies. Farrugia's (2009a, 2009b) research provides a good example: she investigated the use of English and Maltese in mathematics classrooms in Malta. Official policy requires only English to be used, despite the use of a mixture of English and Maltese being widespread in Maltese society. Farrugia found that mathematics classroom interaction reflected this practice, with students drawing on both languages. Her account of this situation clearly illustrates this tension:

> Although codeswitching in many classrooms is a common practice in Malta, the writers of the NMC [the national curriculum] appeared not to agree with it. Hence, they suggested that mathematics [...] be taught through English. Codeswitching was only acceptable when the use of English caused 'great pedagogical problems' (Ministry of Education 1999, p. 82). I cannot exclude the fact that some of the writers may have had a personal perception of English as being in some way 'superior' to Maltese, or, as Baker (2001) implies, that the recommendation is motivated by an ulterior agenda to offer an advantage to certain social groups. However, from my personal acquaintance with some of the writers, and my interpretation of the document, the apparent reasons for the recommendation were to find a way to improve students' competence in English and a disapproval of codeswitching as a pattern of language. The medium of instruction issue for mathematics is a hotly debated topic in Malta. Those in favour of English argue in a similar way to the NMC writers, while those in favour of Maltese (or rather codeswitching) tend to present arguments that prioritise mathematical understanding. (Farrugia 2009b, p. 101)

As a 'hotly debated' subject, the tension is apparently very real in Malta. The terms of the debate can be traced to Bakhtin's centripetal and centrifugal forces. The former are represented in several forms including the supposed superiority of English or the disapproval of code-switching as degenerate, despite its prevalence in the language practices of Maltese society. Similar issues have been reported elsewhere including in Pakistan (Halai 2009), Swaziland (Dlamini 2008) and in Malaysia (Lim and Ellerton 2009). More generally, decisions about the language of textbooks or debates about the suitability of some languages for doing mathematics also reflect this tension. It is noticeable in all these examples that the tension between language policy and mathematics classroom practice is closely interrelated with the preceding tensions—with concerns about code-switching and informal language practices. Language policy generally seems to seek to suppress such practices, or to see them at best as steps on the road to monolingual proficiency in a given high status language. It is rare (at least in my experience) to find policies that explicitly advocate the growth of students' home languages through their work in mathematics classrooms.

6 Tension Between a Language for Learning Mathematics and a Language for Getting On in the World

The final tension I will discuss highlights the political role of language in society. It concerns two, sometimes conflicting roles of language. First, language is used to learn. In many cases, learning mathematics would be more effective if conducted using a language repertoire familiar to students. In some cases, this would mean using a single language (perhaps Urdu, in Pakistan). In other cases, it might mean using a mixture of languages (perhaps French and Spanish for some students in the classe d'accueil). Second, however, language has a socio-political role. The reason for the existence of the classe d'accueil, for example, is, in part, to protect and maintain the French-speaking nature of Québec society. French-medium education is mandatory for the majority of immigrants to Québec, regardless of their language background. At the national level, Canada is an officially bilingual country in which the roles of

French, English and bilingualism are highly politicised—to the extent that consideration of aboriginal languages or immigrant languages is somewhat overshadowed, despite their prevalence.

The appearance of this tension in relation to mathematics classrooms has been investigated by Setati (2008), who drew on interviews with teachers and high-school students in South Africa. The majority of teachers and students in her study expressed a preference for the learning and teaching of mathematics to be conducted in English. Setati attributes their position to an implicit view of English as international and as providing access to higher education, jobs and generally getting on in the world. This preference must, however, be set against the effectiveness of students' learning of mathematics. The two issues are in tension, however, as illustrated by one of Setati's interviews with high-school students:

> Nhlanhla, however, had conflicting cultural models. While she acknowledged the power of English, she also accepted the fact that if she focused on wanting to understand mathematics she would choose her home language as the language of learning and teaching.
>
> Nhlanhla: ...is the way it is supposed to be because English is the standardized and international language.
> MS: Okay, if you had a choice what language would you choose to learn maths in?
> Nhlanhla: For the sake of understanding it, I would choose my language. But I wouldn't like that [English as language of learning and teaching] to be changed because somewhere somehow you would not understand what the word 'transpose' mean, ukhithi uchinchela ngale [that you change to the other side], some people won't understand. They would not understand what it means to change the sign and change the whole equation.

(Setati 2008, pp. 110–111)

In most cases, students and teachers in South Africa appear to prioritise learning English over understanding mathematics. Although Setati's work does not appear to have been replicated elsewhere in the world, the picture it presents is likely to be widespread. In many societies, the possibility of choosing the language of mathematics teaching and learning does not even arise. In England, for example, state schooling is in English: speakers of other languages have little choice but to accept this position. In countries where there is a choice, the tendency is for high-status languages to be preferred. Halai (2009) reports similar issues from Pakistan, where English-medium education is highly valued, to the extent that national policy has shifted towards teaching some high-school subjects, including mathematics, in English rather than Urdu (p. 48). In Pakistan, as in South Africa, there is a trend towards the use of English and away from the official use of national or local languages. This trend and the related tension is again traceable to Bakhtin's centripetal and centrifugal forces. The role of language in getting on in the world seems to lead to a unitary perspective in which high status languages are seen as more suitable for mathematics teaching and learning due to their socio-political value, rather than their educational value. The centripetal force towards a unitary language is not, however, simply an issue of bureaucratic policy-making; it reflects widespread pressure from students, families and communities. The terms of the debate, however, also reflect a unitary language perspective—observation of classrooms in South Africa, Pakistan or elsewhere suggest that language use is much more diverse than Nhlanhla's comments suggest. In reality, the choice is not so clear cut, as illustrated by Nhlanhla's natural use of a mixture of languages in her reply.

7 Discussion: Heteroglossia and Equity

Teaching and learning mathematics in culturally and linguistically diverse class-rooms has been shown to influence the teaching and learning of mathematics through a range of tensions. What I have argued in this chapter is that these tensions all arise from Bakhtin's centripetal and centrifugal forces. The centripetal forces are often in the form of language policies or assumptions on the part of teachers, parents and communities that it is better to learn mathematics in one unitary language. Specifically, the pressure comes from a preference for a single school language, a standardised mathematical register, national and institutional language policies and the high status accorded to some natural languages (e.g. English) and social languages (e.g. the language of the middle classes, the mathematics register). The centrifugal forces arise from the heteroglossia found in all mathematics classrooms, but particularly in culturally and linguistically diverse mathematics classrooms. This heteroglossia is both collective and individual. Each classroom includes speakers of multiple natural languages and multiple social languages, including the languages of class, race, gender and so on. Just as significantly, each individual speaks in diverse ways drawing on a repertoire of natural and social languages to do mathematics. Indeed Bakhtin goes further, arguing that the tension is present in each utterance, which conforms to a relatively rigid language system, while at the same time drawing on multiple natural or social languages and invoking multiple possible meanings (cf. Holquist 2002, p. 48). The tension between these two sets of forces is inherent in human communication, both in general and in mathematics. Without common ways of talking or representing ideas, such communication would be impossible. Without variation, new things could never be said. My presentation of these four tensions is necessarily a simplification. These tensions are abstractions discernible in the heteroglossia of linguistically diverse mathematics classrooms. No classroom can be fully or neatly described in terms of these four tensions alone: the tensions are not things in themselves. Each classroom will swirl with various facets of the tensions I have identified, as well as others, as yet undocumented or unremarked. It is therefore useful to represent these tensions as occupying a wider space, which I have attempted to do in Fig. 1.

For me, such a diagram is an aid for thinking, not a reification. In thinking about the classroom in which I worked in Pakistan, for example, I can readily note the tension arising around the use of English as a medium of instruction for mathematics. How this tension plays out in specific utterances is, however, highly situated. It depends, for example, on the English proficiency of the students and of the teacher, as well as the nature of the English used in the textbooks and so on.

At the start of this chapter, I pointed out that language diversity in mathematics classrooms is widespread but is itself diverse. I suggested several questions that

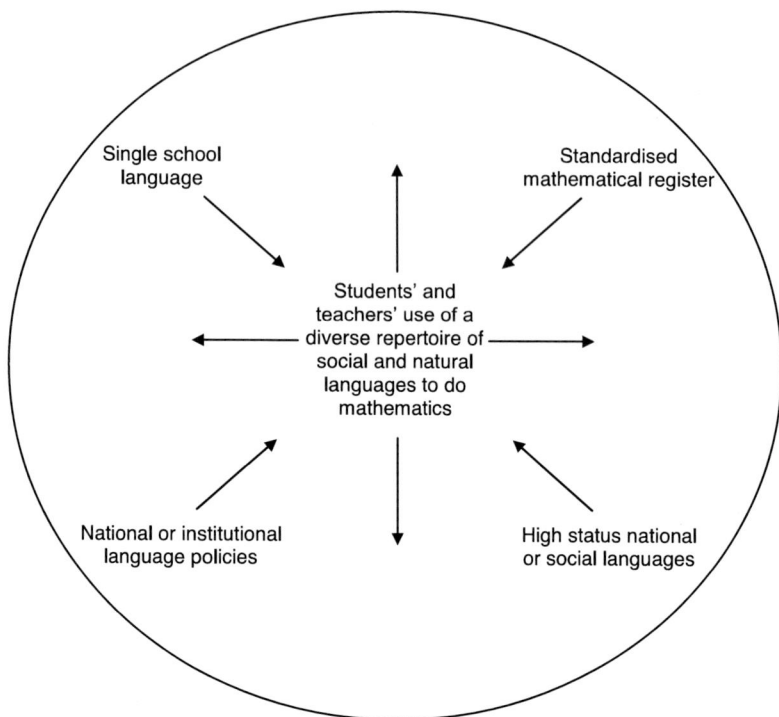

Fig. 1 Centripetal and centrifugal forces in language use in mathematics classrooms

arise in relation to this observation: In terms of mathematics teaching and learning, do these situations and others like or unlike them have anything in common? Are there common issues or challenges that arise? How are language, diversity and mathematics related? In the analysis set out above, I have addressed these questions. Despite the diverse nature of language diversity, teaching and learning mathematics in the context of language diversity does involve various tensions (and associated issues and challenges) that arise in a range of different situations. Four of these tensions are discussed in this chapter. Furthermore, these tensions are all related to the underlying push and pull of the centripetal and centrifugal forces inherent in language itself (indeed, in human communication more generally). This tension is the key nexus in which language, diversity and mathematics interact.

There is one remaining question: How can teachers deal equitably with such diversity? I have argued, following Bakhtin, that the kind of tensions I have described are inherent in human communication. The four tensions I have discussed are unlikely to be the only such tensions that arise; the tension between unitary language and heteroglossia will manifest itself in mathematics classrooms in a variety of different ways. But the underlying tension is ever-present. These tensions are likely to have different impacts on different students. Students whose repertoire of natural and social languages is well aligned with prevailing unitary languages are at an advantage: they do not need to 'crack the code' (Zevenbergen 2000); they already

have the key. Students from the peripheries must either submit to the prevailing unitary language approach or find ways to resist it and assert their own repertoires. In
reality, there is no perfect alignment between a given student's language repertoire
and the language of the mathematics classroom, even if some students' repertoires
are more closely aligned than others. To a greater or lesser extent, all students must
live in the tension between heteroglossia and unitary language, between centrifugal
and centripetal forces. This situation makes consideration of equity somewhat complex. In particular, heteroglossia cannot be eliminated; difference and diversity is
unavoidable. Indeed, diversity accounts for some students' success as much as other
students' underachievement. The challenge, therefore, is to find ways to decouple
diversity from disequity.

8 Conclusion: Shifting the Tension

If the kind of tensions I have described in this chapter are inherent in human communication, they cannot be eliminated by a change in policy or classroom practice.
Nevertheless, my analysis does suggest some ideas that have perhaps not been considered very widely. In particular, much concern with teaching and learning mathematics in diverse classrooms tends to assume a unitary perspective—debates about
whether to use a particular language of instruction, for example. It is apparent from
the research I have referred to and research in linguistics more generally that human
communication is not naturally unitary. I suggest, therefore, that rather than seeking
to eliminate the tensions I have described, a more productive approach would be to
shift the tension more towards heteroglossia and away from a unitary perspective.
I am not advocating a move towards absolute heteroglossia (whatever that would
be), which would be as unhelpful as an approach that is too strongly unitary. But
greater recognition of the heteroglossia of human communication in mathematics
classrooms would better reflect the lives and experiences of students and teachers.
The following two brief examples illustrate what such a move might look like.

In the research I conducted in the classroom in the UK described at the start of
this chapter, I analysed how learners of English engaged with a task set by their
teacher to write arithmetic word problems. My analysis of students' interaction as
they worked together on this task suggests that the task itself has some important
affordances (cf. Barwell 2009). In particular, the task did not require students to interpret an unfamiliar context, as is the case with solving a problem set by the teacher
or a textbook. Rather, the students were able to draw on contexts that were familiar
to them and to draw on their language repertoires to express these contexts. Students
wrote word problems about shopping, buying presents for their families, earning
pocket money, going to concerts or about monsters and morgues. The task also allowed students to engage with the language of word problems, infusing them with
their own forms of expression. And writing word problems provided opportunities
for the students to learn English and to make connections between the mathematics,
context and language of word problems. The word problem task does not eliminate
the tensions discussed in this chapter, but it does allow diversity of expression and

so takes greater account of the heteroglossia of the classroom. In particular, a key feature of the task is that it allows students to bring their own experiences and ways of talking into their mathematics.

Setati et al. (2008) report another kind of task that involves greater recognition of classroom heteroglossia. Their work was conducted in a classroom in South Africa in which there were 36 students, including speakers of Setswana, Xitsonga, IsiZulu and Tshivenda. The official language was English. During work on a complex page-long problem about electricity prices, students were organised into language groups. They were provided with both the English version of the problem and a version of the problem in their home language (e.g. Setswana). Setati et al. report that students drew on both versions as they worked in groups on the task. It is notable that they did not use a single version—they referred to both and drew on a mixture of languages to think about the task. As I have argued above, these kinds of language practices are more typical of students' language use; the difference with this task is that it explicitly recognises these practices. Setati et al. highlight that the use of these multiple languages did not interfere with their work and argue that it enabled a greater focus on the mathematics of the task. Again, this kind of approach does not eliminate the tensions discussed in this chapter, but it does go some way towards recognising and incorporating the heteroglossia of the students and of society into mathematics classroom interaction.

These two examples suggest some possible directions for future work on equity in mathematics classrooms. To develop such strategies requires an awareness of the kinds of tensions I have discussed in this chapter. Of course, such an awareness does not eliminate the tensions: indeed awareness is likely to result in the kind of teaching dilemmas identified by Adler (2001). It is, however, better to work with the tensions than to work in ignorance of them. Whether the setting is broadly characterised in terms of second language learners or of bilingual or multilingual learners, there is a common underlying principle: that more recognition be given to the heteroglossia of the class and of each individual student. To make a start, we simply need to start listening to the diversity around us.

References

Adler, J. (1995). Dilemmas and a paradox: Secondary mathematics teachers' knowledge of their teaching in multilingual classrooms. *Teaching and Teacher Education, 11*(3), 263–274.

Adler, J. (2001). *Teaching mathematics in multilingual classrooms.* Dordrecht: Kluwer Academic Publishers.

Bakhtin, M. M. (1981). *The dialogic imagination: Four essays.* (M. Holquist (Ed.), C. Emerson & M. Holquist (Trans.)). Austin: University of Texas Press.

Barwell, R. (2003). Patterns of attention in the interaction of a primary school mathematics student with English as an additional language. *Educational Studies in Mathematics, 53*(1), 35–59.

Barwell, R. (2005a). A framework for the comparison of PME research into multilingual mathematics education in different sociolinguistic settings. In H. L. Chick & J. L. Vincent (Eds.), *Proceedings of 29th conference of the international group for the psychology of mathematics education* (Vol. 2, pp. 145–152). Australia: PME.

Barwell, R. (2005b). Working on arithmetic word problems when English is an additional language. *British Educational Research Journal, 31*(3), 329–348.

Barwell, R. (2005c). Ambiguity in the mathematics classroom. *Language and Education, 19*(2), 118–126.

Barwell, R. (2005d). Integrating language and content: Issues from the mathematics classroom. *Linguistics and Education, 16*(2), 205–218.

Barwell, R. (2005e). Empowerment, EAL and the national numeracy strategy. *International Journal of Bilingual Education and Bilingualism, 8*(4), 313–327.

Barwell, R. (2009). Mathematical word problems and bilingual learners in England. In R. Barwell (Ed.), *Multilingualism in mathematics classrooms: Global perspectives* (pp. 63–77). Bristol: Multilingual Matters.

Blackledge, A. (2002). The discursive construction of national identity in multilingual Britain. *Journal of Language, Identity, and Education, 1*(1), 67–81.

Bournot-Trites, M., & Reeder, K. (2001). Interdependence revisited: Mathematics achievement in an intensified French immersion program. *The Canadian Modern Language Review/La Revue canadienne des langues vivantes, 58*(1), 27–43.

Canagarajah, S. (2009). The plurilingual tradition and the English language in South Asia. *AILA Review, 1*, 5–22.

Clarkson, P. (1992). Language and mathematics: A comparison of bilingual and monolingual students of mathematics. *Educational Studies in Mathematics, 23*(4), 417–430.

Clarkson, P. C. (2007). Australian Vietnamese students learning mathematics: High ability bilinguals and their use of their languages. *Educational Studies in Mathematics, 64*(2), 191–215.

Clarkson, P. C. (2009). Mathematics teaching in Australian multilingual classrooms: Developing an approach to the use of classroom languages. In R. Barwell (Ed.), *Multilingualism in mathematics classrooms: Global perspectives* (pp. 145–160). Bristol: Multilingual Matters.

Clarkson, P., & Dawe, L. (1997). NESB migrant students studying mathematics: Vietnamese students in Melbourne and Sydney. In E. Pehkonen (Ed.), *Proceedings of 21st meeting of the international group for the psychology of mathematics education* (Vol. 2, pp. 153–160). Lahti, Finland: University of Helsinki.

Clarkson, P., & Galbraith, P. (1992). Bilingualism and mathematics learning: Another perspective. *Journal for Research in Mathematics Education, 23*(1), 34–44.

Department for Children, Schools and Families (DCSF) (2007). *Supporting children learning English as an additional language: Guidance for practitioners in the early years foundation stage.* London: DCSF Publications.

Dlamini, C. (2008). Policies for enhancing success or failure? A glimpse into the language policy dilemma of one bilingual African state. *Pythagoras, 67*, 5–13.

Duranti, A. (1998). *Linguistic anthropology.* Cambridge: Cambridge University Press.

Farrugia, M. T. (2009a). Registers for mathematics classrooms in Malta: Considering the options. *For the Learning of Mathematics, 29*(1), 20–25.

Farrugia, M. T. (2009b). Reflections on a medium of instruction policy for mathematics in Malta. In R. Barwell (Ed.), *Multilingualism in mathematics classrooms: Global perspectives* (pp. 97–112). Bristol: Multilingual Matters.

Gorgorió, N., & Planas, N. (2001). Teaching mathematics in multilingual classrooms. *Educational Studies in Mathematics, 47*(1), 7–33.

Halai, A. (2009). Politics and practice of learning mathematics in multilingual classrooms: Lessons from Pakistan. In R. Barwell (Ed.), *Multilingualism in mathematics classrooms: Global perspectives* (pp. 47–62). Bristol: Multilingual Matters.

Halliday, M. A. K. (1978). *Language as social semiotic: The social interpretation of language and meaning.* London: Edward Arnold.

Holquist, M. (1981). Introduction. In M. M. Bakhtin (Ed.), *The dialogic imagination: Four essays* (C. Emerson & M. Holquist (Trans.)). Austin: University of Texas Press.

Holquist, M. (2002). *Dialogism: Bakhtin and his world* (2nd ed.). London: Routledge.

Khisty, L. L. (1995). Making inequality: Issues of language and meaning in mathematics teaching with Hispanic students. In W. Secada, E. Fennema, & L. B. Adajian (Eds.), *New directions for equity in mathematics education* (pp. 279–297). Cambridge: Cambridge University Press.

Lapkin, S., Hart, D., & Turnbull, M. (2003). Grade 6 French immersion students' performance on large-scale reading, writing, and mathematics tests: Building explanations. *Alberta Journal of Educational Research, 49*(1), 6–23.

Lim, C. S., & Ellerton, N. (2009). Malaysian experiences of teaching mathematics in English: Political dilemma versus reality. In M. Tzekaki, M. Kaldrimidou, & H. Sakonidis (Eds.), *Proceedings of the 33rd conference of the international group for the psychology of mathematics education* (Vol. *4*, pp. 9–16). Thessaloniki, Greece: PME.

Makoni, S., & Meinhof, U. (2004). Western perspectives in applied linguistics in Africa. *AILA Review, 17*, 77–104.

Mendes, J. R. (2007). Numeracy and literacy in a bilingual context: Indigenous teachers education in Brazil. *Educational Studies in Mathematics, 64*(2), 217–230.

Moschkovich, J. (2002). A situated and sociocultural perspective on bilingual mathematics learners. *Mathematical Thinking and Learning, 4*(2&3), 189–212.

Moschkovich, J. N. (2008). "I went by twos, he went by one:" Multiple interpretations of inscriptions as resources for mathematical discussions. *The Journal of the Learning Sciences, 17*(4), 551–587.

Moschkovich, J. N. (2009). How language and graphs support conversation in a bilingual mathematics classroom. In R. Barwell (Ed.), *Multilingualism in mathematics classrooms: Global perspectives* (pp. 78–96). Bristol: Multilingual Matters.

Ní Ríordáin, M., & O'Donoghue, J. (2009). The relationship between performance on mathematical word problems and language proficiency for students learning through the medium of Irish. *Educational Studies in Mathematics, 71*(1), 43–64.

O'Halloran, K. L. (2005). *Mathematical discourse: Language, symbolism and visual images.* London: Continuum.

Parvanehnezhad, Z., & Clarkson, P. C. (2008). Iranian bilingual students' reported use of language switching when doing mathematics. *Mathematics Education Research Journal, 20*(1), 52–81.

Setati, M. (1998). Code-switching in a senior primary class of second-language mathematics learners. *For the Learning of Mathematics, 18*(1), 34–40.

Setati, M. (2005). Teaching mathematics in a primary multilingual classroom. *Journal for Research in Mathematics Education, 36*(5), 447–466.

Setati, M. (2008). Access to mathematics versus access to the language of power: The struggle in multilingual mathematics classrooms. *South African Journal of Education, 28*, 103–116.

Setati, M., & Adler, J. (2000). Between languages and discourses: Language practices in primary multilingual mathematics classrooms in South Africa. *Educational Studies in Mathematics, 43*(3), 243–269.

Setati, M., Molefe, T., & Langa, M. (2008). Using language as a transparent resource in the teaching and learning of mathematics in a grade 11 multilingual classroom. *Pythagoras, 67*, 14–25.

Sfard, A. (2008). *Thinking as communicating: Human development, the growth of discourses, and mathematizing.* Cambridge: Cambridge University Press.

Swain, M., & Lapkin, S. (2005). The evolving socio-political context of immersion education in Canada: Some implications for program development. *International Journal of Applied Linguistics, 15*(2), 169–186.

Turnbull, M., Lapkin, S., & Hart, D. (2001). Grade 3 immersion students' performance in literacy and mathematics: Province-wide results from Ontario (1998–99). *The Canadian Modern Language Review/La Revue canadienne des langues vivantes, 58*(1), 9–26.

van Oers, B. (2001). Educational forms of initiation in mathematical culture. *Educational Studies in Mathematics, 46*, 59–85.

Wells, G. (1999). *Dialogic inquiry: Towards a sociocultural practice and theory of education.* Cambridge: Cambridge University Press.

Zevenbergen, R. (2000). "Cracking the code" of mathematics classrooms: School success as a function of linguistic, social, and cultural background. In J. Boaler (Ed.), *Multiple perspectives on mathematics teaching and learning* (pp. 201–224). Westport: Ablex.

Commentary on the Chapter by Richard Barwell, "Heteroglossia in Multilingual Mathematics Classrooms"

Heteroglossia and "Orchestration" in Multilingual Mathematics Classrooms

Núria Planas

Diria que m'agrada més parlar català a classe de mates, perquè és més elegant, com les mates, les mates també són més elegants! [I would say that I like to speak Catalan better in the mathematics class, because it's smarter, like math, they are also smarter!] (Elena, Spanish-dominant speaker, 13-years-old, May 2011)

In his chapter, Barwell draws on the Bakhtinian notion of co-existing voices, *heteroglossia*, to point to the experience of four tensions in linguistically diverse mathematics classrooms, namely: 1) school versus home languages, 2) formal versus informal language in mathematics, 3) language policy versus mathematics classroom practices, and 4) a language for learning mathematics versus a language for getting on in the world. To further explore Barwell's points, I bring up the metaphor of *orchestration*. My idea of orchestration refers to the manner in which various voices are individually used to produce coordinated results. As in the case with instruments in an orchestra, the final production requires successful individual practices with instruments that successfully interact with each other. My main argument here is that heteroglossia needs adequate practical orchestration, on the part of all members in a class: they all need to learn what the others can and cannot do, what distributions of work sound more adequate at each time, why is it that some of the ideas and contributions sound different depending on who introduces them, etc. By interpreting orchestration as something that includes all participants, not only the actions and choices by the teacher, it becomes clearer that students and teachers have lots in common.

Although all members in a class are responsible for good orchestration, it is not easy to identify the orchestrators that have a role in deciding which voices are asked to play more valuable "notes." In a full orchestra, several instruments alternate their presence throughout the musical representation. Moreover, although the composer may not be physically present, he/she marks the anticipation parts of

N. Planas (✉)
Mathematics and Science Education Department, Universitat Autònoma de Barcelona, Barcelona, Spain
e-mail: nuria.planas@uab.cat

H. Forgasz, F. Rivera (eds.), *Towards Equity in Mathematics Education,*
Advances in Mathematics Education,
DOI 10.1007/978-3-642-27702-3_29, © Springer-Verlag Berlin Heidelberg 2012

the arrangement. Even if there is only one orchestrator for each representation, all contexts are socially and politically charged in such a way that the orchestration cannot be attributed to a single person. In a multilingual mathematics classroom, for instance, a teacher may choose/allow to code switch and promote flexible language practices while teaching. However, other agents also have significant roles in determining the value of these practices. In other words, the understanding of orchestration is not limited to the scenario in which the orchestra with the students and the teacher is placed. It tells a more complex story about broader discourses and relationships within society that highly frame the reading of what happens in a classroom. Throughout the text below, for instance, in my own academic discourse, I use terms like "Catalan teachers" and "late arrival immigrant students" to refer to certain social realities at the risk of unintentionally suggesting monolingualism.

1 Viewing Orchestration in "School Versus Home Languages"

Barwell highlights the stipulation of an official language of schooling as a manifestation of dominant discourses on unitary language in mathematics classrooms. In my research context (Planas 2011), I also elaborate on the institutional emphasis on language, specifically among students whose home language is not Catalan, the language of learning and teaching in the Catalan school system. Despite several discourses on monolingualism all over the world, research has argued in favor of the use of the students' languages in the teaching and learning of mathematics. In Planas and Setati (2009), the analysis of data from small group work in a multilingual mathematics classroom relates the alternate use of the students' languages to move mathematical participation and thinking forward. However, classroom data in situations of whole group discussion (Planas 2011) show fragile author positions for students who are not Catalan-dominant, and who have come to Catalan almost exclusively through school. Changes in some of these students' positions are critical in that opportunities for interaction become altered and affect the overall learning situation. This type of changes in positions from small to whole group work is particularly observable in my data. I have always attempted to work in classrooms with linguistically mixed student population. This option has created a good opportunity to explore differences in the behavior of students of Catalan speaking background and those of Spanish speaking background in different social contexts of the mathematics classroom.

Which orchestrations could help shift the tension more toward heteroglossia and away from unitary perspectives? Which are the potential orchestrators for such a task? Are all members of a class ready to take responsibility for orchestrating particular teaching and learning practices? The research in Tucson, Arizona by Civil and Menéndez (2011) highlights the fact that shifts require complex orchestrations that include teachers, students, and close community members. In home settings, Mexican-American students' voices sound bright while they share their mathematics homework with their parents and when the latter helps the former deal with tasks presented in English. But when the same students are placed in an English-only mathematics classroom with their teachers and other students, their voices sound very different. This suggests that the interpretation of valorizations as expressed in the classroom cannot be isolated from the interpretation of valorizations in other areas of life in which the students' knowledge is highly appreciated where they tend to act as orchestrators on the basis of their language and cultural experience.

2 Viewing Orchestration in Formal Versus Informal Language in Mathematics

Barwell points to a second tension coming from the students' informal expressions of their mathematical thinking and the need to use formal mathematical vocabulary and grammar, which tends to be presented in the official language. In South Africa, Setati (2005) has documented how discussing mathematics in English tends to be more procedural, while sophisticated mathematical thinking is more likely to occur in the students' home languages. If looking at teachers in Catalonia, there are an increasing number of multilingual classrooms where teachers teach mathematics in English, a language that is neither theirs nor the students' dominant language. It is claimed that CLIL (Content and Language Integrated Learning) methods aim for the simultaneous learning of content (i.e. mathematics) and a "foreign" language (i.e. international English). However, preliminary findings based on fieldwork in 2006–2007 from one of my students, M. Gallart, show that Catalan teachers have difficulties with modeling mathematical thinking through English and that they tend to use procedural approaches to mathematical activity.

How could the communication of students' mathematical ideas be facilitated in multilingual classrooms? I agree with Barwell that this question is not easily resolvable. But, again, my position is that at least better orchestrations need to be considered. Which orchestrations can help keep a balance between informal and formal language both in teaching and learning? How students and teachers are able to use language depends on what is interpreted to be convenient in a particular classroom context and which forms of assessment are activated. Research by Enyedy et al. (2008) on practices of revoicing in multilingual mathematics classrooms gives some clues as to how teachers and students can promote a deeper conceptual understanding of mathematics by avoiding an excessive focus on language. In Planas and Morera (2011), practices of revoicing in bilingual mathematics classrooms also appear to be connected with advanced collective argumentation in conversations in which mathematical vocabulary is often reconstructed across languages through a mixture of Catalan and Spanish. It is especially interesting the fact that revoicing often helps move toward more conceptual discourses by means of "repeating" questions in which either grammatical constructions or technical vocabulary have been partially reformulated.

3 Viewing Orchestration in Language Policy Versus Mathematics Classroom Practice

The third tension posed by Barwell relates to language policies penalizing practices of code switching, or at most seeing them as steps on the road to monolingual proficiency in a high status language. Language policies are primarily motivated by political rather than pedagogical issues. Consequently, what is intended in terms of institutional norms may become an obstacle in terms of learning. The interpretation of the effects of language policies on learning has a strong political dimension, too. Based on part of the findings drawn from the PISA 2009 Report for Catalonia-Spain and mathematics (OECD 2010), one of the mainstream journals in my country— *El Mundo, Catalan Edition*—published the rate of failure for Spanish dominant students (42.62%) and compared it with the rate for Catalan dominant students (18.58%). The title of the column in the journal was "Language and school failure in Catalonia." Here, the main reason for the difference was attributed to Catalan being the language of learning and teaching and to "the impossibility for Spanish-speakers to be schooled in their language" (*El Mundo, Catalan Edition*, 07-12-2010, http://www.elmundo.es). The fact that over 80% of 42.62% were Latin American immigrant students and/or children from immigrant families was not mentioned. Moreover, there were no references to the parallel system of "special classes" for late arrival immigrant students, with fewer hours of mathematics per week than those in regular classes. Nowadays, the social debate in the media is still primarily centered on the negative educational effects of having the learning of "key" subject areas such as school mathematics distracted by the learning of language.

If we react to the impossibility for Spanish-speakers to be schooled in their language by changing the language of learning and teaching, then we have other groups facing the challenge of learning mathematics in a language that is not their own. Therefore, it makes sense to search for alternative solutions at the level of school mathematics and classroom practices. Which practical orchestrations can work toward less restrictive interpretations of language policies? Like Setati et al. (2002), I see practices of code switching, especially when used by both teachers and students, as a tool for performing more multilingual exploratory talk and less monolingual procedural thinking. When teachers take on the role of bi/multilingual and share this with students, they can improve their interactions with them and teach in new and distinct ways that foster content instead of unitary language systems. If possible, teachers must be willing not only to talk mathematics with their students but also to model talking mathematics across languages and gain reflexivity. This approach, of course, does not solve the question of how monolingual teachers in multilingual classrooms can gain effective reflexivity with all students.

4 Viewing Orchestration in Language for Learning Mathematics Versus Getting On in the World

The last tension in Barwell's text illustrates the preference of many students for the learning and teaching of mathematics in a language that is not their own in spite of additional personal difficulties that such scenarios may bring. Again, the debate is not only about language and/or mathematics, but it is also about the political role of language and the complexity of the context in which the mathematics is taught and learned. In Setati and Planas (2012), what can be seen is that students and teachers have a clear preference for one language (Catalan in my context, and English in South Africa). From most of the students' and the teachers' perspectives, learning and teaching mathematics in Catalan/English is not so much about choice; it is just how things should be. There is also an implicit message of "wanting the opportunity." Students believe that without fluency in Catalan/English, they would not have access to social goods such as higher education and qualified employment.

It is not surprising that students refer to the social dimension of access and more explicitly to the right to choose the language for their learning of mathematics. This result, nevertheless, tells again a more complex story than the issue of choice itself. It speaks of some languages that are not supported enough within society, while other languages are marked with ideals of social promotion and academic success. In this situation, which are the potential orchestrators, if any? Do they look like 1) particular students, teachers, and close community members or 2) political agents who are not represented by a single person? In my view, we need orchestration from both groups. They work differently as they attempt to arrange significant aspects of the initial "piece" and, thus, need to be analyzed carefully. As researchers and mathematics educators, our task is to make progress in understanding the intentions of all agents and the extent to which certain intentions might better facilitate an efficient orchestration of heteroglossia in multilingual mathematics classrooms (see, e.g., Rivera and Becker 2008). The final ideal representation would not let one language group advance at the expense of the other:

> There is always room for negotiation and re-distribution of identities: any social and po-litical discourse may be contested and become the basis for the construction of alternative identities. Even when institutions behave as monolinguals, monolingualism is supported by law and issues of language identity may be presented as non-negotiable, individuals and groups may struggle for voice by reconstructing what others expect from them—and what they themselves have come to develop as self-representations. (Planas 2011, p. 131)

References

Civil, M. & Menéndez, J. M. (2011). Impressions of Mexican immigrant families on their early experiences with school mathematics in Arizona. In R. S. Kitchen & M. Civil (Eds.), *Transnational and borderland studies in mathematics education* (pp. 47–68). New York: Routledge.

Enyedy, N., Rubel, L., Castellón, V., Mukhopadhyay, S., Esmonde, I., & Secada, W. (2008). Revoicing in a multilingual classroom. *Mathematical Thinking and Learning*, *10*(2), 134–162.

OECD (2010). *PISA 2009 results: What makes a school successful? Resources, policies and practices*. Vol. *IV*. Paris, France: OECD Publishing.

Planas, N. (2011). Language identities in students' writings about group work in their mathematics classroom. *Language and Education*, *25*(2), 129–146.

Planas, N. & Morera, L. (2011). Revoicing in processes of collective mathematical argumentation among students. In M. Pytlak, E. Swoboda, & T. Rowland (Eds.), *Proceedings of the seventh Congress of the European society for research in mathematics education* (pp. 1356–1365). Rzeszów, Poland: ERME.

Planas, N. & Setati, M. (2009). Bilingual students using their languages in their learning of mathematics. *Mathematics Education Research Journal, 21*(3), 36–59.

Rivera, F. & Becker, J. (2008). Ethnomathematics in the global episteme: Quo vadis? In B. Atweh, M. Borba, R. Vithal, & C. Vistro-Yu (Eds.), *Internationalization and globalization in mathematics and science education* (pp. 209–225). New York: Springer.

Setati, M. (2005). Teaching mathematics in a primary multilingual classroom. *Journal for Research in Mathematics Education, 36*(5), 447–466.

Setati, M., Adler, J., Reed, Y., & Bapoo, A. (2002). Incomplete journeys: Code-switching and other language practices in mathematics, science and English language classrooms in South Africa. *Language and Education, 16*(2), 128–149.

Setati, M., & Planas, N. (2012). Mathematics education across two different language contexts: A political perspective. In O. Skovsmose & B. Greer (Eds.), *Opening the cage: Critique and politics of mathematics education* (pp. 120–140). Rotterdam, The Netherlands: Sense Publishers.

Commentary on the Chapter by Richard Barwell, "Heteroglossia in Multilingual Mathematics Classrooms"

Bakhtin, Alterity, and Ideology

Luis Radford

Mathematics classrooms are sites of encounter for different voices, perspectives, and ideas. Those differences become even more visible when the object of difference is language. In his chapter, Barwell draws on Bakhtin's concept of heteroglossia to explore the tensions that underpin multilingual classrooms. He enquires about how those tensions influence the teaching and learning of mathematics and the implications that they may have for equity in mathematics teaching. In my comments, I would like to dwell upon the question of language in the mathematics classroom and on some issues about equity.

1 Language in the Mathematics Classroom

One way or another, for one reason or another, since the time of Babylonian schools, institutional educations have always faced the question of linguistic diversity. However, the manner in which this diversity has been addressed and understood has not always been the same. Contemporary schools seem to be led to address this diversity along the lines of contemporary concerns about equity and social justice. These concerns, of course, are a token of social and political interests in coming to grips with cultural diversity, brought forward by unprecedented migratory movements of a global scale.

In his chapter, Barwell points out four "tensions" that are present in the mathematics classroom, considering them through the lenses of language differences—for instance, tensions between school and home languages or language policy and mathematics classroom practice. We can see, through the illuminating examples he discusses, how difficult it is for teachers and schools to "deal" with cultural diversity. How, in particular, to approach the multiple languages that the students bring

L. Radford (✉)
Laurentian University, Sudbury, Ontario, Canada
e-mail: lradford@laurentian.ca

H. Forgasz, F. Rivera (eds.), *Towards Equity in Mathematics Education*,
Advances in Mathematics Education,
DOI 10.1007/978-3-642-27702-3_30, © Springer-Verlag Berlin Heidelberg 2012

into the classroom? Following Bakhtin, Barwell sees the classroom as immersed in the dynamics of unitary and diverging "forces." Unitary (or centripetal) forces tend towards unified forms of language, often associated with conceptual, cultural, and political centralization. Diverging (or centrifugal) forces stress diversity, often associated with the speakers' cultural, political, and economic background. In the conclusions, he suggests that one of the characteristics of language and communication is the tensions they entail, and goes on to assert "that rather than seeking to eliminate the tensions... a more productive approach would be to shift the tension more towards heteroglossia and away from a unitary perspective."

Although I am in agreement with Barwell's conclusion, it is my contention that the search for "productive approaches" requires us to better understand the interplay between unitary and diverging "forces" in multilingual classrooms and how to take advantage of these forces in school mathematics practices. In particular, I would like to suggest that it might be advantageous for multilingual research in mathematics education to examine the interplay of unifying and diverging forces against the backdrop of two central Bakhtinan ideas: *alterity*—i.e., the relationship of I and Other—and language as necessarily *ideological*.

2 Language as *Ideological*

Bakhtin's concept of language is at odds with most concepts of language developed in the Western tradition—e.g. the empiricist view of Locke or the rationalist one of Leibniz, views that, although different, share nonetheless a common individualist stance: language is, in those views, something lodged in the individual. For Bakhtin, language is not *in* the individual. Language precedes the individual. Language is historical, social and cultural—something that instead of being neutral, is, from the outset, positioned within a larger political context. Thus, in one of Barwell's examples, English appears as the language of elite classes in Pakistan. In general, language operates as a marker of differences in the social, political, and economic arenas of culture. Even more, language is *constitutive* of its subjects whose subjective existence can only be realized through it and the *worldviews* it conveys. To emphasize the fact that language always signifies within particular worldviews—something that makes language much more than a formal channel of communication—Bakhtin and his collaborators referred to the term *ideologya*, understood not as a simple system of ideas, but as a social-cultural human activity (Vološinov 1973). All signs, language included, signify within the sphere of a super-symbolic cultural axis (the axis of *ideologya*). And mathematics' language and ideas are not the exception.

For instance, Lizcano (1993) has shown how the concept of number in Ancient China developed within the symmetries of yin-yang symbolic structures, thereby making it possible to imagine and talk about negative numbers, something that was tremendously difficult to imagine in the West. To come up with a concept of negative number, Western thought had indeed to invent capitalism and its quantifying practice of debts. A more contemporary example of language and *ideologya* is provided by the investigation of forms of knowing in aboriginal communities in Canada. For the Yup'ik people, problem solving is embedded in oral narratives that emphasize intuitions, visions, dreams, and spiritual interaction (Kawagley 1990)—components that are in deep contrast to the analytic rationalist ones that we inherited from the Enlightenment and its emphasis on rigor, deduction, and abstraction. As Barwell reminds us, it is misleading to think that students from other cultures "do not bring different mathematics, only different languages." In the students of multicultural classrooms we already find mathematics (as a plural noun) that speak about different worlds, even if the mathematics is expressed in the same official language. We do not produce accents when talking only. We also produce accents when thinking. All of our thinking and talking is inevitably inhabited by the languages we use and the *ideologyas* those languages unavoidably refer to. A greater sensitivity to the ideological nature of language seems to me to be an aspect to take into account to offer space for learning and to promote justice and equity in multilingual classrooms.

3 Alterity

Naturally, a greater sensitivity to the ideological nature of language is just a step, perhaps the first step to inclusiveness. We still need to know what to do with the various languages the students bring into the classroom. This question can only be answered within the larger context of multiculturalism. One of the imminent risks here is to consider multiculturalism as a benevolent form of tolerance. It might be this form of liberal multiculturalism that informs pedagogical actions that tolerate other languages in the classroom as a trade-in to subtly impose official languages. In this case, cultural and linguistic differences are conceived of as *natural* differences and become engulfed in the official mechanisms of order, centralization, and subjection. We end up forgetting that the differences under consideration have been forged to a large extent through a bloody history of colonialism and domination. Of course, we can *hear* that the person next to us is talking in a different language. But the question is: how are we going to react to this cultural difference? The manner in which we respond to our neighbor's language is not a natural act, but one that has been actively produced on the bases of historically formed social, cultural, and political understandings. This is why a greater sensitivity to the ideological nature of language and a worthy shift of the tensions produced by multicultural encounters towards heteroglossia may not be enough. There is something extremely important in Bakhtin's idea of language that needs to be taken into consideration in our understanding of multiculturalism and multilingualism. And this is the question of *alterity*, the place of the Other in the constitution of the I. Following the Hegelian-Marxist tradition, Bakhtin says:

> All that touches me—beginning with my name and that penetrates into my consciousness—comes from the outside world, from the mouths of others... with their intonation, their affective tonality, and their values. At first I am conscious of myself only through others: they give me the words, the forms, and the tonality that constitute my first image of myself... Just as the body is initially formed in the womb of the mother (in her body), so human consciousness awakens surrounded by the consciousness of others. (Bakhtin 1984, pp. 357–58)

Bakhtin's view of the Other, I want to suggest, is extremely important to understanding multilingualism in the classroom. In this view, the Other is neither an object nor an exotic specimen, but someone who, through his/her languages and *ideologyas*, constitutes me as individual—someone who helps me get out of my tautological confined space.

But we are not out of the woods yet. The two Bakhtinan ideas I have mentioned (language as ideological and the Other as constitutive of the I) have still to be integrated in classroom practices that seek to promote equity and social justice. Such a task could hardly be carried out if we continue to think platonically, that is to say, if we think of mathematics as a discipline dealing with disembodied truths. We might be better off if we think of mathematics as a *situated* process where we come to the public space to think and talk about certain states of affairs. Barwell's heteroglossical shift may mean here the critical encounter of different voices and ideologies unfolding as a historical process against the set of various cultural centripetal and centrifugal traditions, each one becoming enlightened and modified by the others. Naturally, the challenges are colossal. For one thing, we have to cease seeing the mathematics classroom under the model of market economy—a utilitarian space of negotiations and personal promotion. The "banking model" of the classroom—to use Freire's term—should be replaced with a model of genuine human interaction moved by values of solidarity and cooperation. In this new model yet to be built, the language of the classroom would necessarily be collective and polyphonic. It would be a multifarious organ of collective and individual identity formations, a site of struggle and contradictions and, hence, of change and movement.

References

Bakhtin, M. (1984). *Esthétique de la création verbale [The aesthetics of verbal creation]*. Paris: Gallimard.
Kawagley, O. (1990). Yup'ik ways of knowing. *Canadian Journal of Native Education, 17*, 5–17.
Lizcano, E. (1993). *Imaginario colectivo y creación matemática [Collective imaginary and mathematical creation]*. Barcelona: Editorial Gedisa.
Vološinov, V. N. (1973). *Marxism and the philosophy of language*. Cambridge: Harvard University Press.

Part III
Equity and Curriculum Diversity

Ferdinand Rivera and Helen Forgasz

Part III consists of two reprinted ZDM articles and two new chapters that address issues in equity and curriculum diversity. Our working notion of curriculum (diversity) is tied to "the study of educational experience" (Pinar 2004), a view that does not dwell on planned courses of study (unit and lesson planning and assessments, scope and sequence guide construction, effective use of textbooks, etc.) but critically explores and reflects on various underlying power/knowledge issues that "always-already" characterize the development of a curriculum. Thus, a curriculum that is conceptualized in this manner articulates its highly symbolic nature. That is, following Pinar (2004), it is "the site on which generations struggle to define themselves and the world [and that] it is the symbolic character of curriculum that renders debates over the canon" (p. 186), with canon in our case as referring to the valued institutional versions of the mathematics curriculum. The four chapters in this section together address the equity issue that deals with the complex relationship between mathematical knowledge construction, on the one hand, and individual and institutional histories that influence the constructive process, on the other. Thus, the curriculum diversity related concerns that are tackled in this section seek to understand how students' differing levels of access to a mathematics curriculum consequently shape how they understand their own lives and their emerging mathematical identities, and how both are embedded within the complex fabric of society, politics, and culture (Pinar 2004, p. 36).

Skovsmose in his chapter addresses issues concerning the sociopolitical roles of mathematics education and what he calls "mathematics-based rationality" between two opposing lifeworlds (i.e. the progress of science, on the one hand, and technology, on the other). Such rationality, he notes, "brings us into an open space of contingencies, and that development could take very different routes." Skovsmose introduces the concept of "fabrication by mathematics-based rationality," which both enables and complicates our understanding of the different natures of mathematical knowledge. For example, mathematical rationality fabricates: possibilities (with choices that are neither necessarily accessible nor optimal); strategies; facts; contingencies (with risks); and perspectives. The author then explores the implications of such fabrications, including their operational mechanisms, in our societies and especially in the "school mathematics tradition" that cultivates and values "a prescription readiness among students." Prescription readiness involves the (predictable and rather stable) mathematical predisposition of knowing "what it means to operate with given information [that is both necessary and sufficient] within a given space of possible strategies for solution [with possibilities often-times framed within the oppositional binary practice of right and wrong]." In the remaining sections, Skovsmose further explores the implications of fabrication in the school mathematics tradition from the way in which our schools engage in "differentiated labeling of students" to the "ethically filtrated" assumptions we impose in "mathematical transposition" activity. Drawing on several powerful empirical studies in the emerging counterdiscourse of critical citizenship in mathematics education, he provisionally closes the chapter by exploring how it can be put into praxis within/against such fabrications with a cautionary note that responses or conditions of possibility vary depending on the complex context/s of specific life-worlds.

In her chapter, **Valero** uses very recent theoretical and methodological frameworks that enable us to "understand the meaning of mathematics education as sociopolitical practices and the implications of these notions for researching mathematics education." Utilizing two case studies that have been drawn from an empirical study in two schools in Denmark and South Africa, she extrapolates ways in which students' "advantaged and disadvantaged positions" could possibly emerge in a "network of school mathematics practices" (NSMP), which is a reformulation of an institutional system model of mathematics education (ISME). In ISME, the teaching and learning of mathematics in a school system are viewed in the context of a relationship between and among school leaders, mathematics teachers as a group, and the individual teachers. Valero's NSMP encompasses "the network of routines, ideas, shared meanings and ways of talking and conceiving mathematics education among the actors in the school organization, and even outside of it." The two interesting case studies, narrated and interpreted from a critical perspective, illustrate how excluded and marginalized students oftentimes perform their socially constructed roles in the classroom as a consequence of a network in which some "micro-physics of power" may have (unintentionally) contributed to their "differentiated positions," ultimately denying them (full) access to learning. Valero then notes that marginalized students' disadvantages in mathematics classrooms appear to "trespass the boundaries of individual attributes and of instructional organization in the classroom," which means that notions of "disadvantage and equity could be reformulated in terms of positioning within the relations and practices that constitute mathematics education in the school."

Leder in her chapter discusses interpretive findings that she has drawn from studies that have been published in the ZDM journal from 1990 to 2009 in regards to the education and needs of mathematically gifted students and compares them with results drawn from a representative sample of studies taken from the broader mathematics education literature. She notes at the outset that despite the voluminous amount of research on gifted students, no consensus has been reached among scholars in terms of how to conceptualize giftedness as a category. Consequently, her analysis accepts the various definitions of giftedness at face value (e.g. high achieving, mathematically able, expert students, etc.). In her thorough research of published ZDM articles on gifted students, she notes that a range of issues has been covered from promoting special problem-solving programs and developing effective instructional interventions, strategies, and approaches for gifted students to establishing strong (psychological and neural) linkages between and among creativity, intelligence, ability, flexible and adaptive thinking, and gifted students' problem solving proficiency and performance in solving difficult and nonroutine mathematical problems. Additional recent studies describe the quality of their (nonstandard) thinking processes and the significant role of access to practice and appropriate curriculum materials as a way of enhancing their problem-solving strategies. Leder, in one section, underscores important equity issues such as the social value of equality and access to challenging curricula between gifted and "regular" students and possible gender differences among high-ability students (see **Penner** and **CadwalladerOlsker** in this volume for an interesting discussion of gender differences in mathematical performance among high-ability groups across several countries). In their commentaries to Leder's chapter, **Koichu** raises several thought-provoking questions concerning equity issues surrounding gifted students in mathematics, while **Callingham** draws possible research implications of the key themes in Leder's chapter that might inform future empirical investigations with the mathematically gifted.

In his chapter, **Were** writes about his ethnographic experiences with the Nalik people of New Ireland in Papua New Guinea when a federal systemic initiative developed and introduced "a community-based (elementary) mathematics education program in the local vernacular that was sensitive to children's existing knowledge of local cultural traditions." Were's analysis focuses on the properties of cultural forms that are collected, developed, and used in class as tools for learning mathematics, especially in light of the fact that some forms either contain stipulations and secrets of particular clans (e.g. *kirugu*) or are conveyed nonverbally (e.g. the craft of carving canoes) and that, in most other cases, require teachers to work closely with key members of the community who possess the requisite information. Drawing on findings from a few ethnomathematical projects, Were also points out how "studying mathematical ideas in a cultural context" presupposes the necessity of "extending beyond the tangible forms and expressions" and locating them "in abstract relations that order social relations." In his ethnographic case study, he explores how everyday patterns of tangible forms (especially the *kapkap*) are effectively used with the Nalik children in order to help them transition in their mathematical thinking from the concrete (i.e. through the forms) to the abstract (i.e. the relevant systems of geometric transformations). Were also notes how the principles that structure and generate those forms represent systems of paradoxical structures that are "lived out" and "regulate and maintain order in the social domain" among the community. His conclusion provides a materialist view of mathematics, and, in particular, patterns that are loaded with both mathematical and socialization values. In their commentaries to Were's chapter, **Bishop** dwells on several complex issues that arise when "the cultural discourse [moves] into the educational sphere" in which case "the problems revealed are far more complicated than first realized," while **Khan** underscores potential dilemmas such as "desubstantiation" and relevant translational implications when ethnomathematical ideas are appropriated in capitalistic societies.

References

Pinar, W. (2004). *What is curriculum theory?* Mahwah: Erlbaum.

Preface to "Doubtful Rationality"

Ole Skovsmose

In talking about "doubtful rationality" I try to move beyond the *modern conception of mathematics*. This conception includes three elements: the celebration of mathematics as being essential for understanding nature; the celebration of mathematics-based contributions to technological development; and the celebration of mathematics as a pure scientific enterprise.

The conviction that mathematics is crucial for understanding nature emerged as an integral part of the Scientific Revolution. Copernicus used mathematics in order to outline a heliocentric worldview. The results of Kepler's meticulous studies of the orbit of Mercury showed that mathematics could provide an exact description of natural phenomena. Galilei found it vital to present his observations in numbers, as it appeared that the book of nature was written in mathematical terms. Descartes provided a philosophical outline of the worldview that became assumed as part of the Scientific Revolution. He found that nature operated according to some few laws that had been instigated by God. In *The World*, Descartes (1998/1664) formulated these Laws of Nature, three in total, and he announced that God had not implemented other laws. The Laws of Nature only concern the behaviour of material objects, and this way Descartes formulated a strict mechanical worldview. Descartes did not formulate the laws in mathematical terms, but this step was completed by Newton. Furthermore, Newton demonstrated that the laws that Kepler had identified concerning the movement of planets and the laws Galilei had identified with respect of falling bodies could be derived from the same overall mathematically formulated Laws of Natures. This way Newton demonstrated that the same natural laws were operating both on earth and in heaven, and that the whole universe could be considered a unified mechanical structure. Since then mathematics and natural sciences had to be intimately united: Mathematics expressed the rationality of nature.

O. Skovsmose (✉)
Aalborg University, Aalborg, Denmark
e-mail: osk@learning.aau.dk
State University of São Paulo, São Paulo, Brazil

H. Forgasz, F. Rivera (eds.), *Towards Equity in Mathematics Education,*
Advances in Mathematics Education,
DOI 10.1007/978-3-642-27702-3_31, © Springer-Verlag Berlin Heidelberg 2012

The second element of the modern conception of mathematics concerns technology. The initial steps of the Industrial Revolution were not intimately connected to mathematics. Thus, the first weaving machines that symbolise the emerging of this revolution were constructed by means of technical ingenuity more than through mathematical modelling. However, soon after mathematics took up the crucial role in any technical fabrication. This brought a new dimension to the mechanical worldview. Not only nature but also fabricated nature was created on the basis of mathematics. This applies to any construction of houses, bridges, ships, airplanes, lamps, shoes, cigarettes, computes, cell phones, whatever. We can as well think of the fabrication of systems of rationalisation, production, surveillance, tax payment, etc. Mathematics is crucial for the development of our techno-nature. It expresses the rationality of technological development.

The third element of the modern conception of mathematics concerns the academic purity of mathematics, which was clearly formulated by Hardy (1967/1940) in a *Mathematician's Apology*. Hardy claims that what he had done as a mathematician has been without any practical value. Mathematics is simply not any useful science. Instead the work of a mathematician can be compared to the work of an artist. Sublime mathematical theories can be compares to the symphonies of Mozart and Beethoven. Asking what is the use of a number theoretical insight by Fermat is the same as asking about the use of Beethoven's 9th symphony. One can see Euclid's axiomatic approach as a paradigmatic way of removing any empirical element from mathematics. More recently, the formalist philosophy of mathematics has portrayed mathematics as a self-contained formalism. This way mathematics became presented as a pure rationality.

One can naturally ask if the three elements characterising the modern conception of mathematics are consistent. I doubt so. Thus the claim that mathematics is pure hardly squares with the assumption that mathematics expresses the rationality of technological development. Anyway, a paradigmatic conception of mathematics need not be consistent, and the modern conception of mathematics brings together assumptions about nature, technology and purity in eager celebrations. Mathematics becomes seen as representing the rationality of nature, the rationality of progress, as well as a sublime form of human knowledge.

Negating the modern conception of mathematics does not mean to negate the many roles mathematics can play in science, technology and epistemology. Assuming a *critical conception of mathematics* first of all means not to participate in any automatic celebration of the roles that mathematics is playing. Instead a critical conception of mathematics addresses any such roles in all their possible ambiguities.[1]

In trying to present such ambiguities I have pointed out that mathematics are brought in action in many different ways. The crucial point of addressing mathematics in action is that, like any other form of action, so all mathematics in action can assume any kind of qualities. Mathematics-based actions can be expensive, dangerous, solid, doubtful, benevolent, harmful, risky, etc. As a consequence, a critical

---

[1] See the discussion in Skovsmose (2009).

conception of mathematics does not assume any uniform celebration of mathematical rationality, nor does it suggest a uniform dismissal of mathematics. The critical conception of mathematics acknowledges that mathematical rationality is without essence; it can come to operate in very many different ways and serve even contradictory interests.

This brings us to mathematics education. One can develop mathematics education within the perspective of the modern conception of mathematics. The principal aim of any mathematics education, then, becomes to contribute to the celebration of mathematical rationality and to ensure that students come to learn and appreciate mathematics. Assuming that mathematics represents an intrinsic praiseworthy rationality constitutes mathematics teachers as ambassadors of mathematics. Such paradigmatic assumptions constitute what can be referred to as the *modern conception of mathematics education*.

This conception, for instance, dominated the Modern Mathematics Movement, which tried to apply a logical organisation of mathematics as a curriculum structure. This established set theory as a basic discipline. The number of eager supporters of the Modern Mathematics Movement was dramatically reduced as educational problems emerged, but the modern conception of mathematics educations got new directions. Thus the process oriented mathematics educations, inspired by Freudenthal's interpretation of mathematics, joined the universal celebration of mathematical rationality, although it abolished the structuralism associated to the Modern Mathematics Movement. The notion of didactical transposition, crucial in the so-called French tradition in mathematics education, also assumes a celebration of mathematics rationality. In fact I see the French tradition as an important expression of the modern conception of mathematics educations.

How then to characterise a *critical conception of mathematics education?*[2] First thing is to leave behind any assumption about the intrinsic precious qualities of mathematical rationality and instead acknowledge its ambiguity.

However, there are many more elements that make part of a critical conception of mathematics education. Depending on the context, this education can come to exercise many different social, political, economic, and cultural functions. There is no simple set of functions that can be associated to mathematics education. Instead it becomes a highly contested socio-political enterprise. For instance, mathematics education might provide a prescription readiness among students, and this can be considered useful when imposed on a labour force that is going to be submitted to routines that are not supposed to be questioned. Mathematics education might provide some students with an experience of being able to master mathematics, and, simultaneously, a huge number of other students with an experience of defeat and inadequacy. Mathematics education might serve a useful function in today's neoliberal economic order by labelling students according to internationally recognised standards. Such a labelling might be considered important for the smooth functioning of the labour market of a knowledge society. One may also assume that mathematics education can ensure not only the development of functional citizenship, but

---

[2]This conception suggest a range of preoccupations, characteristic of critical mathematics education. See, for instance, Skovsmose (2005).

as well make students ready to read the world in terms of numbers and figures, and reading it as being open to change.

Such observations bring us to acknowledge that mathematics education, just like mathematics rationality, can play both doubtful and commendable functions. Mathematics education can be acted out in many different ways. Mathematics education does not contain some intrinsic qualities due to the fact that we have to do with an education in *mathematics*. Mathematics education is without essence, and as a consequence it is important to address it critically.

References

Descartes, R. (1998/1664). *The world and other writings*. Cambridge, UK: Cambridge University Press.

Hardy, G. H. (1967/1940). *A mathematician's apology*. Cambridge, UK: Cambridge University Press.

Skovsmose, O. (2005). *Travelling through education: Uncertainty, mathematics, responsibility*. Rotterdam: Sense Publishers.

Skovsmose, O. (2009). *In doubt: About language, mathematics, knowledge and life-worlds*. Rotterdam: Sense Publishers.

Doubtful Rationality

Ole Skovsmose

Abstract The paper addresses the following two questions: (1) How can we conceptualise possible socio-political roles of mathematics-based rationality? (2) How can we conceptualise possible the socio-political roles of mathematics education? Together the two questions might help to shed some light on how mathematical rationality might assume different forms and become integrated into social and technological development. It is pointed out how mathematics-based rationality forms part of the fabrication of possibilities, strategies, facts, contingencies, and perspectives. There is, however, no inherent quality to be associated with such mathematics-based fabrications. It is considered to what extent the school mathematics tradition could establish a prescription readiness by submitting students to a school-mathematics absolutism; and to what extent this tradition could facilitate differentiated labelling of the students, as well as endorsement of an ethical filter. These aspects could be important functions of the school mathematics tradition in today's knowledge society. It is considered to what extent mathematics education could prepare people for critical citizenship, which includes a potential for 'talking back' to authority. I do not see such preparation as being related to the school mathematics tradition, nor do I see it as linked to the very nature of mathematics. Rather, I am proposing it as a *possible* function of mathematics education.

In 1916, John Dewey published *Democracy and Education* (Dewey 1966) that overwhelmed readers with an avalanche of ideas and notions. Dewey was inspired by Hegel. Everything can be brought together in an all-embracing analysis, as historical, social, political, cultural, and scientific development form part of a unifying progress. Dewey presented science in the role of an energetic motor of progress. The scientific pattern of thinking has rich and fruitful applications that reach far beyond the limits of science, and when this pattern is incorporated into the field of

O. Skovsmose (✉)
Department of Education, Learning and Philosophy, Aalborg University, Fibigerstræde 10,
9220 Aalborg East, Denmark
e-mail: osk@learning.aau.dk

This chapter is a reprint of an article published in ZDM—The International Journal on Mathematics Education (2007) 39(3), 215–224. DOI 10.1007/s11858-007-0024-5.

education, it, too, will become progressive. Progress includes democracy, and therefore education can represent a powerful move towards democratic life. At the most overall perspective of Hegelian dimensions, Dewey finds that the logic of science brings us in the direction of democracy.

What could it mean to produce a book, in a similar spirit, with the title: *Democracy and Mathematics Education*? It is not difficult to play with this idea. We can see mathematics as condensing the logic of scientific thinking, and in this way mathematics comes to represent the logic of progress. It represents a way of thinking where reason and proper arguments take the leading role, and not personal priorities or ideological idiosyncrasies. In this way, one can see a connection between the mathematical way of arguing and a democratic discourse which must also be free of particular individual interests. This would be the message of the book *Democracy and Mathematics Education*.[1] However, I am afraid that this book might lead us into a dreamworld. We have to consider carefully at least the following two questions: (1) *How can we conceptualise possible socio-political roles of mathematics-based rationality?* (2) *How can we conceptualise possible sociopolitical roles of mathematics education?* Together the two questions might help to shed some light on how mathematical rationality might take different forms and be integrated into social and technological development.

The emergence of the dreamworld has to do with the extent to which we envelop the discussion of mathematics and mathematics education in the assumption that we are basically dealing with an unproblematic rationality which, by its very nature, composes part of progress. My point is that the two questions should not be analysed within Dewey's overall assumption of progress. When we abandon this assumption, we prepare for a different form of analysis. It could become a more sinister one. Thus, inspired by Jacque Ellul (1964), we can see how society is formed by technology propagated by (and propagating) a mathematical rationality. Our lifeworld, to use a notion coined by Edmund Husserl, assumes a technological format, and maybe we[2] should talk about techno-nature as being the ever-changing container of lifeworlds.

I suggest that we not automatically take an optimistic nor a pessimistic approach when addressing the two questions concerning the socio-political roles of mathematics and of mathematics education. We need not assume any deeply based optimism included in Dewey's outlook. Nor do we need to assume that the scientific form of thinking brings about a de-humanised 'brave new world'. I suggest an analysis where we do not make any such assumptions about the nature of mathematics-based

---

[1] It is not difficult to find contributors to such a book. There have been several studies which exaltedly emphasise the link between mathematics education and democracy. Thus, Hannaford (1998) argues that the mathematical way of arguing is intrinsically connected to democracy, by paying no attention to personal priorities, but only to matters of fact. That mathematics in an axiomatic form and Greek democracy developed in the same historical period is not seen as accidental. See also the discussion in Skovsmose (1994, 1998).

[2] For an elaborated discussion of lifeworld see Husserl (1970).

rationality. I suggest that this rationality brings us into an open space of contingencies, and that development could take very different routes.[3]

In light of this uncertainty, I want to address the two questions mentioned.

1 Fabrication by Mathematics-Based Rationality

The possible meanings of 'mathematics-based rationality' depend on how we see mathematics. I try to broaden the notion of mathematics as much as possible. I do not let it refer only to the field as presented in textbooks, intended for schools or universities. Mathematics is also included in disciplines like engineering, economics, and medicine. Furthermore, I refer to mathematics as it is manifested in all kinds of work practices, being in banks, hospitals, shops, industries, where formal calculations play a role. I can see mathematics in very different daily-life practices, shopping being a most obvious example. As a consequence, I am not searching for any universal characteristics of mathematics, as has often been the case in the philosophy of mathematics. I let mathematics remain an open concept.

Knowledge can be acted out in many different ways and contexts. It can be a resource for technological innovations and for establishing new forms of production, automation, and management. What can be said about knowledge in general can also be said about mathematics. Mathematics is a resource for action. It represents a rationality which forms part of our techno-nature and lifeworlds. To illustrate, I will point out how fabrication of *possibilities, strategies, facts, contingencies,* and *perspectives* can be mathematics based.[4]

1.1 Fabricating Possibilities

One of the most general aspects of life emerges from the possibilities one can approach. At a personal level, this is a basic existential parameter, but it is also a general characteristic of our overall life conditions. Possibilities emerge through technological innovations. The information age can be symbolised by the invention of the computer, and the mathematically conceptualised Turing Machine made it possible to analyse in detail the operation of the computer, including its computational limits, even before any real computer was built. We are talking about an analysis based on a mathematical construct. There is an important observation to be made here. The inventions that opened the way for the industrial revolution had

---

[3]This space of contingencies has been addressed in Skovsmose and Valero (2001, 2002) in terms of the critical nature of mathematics education. See also Skovsmose and Valero (2005).

[4]The following presentation of mathematical fabrication draws on an insight obtained through a collaborative project including Keiko Yasukawa, Ole Ravn Christensen, and myself. See also the presentation of mathematics in action in Skovsmose (2005) and in Skovsmose and Yasukawa (2004).

a rather loose relationship to the sciences of the time. Thus, only long after the steam engine was developed was its operation investigated in scientific terms. This is different from the construction of the computer, where the whole conception emerged from mathematical insight. The computer represents a principle example of what a mathematics-based fabrication of possibilities can mean.

Today, it is common for mathematics to help establish technological possibilities. We can think of communication technologies, the conceptualising and construction of the Internet, techniques for cryptography, security technology, and pattern recognition. Nanotechnology and biotechnology operate as resources for a range of technological innovations, and mathematics is indispensable. Modern management, decision making and marketing would not function without mathematical procedures at hand. In all these different areas, technological innovations are based on the fact that computational power is available. To put it simply, in many different fields, the identification of new administrative possibilities depends on the spreadsheet. And more generally: the identification of new possibilities within any area of life relies on mathematics-based rationality.

It is important to observe that mathematics-based rationality cannot be substituted by a commonsense-based rationality or by a natural language formulation of ideas. In this sense, mathematics-based rationality opens a space of possibilities, which is not accessible in other ways. Mathematics-based ways of creating possibilities can take unique forms. This is not only its strength, but also its weakness. Creating 'unique' possibilities does not mean creating the 'best' possibility.

1.2 Fabrication of Strategies

Let us take a look at marketing and the definition of prices. One could think of prices as determined by supply and demand, and the particular price of an item becomes acted out at the market, seen as a bazaar: a negotiation (haggling and bargaining) takes place until the proper price is identified. However, the way of balancing supply and demand is rather different today. In *Travelling Through Education*, I describe an example of a mathematics-based mixed selling strategy. Airlines overbook flights, as statistics show that there are likely to be some 'no shows' for each departure. Naturally, the number of no shows depends on many different parameters, the time of departure being only one of them. However, one can use statistics to estimate the number of no shows for each departure. Such statistics will also reveal a certain variation of no shows. It might be possible to improve the forecasting, for instance by grouping the passengers, determined by different types of tickets and payment conditions. It is important to establish the most accurate prediction of no shows, as 'bumping' a passenger must be prevented.[5]

---

[5] 'Bumping' a passenger means to deny a passenger with a valid ticket access to the plane.

Through grouping of passengers, an airline company can improve the forecasting of no shows, and in this way determine, for a specific departure, the degree of over-booking which maximises revenue, considering also the compensation the company would have to pay a bumped passenger.

This is a phenomenon which I will call *mathematics-based action design*. A tra-ditional principle of selling could be formulated as 'Do not sell any more tickets than the number of seats on the plane.' However, such a principle could be substi-tuted by a more advanced principle 'Overbook, but do it in such a way that revenue is maximised, considering the amount of money to be paid in compensation to the bumped passengers, the destination, the time of departure, the day of the week, as well as the long-term effects of having to sometimes bump a passenger who in fact has a valid booking.' This is a complex principle which only can be managed through a booking model; and this is the point—through mathematical modelling, one can substitute classic principles for doing business with much more elaborated schemes. The airlines business is just one example. This observation applies to any aspect of the market. One has only to think of the many different prices and offers which are operating on the market for mobile phones. Mathematics-based action design becomes a most general aspect of the whole market of the informational society.

1.3 Fabricating Facts

When possibilities are analysed, some may become realised. A company can anal-yse different marketing strategies (for instance by applying mathematical models for epidemics). But when the decision is made about what strategy to apply, the epi-demic becomes real. At this point, the ontological status of the applied mathematical model is changed. The model is no longer a detached description of a possibility. It is not a simulation model any longer. Thus, different booking strategies can be in-vestigated, but once implemented, a particular strategy becomes part of the business reality for both the company and the passengers. This applies to the implementa-tion of any system. When brought into operation, mathematics becomes real. One fabricates facts through mathematics.

However, there are many forms of fabricating facts. The bumping of a passenger can be a most obvious fact, but one can see a different form of fact fabrication. There are many measures of what to count as safe and what not. Standards exist for any kind of building construction with respect to choice of material and dimensions. The construction is safe if it meets some predetermined standards. But in what sense is it safe? In what sense do the standards ensure that the building is safe? How are the standards defined? Or take a different example: there are standards for pollution, and such standards are based on the identification of a degree of pollution below which it is believed that there are no dangers. Our drinking water must not be polluted, which means that the level of pollution of the selected chemicals has to be below certain standards. In autumn of 2006, the news in Denmark reported on a crimi-nal act of somebody adding strychnine to the drinking water in a city in Denmark.

The medical authorities, however, emphasised that the amount of strychnine was so small that there were no health dangers to be concerned about. The possible truth of such a statement is a mathematical fabrication. We can only reach such a statement via statistics. And even the most careful statistical investigation must include assumptions about what to consider. What about the possible effects after 50 years? Surely not! Such a long-term investigation could not be managed. In fact, we cannot make observations of human beings being poisoned by strychnine in different degrees until we are able to make a proper identification of the limit below which there is no danger. We can imagine experimentation with rats. But what logic brings us from observations of rats to safety levels with respect to human beings? All these questions are integrated in a whole model-apparatus which provides safety limits. This is a different example of mathematics-based fact fabrication, and they are facts which are acted upon.

1.4 Fabrication of Contingencies

A pessimistic outlook, like the one presented by Ellul, characterises technology as a determining structure. Society becomes absorbed by the technological development which moves forward according to dynamics that cannot be influenced by particular human decisions, even though the dynamics are a human creation. I do not follow this deterministic line of analysis. Instead, I think of technological development, including its mathematics-based rationality, as a process which opens up for contingencies, although we might be dealing with contingencies which cannot be managed through particular human decisions. Technological development might contain risk production, and risks signify the emergence of man-made contingencies. I find that risk production is grounded in the wider application of mathematical rationality. One can identify technological possibilities, but each possibility might open the way for rather different clusters of risks, which need not happen, but nevertheless could happen.

1.5 Fabrication of Perspectives

Mathematics can be seen as a language, and bringing a new language into operation means bringing about new perspectives and new reasons for decision making. When we describe something in mathematical terms, we provide a new way of looking at things. Language can mean action, and a change of language also signifies a change in forms of action. I find that mathematics can provide new perspectives, including changes related to seeing as well as doing. The scientific revolution was founded on a new way of seeing nature, and mathematics made this revolutionary new perspective possible by providing the language needed to formulate natural laws with a new degree of exactness. Galileo Galilei distinguished between appearance and reality, which turned into a distinction between primary and secondary sense qualities.[6]

While secondary sense qualities provided a human-related perspective on nature, primary sense qualities revealed a non-subjective insight into nature. The primary sense data was the starting point for a proper science of nature, and it is precisely such data that can be expressed though mathematics.

An overall distinction between primary and secondary data turned into a defining assumption behind the broad application of mathematics not only in science, but also in almost all forms of technology. The mathematical way of seeing nature became a way of operating, not only with nature, but also with any possible aspect of techno-nature, or any aspect of our lifeworlds. As a consequence, mathematics became a way of operating with practically anything: economic proposals, strategies for marketing, medical procedures, etc.[7]

What to think of these different forms of fabrication? Relying on Dewey's perspective, we could assume that we are dealing with healthy and progressive fabrications. The fabrication may serve changes, and certainly I think they make changes, but I do not assume that mathematics-based changes ensure progress. The possibilities that are created could be horrifying; they could be wonderful. The strategies could be selfish; they could be highly original. Facts could be trivial; they could appear as inventions. Contingencies could include unwelcome risks; they could open the way for opportunities we have never imagined. New perspectives could be cynical; they could be naïve. There is no inherent quality to be associated with any mathematics-based fabrication.

2 Prescription Readiness

Let us now move from considerations about mathematical fabrications to what might happen in mathematics education. This means that we move away from the question *How can we conceptualise possible socio-political roles of mathematics-based rationality?* towards the question *How can we conceptualise possible socio-political roles of mathematics education?*

What role could we imagine that mathematics, as a subject taught in school, could have in society? In particular, what to think of the *school mathematics tradition*? What functions could we imagine that this tradition might have with respect to overall social, economic and technological development? Let us first briefly characterise the school mathematics tradition. Mathematics lessons follow some standard patterns. It could be the following. First, a new topic is presented by the teacher. This presentation can take different forms, from a teacher monologue to a dialogue between teacher and students. The presentation is based on the selected textbook, and the students are asked to study a particular part of the textbook for the next lesson.

---

[6]See in particular the essay *The Assayer*, which is reprinted in Galileo (1957, pp. 229–280).

[7]I refer to this phenomenon as a mathematical transposition, which I will discuss later.

Next, the students will be asked to solve some exercises. They may work individually or in groups. The teacher can discuss particular problems with the students with respect to the exercises, and normally the students are assigned some exercises to solve as homework. Finally, part of the lesson is spent on the teacher's control of the students' work. It could be of the results of the exercises done as homework, and some difficulties might be clarified at the blackboard. Students might be asked about what methods they have used. If the previous lesson included some principal mathematical ideas, like a proof, then the teacher may check to see whether or not a student is able to reproduce the proof. In fact, proof reproduction is a principle element of the school mathematics tradition, at least from secondary school and up. In short, the school mathematics tradition emerges as a combination of teacher presentation, students solving exercises, and teacher's control of students' work.

No research indicates that this tradition brings about any adequate understanding or learning of mathematics to a majority of students. It appears that we are dealing with a serious dysfunction incorporated into the educational programme. Explanations of this stubborn perpetuation of the school mathematics tradition have been suggested along the following line. Teachers have beliefs and schools have traditions which result in the survival of educational patterns even after they have lost their functionality. The task, therefore, is to struggle with outdated beliefs and traditions. I do not appreciate this line of explanation that seems condescending to teachers.

One could, however, assume that the dysfunction is only apparent. It could be that when we take a closer look at things, it becomes revealed that the school mathematics tradition serves some important functions in today's society. But what function could we imagine this tradition to have? One of its characteristics is the many exercises. They have the form "Calculate the length of. . . !", "Find x from the following equation. . . !", "Construct the triangle where. . . !" The sequence of exercises appears like a sequence of orders.[8]

Commands do not seem to facilitate an understanding of mathematics; they seem to have nothing to do with mathematical thinking and reasoning. But could it be that the school mathematics tradition takes on significance by cultivating a *prescription readiness* among students, exactly through the sequence of orders it administers so carefully? Could the tradition endorse a disciplining of the students which is a relevant response to the demands of today's labour market?

When we take a look at the exercises in mathematics textbooks, some features are characteristic. The exercise provides some information which is *necessary* but also *sufficient* for solving the exercises. Students need not look up any additional information in order to solve an exercise. They need not go to the nearest shopping centre to find out prices of a sale, for instance. The students can remain in the classroom, and the teacher can be sure that they have all the relevant information. The

---

[8]In Alrø and Skovsmose (2002), the notion of bureaucratic absolutism is suggested in order to clarify this possible aspect of the school mathematics tradition. The point is that the nature of command comes to include the pattern of communication in the classroom, and in this way comes to frame the learning of mathematics.

second thing is that the information is *exact*. If the task is to find out how long it takes for a car to drive from *A* to *C* via *B*, and the distances between *A* and *B* and between *B* and *C* are given and it is stated that the car has a speed of 70 km/h from *A* to *B*, and then 80 km/h form *B* to *C*, then this information is exact. We need not consider what in fact is happening at point *B*. If *B* is considered a point, and in an exercise *B* would normally be considered as such, then it is completely incomprehensible from a physical point of view what the driver and the car are doing at point *B*. But any such considerations are irrelevant for solving the exercise. And if a student were to raise such questions, it could be experienced as a kind of sophisticated obstruction of the classroom order, defined through the logic governing the imaginary world of exercises. However, working with exercise after exercise, the students will learn what it means to operate with given information within a given space of possible strategies for solution. In this way, they assimilate a *prescription readiness*.

The answer to an exercise can be right or wrong; there are no other possibilities. Within the school mathematics tradition, exercises are supposed to have one-and-only one correct answer. Absolutism governs the mathematics classroom. Interesting, however, is that the right–wrong duality becomes extended to all kinds of activities in the classroom. The duality becomes applied far beyond the scope of answers to exercises. One of the classic mistakes that the teacher has to correct again and again is that students often copy the exercise incorrectly from the textbook. Very often the teacher has to emphasise that when one copies the figures incorrectly, then everything comes out wrong. Students can also solve the wrong exercises, or use the wrong method. There are so many things that get caught up by the right–wrong dichotomy. I see the submission to a broadly applied right–wrong dichotomy, which can be referred to as the *school-mathematics absolutism*, as an important part of establishing a prescription readiness.[9]

This could be the principle function of the school mathematics tradition.

3 Differentiated Labelling

Let us consider the labour market from the perspective of the knowledge society. I am well aware that the label 'knowledge society' is contested. Very many other notions have been presented in order to characterise our present society: informational society, risk society, network society, etc. Let me, however, stay with the notion of knowledge society, being aware that the application of this notion might falsely assume some degree of universal homogeneity: we are entering a knowledge society, but who are the 'we'? Processes of globalisation include processes of ghettoising. Globalisation means not only inclusion, but also exclusion. Keeping this in mind, I continue using the notion of knowledge society.

---

[9]This exaggerated use of the right–wrong distinction has been addressed in terms of an ideology of certainty. See Borba and Skovsmose (1997).

In the knowledge society, the labour market might take the form of a knowledge market. One could assume that the more there are people with more knowledge, the better the knowledge market. But this may not necessarily be so. In fact, I would suggest that for today's liberal-capitalist knowledge societies the most important thing for the functioning of the knowledge market is that the 'goods' be clearly labelled. People in the knowledge market have different competencies, and it is important that the competencies possessed by each and every person be as easily identifiable as possible. Naturally, there is a demand for people to provide innovations; but there is also a need for people who can manage the many different routine functions in society. And most important is to be able to select the persons with the competence profile in question. So my point is that differentiated labelling is a precondition for a well-functioning knowledge market, assuming the economic order of today.

Differentiated labelling presupposes that identification of competencies and procedures for testing and labelling students are taking place in school. This could explain the emerging preoccupation with tests, assessments and measuring as well as the meticulous development of the notion of competencies. I do not see this as a simple political twist of today's educational discourse. Instead, I see it as linked to a basic need of the knowledge society. The point is not that every student should master all possible competencies. If that were the case, there would in fact be no great need for tests and measuring. But if we have in mind that the supply to the knowledge market should be synchronised with the demands of the market, then the education programmes appear in a new light. Providing higher education to too many might be considered a wasteful competence production. It calls for a better adjustment of the educational production line, which has to operate according to a just-in-time production, where people with the competencies in demand appear exactly when they are in demand.

On a liberal market, students can choose among different educational possibilities. But how could this result in a just-in-time production of competencies for the labour market? We have to do with an apparent complication. However, highly elaborated tests and access conditions may establish sufficient regulations of students' possibilities for making choices. And as part of such regulations, a careful labelling serves its functions.

With such considerations in mind, we can see a new function associated with mathematics education, which the school mathematics tradition certainly complies with. This tradition operates with meticulous labelling, which is easy to adjust and expand to new forms of testing and labelling. So if we consider the emergence of the knowledge society, then the school mathematics tradition appears less obsolete. Rather, it seems to include educational procedures that can easily provide testing and labelling for the market.

4 Ethical Filtration

Mathematics forms part of many different programmes in higher education. One finds mathematics in all kinds of study programmes having to do with business, marketing, engineering, computing, economics, medicine, land surveying, architecture, etc. Mathematics courses may take the form of lectures; they may appear as upgraded variations of the school mathematics tradition; they could also operate as integrated elements of project work.

In all such cases, we can consider the mathematics course to be part of the development of professionalism within a particular field. And if we consider the great many ways in which mathematical rationality can be brought into operation, the relevance of mathematics as part of professionalism seems obvious. The mathematical dimension of professionalism includes many ingredients. Here, I want to address what I call an *ethical filtration*, which can be experienced each and every time mathematics is brought into operation.

This filtration concerns the relationship between formal and natural language. Natural language is rather tolerant. It is constructed over a longer period and across many different stages of human insight and understanding, so certainly all kinds of assumptions, superstitions, traditions, etc., are engraved in linguistic conventions and grammatical structures. Natural language represents a grid on the basis of which we formulate ideas, interpret our situation, make descriptions, formulate priorities, and carry out decisions. It might appear consequential, then, to try to establish a language which is not overburdened by all kinds of cultural artefacts. Mathematical language could be considered filter for all such kinds of imprecise formulations and preconceptions.[10]

When mathematics is brought into operation, we establish a *mathematical transposition*, which can be seen as a change of discourse that leaves behind an overloaded natural-language metaphysics. But could it be that this transposition is also entangled in a different, but not less overloaded metaphysics? An example could be the establishment of a mechanical worldview which, in modified form, is so dominant in science and engineering.

We have to raise some questions: Does mathematics provide an apt way of seeing the world, as material structures become reflected by mathematical structures? Or, are we dealing here with a sophisticated example of projecting mathematical structures into the world so that the world appears as material structures? Do we confuse projected properties of mathematics with properties of the material world? Do we find the roots of the mechanical worldview in the grammar of mathematics or in reality?

Whichever the case, mathematical transpositions have spread through the natural sciences and taken hold in many engineering activities, economic approaches, management strategies, etc. The mathematical perspective highlights some things as important and ignores others. It makes something 'primary' and other things 'secondary'. This applies when mathematics is used in describing any kind of physical phenomenon, but it applies as well when mathematics is used to describe aspects of a market, of investigating new forms of medicine, of identifying new forms of

---

[10]This possibility was celebrated by logical positivism, which found Galilei's insight important. What could be counted as 'secondary sense qualities', generally speaking, were secondary from a scientific point of view.

rationalisation of production, of providing new forms of promotion (the booking model being just one example). In any situation where we use mathematics, we witness transposition that provides a new way of seeing and acting in the world. A mathematical transposition is basic to the different forms of fabrication we have mentioned previously.

These processes of fabrication also include an *ethical filtration*. A mathematical transposition provides the illusion that we are dealing with things in an objective and neutral way, and that we have left behind the domain of values, political interest, personal priorities, etc. The idea of the transposition as being neutral and objective in and of itself hinges on the assumption that the transposition brings problems and actions into a unique and adequate perspective. My point is, however, that the mathematics transposition brings about a change of perspective. It substitutes some ways of seeing and doing with different ways of seeing and doing. I find that a mathematical transformation includes numerous assumptions which are in need of careful consideration. A mathematical transposition brings about a new outlook. It could be, say, a more or less mechanical worldview. But this is not an unproblematic perspective. It could include many assumptions and create specific forms of fabricating possibilities, strategies, facts, contingencies, and perspectives. (Yes, also perspectives, meaning that a mathematics perspective, when brought into operation, can be proliferated into many other perspectives.) The whole spectrum of fabrication becomes an ethical challenge.

And now coming back to mathematics education, as already emphasised, many programmes in science and technology education include elements of mathematics. A general characteristic of such programmes is that the mathematical transposition is taken as both necessary and unproblematic, or problematic only in a technical sense, meaning that if one masters the mathematical techniques, then one could exercise adequate professionalism within one's field. By supporting such an assumption, the mathematical part of the educational programme comes to include an ethical filtration. I find such a filtration highly problematic. It could lead to the assumption that an expert is neutral and somehow remains beyond the particular interest that might be associated with a mathematical rationality. In this sense, an ethical filtration is just a different variation of prescription readiness.

5 Citizenship and Critical Citizenship

We have moved rather far away from Dewey's optimistic perspective of seeing science and also mathematics as intimately connected to progress in all aspects of life. But there are several other stories to be told, and let me now consider the notions of citizenship and critical citizenship.

Citizenship can be interpreted in a broad sense as membership, both formally and informally, in whatever kind of society we might have in mind. There could be many different forms of membership. One could be a member by virtue of being submitted to decisions of the government, whatever form the government might take—democratic or otherwise. If citizenship includes preparation for entering the labour market of society, prescription readiness can be seen as preparation for citizenship. In a similar way, we can see the meticulous labelling as ensuring citizenship to the extent that citizenship means becoming included in society in accordance with one's capacities. If we consider the economic order of today's liberal-capitalist society, such a position depends on one's competence profile as it has been measured and pigeonholed. In other words, when we take citizenship in its broadest terms, the previous considerations seem to indicate that mathematics education, also in its traditional form, could serve adequately to prepare people for citizenship.

Critical citizenship, however, includes a potential for 'talking back' to authority. It includes opposition to taking decisions for granted. Paola Valero (1999) has characterised democracy in terms of collectivity, transformation, deliberation, and coflection.[11]

The notion collectivity emphasises that democracy is a social and political concern; the notion of transformation makes clear that democracy involves changes in social affairs—deliberation points towards the dialogical nature of democracy—while the construct of coflection, which could be thought of as collective reflection, stresses that democracy has to be learnt and re-learnt, reflected upon and changed.[12]

This characteristic of a democratic form of life includes the idea of critical citizenship, and it is radically different from the conception presented by Dewey in terms of a 'logic of progress'.

The important question now is to what extent mathematics education could prepare for critical citizenship. I do not see such preparation as related to the school mathematics tradition. Nor do I see it as linked to the very nature of mathematics. It has to do with a *possible* function of mathematics education. Let me provide some indication of how such a function can be conceptualised.

Often the notions of 'empowerment' and 'critical mathematics education' have been exemplified with reference to primary and secondary education or to basic adult education, and certainly these are important areas to discuss. To illustrate that there is something to be obtained here, I will refer to work carried out by Erik Gutstein in a Latino neighbourhood in Chicago. One of the students made the following comment after being involved in the programme:

> With every single thing about math that I learned came something else. Sometimes I learned more of other things instead of math. I learned to think of fairness, injustices and so forth everywhere I see numbers distorted in the world. Now my mind is opened to so many new things. I'm more independent and aware. I have learned to be strong in every way you can think of it. (Gutstein 2003, p. 37)

This is a clear statement of an empowering experience, and I see it as an indication of how mathematics education could help to bring about critical citizenship. This idea has been elaborated by Gutstein in reference to the work of Paulo Freire, who proposed that literacy could be developed in such a way as to include a capacity to read the world and to see it as being open to change.[13]

---

[11] See also Valero (2002, 2004).

[12] See also Skovsmose and Valero (2001), for a discussion of the notion of democracy.

[13] See also Gutstein (2006).

However, mathematics could come to serve a similar purpose. Also, through mathematics one can come to read the world as being open to change.

The idea of the project 'Terrible Small numbers', described in Alrø and Skovsmose (2002), is to address decisions and actions which are carried out with reference to mathematics. The project presents what it could mean in secondary school to deal critically with a mathematical transposition. Thus, the project placed the students in a situation where they had to make decisions based on mathematical calculations (which could be more or less reliable). In this way, they experienced what mathematics-based actions could mean and perceived the relevance of reflection. The project tries to show that what might appear as ethical filtration is really just a change in perspective, and that this change must be addressed through ethical considerations.[14]

The project, presented and carefully analysed in Renuka Vithal (2003), took place in post-apartheid South Africa, and it demonstrates what mathematics education for critical citizenship could mean in this situation. Here the issue of cultural conflicts became part of an educational concern and was addressed through different activities in the mathematics classroom. Cultural complexities pointed to new challenges regarding what *reading the world as open to change* could mean. When we consider multicultural issues, we might realise, in a more profound way, what it could mean to be concerned with critical citizenship.

It must be emphasised that critical citizenship also concerns the preparation of future experts. Here I can refer to many attempts to bring a new perspective to university education. One can view attempts to address the more profound aspects of the mathematical transposition as an aspect of critical citizenship for experts. Here I can refer to a project carried out by Otavio Roberto Jacobini (2004). It exemplifies how university students can be involved in mathematical modelling not as a technique, but as a transition between a natural language environment and mathematics language environment. When the very nature of a mathematical transposition becomes part of university education, I consider this to be an example of what education for critical citizenship could mean for the preparation of future experts.[15]

The idea that mathematics education could serve to prepare people for citizenship is rather straightforward. The interesting point with respect to the question *How can we conceptualise the socio-political role of mathematics education?* is, however, to what extent it is possible for mathematics education to prepare people for critical citizenship, and that this preparation concerns all stages of the educational systems, from primary school to the university level. This possibility, however, is not simple to realise, and it must be sought outside the school mathematics tradition.

---

[14] See also Skovsmose (2006a, 2006b).

[15] See also Christensen (2003) for a discussion of the role of reflections in university mathematics and science education, and (Vithal et al. 1995) for a discussion of project work in university-level mathematics education.

6 Conclusions

The question *How can we conceptualise possible sociopolitical roles of mathematics-based rationality?* brought us to the notion of fabrication. If we share Dewey's perspective, the notion of development means progress. It has a sound significance. According to the perspective developed in this article, the notion of development only signifies change. There is no assumption of progress inherently associated with such change. When we talk about mathematics-based fabrications of possibilities, strategies, facts, contingencies, and perspectives, possibilities can be horrifying, strategies can be wasteful, facts can be suspicious, contingencies can be threatening, and perspectives can be limited. But they need not be so.

The question *How can we conceptualise possible sociopolitical roles of mathematics education?* leads to the insight that mathematics education can play many different roles. We have referred to prescription readiness, differentiated labelling and ethical filtration. We related this filtration to a mathematical transposition, and mathematics education can support the assumption that such a transposition brings the issue one is addressing into a natural and objective landscape, leaving behind values and personal preferences. Furthermore, it was emphasised that a possible function of mathematics education could be conceptualised in terms of critical citizenship.

There is one overall observation I want to make with respect to the analysis of the two questions about conceptualising the functions of mathematics and mathematics education. The answers to the questions cannot be formulated in a uniform way. One has to be specific with respect to the context one has in mind. The analysis could take one form if we consider mathematics education for technicians, including future experts. We could also consider the questions with respect to those many people who will come to occupy routine-like job functions. The question could be discussed with reference to people as consumers. We could ask what could be the functions of mathematics and of mathematics education in constructing the lifeworld of different groups of people. My point is that to address the two overall questions, it is necessary to refer to specific lifeworlds. And here we really have to consider what different worlds we are dealing with. There are many groups who are partially excluded from what could be called the informational economy. What does mathematics and mathematics education mean for such groups of people? My point is that notions such as citizenship and critical citizenship might be interpreted very differently with respect to the variety of lifeworlds. As a consequence, one should expect a variety of answers to the two questions. I have no doubt that the analysis at this point must be substituted by a much more context-sensitive analysis. Let this be my conclusion so far.

Acknowledgements This paper makes part of the project 'Mathematics Education and Democracy', which is based on a cooperation between Paola Valero, Department of Education, Learning and Philosophy, Aalborg University and me. I want to thank Anne Kepple for completing a careful language revision of the manuscript.

References

Alrø, H., & Skovsmose, O. (2002). *Dialogue and learning in mathematics education: Intention, reflection, critique.* Dordrecht: Kluwer Academic.

Borba, M., & Skovsmose, O. (1997). The ideology of certainty in mathematics education. *For the Learning of Mathematics, 17*(3), 17–23.

Christensen, O. R. (2003). *Exploring the borderland: A study on reflections in university science education.* Ph.D. thesis, Department of Education, Learning and Philosophy, Aalborg University, Aalborg.

Dewey, J. (1966). *Democracy and education: An introduction to the philosophy of education.* New York: Free Press. (First published 1916.)

Ellul, J. (1964). *The technological society.* New York: The Random House. (Original French edition 1954.)

Gallilei, G. (1957). *Discoveries and opinions of Galileo.* Translated with an introduction and notes by Stillman Drake. New York: Anchor Books.

Gutstein, E. (2003). Teaching and learning mathematics for social justice in an urban, Latino school. *Journal for Research in Mathematics Education, 34*(1), 37–73.

Gutstein, E. (2006). *Reading the world with mathematics.* New York: Routledge.

Hannaford, C. (1998). Mathematics teaching is democratic education. *Zentralblatt für Didaktik der Mathematik, 98*(6), 181–187.

Husserl, E. (1970). *The crisis of European sciences and transcendental phenomenology.* Evanston: Northwestern University Press. (First published 1936.)

Jacobini, O. R. (2004). *A Modelagem Matemática como Instrumento de Ação Política na Sala de Aula.* Doctoral thesis, Instituto de Geociências e Ciências Exatas, Universidade Estadual Paulista, Rio Claro.

Skovsmose, O. (1994). *Towards a philosophy of critical mathematics education.* Dordrecht: Kluwer.

Skovsmose, O. (1998). Linking mathematics education and democracy: Citizenship, mathematics archaeology, mathemacy and deliberative interaction. *Zentralblatt für Didaktik der Mathematik, 98*(6), 195–203.

Skovsmose, O. (2005). *Travelling through education: Uncertainty, mathematics, responsibility.* Rotterdam: Sense Publishers.

Skovsmose, O. (2006a). Reflections as a challenge. *Zentralblatt für Didaktik der Mathematik, 2006*(4), 323–332.

Skovsmose, O. (2006b). Challenges for mathematics education research. In J. Maasz & W. Schloeglmann (Eds.), *New mathematics education research and practiced* (pp. 33–50). Rotterdam: Sense Publishers.

Skovsmose, O., & Valero, P. (2001). Breaking political neutrality: The critical engagement of mathematics education with democracy. In B. Atweh, H. Forgasz & B. Nebres (Eds.), *Sociocultural research on mathematics education* (pp. 37–55). Mahwah (USA): Lawrence Erlbaum.

Skovsmose, O., & Valero, P. (2002). Democratic access to powerful mathematical ideas. In L. English (Ed.), *Handbook of international research in mathematics education* (pp. 383–407). Mahwah (USA): Lawrence Erlbaum.

Skovsmose, O., & Valero, P. (2005). Mathematics education and social justice: Facing the paradoxes of the informational society. *Utbildning & Demokrati, 14*(2), 57–71.

Skovsmose, O., & Yasukawa, K. (2004). Formatting power of 'Mathematics in a package': A challenge for social theorising? *Philosophy of Mathematics Education Journal, 18.* http://www.ex.ac.uk/~PErnest/pome18/contents.htm.

Valero, P. (1999). Deliberative mathematics education for social democratization in Latin America. *Zentralblatt für Didaktik der Mathematik, 1998*(6), 20–26.

Valero, P. (2002). *Reform, democracy, and mathematics education: Towards a socio-political frame for understanding change in the organization of secondary school mathematics.* Doctoral thesis, Department of Curriculum Research, Danish University of Education, Copenhagen.

Valero, P. (2004). Socio-political perspectives on mathematics education. In P. Valero & R. Zevenbergen (Eds.), *Researching the socio-political dimensions of mathematics education* (pp. 5–23).
Dordrecht: Kluwer Academic Publishers.

Vithal, R. (2003). *In search of a pedagogy of conflict and dialogue for mathematics education.*
Dordrecht: Kluwer.

Vithal, R., Christiansen, I. M., & Skovsmose, O. (1995). Project work in university mathematics education: A Danish experience: Aalborg University. *Educational Studies in Mathematics,*
29(2), 199–223.

Preface to "A Socio-Political Look at Equity in the School Organization of Mathematics Education"

Paola Valero

Whenever we place our attention on the issue of equity in mathematics education, we are addressing a broad social, political, economic, and cultural problem that is related to how societies create and set in place mechanisms that differentiate, include, and exclude certain peoples from valued resources, either material or symbolic. My main point in this paper is to provide evidence that it is not possible to assume that "equity" is a matter that only and exclusively emerges and is reproduced inside mathematics classrooms. Since "problems of equity" in mathematics are a part of general problems of equity, inclusion, and exclusion in society at large (Pais and Valero 2011), I intend to focus on how the notions of ability/disability associated with the exclusion of certain children emerge in the practices and discourses inside the school organization. Of course, the school organization, again, is only one site in which mechanisms of inclusion and exclusion operate in society.

In the ZDM article there were two central arguments. The first had to do with a particular understanding of the term "mathematics education". My proposal of thinking about mathematics education from a social and political perspective revolved around the notion of a *network of mathematics education practices*. Such a perspective highlights the idea that whatever counts as the meaning of "mathematics education" in a given historical time is a complex construction that involve many participants, practices, and discourses. I further elaborated on the notion of a network of mathematics education practices in Valero (2010). In this paper, I examined the idea that in the historical constitution of mathematics education research, there has been a different understanding of the practices of mathematics education. My proposal tried to open up a space so that the field of research embraces as legitimate objects of study the links and connections between micro-processes of teaching and learning and the rest of contextual or macro-processes, factors, and elements that contribute to the constitution of those practices. In the particular case of research with a concern for issues of equity, the view of a network of mathematics education

P. Valero (✉)
Aalborg University, Aalborg, Denmark
e-mail: paola@learning.aau.dk

H. Forgasz, F. Rivera (eds.), *Towards Equity in Mathematics Education,*
Advances in Mathematics Education,
DOI 10.1007/978-3-642-27702-3_33, © Springer-Verlag Berlin Heidelberg 2012

practices would precisely provide a possibility of seeing equity in mathematics education as being strongly connected to larger processes of inclusion and exclusion in society.

My second argument in the ZDM article had to do with the illustration of the first argument through the examination of two concrete cases in Denmark and in South Africa. I would like to come back to one particular element that has bothered me from the time this paper was first published. It was the working definition that I provided for the term "power" and how it could be seen in relation to the cases examined. I wrote that in a view of power inspired by the work of Foucault, power could be view as "a relational capacity of social actors to position themselves in different situations, through the use of various resources. Power is not an intrinsic and permanent characteristic of social actors; rather, it is their capacity to participate by taking and defining the positions and conditions for engaging in social practice" (Valero 2007, p. 226). This formulation was not accurate and now needs further clarification. In discussions about how the notion of "power" has been used in the mathematics education research literature (Valero 2009), I have identified three main theoretical trends, namely: the liberal tradition that views power as an intrinsic capacity and characteristic of both mathematics and of those who learn mathematics successfully; a Marxist tradition that posits power also as a capacity but that depends on a structural imbalance of knowledge control; and a post-structural tradition that views power as a distributed positioning in the network of social practices and related systems of reason that regulate what is possible to do and think in mathematics education. It was the third trend that I tried to inscribe my analysis in the ZDM paper.

However, the fundamental difference that I now see in relation to the type of analysis that I showed in the ZDM article has to do with *the construction of subjectivities* in and through the practices of mathematics education. In other words, by constructing particular positioning of students in relation to school mathematics, the practices of mathematics education set in place categories of ability/disability and processes of inclusion/exclusion. As students face and insert themselves in those practices, they are not only learning "school mathematics", they are mostly learning to become certain type of people and take as theirs the categories available in those practices. Being excluded from school mathematics and fitting in a category of exclusion—either in terms of one's ability, gender, race, religion, ethnicity, etc.—is a social and organizational construction that subjectifies students in particular ways.

Back to the work of Foucault, the importance of studying the micro-physics of power has to do *not* in recognizing structures of power but rather in understanding the effects of power in the constitution of subjectivities (Foucault 1982). Looking back to the ZDM paper in these terms, I tried to show how in a network of mathematics education practices systems of reason are created that make possible the shaping of certain subjectivities in relation to school mathematics. A very important form of becoming in relation to mathematics is that of being "school mathematics incapable". Not being able to cope, participate, or succeed in the demands of school mathematics is in no way a characteristic of the individual learner, but it is a result of how the whole set of participants in the practices and discourses of school mathematics subjectify certain students.

After the seminal work of Walkerdine in the late 1980s (see, for e.g., Walkerdine 1988), many studies in recent years have pointed to similar directions and have helped in substantiating the implications of school mathematics in the constitution of students' identities and subjectivities (Knijnik and Wanderer 2010; Lange 2009; Martin 2010; Mendick 2008; Moreau et al. 2010; Stentoft and Valero 2010). These studies, from my point of view, contribute to a deeper understanding of how and why mathematics education touches many people in deep ways and not all of them in the most positive sense. Studies of equity in mathematics education that follow these lines enable us to go deeper in unraveling the political constitution of mathematics education.

References

Foucault, M. (1982). The subject and power. *Critical Inquiry, 8*(4), 777–795.

Knijnik, G., & Wanderer, F. (2010). Mathematics education and differential inclusion: A study about two Brazilian time–space forms of life. *ZDM, 42*(3–4), 349–360.

Lange, T. (2009). *Difficulties, meaning and marginalisation in mathematics learning as seen through children's eyes.* Aalborg: Aalborg University—Department of Learning and Philosophy.

Martin, D. B. (2010). Not so strange bedfellows: Racial projects and the mathematics education enterprise. In U. Gellert, E. Jablonka & C. Morgan (Eds.), *Proceedings of the sixth international mathematics education and society conference (MES6)* (pp. 57–79). Berlin: Freie Universität Berlin.

Mendick, H. (2008). Subtracting difference: Troubling transitions from GCSE to AS-level mathematics. *British Educational Research Journal, 34*(6), 711–732.

Moreau, M.-P., Mendick, H., & Epstein, D. (2010). Constructions of mathematicians in popular culture and learners' narratives: A study of mathematical and non-mathematical subjectivities. *Cambridge Journal of Education, 40*(1), 25–38.

Pais, A., & Valero, P. (2011). Beyond disavowing the politics of equity and quality in mathematics education. In B. Atweh, M. Graven, W. Secada & P. Valero (Eds.), *Mapping equity and quality in mathematics education* (pp. 35–48). New York: Springer.

Stentoft, D., & Valero, P. (2010). Fragile learning in mathematics classrooms: How mathematics lessons are not just for learning mathematics. In M. Walshaw (Ed.), *Unpacking pedagogies. New perspectives for mathematics* (pp. 87–107). Charlotte, USA: IAP.

Valero, P. (2007). A socio-political look at equity in the school organization of mathematics education. *Zentralblatt für Didaktik der Mathematik. The Intentional Journal on Mathematics Education, 39*(3), 225–233.

Valero, P. (2009). What has power got to do with mathematics education? In P. Ernest, B. Greer & B. Sriraman (Eds.), *Critical issues in mathematics education* (pp. 237–254). Greenwich, USA: IAP.

Valero, P. (2010). Mathematics education as a network of social practices. In V. Durand-Guerrier, S. Soury-Lavergne & F. Arzarello (Eds.), *Proceedings of the sixth congress of the European society for research in mathematics education* (pp. LIV–LXXX). Lyon: Institut National de Récherche Pédagogique.

Walkerdine, V. (1988). *The mastery of reason: Cognitive development and the production of rationality.* London, New York: Routledge.

A Socio-Political Look at Equity in the School Organization of Mathematics Education

Paola Valero

Abstract This paper presents some theoretical tools to help understand the meaning of mathematics education as socio-political practices and the implications of these for researching mathematics education. Taking two cases of schools and students in Denmark and South Africa, the paper illustrates how the theoretical and methodological ideas come into operation when illuminating issues of equity. It is contended that the disadvantaged positioning of some students for participating in mathematics teaching and learning is the result of the routines, ideas, shared meanings, and ways of talking and conceiving mathematics education among the actors in the school organization, inside as well as outside the classroom.

1 Introduction

That mathematics education is political should not surprise many nowadays. In the last two decades, several mathematics education researchers have formulated different thesis about how the teaching and learning of mathematics, a very important discipline in the construction of the modern and contemporary world, is deeply involved in a differentiated positioning of diverse groups of students and people (see, e.g., Keitel 2006). In the practices of teaching and learning mathematics in classrooms at all educational levels, people may construct more or less influential positions for their action in relation to their successful (or less successful) exercise of mathematical competence. In further social action outside of educational settings, the advantaged (or disadvantaged) positioning may persist. Formulations about mathematics education as a means to empower individuals and consolidate democracy, as well as to discriminate and exclude particular groups of children and people, have not only been part of the concern of researchers. Practitioners who care

P. Valero (✉)
Department of Education, Learning and Philosophy, Aalborg University, Fibigerstræde 10, 9220 Aalborg East, Denmark
e-mail: paola@learning.aau.dk
url: http://www.learning.aau.dk/en/department/staff/paola_valero.htm

This chapter is a reprint of an article published in ZDM—The International Journal on Mathematics Education (2007) 39(3), 225–233. DOI 10.1007/s11858-007-0027-2.

H. Forgasz, F. Rivera (eds.), *Towards Equity in Mathematics Education*,
Advances in Mathematics Education,
DOI 10.1007/978-3-642-27702-3_34, © Springer-Verlag Berlin Heidelberg 2012

about the effects and consequences of mathematics education on the improvement (or worsening) of students' lives have also shown interest in thinking about how power is in operation in their teaching practices. Even educational policy makers around the world, for whom all school subjects should contribute actively in the education of citizens, have clearly formulated a desire of making mathematics education strengthen the general competencies of citizens in the twenty-first century. That mathematics education is highly political is part of the current way in which this social practice is conceived at this time in history.

It is not my intention in this paper to go into the details of the general formulations about mathematics education being socio-political practices, nor about how and why mathematics education is related in a positive or in a negative way with the construction of democratic social relations and of critical citizens. Others previously have done this in depth.[1] My intention in this paper is twofold. On the one hand, I will present some theoretical tools to understand the meaning of mathematics education as socio-political practices and the implications of these notions for researching mathematics education. On the other hand, I will illustrate how the theoretical and methodological ideas come into operation in an empirical study on equity issues in two schools in Denmark and South Africa.

2 Mathematics Education as Social and Political Practices: A Theoretical and Methodological Framework

I will start by addressing a series of central theoretical and methodological points when researching mathematics education from a socio-political perspective. First, it is important to clarify the meaning I give to the term "mathematics education." It refers to the whole series of practices where the meaning of the teaching and learning of mathematics is constituted. Such practices are not limited to the space of the classroom where teachers and students put into play forms of communicating around a particular mathematical content, but also include other spaces where decisions are made and actions are taken around the teaching and learning of mathematics. Sites of practice such as international or national educational policy making in mathematics, teacher education, textbook production, the labor market, and even the very same research on all these practices, among others, are part of the practices of mathematics education. Such a broad and complex collection of sites is what I have called the *network of mathematics education practices* (Valero 2002b).

Second, mathematics education practices are social not only because the act of learning is social and because mathematics is a socially constructed knowledge (Lerman 2006), but also and foremost because the learning and teaching of the subject are embedded in that broad network of practices. Being so, their meaning is constantly created and recreated though history, in particular social and cultural conditions, by all those people involved in shaping what is understood at a particular

---

[1] Ole Skovsmose's paper here in this number provides a good overview on this issue.

time and in a particular site by "mathematics education." This view implies that, for example, the activity of policy makers or the demands of the labor market at particular times in history have an important role to play in providing frames for action in schools when mathematical instruction is being planned and implemented. The actions and meanings constructed in the classroom around mathematics cannot be completely isolated from the actions and meanings constructed in other sites of the network.

Third, mathematics education practices are also political because they are implicated in the functioning and distribution of power in social relations. Among the different ways of conceiving power and relating it to mathematics education, poststructuralist definitions inspired by the work of Michael Foucault (e.g., Foucault 1972; Foucault and Faubion 2000) have been illuminating for many researchers such as, for example, De Freitas (2004), Hardy (2004), Meaney (2004), and Popkewitz (2002). In this view, power is a relational capacity of social actors to position themselves in different situations through the use of various resources. Power is not an intrinsic and permanent characteristic of social actors; rather, it is their capacity to participate by taking and defining the positions and conditions for engaging in social practice. Thus, power is not monolithic; it is distributed in social relations and is in constant transformation. This transformation does not necessarily happen directly in open struggle and resistance, but through the everyday participation of actors in social practices, in the creation of their meaning, and in the constitution of their associated discourses. Power is not openly overt and easy to identify but subtly exercised in and through social action.

Viewing power in this way invites analysis of the microphysics of power in mathematics education practices; that is, investigating how the everyday interaction among the different social actors that participate in the practices of mathematics education, inside and outside educational institutions, bring into life ways of talking and viewing the teaching and learning of mathematics. Such an analysis makes evident the mechanisms through which different actors get positioned as more or less influential participants and thereby get included and excluded.

Fourth, all the previous considerations about the nature of mathematics education practices have implications in the way the researcher approaches the study of those practices. A very first implication has to do with the delimitation of the site of study. In mathematics education research, the individual learners and teachers and the classroom have been privileged sites of study. Few studies have considered, for example, the school organization, the sites of policymaking, the working place, or the family. Mathematics education research with a serious concern and commitment with evidencing the political and social nature of mathematics education practices needs to open up its focus of study; and the school organization is an important site and unit of research (Skovsmose and Valero 2001, 2002). A second implication that follows is the necessity of considering the site as part of the network of mathematics education practices. This means that the researcher is invited to analyze the relations among different actors within the site, as well as the connections between it and its context. The complexity of mathematics education practices at any site can only be understood through an analysis that links in significant ways the construction of meaning for mathematics education at micro- and macro-levels of social practice. Such methodological challenges, among others, are some of those facing researchers adopting a socio-political perspective on mathematics education.

In what follows, I intend to illustrate the previous points by concentrating on an empirical study that allowed discussing student's participation in mathematics education and disadvantage, one of the core issues addressed by research concerned with equity.

3 Studying Equity in the Network of School Mathematics Education Practices

In Valero (2002b), I engaged in formulating a language of interpretation to talk about mathematics education reform within the school organization and from a socio-political perspective. I built on the work of Perry and collaborators (Perry et al. 1998; Perry et al. 1996) who developed the notion of *institutional system of mathematics education* (ISME). The ISME is a model that views the functioning of the teaching and learning of mathematics inside a school as a system in which at least three types of actors intervene: the school leaders, the group of mathematics teachers in the school, and the teacher as an individual in his or her classroom.[2]

Three case studies in three schools—Nyspor School, a primary school in Denmark, Rajas Secondary School in South Africa, and Esperanza Secondary School in Colombia—were carried out with the intention of being an empirical ground for triggering reflection, challenging, and reformulating the ISME model. The three qualitative case studies (Stake 1995, 2005) were conducted in order to get a detailed understanding of the functioning of the ISME in the three schools selected to be the cases. A number of methods—such as classroom observation of the group of mathematics teachers in each school, observation-based interviews with teachers on their teaching practice, observations of staff meetings, observation-based discussion with school leaders and other relevant staff members, documentation of national and local educational policy, and mathematics education related policy in the school, among others—were used for collecting information about the functioning of the ISME in the three schools. The information was used to construct a series of critical episodes, presented in the form of vivid stories of practice, addressing relevant aspects of the functioning of school mathematics in the school organization.[3]

The analysis of the critical episodes concentrated around the exploration of five clusters: (1) the significance of *the context of schooling* in mathematics education change (for details see Valero 2007); (2) the conception of *the mathematics learner* as a fully social and political human being constituted within the practices of school

---

[2]Such an approach shares many points in common with the recent work of Paul Cobb and collaborators (Cobb et al. 2003) on viewing mathematics teachers' instructional practices as part of the school community of practice.

[3]More details on the methodology and analysis of data in the empirical studies can be found in Valero (2002b).

mathematics education (see Valero 2002a, 2004); (3) the role of *school leaders* as navigators in organizational contradictions when dealing with the transformation of mathematics education practices in the school; (4) the constitution of *professional communities of mathematics teachers*; and (5) the realization of the *school mathematics teachers* as socio-political beings who contribute to the creation of the meaning of school mathematical in the classroom and in the whole organizational environment outside of it. The overall analysis led to the formulation of the notion of the *network of school mathematics practices* as presented in the previous section of this paper.

Part of the discussion about the conception of the mathematical learner emerged from the observations in the three schools from students who seemed to be marginalized from active participation in classroom activities. Processes of inclusion and exclusion of some students have been the central concern of most of the research studies and policy formulations arguing in favor of more equitable and democratic mathematics teaching and learning practices. Digging in the network of school mathematics practices and discussing existing research literature and policy documents opened the way for allowing insights into how these processes operate and how advantaged/disadvantaged positioning of students come into being in mathematics education. In what follows, I will enter that discussion, starting with two snapshots from Nyspor School in Denmark and Rajas Secondary School in South Africa.

3.1 The Lonely Girl

Mette, a young, novice female Danish teacher, and I walked to her eighth grade classroom together. Carrying a pile of photocopies with the activities she had planned, she entered the classroom, which was organized in a typical matrix of three rows and three columns of students' desks. Students sat in pairs. There was only one individual desk, right in front of the teacher's table. As students slowly moved to their places, the shelves with books, dictionaries, pencils, paper, and notebooks became more visible and suddenly seemed to get ready for doing some mathematics. It took some time before the 18 students got ready to work. Mette's soft voice became clearer as there was silence enough for her to deliver the photocopies with the tasks for the whole week. Students complained but finally got started, solving the problems in pairs, as they were sitting in the class. Mette attended those who called for assistance, and I concentrated on observing and talking to the girls sitting at the front.

Quite soon, Gitte, a Danish teenager with colored hair and fashionable clothes, attracted my attention. Two girls sitting by the window needed a sharpener and a ruler. "Gitte, give us the sharpener and the ruler," said one of them. "Here they are," Gitte replied. Other girls dropped a pencil on the floor. "Gitte, can you pick up the pencil?" "Yes," Gitte replied, picking it up and passing it to them. From time to time the girls behind her also asked her similar things. Gitte was solving a different working sheet. She was sitting alone in front of the teachers' desk. I started wondering... I offered her assistance but she refused it. The rest of the class was chatting, sometimes more about the weekend than about drawing lawns and fences, and calculating areas, and perimeters. Gitte, the lonely girl, suddenly stood up and left the room. After a while she came back and sat staring at the girls sitting by the window. "Stop looking and do your work" was the answer to Gitte's friendly smile. She turned back to her desk and did her exercises until the class finished.

Back in the staffroom, Mette commented on her difficulties with this class. As a novice teacher, it had been hard for her to convince them to engage in doing some work in mathematics. She had to adopt a low profile in front of these problematic students. I asked Mette directly about Gitte. She explained that this girl had some learning problems. She was allowed to be in the school and do as much as she wanted. Mette had tried to help her, but it was time consuming and the other students also demanded attention. "It is important to give her a hand," I insisted. "But you see, it is difficult, with the messy class, I cannot give her more time. I make special copies for her. Besides, she has serious problems, some members of her family have also studied here and they have also had difficulties in learning, so when there is a case like this, what can you do?"

3.2 They Are Creeping into...

African students were together, always together with their African peers in Rajas Secondary School. In the mornings, groups of uniformed students walked some kilometers up the hill to reach the school. Many of them started walking from the main avenue where the white "black taxis" dropped them. In the schoolyard, they walked and talked only with other African students. In the classrooms they also sat next to each other, in pairs or in groups of four, scattered in the room, never all in the same spot. Very seldom did they ask a question or raise their hand to answer the teachers' questions. Only few times I heard them speaking in their native languages in the midst of the sea of English words that flooded the school. Teachers seemed not to pay much attention to them, even though teachers recognized that many of these students had difficulties. At least their marks in mathematics assessments and exams were systematically low. For an outsider like me, African students constituted a silent minority inside the school, with the same formal rights to be members of it, but also with high and thick invisible walls around them. The walls did not let them approach others neither to be approached.

Breaking those walls was one of the challenges of Rajas School in the era of the "New South Africa." Mr. Ruikar, the acting school principal, when talking about the composition of the student body, referred to the change from an Indian segregated school to an integrated school with a mixed racial composition. Legally, the school was compelled to open places for students from other racial groups. In a new, democratic South Africa all should have the same opportunities, and accessing better schooling constituted one of the big achievements for the most disadvantaged although larger group, the African population. But making democracy is not only a matter of opening seats in a more advantaged school. For Mr. Ruikar, there were problems of access to resources, but especially the "Blacks' learning culture" needed transformation. For Arun, one of the mathematics teachers, language was the greatest barrier.

The invisible wall around African students was a double-sided fortification built with bricks of misunderstandings along the very many years of racial segregation and struggle. African students in Rajas resembled intruders who are slowly creeping into a forbidden place. I wondered about how and when they could become legitimate insiders...

3.3 Entering the Discussion of Equity in Mathematics Education Research

These two episodes invited me to take up the issue of equity in consideration. In "The lonely girl," Mette tries to do her teaching in the best possible way. In many respects, she brings into practice some of the curricular recommendations of the moment (Undervisningsministeriet 1995) such as engaging all students in learning, by meeting them at their own individual level. The principle of "differentiated teaching" encourages teachers to deal with the diversity of students' competency and to provide support to students' individual development. Although Mette does her best, she finds difficult to deal with Gitte, who has the right to basic education and to be taken care of by teachers in a school. She gets some attention from Mette and she does special mathematics tasks at her own pace. She interacts with her classmates in a condition of "inferior": While they do the mathematics tasks, Gitte picks up pencils and rulers from the floor. The episode was shocking since it illustrated the predicaments of equity: disadvantage is being built even at the heart of an educational system that has inclusion and democracy as an organizational principle; the predicament lives right in the middle of the mathematics classroom.

"They are creeping into..." shows the situation of African students in a predominantly Indian school in South Africa. In a time of national reconstruction and democratization, changes in policy allowing access of African students to more advantaged schools such as Rajas Secondary School was a challenge. The historical walls of apartheid were thrown down in laws and regulations but pretty much existed in the experience of people when meeting other racial groups. School administrators and teachers recognized that there were "problems" with the African students but in spite of more or less conscious efforts the walls were still there, even in the mathematics classrooms. The episode called my attention and invited thinking about the actual possibilities of throwing the thick walls down. Was there ever a chance to do it?

Gitte in Nyspor and the African students in Rajas represent students at risk of disadvantage in mathematics education. How to interpret the students and their apparent disadvantage? Research literature dealing with issues of equity and disadvantage in mathematics education had identified gender, ethnicity, language, social class, and ability as key factors related to the differentiated access and success in mathematics learning (e.g., Burton 2003; Keitel 1998; Rogers and Kaiser 1995; Sriraman 2007; Zevenbergen and Ortiz-Franco 2002). There have been different approaches in researching why and how mathematics education becomes a gate-keeper for particular social groups. A first trend of essentialist explanations related differentiated mathematical performance to genetic factors (for a critique of this type of studies see Ginsburg 1997). In a second trend, attention has been then given to students' psychological characteristics—such as affection, self-esteem, confidence, motivation, and persistence—and how these influence low performance and participation when engaging in mathematical learning (e.g., Fennema 1995; Leder and Forgasz 1998). This trend has been criticized for being "attributionist" (Boaler 1998) and not going into the mechanisms for the construction of exclusion. A third trend has built on the assumption that equity is a socially constructed phe-nomenon; that is, differentiation is not the result of intrinsic, genetic deficiency, or personal characteristics, but of larger social inequalities that are also visible at the level of individuals and of collectivities. Studies have recognized that the problems of inequity reside in the organization of mathematical instruction in the classroom. Patterns of interaction and teachers' attention to particular students, teachers' dif-ferentiated mathematical demands, the learning styles privileged by different types of instruction, the pedagogical discourses in mathematics classrooms among others are factors that give an account of the creation of disadvantage (e.g., Boaler 1997; Licón-Khisty and Chval 2002; Zevenbergen and Flavel 2007).

An examination of the situations described above suggested that students' dis-advantage in mathematics trespassed the boundaries of individual attributes and of instructional organization in the classroom. Disadvantage and equity could be re-formulated in terms of positioning within the relations and practices that constitute mathematics education in the school.

3.4 Examining the Construction of Disadvantage in the School Organization of Mathematics Education

How is an advantaged and disadvantaged position for participation in mathematics education constructed in the network of mathematics education practices within the school organization? Let us consider the episodes again.

Gitte, the "lonely girl" in Nyspor School, seems to have a position of a student with learning difficulties in mathematics. Such position, however, is not given by the fact that she may have some kind of genetic or psychological deficiency.[4] Whether she has it or not is in fact unknown to the teachers. However, the series of inter-actions between Gitte and other actors in the school clearly isolate her more than her own possible difficulties. How is it then that Gitte's environment constructs her "deficient" position? First of all, Gitte's position is evidently marked physically:

---

[4]The case of Gitte could be interpreted within the research on mathematics education for children with special needs. For a discussion of this trend see Magne (2001).

Her desk is placed just in front of the teacher's desk. She sits alone while all other students sit in pairs. This very physical setting marks her as being different. Her classmates also relate to her in a particular way. Gitte can be disturbed to pick up pencils from the floor, and passing sharpeners and rulers. Given the organization of the primary and lower secondary school in Denmark where students stay in the same class and with the same teacher for many years, students in the class have known Gitte for some years, and in their everyday interaction with her have also contributed to reinforce her being "lonely." Mette, the teacher, has identified Gitte's difficulties. She has created a way of interacting with girl that clearly gives her the place of a lower achiever and a slow learner in mathematics. The sets of special and easier problems given to Gitte function both as a help and as a marker of Gitte's difficulties. All these interactions in the classroom are only one of the elements in constructing Gitte's position as a less able student in general and in mathematics. Mette mentioned Gitte's belonging to a family with learning difficulties. Mette has been in the school for only 2 years, and it is the first year that she had that class. In the classroom, she may see that the girl does not perform as the rest of the students. But how does she know that Gitte's problems are common to other members of her family? In her socialization as a new teacher in that school she came to know from colleagues that Gitte had problems, but also that her learning difficulties was some kind of "genetic" deficiency. Gitte's previous mathematics teacher had identified a lack of (mathematical) ability in Gitte, in her passing from first to eighth grade, which coincided with the deficiency of her relatives. Other teachers, in diverse spaces of interaction, have also commented about both the girl and her family. Probably not only Mette but the teachers in other school subjects also position Gitte in this way, given their shared idea on her disability.

The municipal and school policy on support to students with special needs have also impacted the way in which the school master and teachers have organized the possibilities for providing students with difficulties—and Gitte in particular—with additional help. The decisions on how to deal internally in the school with the support to mathematically weak students concern the school leaders, the teachers, and the students' parents. The emphasis that local authorities put on bilingual students leaves few resources for students such as Gitte. Therefore, most of the help given to her has to go through Mette and her strategies for differentiation of teaching and meeting students' different ability levels in the same class. Besides, the fact that Gitte is a quiet, well behaved girl may distract the attention of the teacher to other students whose evident misbehavior is considered in the school culture as a sign of being "problematic" and, therefore, in need of special attention.

It is clear that in the interaction among leaders, other teachers, and Mette, collective ideas about ability, in general, and of mathematical competence in particular students such as Gitte, are built. The resulting decisions made and actions taken in the school organization and in the classroom contribute in constructing Gitte's positioning as a low-ability, female (mathematics) learner.

In the case of the African students in Rajas Secondary School, how is it that the walls around African students come in place? Are they only built in the classroom? A very first answer to this question could be that the position of African students in an ex-Indian school in South Africa is the result of apartheid and the deep structural inequalities of that regime. However, such an answer does not allow seeing the instauration of mechanisms of inclusion/exclusion in the everyday functioning of schools and mathematics education in them. It is necessary to dig into the constitution of practices in the school about mathematics education and how those position African students.

In *Rajas School Constitution*, it is stated that one of the aims of the school is to "educate and prepare learners to accept and manage the challenges and demands of a changing society based on equality and freedom and the eradication of all other forms of unfair discrimination" (Rajas Secondary School 1997, p. 2, unpublished data). Formally, there is an institutional commitment against discrimination and racism, in accordance with the national democratic transformation in South Africa and with specific policies of affirmative action, which intend to favor African population as a way of balancing the historical racial unbalance and discrimination. Although in Rajas, a predominantly Indian school, there is no explicit policy for the admission of children belonging to other racial groups, the criterion of "first come, first served" has opened the doors to African students who make up 20% of the school population. Most of the African students live in townships far from the school area and must find and pay for their transportation to and from school. Although these students are allowed in the school, it is also evident that they perform poorly. Mr. Ruikar, the school's principal, mentions the difficulty of these students to cope with school mathematics. The difference in "culture of learning" between the school and the African students is the reason of the poor results:

> [...] if you take like specially the Blacks, you will find the Blacks are in a fairly quite hopeless community when it comes to mathematics results or in fact in any other subjects. In their culture of learning [unrecognizable] whatever they can do, they use to do it in the school. Now, when these Black students are coming to Indian schools, for us, homework is always there. In fact we've got a special homework timetable, where they [students] are suppose to have about 45 minutes of homework per day or every second day. And we insist on homework because the time that is allocated to the subjects during normal class hours is not sufficient. So we do the basics in class and they have to go home and consolidate in the form of examples and so on. But the Blacks are not doing the homework. And only now, in these Indian schools by being forced, they are slowly attempting or learning to do the homework (Mr. Ruikar, Interview 1).

Mr. Ruikar recognizes that the "lack of a learning culture" is associated with the lack of resources in many African families and even the lack of parental support:

> Mr. Ruikar: [...] when you give homework to a child, one from a deprived community and one from the elite, that child has got computers at home. When he goes and speaks to the parents and looks at the questions and what they are expected to do, the child can afford to go to the bazaar and buy certain types of aids, and pictures and other things in order to enhance the assignment. Whereas in a poorer home, the child just is being assisted. He hasn't got even the daily news or the daily paper from which he can take certain cutting and he can paste it.
> Paola: And the parents probably don't even have the time to check the homework.
> Mr. Ruikar: Maybe they didn't even go to school. They haven't got the ability to help the child even if they want to help the child they haven't got the knowledge. So this is the situation that we are facing (Mr. Ruikar, Interview 1).

Mr. Ruikar's explanations resemble the attributionist views of equity mentioned in the previous section. Nevertheless, in the first quotation, he pointed to the fact that the organization of instruction in the school has settled the practice of homework making it an important part of the activities that students are expected to do in order to learn. In this fragment of conversation with Mr. Ruikar, it is evident that the organization of teaching and learning installs a practice, central to the learning of mathematics, which disadvantages African students. Disadvantage is partly constructed by the way in which instruction happens, although it is perceived as a problem of the students for not sharing an important part of the "learning culture" of the dominant Indian community in the school. Mr. Ruikar identifies this as a problem, but still does not foresee possibilities for overcoming it with introducing changes to the "Indian learning culture."

A disadvantaged position is also constructed from the head of the mathematics and science department, Seema. She thinks that hard work is the key of success in mathematics:

[...] And so my experience with all the people we teach is that they realize they must work hard. And the mark gets better, you would never believe they did badly in grade 9 because they reached that level where they know. And the mark has got very good [...] But some of them think that automatically they go for tuition, automatically they will get good marks. It doesn't happen, they have to work hard (Seema, Interview 1).

Other teachers in the department also share this opinion. For example, Mahesh, referring to moments of discouragement when teaching in stressful situations, values the effort and hard work that some students have invested in their mathematics lessons. These students are his motivation to continue with his teaching:

Because you may reach a point in your life, even in school, when you get highly frustrated of students not cooperating [...] But at the end of the day when you look at what is important, the pupil is important and they need to produce a result based on knowledge and your duty is to impart that knowledge and also take care of them for their future [...] So you find that you come to a class when half of them may not do the work. At first, your initial reaction is to yield. But then you look at the other and those are the students we work for. Don't worry about the other half. And then the other half who don't do it, they get the message eventually [...] But then I usually look at the kid who sat until 10 last night and did the work. So I always maintain that there is at least one kid who benefits form it, that is enough. But that shouldn't be the goal. You should try to get everyone to that level (Mahesh, Interview 1).

Both Arun and Vijay, the other mathematics teachers, also expressed their expectations about students doing their homework, reading the textbook, doing the exercises in it, and completing additional work that they prepared about topics they recognized as difficult. Students have to study hard in order to successfully cope with the examinations. If they fail it is because, as Vijay said, "they don't want to learn anymore." The group of mathematics teachers led by Seema value hard work and connects it with success in mathematics. They all expect that students are able to do a considerable amount of additional activities so that they can cope with the demands. As Mr. Ruikar pointed out, more "well-off" students have the resources—material and parental—to do so. But the possibilities of doing all the extra work needed may not be within the reach of most African students. Lack of material resources may restrict access to textbooks, additional materials, and free

time—some students have to do work at home or outside to help supporting their
families. Parental support is also a restricted resource that limits students' possibilities to fulfill all expectations concerning hard work.[5]

The teachers and the principal in Rajas identified private tuition—see Seema's
comment above—as an important support to perform well in examinations, particularly when approaching the end of secondary school and the national tests when
leaving school. Even some of these teachers service students from other schools
as private tutors in their free time. An organization of mathematics education in a
school which requires additional, paid tuition may certainly be in the disadvantage
of less affluent students, among them most of the African youngsters attending this
school.

The fact that English is the language of instruction in Rajas contributes to the
disadvantage of students whose main language is not English.[6] For Arun, one of the
mathematics teachers, language was a serious problem:

> Arun: [...] Now from 94 with all the transformations, you have different cultures in the
> classroom.
> Paola: And which kinds of problems do you see in this integration of cultures, in your
> classroom, for example, with Black children?
> Aarun: The main problem is really communication. Black students have problems because
> the medium of instruction here is English and they have been used to a culture where the
> medium of instruction is Zulu or one of their languages, and now they come to an English-
> Indian school, and the main problem is the communication. For example, in maths, you do
> a lot of things with words, and if they may not follow the problem, they won't pick it [...]
> (Aarun, Interview 1).

It was evident that teachers recognized the difficulty of having several cultures
and languages in the classroom. However, there was not an explicit institutionalized
effort in addressing the difficulties. On the contrary, the decisions made concerning
instruction in a time of shortage of resources accelerated the pace of teaching in
order to cover the mandated syllabus; the result of that being a very compact and
pressed school day which made it difficult for students who are not very proficient
in English to cope with the teaching.

The organization of school demands, the valuing of hard work, the expectations
of extra home work and eventual private tuition, and the language of school mathematics are part of the institutional practices of school mathematics in Rajas Secondary School. These practices have implications for teaching in the classroom, but
are definitely not limited to it. They are built and exercised in the school organization where mathematics teachers as a collectivity, in relation with other teachers
and with school leaders, create ways of talking about what is valued in teaching and
learning mathematics. These ideas are many times the grounds for decision making
about the organization of a type of instruction that construe a disadvantaged position
for African students.

---

[5]These problems have been documented in under-resourced African communities in South Africa
(e.g., Adler 2001a; Vithal 2003).

[6]The discussion of multilingualism and disadvantage in South African classroom has been broadly
documented in the work of, for example, Adler (2001b) and Setati (2005).

4 Power and Equity in Mathematics Education Research

One of the tasks of mathematics education research as an international field of study is bringing an understanding of the complexity of the practices of mathematics education. Identifying key predicaments in those practices is part of the scientific endeavor. In my trajectory as a researcher in this field, it has been important to open the research scope in an attempt to contextualize mathematics teaching and learning practices in the broader social space in which they live. In this paper, I have outlined some of the key ideas that make part of my socio-political perspective and have illustrated how those ideas are the theoretical and methodological tools for researching mathematics education practices.

I used two cases, Gitte in Nyspor School and the African students in Rajas Secondary School, to bring forward a discussion of equity in mathematics education. Issues of equity, social justice, and democracy have been a concern of many mathematics education researchers who acknowledge the fact that performance in mathematics is used in many school systems as a gate-keeping and selection mechanism from entering into socially valued educational and working possibilities. It is argued that not being able to learn mathematics effectively diminishes students' capacities to act in powerful ways in the current global society. From a broader socio-political perspective, understandings of exclusion and inclusion are incomplete if no attention is paid to the practices of mathematics education outside of the classroom, but that make part of the network of routines, ideas, shared meanings and ways of talking and conceiving mathematics education among the actors in the school organization, and even outside of it. Digging the network of school mathematics education practices in Nyspor School and Rajas Secondary School allowed evidencing that the possibilities of participation that Gitte and the African students have in their mathematics classrooms are constructed in the school. The micro-physics of power of mathematics education point to the school as the arena where differentiated positioning, and thereby access, is constructed.

Addressing equity issues in mathematics education research and in practice demands the broader type of understanding presented here. Without it, any attempt at tackling exclusion within and through mathematics teaching and learning will fall short in making a substantial difference for students such as Gitte or the African children in South Africa.

Acknowledgements   This paper makes part of the project "Mathematics Education and Democracy" which Ole Skovsmose and I are engaged with, at the Department of Education, Learning and Philosophy, Aalborg University.

References

Adler, J. (2001a). Resourcing practice and equity: A dual challenge for mathematics education. In B. Atweh, H. Forgasz, & B. Nebres (Eds.), *Sociocultural research on mathematics education. An international perspective* (pp. 185–200). Mahwah: Erlbaum.
Adler, J. (2001b). *Teaching mathematics in multilingual classrooms.* Dordrecht, Boston: Kluwer.

Boaler, J. (1997). *Experiencing school mathematics. Teaching styles, sex and setting.* Milton Keynes: Open University.

Boaler, J. (1998). Nineties girls challenge eighties stereotypes: Updating gender perspectives. In C. Keitel (Ed.), *Social justice and mathematics education. Gender, class, ethnicity and the politics of schooling* (pp. 278–293). Berlin: IOWME—Freie Universität Berlin.

Burton, L. (Ed.). (2003). *Which way social justice in mathematics education?* Westport: Praeger.

Cobb, P., McClain, K., de Silva Lamberg, T., & Dean, C. (2003). Situating teachers' instructional practices in the institutional setting of the school and district. *Educational researcher: A publication of the American Educational Research Association, 32*(6), 13–25.

De Freitas, E. (2004). Plotting intersections along the political axis: The interior voice of dissenting mathematics teachers. *Educational Studies in Mathematics, 55,* 259–274.

Fennema, E. (1995). Mathematics, gender and research. In B. Grevholm & G. Hanna (Eds.), *Gender and mathematics education* (pp. 21–38). Lund, Sweden: Lund University Press.

Foucault, M. (1972). *The archaeology of knowledge* (1st American ed.). New York: Pantheon Books.

Foucault, M., & Faubion, J. D. (2000). *Power.* New York: New Press; Distributed by W.W. Norton.

Ginsburg, H. (1997). The myth of the deprived child: New thoughts on poor children. In A. B. Powell & M. Frankenstein (Eds.), *Ethnomathematics: Challenging eurocentrism in mathematics education* (pp. 129–154). Albany: State University of New York Press.

Hardy, T. (2004). "There's no hiding place". Foucault's notion of normalization at work in a mathematics lesson. In M. Walshaw (Ed.), *Mathematics education within the postmodern* (pp. 103–119). Greenwich: Information Age.

Keitel, C. (2006). Mathematics, knowledge and political power. In J. Maasz & W. Schloeglmann (Eds.), *New mathematics education research and practice* (pp. 11–22). Rotterdam: Sense.

Keitel, C. (Ed.). (1998). *Social justice and mathematics education. Gender, class, ethnicity and the politics of schooling.* Berlin: IOWME—Freie Universität Berlin.

Leder, G., & Forgasz, H. (1998). Single-sex groupings for mathematics: An equitable solution? In C. Keitel (Ed.), *Social justice and mathematics education. Gender, class, ethnicity and the politics of schooling* (pp. 162–179). Berlin: IOWME—Freie Universität Berlin.

Lerman, S. (2006). Cultural psychology, anthropology and sociology: The developing 'strong' social turn. In J. Maasz & W. Schloeglmann (Eds.), *New mathematics education research and practice* (pp. 171–188). Rotterdam: Sense.

Licón-Khisty, L., & Chval, K. (2002). Pedagogic discourse and equity in mathematics: When teachers' talk matters. *Mathematics Education Research Journal. Special issue: Equity and Mathematics Education, 14*(2), 154–168.

Magne, O. (2001). *Literature on special educational needs in mathematics. A bibliography with some comments.* Malmö, Sweden: Department of Educational and Psychological Research—Malmö University.

Meaney, T. (2004). So what's power got to do with it? In M. Walshaw (Ed.), *Mathematics education within the postmodern* (pp. 181–200). Greenwich: Information Age.

Perry, P., Valero, P., Castro, M., Gómez, P., & Agudelo, C. (1998). *Calidad de la educación matemática en secundaria. Actores y procesos en la institución educativa.* Bogotá: Una empresa docente.

Perry, P., Valero, P., & Gómez, P. (1996). La problemática de las matemáticas escolares desde una perspectiva institucional. In P. Gómez (Ed.), *La problemátca de las matemáticas escolares. Un reto para directivos y profesores.* México: Una empresa docente y Grupo Editorial Iberoamérica.

Popkewitz, T. S. (2002). Whose heaven and whose redemption? The alchemy of the mathematics curriculum to save (please check one or all of the following: (a) the economy, (b) democracy, (c) the nation, (d) human rights, (d) the welfare state, (e) the individual). In P. Valero, & O. Skovsmose (Eds.), *Proceedings of the third international conference on mathematics education and society* (pp. 29–45). Copenhagen: Center for research in learning mathematics.

Rajas Secondary School (1997). *Constitution of Rajas Secondary School.* Unpublished manuscript, Durban.

Rogers, P., & Kaiser, G. (1995). *Equity in mathematics education: Influences of feminism and culture*. London, Washington: Falmer Press.

Setati, K. (2005). *Power and access in multilingual mathematics classrooms*. Paper presented at the fourth international conference on mathematics education and society, Gold Coast, Australia.

Skovsmose, O., & Valero, P. (2001). Breaking political neutrality: The critical engagement of mathematics education with democracy. In B. Atweh, H. Forgasz, & B. Nebres (Eds.), *Sociocultural research on mathematics education. An international perspective* (pp. 37–55). Mahwah: Erlbaum.

Skovsmose, O., & Valero, P. (2002). Democratic access to powerful mathematical ideas. In L. D. English (Ed.), *Handbook of international research in mathematics education. Directions for the 21st century* (pp. 383–407). Mahwah: Erlbaum.

Sriraman, B. (Ed.). (2007). *The montana mathematics enthusiast*. Monograph *1*: *International perspectives on social justice in mathematics education*. Missoula: Department of Mathematical Sciences—The University of Montana.

Stake, R. (1995). *The art of case study research*. Thousand Oaks: Sage.

Stake, R. (2005). Qualitative case studies. In N. K. Denzin & Y. S. Lincoln (Eds.), *The sage handbook of qualitative research* (3rd ed., pp. 443–467). Thousand Oaks: Sage Publications.

Undervisningsministeriet (1995). *Matematik. Faghæfte 12*: Undervisningsministeriet.

Valero, P. (2002a). The myth of the active learner: From cognitive to socio-political interpretations of students in mathematics classrooms. In P. Valero & O. Skovsmose (Eds.), *Proceedings of the third international conference on mathematics education and society* (2nd ed., Vol. *2*, pp. 489–500). Copenhagen: Center for research in learning mathematics.

Valero, P. (2002b). *Reform, democracy and mathematics education. Towards a socio-political frame for understanding change in the organization of secondary school mathematics*. Unpublished Ph.D. thesis, Danish University of Education, Copenhagen.

Valero, P. (2004). Postmodernism as an attitude of critique to dominant mathematics education research. In M. Walshaw (Ed.), *Mathematics education within the postmodern* (pp. 35–54). Greenwich: Information Age.

Valero, P. (2007). In between the global and the local: The politics of mathematics education reform in a globalized society. In B. Atweh, A. Calabrese Barton, M. Borba, N. Gough, C. Keitel, C. Vistro-Yu, & R. Vithal (Eds.), *Internationalisation and globalisation in mathematics and science education* (pp. 421–439). New York: Springer.

Vithal, R. (2003). *In search of a pedagogy of conflict and dialogue for mathematics education*. Dordrecht, Boston: Kluwer.

Zevenbergen, R., & Flavel, S. (2007). Undertaking an archaeological dig in search of pedagogical relay. In B. Sriraman (Ed.), *The montana mathematics enthusiast*. Monograph *1*: *International perspectives on social justice in mathematics education* (pp. 63–74). Missoula: Department of Mathematical Sciences—The University of Montana.

Zevenbergen, R., & Ortiz-Franco, L. (Eds.). (2002). *Mathematics education research journal. Special issue: Equity and mathematics education*, 14(2). Melbourne: MERGA.

Looking for Gold: Catering for Mathematically Gifted Students Within and Beyond ZDM

Gilah C. Leder

Abstract ZDM—The International Journal on Mathematics Education has a forty year long history of sustained publications. There is pride in the Journal's tradition of "publication of themed issues that aim to bring the state-of-the-art on central sub-domains within mathematics education" (Kaiser and Sriraman 2010, p. 143). In this chapter I trace the scope and themes of ZDM publications in which the education and needs of mathematically gifted students are discussed and compare the findings with those reported in the broader mathematics education research literature.

1 Introduction

ZDM—The International Journal on Mathematics Education is internationally one of the oldest research journals in mathematics education. It was published originally under the name Zentralblatt für Didaktik der Mathematik... (It) is the only research journal in mathematics education which strictly follows the approach of publishing thematic issues. The themes of the issues consider broad topics distinct for mathematics education all over the world such as problem solving or proof or features of research in mathematics education as well as more specialized topics such as development of algebraic thinking, gender issues, semiotics etc. With this approach the journal meets the interest of the entire spectrum of ongoing mathematics education debates as well as specialized perspectives... (ZDM) has broad international governance through its editorship and editorial board and this international breadth is reflected in the choice of theme editors and paper authors as well as the readership. (ZDM n.d.)

Quantifying the amount of research in a specific field in education is no simple matter. "The number of special journals, monographs, edited books, and essays on giftedness research as well as on practical interventions with gifted children both inside and outside schools is now almost incalculable. Hundreds of new publications are appearing every year" (Weinert 2000, p. xii). Many of these contain confirmation or refinements of findings from earlier research.

In the remainder of this chapter I trace the extent to which issues related to the education and development of mathematically gifted students have been examined

G.C. Leder (✉)
Monash University, Melbourne, Australia
e-mail: gilah.leder@monash.edu

H. Forgasz, F. Rivera (eds.), *Towards Equity in Mathematics Education*,
Advances in Mathematics Education,
DOI 10.1007/978-3-642-27702-3_35, © Springer-Verlag Berlin Heidelberg 2012

in ZDM, and in particular on material published in the journal over the past two decades. Whether this material reflects the work reported in other recent and relevant education research literature is also of interest. For the latter I have relied heavily on the contents of three major *Handbooks* published between 2000 and 2009: Colangelo and Davis (2003), Heller et al. (2000), and Shavinina (2009), as well as on material published in recent years in the *Gifted Child Quarterly [GCQ]*[1], a journal that "publish(es) articles that offer research findings and new and creative insights about giftedness and talent development in the context of the school, the home, and the wider society" (Gifted Child Quarterly n.d.).

The choice of material from the body of research beyond ZDM has, inevitably, an idiosyncratic element. Undoubtedly, other handbooks could well have served as useful sources, for example, Sternberg (2000) and Davis and Rimm (2004). Word limits imposed on the length of contributions to this volume allowed citation of only selected sources. It is worth noting that two of these editors, Sternberg and Davis, are represented among the team whose edited volumes I have used for comparative purposes. Space constraint considerations similarly led to the decision to focus on the contents of only one of the major journals concerned with gifted education[2]. Collectively, the sources selected enable a representative, rather than exhaustive, picture to be constructed of relevant research published outside ZDM.

No serious attempt is made in this chapter to provide a unique definition of mathematically gifted students or its pseudonyms. Indeed, Stoeger (2009, p. 17) has argued that "there has not been, and still is no uniform conception of giftedness". In keeping with the spirit of this chapter, the diversity of definitions used in different publications is accepted at face value[3].

2 ZDM

Although various changes have been implemented since the journal was first published in 1969, the English title of *International Reviews on Mathematical Education* was associated with ZDM from the outset. Initially the journal comprised two parts: "the documentation section (which) gave an overview about new publications... and the analysis section (which) was meant to reflect on the current discussion in mathematics didactics" (Kaiser 2007, p. 1). By 2001 financial constraints lead to the abandonment of the printed version of the latter section which,

---

[1]The GCQ is considered "the premier scholarly journal of the National Association for Gifted Children" (Gifted Child Quarterly n.d.). Launched in 1957 it, too, has a long publication history.

[2]Inspection of the *Gifted Child Quarterly* and, for example, the *Roeper Review*, another journal concerned with multiple aspects of gifted education, indicates that there is much overlap in the core issues covered in the two publications.

[3]A multitude of definitions can, in fact, be found in the research literature. Brief, readily accessible and practice oriented discussions can be found in Borland (2009) and Friedman-Nimz (2009). A more detailed overview can be found in Ziegler and Heller (2000), as well as in chapters in the other *Handbooks* listed above.

however, remained accessible electronically. Further changes, described in some detail by Kaiser (2006, 2007) followed. Significantly, the data base (documentation section) and original articles (analysis section) were separated.

In 2007 ZDM began to be published by the Springer Publishing House. At the same time, the journal was renamed *ZDM—The International Journal on Mathematics Education*. Original articles, still clustered in themes, once again became available in printed and electronic format.

2.1 ZDM—1990 to 1999

Inspection of the themes examined in the journal in the decade spanning 1990 to 1999 proved a rewarding exercise. A rich diversity of topics was investigated. These included—listed alphabetically here to illustrate, but not portray exhaustively, the wide range covered—"cross curricular activities", "environmental mathematics education", "ethnomathematics", "intelligent tutoring systems", "interpretative research in Primary mathematics Education", "mathematical beliefs", "mathematics and art", "mathematics, peace and ethics", "multicultural issues in the teaching and learning of mathematics", "problem solving in mathematics", "semiotics", and "women and mathematics". The themes of particular relevance to the topic of 'catering for the mathematically gifted' were considered in two issues: *Promotion of mathematically gifted primary school children* and *Fostering mathematical creativity*. The content of each of these is described in more detail below.

2.1.1 Promotion of Mathematically Gifted Primary School Children [1996: Volume 28(5)]

Although the abstracts for the first three articles listed in this issue were written in both German and English, the full articles appeared only in German. In line with the theme of the issue, each had a focus on primary school students. In the first, Sprengel (1996) discussed mathematical competitions at primary school. These, he reported, have been held in the former German Democratic Republic, for students from grade 3 onwards, for at least 30 years. Throughout that time competition-related experiences were largely positive, provided the problems used were age and achievement appropriate. To illustrate material deemed suitable, six exemplary problems and one strategic game were included in the article. The aims, organization, and selection procedures of a project which provided out-of-school support for "mathematically interested and gifted students" in grades 3 and 4 were the subject of Käpnick's (1996) contribution. Examples of the topics chosen in the project and their possible extension for in-school work were again provided. Reference was also made to any "strange and unusual behavior and school performance" of the participating students. Whether mathematically gifted students could and should be identified and given special programs was examined in the third article by two authors (Nolte and

Kießwetter 1996) with quite different prior experiences. The former was described as having extensive experience in working with mathematically weak students; the latter with "vast experience in special training for mathematically gifted children". After collating information gathered from observing students of different abilities, working under different conditions, and on carefully selected problems, the authors gave a qualified "yes" response to the research questions posed, that is, they concluded that there were benefits to be gained from identifying gifted students and providing special program for them. König's (1996) annotated bibliography covering some 60 publications, printed between 1980 and 1995, on work primarily concerned "with giftedness and giftedness research especially in the area of mathematics education at primary level" (p. 158) completed this issue. Most of the abstracts included in this bibliography were written in both English and German. Collectively the work covered in this issue served as a rich source of information, gathered in diverse geographic, research, and practical settings, on work spanning the identification and development of mathematically gifted students.

From the contents of the first three papers, with their focus on mathematics competitions, in- and out-of-school enrichment, and identification of development of gifted primary school students, it could be inferred, in the words of the issue's editors, "that success is strongly determined by the professional skill and competence of the teachers (for example by their ability to empathize), and by their interpersonal response (the way mistakes are handled, for example)" (Kießwetter and Nolte 1996, p. 130). From later contributions in ZDM, discussed in subsequent sections of this chapter, as well as from the literally hundreds of articles in these areas identified in the general educational research literature through search engines such as Google Scholar, and published just in 2009 alone[4], it can be further inferred that these topics have continued to be of interest to researchers.

The lack of articles in ZDM with a focus on older students is a notable omission, not replicated in the wider research literature where programs such as Julian Stanley's *Study for Mathematically Precocious Youth*, started in the 1970s, had spawned reports of a variety of other projects and investigations (Sowell 1993). Reflecting on the contents of their issue and the field more broadly, the editors concluded that "the field of work presents a still rather poorly defined picture... There is a lot of work ahead!" (Kießwetter and Nolte 1996, p. 130).

2.1.2 Fostering Mathematical Creativity [1997, Volume 29(3)]

The close link, explicit or strongly implied, between creativity and problem solving proficiency in the majority of the contributions published in this particular issue served as a justification for the attention given to it in this chapter. The perceived link is captured effectively by Haylock (1997), who noted, approvingly, that "Krutetskii

---

[4]In work listed for 2009 under the grouping of *Social Sciences, Arts and Humanities*, about 3450 articles were identified as dealing with aspects of both mathematics education and gifted students. (Data yielded from search on June 3, 2010.)

(1976) seems to equate mathematical creativity in school children with mathematical giftedness, using the two terms synonymously" (p. 68). Continuing support for this position has been given by authors publishing in later issues of ZDM. For example, more than a decade later Plucker and Zabelina (2009) discussed at some length whether "creativity and problem solving are domain-general—applicable to all disciplines and tasks—or domain specific—tailored to specific disciplines and tasks" (p. 5) and concluded that there is credible theoretical and empirical support for both propositions. That there is a strong link between mathematical creativity and mathematical problem solving was not disputed by these authors. Sriraman (2009) also considered it to be "in the best interest of the field of mathematics education that we identify and nurture creative talent in the mathematics classroom" (p. 25), and added that "between the work of a student who tries to solve a difficult problem in mathematics and a work of invention (creation)... there is only a difference of degree"[5] (p. 26). Outside ZDM the strong link between creativity and giftedness has also been emphasized: "A combination of intelligence and creativity defines giftedness" (Cropley and Urba 2000, p. 484), while Phillipson and Callingham (2009) unequivocally declared: "clearly, there is a link between ability and mathematical creativity" (p. 685). Leikin, Berman, and Koichu's (2009) edited collection, with the dual theme of mathematical creativity and the education of mathematically gifted students, is illustrative of the diverse theoretical perspectives of continuing explorations within and between these fields.

In the opening contribution of the themed issue on creativity, Pehkonen (1997) reflected on "the state-of-art in mathematical creativity" (p. 63). From the outset he maintained that there is a strong link between creativity and mathematics performance: "flexible thinking which is one component of creativity is one of the most important abilities—perhaps the most important—which a successful problem solver ought to have" (pp. 63–64). Flexibility and strategy adaptation in mathematics education, it is worth noting, have been addressed at some length in a more recent issue of ZDM.

"In successful problem solving", Pehkonen (1997) wrote,

> both hemispheres will be needed: First, the right hemisphere has a leading role as this is where holistic data processing takes place. The left hemisphere is better in logical tasks, therefore it dominates the work in the second stage of problem solving. When the solution has been reached, the solver will again consider the situation in a holistic manner (the right hemisphere) in order to check the reasonableness of the constructed solution... [But] a computer always works within its programming, its thinking is similar to "left-hemisphere thinking", not "right-hemispheric thinking". (p. 65)

The emphasis on neurological research is of particular interest, given the present burgeoning concern and activity in this area. After an overview of relevant research—much, though certainly not all, published after Pehkonen's article—Geake (2009) concluded that "the network can be conceptualized as providing a dynamic workspace in which information is processed; the greater efficiency and extent of the network in the brains of gifted people support their superior capacity

---

[5]This quotation was drawn from Polya (1954).

for information processing" (p. 271). In a paper in which contemporary trends in gifted education were discussed, Shaughnessy and Persson (2009) in fact speculated that "the practical application of our increasing knowledge of brain functions may well be the most important concern over the next two decades" (p. 1286).

Pehkonen (1997) further argued that students benefitted most if they were allowed to develop their own solutions to problems rather than to have structured and mechanistic procedures imposed on them. The increased reliance on computer usage within the mathematics classroom may thus well have constraining and negative consequences on building optimum problem solving strategies. Further research, it seems, might well be directed at this issue given Phillipson and Callingham's (2009) recent observation that "(m)any programs for mathematically gifted students advocate the use of technology as a motivator but there appears to be little research on their effectiveness with groups of gifted students" (p. 690). In a similar vein, Subhi-Yamin (2009) argued that "there are still many debates on how, and to what degree, the virtual environment can enhance schooling outcomes and how this environment can provide special provisions for gifted and talented children" (p. 1481).

Mathematical tasks designed to recognize and develop creative thinking in 11–12-year old students were the focus of the contribution by Haylock (1997). Theoretical justifications were provided for the assumption that critical features linked to creative problem solving behaviors include flexibility, independence, divergent thinking, looking beyond well known algorithms, and avoiding 'fixation'. The author ultimately concluded that "mathematical attainment *limits* the pupil's performance on both overcoming fixation and divergent production tasks, but does not *determine* it" (p. 73, emphases in the original). High mathematical achievement, Haylock further argued, is a necessary but not sufficient condition for avoiding 'fixation', to be willing to take risks, and to be able to think divergently. The concluding words to this paper have a remarkably current tone:

> A challenge for further research in this area is to identify teaching approaches that are effective in moving... pupils who have good mathematical knowledge and skills away from their overreliance on routines and stereotypes and their rigidity in thinking about mathematical situations towards the kinds of thinking that have been identified... as representing creativity in the school context. (Haylock 1997, p. 73)

What comprises optimally effective practices and approaches for the mathematical development of all students, including the gifted, remains tantalizingly elusive to judge from the National Council of Teachers of Mathematics [NCTM]'s current research agenda, outlined in Arbaugh et al. (2010). Among the issues highlighted as in urgent need of further investigation is the perennial question: "In what ways do different curricular approaches and/or combinations of those approaches support or impede students' development for mathematical proficiency?" (Shaughnessy 2010, p. 213).

Although Silver (1997) clearly recognized that "problem posing, or problem finding, has long been viewed as a characteristic of creative activity or exceptional talent" (p. 76) and that "creativity is often viewed as being associated with the notions of 'genius' or exceptional ability" (p. 75), he argued in this article that mathematical creativity can be fostered in a much wider range of school students. Using "inquiry-oriented mathematics instruction that includes opportunities for problem posing and problem solving", he urged, "teachers can assist students to develop greater representational and strategic fluency and flexibility and more creative approaches to their mathematical activity" (p. 79). References to the implementation of different programs "in a variety of classroom settings around the world" (p. 79) were used as support for this assertion. Silver's views resonate with those expressed years later by the contributors to Barbeau and Taylor's (2009) edited collection, with its strong theme that challenging activities in and beyond the classroom, typically judged suitable for high achievers in mathematics, can be as appropriate for students of all standards.

In her study Leung (1997) examined the link between creativity and problem posing tasks. The sample comprised 96 Taiwanese grade 5 students. The tasks, 18 items administered over two days, were presented in either text or picture format, or as a known answer to an unknown problem. Number, quantity, space, and statistics contexts were used. Both across- and within-task creativity were examined. Based on her findings Leung (1997) concluded that "a general, rather than specific, problem posing competence exists in children [which] can be measured by the test [she devised]" (p. 81). A cautionary note against too dogmatic an acceptance of this conclusion is implied by Worrell's (2009) contention "that to predict creativity in the verbal and math domains, one needs to use both verbal and math scores as well as the relative strengths in math and verbal scores—that is, three predictors rather than one" (p. 242).

An experiment to discover mathematical talent in a primary school in Kampong Air was the topic of Ching's (1997) contribution. To test whether students may often indeed be unable to show their full mathematical potential in regular classroom activities, five non-routine problems were given, one at a time and in increasing order of difficulty, to students in a standard grade 5 class. Ching focused particularly on the responses of the one student in the class of 25 students who "consistently exhibited mathematical creativity and talent in answering the questions" (p. 94). Such students, the author recommended, should be nurtured carefully to ensure that their mathematical talents are fully developed. That this is not an easy task is seemingly implied by Hertberg-Davis' (2009) conclusion, written more than a decade later, about the pressures faced by elementary school teachers responsible for teaching different curriculum areas:

> For all these reasons—lack of sustained teacher training in the specific philosophy and methods of differentiation, underlying beliefs... that gifted students do fine without any adaptations to curriculum, lack of general education teacher training in the needs and nature of gifted students, and the difficulty of differentiating instruction without a great depth of content knowledge—it does not seem that we are yet at a place where differentiation within the regular classroom is a particularly effective method of challenging our most able learners. (p. 252)

2.2 ZDM—2000 to 2009

As in the previous decade, a great variety of topical subjects was explored in the pages of ZDM over this period. However, there was no specific themed issue that focused explicitly on mathematically gifted students. A different approach—reliance on explorations with relevant search terms—was therefore required to determine the thrust and scope of this topic in the journal. The search terms used were: *accelerate[d]*, *creativity*, *exceptional performance*, *exceptional talent*, *exceptionally able*, *gifted students*, *high ability*, *high achievement*, *high performance*, *streaming*, and *talent*[6]. The information yielded through this approach, much of it resonating with topics already surveyed, is described next.

2.2.1 Mathematically Gifted Students—Problem Solving

Articles concerned with aspects of problem solving were the most fertile, though not the only source of references to the development and needs of gifted students. In several of these the strategies and capabilities of mathematically gifted students were given some prominence.

After a review of relevant literature Christou et al. (2005) described the four stages inherent in their hypothesized model of problem solving: editing, selecting, comprehending, and translating. They also presented data from a study with 143 6th grade students in Cyprus in which they tested the model's validity and applicability. They concluded that three categories of students could be identified. For some, category 1 students, the comprehension tasks were all they could handle. Others, category 2 students, were also able to respond correctly to the translation tasks. Students who were able to manage all four stages—editing, selecting, comprehending, and translating—were classified as Category 3 students. Of particular relevance for this chapter was their assertion that "*Both the editing and selecting processes characterized the most able students.* This finding is related to and confirms numerous qualitative studies on the mathematical thinking of gifted students" (Christou et al. 2005, p. 156, emphasis in the original).

Using data from the Australian Mathematics Competition, a large national data base, Leder (2007) examined the suitability of mathematics curriculum content for students with different levels of mathematical proficiency. By comparing group responses on items attempted by students at four different grade levels, characteristics of mathematical problems able to be answered correctly by students at the different grade levels were identified. "Chronological age per se... (was found to be) an inadequate predictor of a student's likely success on cognitively demanding problems" (Leder 2007, p. 93). Confirmation that items requiring highly abstract reasoning were often able to be solved by the best students in grade 7, even though they were found to be too difficult by average students in grade levels up to four years higher

---

[6]The words "gifted" or "giftedness" do not occur in the ZDM Subject Classification Scheme. The classification code which is most closely related is C, *Psychology of mathematics education. Research in mathematics education. Social aspects.* Within it, C40 is the most relevant category. It comprises the words: intelligence and aptitudes, personality, talent, intelligence, abilities and skills, creativity, behavior, personality traits, personality, development. There is no relevant elaboration for C90, special education. This coding system is no longer used in the most recent issues of ZDM.

could, Leder argued, serve as useful guidelines for those planning extension work for mathematically able students.

Prompted by a request from researchers outside the Netherlands, Doorman et al. (2007) designed a study to explore problem solving skills and strategies of high achieving grade 4 students. The sample comprised 152 students, selected by their teachers on the basis of their mathematics performance, from 22 different schools in the Netherlands. "Most of the (15) tasks were puzzle-like problems, such as number riddles" (Doorman et al. 2007, p. 410), and unlike the mathematical tasks found in the regular textbooks or tests. "In reviewing all the students' responses and the experiences from the interviews, three tendencies were found: many students did not write anything down, many students did even not start, and if they started quite a number showed lack of persistence" (Doorman et al. 2007, p. 410). Non-routine problem solving, the authors concluded, needed more attention at both the primary and secondary school levels. In the remainder of the paper they discussed problems posed in the Mathematics A-lympiad competition. These they described as a fruitful source and model for open-ended problems that require higher order thinking skills. They were deemed suitable for all secondary students and not only for the highest achievers.

In a study, with the same sample as that described by Doorman et al. (2007)[7], but designed "to give insight into the strategy use and strategy flexibility of high achievers in primary school mathematics in non-routine problem solving" (Elia et al. 2009, p. 605), Elia et al. (2009) relied on students' written material and explanations for their inferences about the students' problem solving strategies. Students were asked to complete three tasks, without outside help, and to show all their working. The authors noted several limitations of the approach adopted. Only half the students, they found, made use of the space allocated for showing their working. This, they speculated, might be because students found it difficult to write down their thinking or solution steps, or because they believed:

> that it is better not to use the paper, because solving the problems mentally indicates a higher level of mathematics... (which) is especially true for high achievers who are not accustomed to use scrap paper when they deal with tasks during regular mathematics class work. (p. 615)

The data obtained from the study were thus interpreted with caution and various suggestions for further research concluded the article.

2.2.2 In Brief

Confirmation was provided in several of these articles that mathematically able students are often shown to have a superior approach to problem solving. In her synthesis of research about educating the gifted and talented, Rogers (2007) also reported

---

[7]This assumption is made on the basis of the description of the two samples and not on any clear indication of this in the paper by Elia et al. (2009).

that "gifted learners are more likely to switch to an alternative strategy when faced with a mathematics challenge they cannot resolve, than to resort to trial and error" (p. 391). At the same time it nevertheless seemed that practice and appropriate curriculum materials are needed to enhance strategy development. A range of possible options to achieve this are included in Roger's (2007) synthesis as well as in Barbeau and Taylor (2009). Absent from the discussion within ZDM are the many references in the wider research literature to the benefits of acceleration for strategically selected mathematically gifted students found in the work, among others, of Colangelo and Assouline (2009), Gallagher (2009), and Rogers (2007). Rogers, for example, noted that "the academic effects of compacting are powerful, especially in mathematics and science when the replacement activities have been accelerated and advanced in complexity" (p. 386), and summarized evidence that "studies of subject acceleration, especially pertaining to science and mathematics, show dramatic achievement gains" (p. 387) for elementary and secondary students, as well as "remarkable affective (social and self esteem) gains" (p. 387).

2.3 *Mathematically Gifted Students—A Matter of Convenience in Problem Solving*

Close inspection of other articles published in ZDM over the past decade revealed that mention of gifted students was typically a matter of expediency, that is, of peripheral rather than of core interest, and was used as a platform for discussing suitable content and strategies for the broader school population. For convenience such examples are listed chronologically.

In an article expounding the principles of "the didactical technology developed and applied for mathematics learning in school during 1994–2000 in Romania" (Singer 2001, p. 205), Singer argued, but did not develop the assertion, that only "an elite of gifted children is able to achieve the high theoretical standards; for the other students learning mathematics is done by successive accumulation of algorithms without any perspective or clear goal" (p. 204). Schumann (2003) considered that the teaching of 'Algebraic Curves' should be re-evaluated and considered "for the purposes of general teaching projects, course work projects, working groups and extracurricular mathematics e.g. for gifted students in consideration of newly developed and adequate dynamic geometry computer tools" (p. 301). In the remainder of the article, characteristics of the program, but not its suitability for gifted students, were addressed by the author. In their paper about creating interesting problems for all students Anderson et al. (2004) argued that it is typically only the more gifted students who benefit from grappling with the notion of proof in geometry. Instead, they suggested "maybe the field of discrete mathematics can offer other, more challenging and accessible problems to get high school students interested in mathematics" (Anderson et al. 2004, p. 105). Although Freiman et al. (2005) noted that productive usage of technology could "not only foster a development of mathematical potential in gifted students in early grades, but it could also be beneficial for all students"

(p. 180), they made no further reference to mathematically gifted learners in their description of a project in which school students and pre-service teachers solved mathematics problems interactively via a specific WEB site. Sriraman and Pezzuli (2005), "a university mathematics educator and an idealistic pre-service elementary teacher" (p. 431) respectively, combined to "try to resolve the dilemma of balancing the (American NCTM) Standards with research and personal beliefs about the teaching and learning of mathematics" (p. 431). In the final reflections on the material prepared and attempted by students, the needs of gifted students were given some attention:

> The lessons presented (in this article) I believe are quite challenging to the students, even those who are gifted. They require a lot of thinking and problem solving. However, I could add some more problems on to the lessons for more talented math students. I could have them find the area of a more complex polygon, even a concave polygon. They could do this by breaking the concave figure up into recognizable convex polygons and find the area of each region, then add those up. Or I could alter their lab activity to allow for more free-thinking. ...I am a huge supporter of having the students learn as much as possible from themselves or each other, rather than from the teacher. (Sriraman and Pezzuli 2005, p. 435)

Whether constraints imposed by traditional assessment tasks might prevent students from showing the full extent of their mathematical capability was explored by Iversen and Larson (2006). Modeling tasks, they concluded, enabled students who are not normally recognized as high-achievers to show characteristics typically associated with the latter, namely that they were capable of "using creative, critical problem-solving, decision-making and innovative thinking processes, and... (were) able to evaluate, judge and predict consequences" (Iversen and Larson 2006, p. 290). Apart from this listing of their putative characteristics, no further mention was made of gifted students. In their overview of the role of problem solving in mathematics education, Reiss and Törner (2007) explained that in Germany, as in many countries, it was increasingly recognized that all students could benefit from problem solving activities: "the mathematics competitions which formerly addressed only mathematically gifted students have been supplemented by competitions that address (nearly) all students in a classroom" (p. 437). How gifted students *per se* had benefitted from participation in these competitions was, apparently, deemed outside the scope of the paper. "Mathematics education in Hungary for the gifted and talented looks enviable. This is because the teachers have always put extra energy into the work of interest groups for talented students", according to Szendrei (2007, p. 453). In her article she paid homage to a group of committed researchers who introduced new topics and instructional strategies into the mathematics school curriculum, strengthened the emphasis on problem solving, and built on work done with gifted and talented students to open up mathematical opportunities "for more average students" (Szendrei 2007, p. 453). Once again, however, the focus of the paper was on the latter group for whom, Szendrei concluded, particularly at the upper school level, the "dream where problem solving is not fear inducing but rather a sound intellectual sport enjoyed by everyone involved and a thrilling success story for much more than the favored few—has not come true" (p. 454).

2.3.1 In Brief

The seemingly rich and diverse field of material on the education of mathematically gifted students that emerged from the search of the more recent ZDM content, proved, on closer inspection to lead to many dead ends on this topic and at best yielded a meagre harvest of productive information about gifted students. In many of the articles identified, the authors highlighted the merit of adapting or drawing on programs and strategies found fruitful with gifted students for implementation with a wider group of students. There seems a marked degree of congruency between the thrust of this work, and that of those who argue, like Gentry (2009) that "to recognize whether talent exists there must be opportunities for talent to emerge" (p. 265) or Wallace and Maker (2009) whose framework for curriculum development included in a book primarily focused on gifted students, embodied their belief that "all learners are capable of improving their problem-solving processes" (p. 1113). Significant, too, are reports such as those by Swanson (2006) of a project in which all students in three elementary schools were exposed to curriculum and teaching strategies generally reserved for gifted and high-achieving students. "It was evident that many project teachers changed their practice, with changes ranging from stronger questioning to teaching students to reason critically through an issue. …(The project's) results provide additional evidence that achievement improves when students experience rich and challenging curriculum" (Swanson 2006, p. 23).

2.4 Mathematically Gifted Students—Miscellaneous Issues

As mentioned above, attention to gifted students was not confined solely to papers with a focus on problem solving. Other examples of work in which authors referred to gifted mathematics students are discussed in this section. Once again, these references can most aptly be described as tangential and somewhat gratuitous rather than indications of a central concern with the development of mathematically gifted students.

Whether students should be streamed into like ability groups was discussed in several papers. Braathe and Ongstad (2001), though not directly addressing the needs of mathematically gifted students, raised some interesting issues. Prompted by comparisons of the performance of Norwegian students with that of students in other countries also involved in TIMSS, these authors examined how the strong societal value of equality in Norway affected the content and teaching of mathematics, and concluded that the goal of mathematics education of "rational and reflexive *understanding*" (Braathe and Ongstad 2001, p. 5, emphasis in the original) is at variance with Norway's egalitarian ideology. "Whether educationalists in mathematics in Norway like it or not, most mathematics classrooms seem to convey these tensions: a Norwegian student in mathematics has to be slow *and* equal or bright *and* equal etc." (p. 5, emphasis in the original) they argued. Elsewhere in the article they concluded that "within the Norwegian context, egalitarianism has to give in, since mathematics as a subject individualizes students and makes collectivity secondary to personal understanding, success and progress" (p. 10).

Australian teachers' views on effective mathematics teaching and learning were examined by Perry (2007). Several brief excerpts which reflected the group's beliefs about gifted students were included in the report. "As a teacher of gifted and talented children" Perry wrote about one participant, "she went on to espouse a belief that 'bright children don't need to construct anything, they just do it abstractly in the first place'" (p. 275) and elsewhere: "One teacher clearly felt that her gifted and talented students had no need for concrete materials because they were already capable of structured logical thinking" (p. 278). The reasoning skills of mathematically gifted students were also referred to by Presmeg (2009) in a paper concerned with the relevance to mathematics education research of the quantitative and qualitative methods embodied in the sciences and the arts. Presmeg (2009) drew briefly—but did not elaborate—on research involving mathematically gifted students:

> A "mathematical cast of mind" may be a characteristic of students who are gifted in mathematics (Krutetskii 1976). This mathematical cast of mind enables these students to identify and reason about mathematical elements in all their experiences; they construct their worlds with mathematical eyes, as it were. But unless teachers are aware of the necessity of encouraging students to recognize mathematics in diverse areas of their experience, only a few students will develop this mathematical cast of mind on their own. (p. 133)

In a later issue published the same year, devoted to *flexible and adaptive use of strategies and representation in mathematics education*, there were several articles that touched, indirectly rather than directly, on high achievers in mathematics. For example, Star and Newton (2009) explored strategies used and favored by experts. They concluded "that our experts did exhibit strategy flexibility on the tasks [solving equations] that they were asked to complete: experts showed knowledge of multiple strategies and the ability to select appropriate strategies for given problems" (p. 563). Although they did not refer explicitly to gifted or talented mathematics students, many readers might see a link between optimum use of flexible strategies and outstanding mathematical achievement[8]. Heinze et al. (2009) reported the findings of a study conducted with much younger students, 245 third grade students in Germany. These authors, too, were concerned with flexibility and adaptive strategy use in problem solving, and devoted the bulk of their paper to reporting the impact of students being exposed to textbooks using either an investigative or problem solving approach. As part of setting the context for their own study they noted:

> However, a question that is still not answered satisfactorily is how teaching and learning processes can be organized in such a way that students will acquire an adequate competence in an adaptive use of computation strategies. This encompasses the open question whether there exists an approach which is beneficial for all students or whether different teaching

---

[8]For example, Wieczerkowski, Cropley, and Prado (2000,) noted that "ability in abstracting concrete problems, the ability to generalize, flexibility,… fluency of thought, or strategic decision-making have been regarded as core variables for mathematical giftedness" (p. 414).

approaches should be implemented for high-achieving and low achieving students. (Heinze et al. 2009, p. 594)

In their own paper they defined high achieving students as those who scored in the top 50%—a particularly blunt definition—and did not resolve the open question outlined in the above quotation.

In work concerned with possible gender differences in mathematics learning, the additional problems faced by mathematically gifted females have often been highlighted. Typical of such concerns within ZDM are those voiced by Piatek-Jimenez (2008):

> even women at the undergraduate and graduate level who have demonstrated both interest and talent in mathematics continue to endure messages from their professors that they are not being taken seriously as mathematics students or that they do not belong in the field of mathematics. (p. 634)

However, in the remainder of the article Piatek-Jimenez focused primarily on the career decisions made by female students in general, rather than on those made by the highest achievers in mathematics.

2.4.1 In Brief

As has been evident throughout this chapter, references to work with, or by, high achieving students can be found in reports of diverse research in mathematics education published in ZDM. At times such references seem somewhat forced; at others they are highly pertinent to the issues being explored.

The topics potentially affecting the education and development of mathematically gifted students explored at some level in the ZDM papers listed in the section above have also attracted attention from the wider research community, as well as from practitioners, and relevant reports have been published outside ZDM. Space constraints allow only a few examples to be listed here.

Like Braathe and Ongstad (2001), Cooper (2009) explored whether and how the notions of fairness and equality can be reconciled if the needs of all students are to be met. Rogers (2007) and VanTassel-Baska (2000) offered a useful synthesis of the large body of research on, respectively, teachers' beliefs and practices and curriculum development, and their impact on the progress of gifted and talented students in general, while issues central to the development of such students in mathematics were discussed by Wieczerkowski, Cropley and Prado's (2000). Gender differences in mathematics performance, another issue touched on in work published in ZDM, is discussed at some length in each of the three already mentioned *Handbooks* (Colangelo and Davis 2003; Heller et al. 2000, Shavinina 2009). In brief, the overlap in the performance of males as a group, and females as a group, is invariably described as substantial. Relatively larger differences, in favor of males, are frequently reported when students at the highest levels of achievement and ability are compared.

3 Concluding Comments

A careful reading of issues of ZDM published over the past two decades provided tantalizing allusions to the needs and development of mathematically gifted students, but there were few in-depth discussions of these issues. For a more comprehensive treatment resources outside ZDM needed to be consulted. At the same time extensive reports of studies about exceptional mathematics students proved to be relatively scarce in the eminent texts on giftedness on which heavy reliance was placed in this chapter; few chapters in the strategically selected *Handbooks* were primarily devoted to mathematics students. Supplementary evidence needed to be sourced from recent books specifically concerned with mathematically gifted students, for example, Barbeau and Taylor (2009) and Leikin et al. (2009).

A surprisingly small number of articles published in the *GCQ* had mathematically gifted students as their core focus. Relevant work can, from time to time, be found in other journals, including mathematics education research journals, educational psychology journals, and other gifted education journals including the *Roeper Review*, mentioned earlier in this chapter. A comprehensive overview of such articles was, however, beyond the scope of this chapter.

Combining material from the various sources reviewed in this chapter yielded a long list of perceptions about exceptional mathematics students but fewer evidence-based insights. The need for further, rigorously planned and methodically executed research was often among recommendations made. That areas in need of further investigation go well beyond those to which attention was drawn in the chapter can be gauged from Monks et al. (2000) challenging observation that,

> as in other fields, the more that is known, the more issues are raised and controversies fueled. Issues concerning the gifted field have never been confined to a small group of children and youth identified as 'gifted or talented' but have an impact on the whole of education. For example, the issues surrounding questions of **excellence and equity** affect all educational decisions, not just provisions made for the gifted. (p. 839, emphasis in the original)

Within the education research and broader communities there is widespread concern about the declining popularity of mathematics, the drift away from advanced mathematics subjects, and the limited pool of students willing to enter careers requiring a high level of mathematical expertise. Supporting the development of talented mathematics students and exploring optimum strategies for nurturing talent undoubtedly warrant ongoing attention.

References

Anderson, I., van Asch, B., & van Lint, J. (2004). Discrete mathematics in the high school curriculum. *ZDM, 36*(3), 105–116.

Arbaugh, F., Herbel-Eisenmann, B., Ramirez, N., Knuth, E., Kranendonk, H., & Quander, J.R. (2010). *Linking research & practice. The NCTM research agenda conference report.* Reston: National Council of Teachers of Mathematics.

Barbeau, E. J., & Taylor, P. J. (Eds.) (2009). *Challenging mathematics in and beyond the classroom.* New York: Springer.

Borland, J. H. (2009). Myth 2: The gifted constitute 3% to 5% of the population. Moreover, gift-edness equals high IQ, which is a stable measure of aptitude. *Gifted Child Quarterly, 53*(4), 236–238.

Braathe, H. J., & Ongstad, S. (2001). Egalitarianism meets ideologies of mathematical education-instances from Norwegian curricula and classrooms. *ZDM, 33*(5), 1–11.

Ching, T. P. (1997). An experiment to discover mathematical talent in a primary school in Kampong Air. *ZDM, 29*(3), 94–96.

Christou, C., Mousoulides, N., Pittalis, M., Pitta-Pantazi, D., & Sriraman, B. (2005). An empirical taxonomy of problem posing processes. *ZDM, 37*(3), 149–158.

Colangelo, N., & Assouline, S. (2009). Acceleration: Meeting the academic and social needs of students. In L. V. Shavinina (Ed.), *International handbook on giftedness* (pp. 1085–1098). Dor-drecht, The Netherlands: Springer.

Colangelo, N., & Davis, G. (Eds.) (2003). *The handbook of gifted education*. Boston: Allyn & Bacon.

Cooper, C. R. (2009). Myth 18: It is fair to teach all children the same way. *Gifted Child Quarterly, 53*(4), 283–285.

Cropley, A. J., & Urban, K. K. (2000). Programs and strategies for nurturing creativity. In K. A. Heller, F. J. Mönks, R. J. Sternberg, & R. F. Subotnik (Eds.) *International handbook of giftedness and talent* (pp. 485–498). Oxford: Elsevier.

Davis, G. A, & Rimm, S. B. (2004). *Education of the gifted and talented*. Boston: Pearson Education Press.

Doorman, M., Drijvers, P., Dekker, T., van den Heuvel-Panhuizen, M., de Lange, J., & Wijers, M. (2007). Problem solving as a challenge for mathematics education in the Netherlands. *ZDM, 39*, 405–418.

Elia, I., van den Heuvel-Panhuizen, M., & Kolovou, A. (2009). Exploring strategy use and strategy flexibility in non-routine problem solving by primary school high achievers in mathematics. *ZDM, 41*(5), 605–618.

Friedman-Nimz, R. (2009). Myth 6: Cosmetic use of multiple selection criteria. *Gifted Child Quarterly, 53*(4), 248–250.

Freiman, V., Vézina, N., & Gandaho, I. (2005). New Brunswick pre-service teachers communicate with school children about mathematics problems: CAMI project. *ZDM, 37*(3), 178–189.

Gallagher, S. A. (2009). Myth 19: Is advance placement an adequate program for gifted students? *Gifted Child Quarterly, 53*(4), 286–288.

Geake, J. G. (2009). Neuropsychological characteristics of academic and creative giftedness. In L. V. Shavinina (Ed.), *International handbook on giftedness* (pp. 1085–1098). Dordrecht, The Netherlands: Springer.

Gentry, M. (2009). Myth 11: A comprehensive continuum of gifted education and talent develop-ment services. Discovering, developing, and enhancing young people's gifts and talents. *Gifted Child Quarterly, 53*(4), 262–265.

Gifted Child Quarterly (n.d.). *About the journal*. Retrieved May 26, 2010 from http://gcq.sagepub.com/.

Haylock, D. (1997). Recognising mathematical creativity in schoolchildren. *ZDM, 29*(3), 68–74.

Heinze, A., Marschick, F., & Lipowsky, F. (2009). Addition and subtraction of three-digit numbers: Adaptive strategy use and the influence of instruction in German third grade. *ZDM, 41*(5), 591–604.

Heller, K. A., Monks, F., Sternberg, R. J., & Subotnik, R. F. (Eds.) (2000). *The international handbook of giftedness and talent*. Oxford, UK: Elsevier Science.

Hertberg-Davis, H. (2009). Myth 7: Differentiation in the regular classroom is equivalent to gifted programs and is sufficient: Classroom teachers have the time, the skill, and the will to differen-tiate adequately. *Gifted Child Quarterly, 53*(4), 251–253.

Iversen, S. M., & Larson, C. J. (2006). Simple thinking using complex maths v complex thinking using simple math—A study using model eliciting activities to compare students' abilities in standardized tests to their modelling abilities. *ZDM, 38*(3), 281–292.

Kaiser, G. (2006). On the occasion of Gerhard König's retirement as editor-in-chief of ZDM and MATHDI. *ZDM, 38*(1), 79–81.

Kaiser, G. (2007). Editorial. *ZDM, 39*, 1–2.

Kaiser, G., & Sriraman, B. (2010). Advances in mathematics education: New book series connected to ZDM—The International Journal of Mathematics Education. *ZDM, 42*, 143–144.

Käpnick, F. (1996). Mathematically interested and talented primary school children—The Neubrandenburg project. *ZDM, 28*(5), 136–142.

Kießwetter, K., & Nolte, M. (1996). Introduction to the following series of articles discussing furthering of the gifted in primary education. *ZDM, 28*(5), 129–130.

König, G. (1996). Bibliography "Giftedness and promotion of gifted primary grade students". *ZDM, 28*(5), 158–163.

Krutetskii, V. A. (1976). *The psychology of mathematical abilities in school children.* Chicago: The University of Chicago Press.

Leder, G. C. (2007). Using large scale data creatively: Implications for instruction. *ZDM, 39*(1–2), 87–94.

Leikin, R., Berman, A., & Koichu, B. (Eds.) (2009). *Creativity in mathematics and the education of gifted students.* Rotterdam, The Netherlands: Sense Publishers.

Leung, S. S. (1997). On the role of creative thinking in problem solving. *ZDM, 29*(3), 81–85.

Monks, F. J., Heller, K. A., & Passow, A. H. (2000). The study of giftedness: Reflections on where we are and where we are going. In K. A. Heller, F. J. Mönks, R. J. Sternberg, & R. F. Subotnik (Eds.), *International handbook of giftedness and talent* (pp. 839–863). Oxford: Elsevier.

Nolte, M., & Kießwetter, K. (1996). Can and should mathematically gifted children be identified and promoted already at primary school? *ZDM, 28*(5), 143–157.

Pehkonen, E. (1997). The state-of-art in mathematical creativity. *ZDM, 29*(3), 63–67.

Perry, B. (2007). Australian teachers' views of effective mathematics teaching and learning. *ZDM, 39*(4), 271–286.

Phillipson, S. N., & Callingham, R. (2009). Understanding mathematical giftedness: Integrating self, action repertoires and the environment. In L. V. Shavinina (Ed.), *International handbook on giftedness* (pp. 1085–1098). Dordrecht, The Netherlands: Springer.

Piatek-Jimenez, K. (2008). Images of mathematicians: A new perspective on the shortage of women in mathematical careers. *ZDM, 40*(4), 636–646.

Plucker, J., & Zabelina, D. (2009). Creativity and interdisciplinarity: One creativity or many creativities? *ZDM, 41*(1–2), 5–11.

Polya, G. (1954). *Induction and analogy in mathematics.* Princeton: Princeton University Press.

Presmeg, N. (2009). Mathematics education research embracing arts and sciences. *ZDM, 41*(1–2), 131–141.

Reiss, K. & Törner, G. (2007). Problem solving in the mathematics classroom: The German perspective. *ZDM, 39*(5–6), 431–441.

Rogers, K. B. (2007). Lessons learned about educating the gifted and talented: A synthesis of the research on educational practice. *Gifted Child Quarterly, 54*(4), 382–394.

Schumann, H. (2003). A dynamic approach to 'simple' algebraic curves. *ZDM, 35*(6), 301–316.

Shaughnessy, J. M. (2010). Linking research and practice: The research agenda project. *Journal for Research in Mathematics Education, 41*(3), 212–215.

Shaughnessy, J. M., & Persson, R. (2009). Observed trends and needed trends in gifted education. In L. V. Shavinina (Ed.), *International handbook on giftedness* (pp. 1285–1291). Dordrecht, The Netherlands: Springer.

Shavinina, L. V. (Ed.) (2009). *International handbook on giftedness.* Dordrecht, The Netherlands: Springer.

Silver, E. A. (1997). Fostering creativity through instruction rich in mathematical problem solving and problem posing. *ZDM, 29*(3), 75–80.

Singer, M. (2001). Information structuring—A new way of perceiving the content of learning. *ZDM, 33*(6), 204–217.

Sowell, E. (1993). Programs for mathematically gifted students: A review of empirical research. *Gifted Child Quarterly, 37*, 124–129.

Sprengel, H. (1996). Promotion of mathematically gifted primary school children. *ZDM, 28*(5), 131–135.

Sriraman, B. (2009). The characteristics of mathematical creativity. *ZDM, 41*(1–2), 13–27.

Sriraman, B., & Pezzuli, M. (2005). Balancing mathematics education research and the NCTM standard. *ZDM, 37*(5), 431–436.

Star, J. R., & Newton, K. J. (2009). The nature and development of experts' strategy flexibility for solving equations. *ZDM, 41*(5), 557–567.

Sternberg, R. J. (Ed.). (2000). *Handbook of intelligence*. New York: Cambridge University Press.

Stoeger, H. (2009). The history of giftedness research. In L. V. Shavinina (Ed.), *International handbook on giftedness* (pp. 17–38). Dordrecht, The Netherlands: Springer.

Subhi-Yamin, T. (2009). Gifted education in the Arabian gulf and the middle Eastern regions: History, current practices, new directions, and future trends. In L. V. Shavinina (Ed.), *International handbook on giftedness* (pp. 1463–1490). Dordrecht, The Netherlands: Springer.

Swanson, J. D. (2006). Breaking through assumptions about low-income, minority gifted students. *Gifted Child Quarterly, 50*(1), 11–25.

Szendrei, J. (2007). When the going gets tough, the tough gets going problem solving in Hungary, 1970–2007: Research and theory, practice and politics. *ZDM, 39*(5–6), 443–458.

VanTassel-Baska, J. (2000). Theory and research on curriculum development for the gifted. In K. A. Heller, F. J. Mönks, R. J. Sternberg, & R. F. Subotnik (Eds.), *International handbook of giftedness and talent* (pp. 345–365). Oxford: Elsevier.

Wallace, B., & Maker, C. J. (2009). DISCOVERY/TASC: An approach to teaching and learning that is inclusive and that maximises opportunities for differentiation according to pupils' needs. In L. V. Shavinina (Ed.), *International handbook on giftedness* (pp. 1113–1141). Dordrecht, The Netherlands: Springer.

Worrell, F. C. (2009). Myth 4: A single test score or indicator tells us all we need to know about giftedness. *Gifted Child Quarterly, 53*(4), 242–244.

Weinert, F. E. (2000). Foreword. In K. A. Heller, F. J. Mönks, R. J. Sternberg, & R. F. Subotnik (Eds.), *International handbook of giftedness and talent* (pp. xi–xiii). Oxford: Elsevier.

Wieczerkowski, W., Cropley, A. J., & Prado, T. M. (2000). Nurturing talents/gifts in mathematics. In K. A. Heller, F. J. Mönks, R. J. Sternberg, & R. F. Subotnik (Eds.), *International handbook of giftedness and talent* (pp. 413–425). Oxford: Elsevier.

ZDM (n.d.). *Detailed aims and scope of the journal*. Retrieved May 26, 2010 from www.springer.com/11858.

Ziegler, A., & Heller, K. A. (2000). Conceptions of giftedness from a meta-theoretical perspective. In K. A. Heller, F. J. Mönks, R. J. Sternberg, & R. F. Subotnik (Eds.), *International handbook of giftedness and talent* (pp. 3–21). Oxford: Elsevier.

Commentary on the Chapter by Gilah Leder, "Looking for Gold: Catering for Mathematically Gifted Students Within and Beyond ZDM"

Some Gold Is Found—Much More Is in the Mine

Boris Koichu

Gilah Leder thoroughly reviews, within the constraints imposed by the structure of this volume, what seems to be a very rich but not fully institutionalized area of research enterprise. An important (and, probably, inevitable) decision about the reviewing strategy was "[not to attempt] to provide a unique definition of mathematically gifted students or its pseudonyms, [but to accept] the diversity of definitions used in different publications... at face value" (Leder, this volume, p. 390). As a result, the review includes publications that, generally speaking, are concerned with *supporting* or *characterizing* those students who seem to be insufficiently challenged by a regular mathematical curriculum or show at least above-average achievements when studying mathematics in different educational settings. In fact, the chapter invites the reader to critically consider whether such a broad population can be referred to as "mathematically gifted students" and, more importantly, whether results of numerous studies utilizing the term "mathematically gifted students" or its pseudonyms can be perceived as a body of accumulated knowledge or as "tantalizing allusions to the needs and development of mathematically gifted students" (Leder, this volume, p. 403).

Leder utilizes the very precise metaphor of "looking for gold" in the title of her chapter. This metaphor is stretched in the remainder of my commentary. In the "What counts for gold?" section I comment on one well-substantiated conclusion that can serve as a starting point for further research and handling the equity issue. In the "More gold is in the mine" section I reflect on selected studies in light of quality criteria for judging research about intellectual abilities.

B. Koichu (✉)
Technion—Israel Institute of Technology, Haifa, Israel
e-mail: bkoichu@technion.ac.il

H. Forgasz, F. Rivera (eds.), *Towards Equity in Mathematics Education,*
Advances in Mathematics Education,
DOI 10.1007/978-3-642-27702-3_36, © Springer-Verlag Berlin Heidelberg 2012

1 What Counts for Gold?

One of the most robust conclusions that can be drawn from Leder's review is about ways of supporting mathematically gifted students. Specifically, there is overwhelming evidence that *all* students benefit from studying mathematics through challenging problem-solving tasks, but there are *some* students within every age cohort who require more mathematically advanced tasks than others in order to be adequately challenged. This conclusion can be taken as a starting point for transforming the main question associated with the equity issue and giftedness into a more operational one. Namely, the question "Should mathematically gifted students be taught differently and if yes, how?" can be re-stated as follows: How can the right of all students, including mathematically gifted, to be engaged in challenging mathematical activities be realized within a particular educational system? The latter question can be split into a number of researchable questions, such as:

i.   What are the consequences of explicitly offering a group of students of a particular age mathematical tasks that are challenging and feasible for some and not feasible for the rest?
ii. Is it possible, and if yes how, to challenge the most mathematically able students of a particular group by tasks that would also be challenging and feasible for the rest?
iii. What are the differences, if any, in pedagogical approaches needed to support the use of more mathematically advanced tasks (with more mathematically advanced students) and the use of less mathematically advanced tasks (with less mathematically advanced students)?

Each of these questions presumes a complex, conditional answer. It is encouraging, however, that the questions can be explored within readily available research arenas, that is, with relatively low risk for students. For instance, the first question can be approached by studying mathematics competitions. Exploring the second question may involve tasks which can be offered to all, but probably taken further by some students than by others. Characteristic examples of such tasks include multiple-solution tasks, problem-posing tasks, modeling tasks, and multi-step explorations—see Sheffield (1999) and Leikin et al. (2009) for elaborated examples.

The third question calls for studies focusing on a particular pedagogical approach when it is utilized with students of different mathematical aptitudes. Consider two examples: a. A study by Diezmann and Watters (2001) examined middle-school mathematically gifted students' preferences for collaboration in relation to the difficulty of the tasks being undertaken. An interesting finding was that the students preferred working with peers only when the task was sufficiently challenging. The study is relevant to the third question because it calls for reconsideration of the popular belief that collaborative problem solving is more appropriate for regular students than for the gifted. b. A study by Koichu and Orey (2010) concerned the opposite side of the aptitude spectrum, mathematically disadvantaged high-school students. The main finding of the study was that the students were capable of demonstrating creative behaviors when given feasible tasks, adequately challenging for them, in an emotionally safe and encouraging environment. This study, as well as several studies mentioned in Leder's review, suggests that pedagogical approaches for creativity promotion in students of different mathematical aptitudes may have certain common features.

Systematic exploration of the above questions may inform policy-making concerned with the following practical question:

Under which conditions it is beneficial to teach mathematically gifted students separately from the rest of the students of the same age, for instance, in a special class, program, or with older students?

Grounded suggestions addressing this question are offered in several studies included in Leder's review. More comprehensive answers may be obtained based on questions (i)–(iii) above and consideration of additional factors. One of the most influential factors is a perspective taken on the equity issue. Indeed, the above comments implicitly rely on an opportunity-oriented perspective on the equity issue, which presumes "high expectations and worthwhile opportunities for all" (NCTM 2000, p. 12). However, alternative perspectives exist and are being debated in the gifted education research community. For instance, a recent issue of *Talent Development & Excellence* is devoted to the discussion of the equity issue conceptualized by Gagné (2011) as "expressed judgments by many professionals and scholars that members of disadvantaged groups suffer from unfair selection practices, which leads to their significant underrepresentation in gifted programs" (p. 3). Gagné argued that a "focus on [demonstrated] performance as the main entry requirement to a talent development program offers the best guarantee of equity and objectivity" (p. 18) and that the observed disproportions should not be seen as concerning. His paper triggered a hot debate among many gifted education researchers all over the globe. No attempt to summarize this discussion can be made here. The point is that a performance-oriented perspective on the equity issue, if taken, would imply a very different research agenda than that offered in Leder's chapter and by the comments above.

2 More Gold Is in the Mine

A sizable portion of the studies reviewed by Leder deals with identification of core variables for mathematical giftedness. Such studies seek to relate the highest levels of mathematical performance to underlying intellectual abilities or cognitive processes. This important research venue has recently been scrutinized in the field of general giftedness. For instance, Wenke and French (2003) pointed out various contradicting or elusive results related to the relationships between intellectual ability and competencies in solving complex problems. The scholars concluded that empirical support for the popular belief that intellectual ability is related to complex problem solving is rather poor; some positive relationship was found only for narrowly defined types of problems. Their analysis was driven by the following criteria:

Criterion 1: Both the intellectual ability presumably underlying problem-solving competence and problem-solving competence itself need to be explicitly defined and must not overlap at theoretical and/or operational levels... If the latter is not the case, then any attempt to explain problem-solving competence in terms of an underlying intellectual ability is necessarily circular and redundant...

Criterion 2: The presumed relations between intellectual ability and problem-solving competence must have a theoretical explanation (Wenke & French, 2003, p. 92).

I believe that acceptance of these criteria in research aimed at better understanding core variables for mathematical giftedness would be beneficial for securing further accumulation of knowledge in the field—see Koichu (2011) for an elaborated argument. It seems, however, that having a broadly accepted theoretical definition of mathematical giftedness (which is still lacking) is now more important for designing studies focusing on *characterizing* our capable students than on *supporting* them.

References

Diezmann, C., & Watters, J. (2001). The collaboration of mathematically gifted students on challenging tasks. *Journal for the Education of the Gifted*, *25*(1), 7–31.

Gagné, F. (2011). Academic talent development and the equity issue in gifted education. *Talent Development & Excellence*, *3*(1), 3–22.

Koichu, B. (2011). Overcoming a pitfall of circularity in research on problem solving by mathematically gifted schoolchildren. *Canadian Journal of Science, Mathematics and Technology Education*, *11*(1), 67–77.

Koichu, B., & Orey, D. (2010). Creativity or ignorance: Inquiry in calculation strategies of mathematically disadvantaged (immigrant) high school students. *Mediterranean Journal for Research in Mathematics Education*, *9*(2), 75–92.

Leikin, R., Berman, A., & Koichu, B. (Eds.) (2009). *Creativity in mathematics and the education of gifted students*. Rotterdam, The Netherlands: Sense Publishers.

NCTM (National Council of Teachers of Mathematics) (2000). *Principles and standards for teaching mathematics*. Reston: NCTM.

Sheffield, L. J. (Ed.) (1999). *Developing mathematically promising students*. Reston: NCTM.

Wenke, D., & French, P. (2003). Is success or failure at solving complex problems related to intellectual ability? In J. Davidson & R. Sternberg (Eds.), *The psychology of problem solving* (pp. 87–126). Cambridge, UK: Cambridge University Press.

Commentary on the Chapter by Gilah Leder, "Looking for Gold: Catering for Mathematically Gifted Students Within and Beyond ZDM"

Rosemary Callingham

The title of Leder's article "Looking for Gold" seems particularly apt given the difficulties she identified in finding recent material that focussed specifically on mathematically gifted and talented students. Over the past 20 years in ZDM, as Leder states, there have been "...tantalizing allusions to the needs and development of mathematically gifted students but few in-depth discussions of these issues." Leder's review is comprehensive and canvasses many concerns. The paucity of topical research, however, into the nature of mathematical giftedness, the effects of being mathematically gifted, or ways in which students with mathematical talent should be taught, especially at the high school level, seems surprising given the importance and documented need for quality mathematics students.

Outside the mathematics education community there seems to be a vast amount of material, much of it addressing students who appear to be gifted and talented in mathematics. This raises the issue of why mainstream mathematics education researchers are not taking up the challenge of some potentially interesting questions in relation to gifted students in mathematics. The Norwegian work cited by Leder—Braathe and Ongstad (2001)—may provide some clues about the relative lack of focus on mathematically gifted students. Over the past 20 years or so, there has been an increasing emphasis on both international comparative studies such as Trends in Mathematics and Science Study (TIMSS), Program for International Student Assessment (PISA) and national studies such as the National Assessment Program—Literacy and Numeracy (NAPLAN) in Australia and Key Stage tests in England. The centre of attention arising from these programs is improving overall performance which is generally interpreted as improving outcomes for lower achieving students. As a result, those students at the high achieving end of the scale are ignored in favour of their less capable peers. No one would suggest that improving lower performances is unimportant but it is equally inequitable to ignore gifted and talented mathematics students. Mathematics education research, however, is captive

R. Callingham (✉)
University of Tasmania, Tasmania, Australia
e-mail: Rosemary.Callingham@utas.edu.au

H. Forgasz, F. Rivera (eds.), *Towards Equity in Mathematics Education*,
Advances in Mathematics Education,
DOI 10.1007/978-3-642-27702-3_37, © Springer-Verlag Berlin Heidelberg 2012

to available funding, which has been directed at lower achievers. Further studies into the policy and social environments that have led to this situation seem warranted.

Work on brain functioning referred to by Leder (e.g., Geake 2009) may provide the next research breakthrough into mathematical giftedness. Limited studies have already taken place and understanding better how talented mathematics students go about problem solving, especially non-routine problems, at a brain function level could shed light on the nature of mathematical giftedness. The difficulties associated with undertaking research into brain functioning should not be underestimated, however. Not only do studies require access to expensive equipment, they are also physically intrusive and cannot come close to modelling a classroom environment. At the present time, this kind of research can have only limited direct application to teachers, and this is likely to remain the situation for the foreseeable future.

Nevertheless, there are many studies that indicate that providing a problem solving environment does allow mathematical creativity to emerge and, as Leder says, such thinking is more likely to be associated with highly achieving students. Studies around problem solving abound, and all appear to suggest that this approach has the potential to develop every student's capabilities, and particularly to provide opportunities for the mathematically gifted to widen their repertoire of mathematical thinking. Recommendations about mathematical pedagogy that can nurture mathematical talent are, as Leder suggests, similar to those that currently characterise quality mathematics pedagogy for all students. The extent to which this is true appears uncontested, and there is a need for further work into the provision of appropriate pedagogy for gifted and talented mathematics students in school, based on sound research. One aspect that appears to have been neglected in the literature reviewed is the nature of teachers' interactions and dialogue with gifted students. It may not be enough to provide rich, open-ended tasks; the quality of teachers' questions and scaffolding seem also likely to play a role in helping mathematically gifted students develop their talents and interests in mathematics.

Leder refers to the use of technology, but rigorous studies focussing on the ways in which technology can enhance mathematically able students' learning are remarkably few. There are now many types of educationally focussed software available, including dynamic geometry (DGS) and computer algebra (CAS) systems. There has been considerable rhetoric about the removal of the cognitive load associated with routine procedures, freeing up the capacity for creative problem solving and concept development. This area appears to be open for good quality studies, which could also shed further light on the nature of mathematical giftedness as it is revealed through interaction with technological tools. For example, it could be worth knowing whether students identified as mathematically gifted use such tools in different ways from less able students.

There is also a gap in the literature surrounding the social aspects of mathematical giftedness and the consequent effects on the individual, and this is only tangentially considered in Leder's review. Karlsson (1999), for example, found that mathematically gifted individuals in Iceland have a higher than expected history of mental illness. There is also evidence that the mathematically gifted may have more social difficulties because of the perceptions of mathematics among the community, and that these views may impact more on mathematically gifted girls (Damarin 2000). These considerations are not trivial, particularly when linked to the falling participation rates in high level mathematics and science courses.

The real difficulty with all of these studies is that of "frame of reference" (Marsh 1986). In any group of students, some will be more able than others and may thus be perceived as gifted and talented. Putting these students together provides a different frame of reference, for the students as well as the teacher. The question is whether such ability grouping in mainstream schooling fosters gifted students' mathematical thinking. In extra-curricular activities the provision of specialist mathematics programs, such as Mathematics Olympiads, is analogous to training elite sportspersons, and assumes that gifted students will thrive in a competitive environment. Indeed the early part of Leder's review has a focus on competitions. This assumption may not hold true for all gifted students, and may, in part, explain the relative lack of girls among the very top mathematics achievers. The intriguing comment from Braathe and Ongstad (2001) about equity in the Norwegian context suggests that mathematics may be somewhat different from other subjects, requiring a more individual approach. Certainly it seems that Krutetskii (1976) would concur with this view from his description of a "mathematical cast of mind". Many potential research questions associated with equity in relation to gifted and talented mathematics students could be developed.

Leder's review is comprehensive, and covers the current state of play in mathematics education research. The gaps identified are those where new directions are needed for both pragmatic and philosophical reasons. Pragmatically, knowing more about mathematically gifted and talented students should lead to improved provision for these individuals, with anticipated enhanced participation rates in higher level mathematics courses. Philosophically, it appears inequitable that the particular needs of an identified group of students are not being identified adequately, let alone met. It is to be hoped that Leder's review will lead to an increased research focus on conditions that will nurture those students capable of further developing mathematics itself.

References

Braathe, H. J., & Ongstad, S. (2001). Egalitarianism meets ideologies of mathematical education-instances from Norwegian curricula and classrooms. *ZDM, 33*(5), 1–11.

Damarin, S. (2000). The mathematically able as a marked category. *Gender and Education, 12*(1), 69–85.

Geake, J. G. (2009). Neuropsychological characteristics of academic and creative giftedness. In L. V. Shavinina (Ed.), *International handbook on giftedness* (pp. 1085–1098). Dordrecht, The Netherlands: Springer.

Karlsson, J. L. (1999). Relation of mathematical ability to psychosis in Iceland. *Clinical Genetics 56*, 447–449.

Krutetskii, V. A. (1976). *The psychology of mathematical abilities in school children*. Chicago: The University of Chicago Press.

Marsh, H. W. (1986). Verbal and math self-concepts: An internal/external frame of reference model. *American Educational Research Journal, 23*, 129–149.

From the Known to the Unknown: Pattern, Mathematics and Learning in Papua New Guinea

Graeme Were

Abstract During the late 1990s, the Papua New Guinean Department of Education introduced a new elementary school mathematics curriculum that utilised the country's rich and diverse cultural traditions. The resulting changes saw patterns, one of a family of practices related to the decorative arts, take on a prominent role as a tool for understanding number, space, time, measurement, all of which form the basis of mathematics. Drawing primarily on the author's own anthropological fieldwork, this chapter examines the culture of pattern in community life in order to understand its selection as a cultural resource for mathematics learning. It will demonstrate that while pattern is not spoken about, people are nevertheless especially adept at engaging with it. Since Papua New Guinea is full of patterns, and pattern plays such a robust role in the mathematics curriculum, the chapter demonstrates how pattern can be understood as an expression of the mathematical mind at work.

During the late 1990s, the Papua New Guinean Department of Education introduced a new elementary school mathematics curriculum that utilised the country's rich and diverse cultural traditions. This initiative was introduced in order to deliver a community-based mathematics education programme in the local vernacular that was sensitive to a child's existing knowledge of local cultural traditions (Department of Education 1998, p. iii). It required that schoolteachers consult with local communities, such as knowledgeable men and women, in order to identify local cultural resources that could be used to express mathematical ideas. The identified forms would then be included as content of a curriculum to allow schoolchildren to work with mathematics in a concrete way. This process of cultural engagement has been expressed by the then Secretary for Education Baki as an exercise in going "from the known to the unknown." In other words, children come to school with a mathematics of their own (D'Ambrosio 1989, p. 7) that is then harnessed, formalised, and built on.

G. Were (✉)
University of Queensland, Brisbane, Australia
e-mail: g.were@uq.edu.au

H. Forgasz, F. Rivera (eds.), *Towards Equity in Mathematics Education,*
Advances in Mathematics Education,
DOI 10.1007/978-3-642-27702-3_38, © Springer-Verlag Berlin Heidelberg 2012

The reason why I begin this chapter by drawing attention to the cultural mathematics programme in Papua New Guinea is that it demonstrates how culture provides a resource for the learning of mathematics within specific contexts. This cultural approach relies on the recognition by educationalists that mathematics is all around us and not just in the formal settings of school classrooms but is also located in the mundane and everyday routines of our daily lives as well as more specialised tasks of doing and making (e.g. Lave 1988). But what is especially interesting, and is the focus of this chapter, are the properties of the cultural forms that are selected for use within the school classroom as tools for learning. Culture is, it appears, for educationalists, a resource there to be utilised: it is tangible and expressive in practices, performances and objects. But given the rich diversity of cultural expressions in Papua New Guinea, what then do these forms have in common and what kind of issues, if any, arise in their usage?

In Papua New Guinea, such questions are especially pertinent as assumptions about liberal access to cultural knowledge raise difficult challenges not only for educationalists but also for local people, who may face restrictions on access to specific knowledge due to secrecy or taboos. For example, anthropologist Wassmann shows how the Yupno people of Papua New Guinea use a special cord device, *kirugu*, during ancestral song cycles recitals (Wassmann 1991). Its design of husks and knots tied into the main cord at regular intervals suggests its mathematical complexity. The cord is run through the fingers of a knowledgeable man who sings verses as he feels the husks and knots.[1] The use of the cord is restricted to knowledgeable men and access to recitals is only to male initiates. This is because the verses sung aloud contain secret knowledge about a clan's history and the singing of elements of the entire song allows other men to piece together elements of it during the private recitals. Such an example raises the question of how to develop a cultural curriculum that is sensitive to stipulations placed on traditional knowledge.

Another problem is the language of indigenous mathematics. Whilst Pacific society is made up of many practices, performances and objects that express mathematical concepts, much of the way this knowledge is communicated is in nonverbal form. For example, Borofsky describes how, on the island of Pukapuka in the Cook Islands, young men observe craftsmen carving canoes (Borofsky 1987). The relationship between master craftsman and apprentice is one of respect. Even though canoe construction involves many mathematical calculations, transmission of this knowledge is not expressed through verbal instruction. Young men simply observe the master craftsman: asking questions is considered disrespectful. It appears, therefore, that not only is there a conspicuous lack of verbal communication when performing mathematical activities, but these activities are situated within cultural contexts of respect, status and hierarchy. Both these examples thus highlight

---

[1] The Yupno cord, *kirugu*, could be compared to the Inca knotted cord, *quipu*, a focus of study in the field of ethnomathematics by Ascher and Ascher (1981) and Urton (2003). Both cord devices use mnemonic devices tied into the main thread to schematise and give order to cultural knowledge. They both play important roles in encoding and transmitting cultural memories.

the potential difficulties in working with local culture in Papua New Guinea and in the development of a cultural mathematics programme.[2]

Indeed, Vagi and Green (2004) state that the Papua New Guinean cultural mathematics programme is not without its challenges. They outline two main reasons: firstly, there are practicalities of training. Teachers are only given a short period of training in mathematics which leaves them with limited opportunities to develop more complex mathematics concepts in their own language or within the context of their own communities. Secondly, issues of cultural knowledge and access arise. Teachers are required to develop their own curriculum materials which requires closely working with community members who share the knowledge and experiences of cultural traditions. The development of the curriculum alongside community participation evidently takes time, openness, and flexibility, especially in terms of how best to work with people (Vagi and Green 2004, p. 317). This latter point seems to suggest, as has been shown amongst the Yupno and the Pukapukans, that there is a culture of mathematics to be taken into account in the gathering of cultural resources for the local curriculum. This, I will demonstrate, raises important issues about the types of cultural forms selected in the cultural mathematics programme in Papua New Guinea.

1 Mathematics in the Pacific

Studying mathematical ideas in a cultural context may now seem common practice amongst social scientists and educationalists—as the wealth of publications in the field of ethnomathematics testifies—but thirty years ago, mathematics in culture was much harder to find. As Zaslavsky (1973) states in her pioneering book, *Africa Counts*, "on the basis of references to Africa in books about the history of mathematics and the history of number, one would conclude that Africans barely knew how to count" (p. 2). Even as recently as the 1990s, the anthropologist Crump also observes how the use and understanding of numbers has been virtually overlooked by anthropologists in the study of culture even in cases where their role is a central factor in everyday practice (Crump 1990, p. vii). Where anthropologists have approached mathematics in culture, their work has tended to dwell on metaphoric interpretations rather than any kind of mathematical analysis of objects, practices or ideas (Kuechler 1999). Their reduction to language-like systems of thought is hardly surprising, given as Stafford states in her analysis of the history of visual thinking, cultural forms seem to have been treated as lesser or illusory forms of written communication (Stafford 1999). Even when anthropologists have documented mathematics in culture, such as the symmetry analysis of pots, baskets and other objects (e.g. Wash-

---

[2]Bill Barton has discussed some of the issues around the development of mathematics programmes in indigenous languages, specifically with regards to Maori learners in New Zealand (Barton et al. 1998, Barton 2008).

burn and Crowe 1988), the approach has been to use symmetry as a classificatory tool rather than to understand why such complex forms arise.[3]

Some of the earliest work that documents mathematical ideas in the Pacific is by the anthropologists Bell and Deacon who both focus on the practice of sand drawing in Melanesia. Bell (1935) examines the geometrical sand drawings on the small island of Tanga in eastern Papua New Guinea whilst Deacon describes the different types of sand drawings practised by men on the island of Malekula and other islands in the New Hebrides (Deacon and Wedgwood 1934). Both these works, while showing an appreciation of the technique of drawing and associated myths, provide a descriptive analysis rather than any mathematical focus.

Recently, much of this descriptive ethnographic work has been revisited in a series of case studies by Ascher (2002) in the emerging field of ethnomathematics[4]. Her analysis of the Marshall Island navigation charts of Micronesia demonstrates how what appear seemingly incongruous arrangements of sticks in a regular grid formation are devices that teach seafarers how to navigate across the open ocean by an outrigger canoe. Ascher shows how a type of stick chart called 'mettang' are prototypical devices for representing the topography of wave reflection and refraction around clusters of tiny coral atolls. Two other classes of stick charts, 'rebbelith' and 'meddo' are maps of entire archipelagos or smaller sub-regions with small shells marking the position of atolls while straight or curved sticks represent the wind and sea interaction in and around known groups of atolls (Ascher 2002, pp. 89–126). Ascher's work is important because it shows how an analysis of geometric configurations of sticks and shells reveals how people schematise and express complex mathematical relations that sustain their lives and communities.

The centrality of mathematics in people's lives is taken up directly by the anthropologist Mimica who authors a compelling critique of universalistic ideas about mathematics. His work challenges a prevailing perception that the notion of infinity is unbounded and unknowable (Mimica 1988). Focusing on his ethnography of the Iqwaye people of Papua New Guinea, he documents how they believe that their creation hero encompasses the entire Iqwaye society. According to Iqwaye understandings, all people—both living and dead—are equal to one. The concept of 'one' is all encompassing and representative of the Iqwaye creation hero. Drawing out the social and cosmological significance of the local counting system, Mimica shows that what emerges is the Cantor set expressed through kinship and cosmological understandings whereby everything, when added together, always equals one.

While Mimica's work rests on the theoretical position that mathematical relations encapsulate the Iqwaye people's life-world which is expressed through myth

---

[3]One interesting exception—although not in the field of anthropology—is Schiralli's work on the meaning of pattern (Schiralli 2007). Drawing on the work of anthropologist Bateson and art historian Gombrich, Schiralli argues that because of pattern's ubiquity and its significance in various disciplines, it merits further attention within mathematics, one that returns to the roots of the subject. He does so by examining the relation between pattern and number in the school of Pythagoras.

[4]Goetzfridt (2008) has compiled a bibliography of Pacific ethnomathematics.

and number; other anthropologists working in the Pacific have drawn attention to the socio-historical processes involved in the learning of ritual categories, through an examination of the formation of spatial concepts. In her study of social hierarchy in Fiji, Toren (1990) examines the acquisition of spatial relations and categories in social practice, expressed through the *kava* drinking ceremony. She states that a person's concept of hierarchy is related to spatial reference points during *kava* drinking. The positions in which persons acquaint themselves spatially are concretised through reference points within buildings and the ritual space. Toren states (1990) that "persons are at the same time both products and producers of... group processes" (p. 2). Utilising drawings made by children, she demonstrates that ideas about hierarchy are constructed temporally in social processes and manifested through sociality (Toren 1990, p. 17). Toren's key point is to argue that representations are by no means ready-made but take shape in practice.

Both the work of Toren and Mimica underline how culture takes on concrete and abstract expression and mathematics lies at the heart of this. In Fiji, spatial relations are vital to the articulation and expression of social and ritual hierarchies. Amongst the Iqwaye, numerical relations articulate cosmological relations, providing an ontological explanation of the relation between a mythical hero and a people's life-world. Mathematics, it appears, forms a vital basis for allowing people to fully participate in their life-worlds in the Pacific, and yet, neither case suggests that mathematical ideas are classified by local people as 'mathematical'; they are simply lived out through practices and ideas in the course of their daily lives.

So far, I have pointed to how mathematics extends beyond tangible forms and expressions and is also located in abstract relations that order social relations. Returning to the Papua New Guinean cultural mathematics programme, I want to examine to what extent this holds true in the context of the cultural forms selected in the cultural mathematics programme, and what are the implications of this. I will take as my example, pattern, one of a family of practices related to the decorative arts, as it plays a prominent role in the cultural mathematics programme in Papua New Guinea. School teachers everywhere are required to find patterns in their communities as within the curriculum, it is utilised as a tool for understanding number, space, time, measurement, all of which form the basis of mathematics. Drawing primarily on an ethnographic case study of the Nalik people of New Ireland, Papua New Guinea—a group of about 4000 non-literate Austronesian speakers—I intend to examine the culture of pattern in community life in order to understand its selection as a cultural resource for mathematics learning. My aim is to demonstrate that while pattern is not spoken about, Nalik people are nevertheless especially adept at engaging with it. By investigating the technical properties of pattern—its mathematical basis composed of numerical sequences and symmetrical transformations—my approach will show how pattern permeates beyond the visual and material world, structuring ideas and actions through processes of resemblance and difference. Since Papua New Guinea is full of patterns, and pattern plays such a robust role in the mathematics curriculum, I end by arguing that pattern can be understood as an expression of the mathematical mind at work.

2 Pattern in New Ireland, Papua New Guinea[5]

On the island of New Ireland, just east of the New Guinea mainland, the Nalik speaking people are constantly immersed in a variety of patterns from the moment of birth through to the time of death. Infants sleep close to woven bamboo walls that radiate regular crisscross patterns and provide protection from the night; at school, Nalik children line up in rows for morning assembly or learn traditional dances following carefully choreographed patterns and movements. Adolescents become increasingly involved in the ritual and everyday activities that regulate the lives of adults, such as learning the cultivation of fruit trees and root crops or fishing seasons that follow recognisable movements of the sun, moon and sea.

Pattern also plays a significant role in the new cultural mathematics programme in the region. Those patterns that structure the temporal, material and spatial aspects of village life are utilised in New Ireland elementary schools. "Working with patterns helps students to think about patterning. When working with patterns they can see and talk about relationships between mathematics and their daily lives" (Department of Education 2006, p. 7). Nalik schoolchildren are required to learn the skills to recognise patterns, make patterns, and compare patterns. For instance, pupils are challenged to find names for traditional patterns woven into locally produced mats, baskets, fishing traps, house structures and cooking pots in order to express mathematical ideas such as number sequences, geometry and symmetry in a concrete way. Children also learn to make a village calendar, which includes knowledge of gardening, planting and harvesting times. They are also taught how to make patterns with leaf fibres and paints, and then encouraged to elicit the underlying rules for generating the designs. Knowledgeable men and women regularly visit the elementary school to show patterns in village life. This may involve teaching schoolchildren traditional dances or playing local music. It may even involve talking about ritual activities involving designs on traditional objects or seasonal activities that are governed by the sun, sea, and stars.

If pattern appears at the forefront of the cultural mathematics programme in rural villages, then it also plays a vitally prominent role during ritual events following the death of a clansman or woman in the village. Mortuary feasts are organised in the village hamlet in order to capture and release the deceased's soul and allow it to join the other dead in the ancestral world beyond the horizon (Kuechler 1987). Groups of people related to the clan of the dead converge on the hamlet and engage in ceremonial exchange (of pigs, shell money, taro and cash) and feasting. Then, patterns

[5]The author conducted anthropological fieldwork amongst the Nalik people in New Ireland for twelve months in 2000–2001 towards his doctoral research. The principal aim of this research was to conduct an ethnographic study of the production, circulation and significance of the *kapkap* shell valuable, which number in their thousands in western museum collections. What is striking is how little is known about this cultural object, other than its relation to social status, warfare, and mortuary feasting, most probably because colonial collectors were not interested in conducting in-depth research. Hence, the author carried out sustained research in the region in order to gain insights into the workings of pattern through the establishment of close relations with the Nalik people. His research was kindly funded by the Economic & Social Research Council (UK).

Fig. 1 Detail of New Ireland *kapkap*. Photography by Graeme Were

are fleetingly revealed in controlled ways so as to heighten one's attention to them. These patterns appear on wooden carvings, *malanggan*, as geometric shapes, and most conspicuously, clam- and turtle-shell breastplates, *kapkap*, that are worn by *maimai* (at no other time are *malanggan* or *kapkap* revealed publicly). *Maimai* are chiefly men or women, knowledgeable of tradition, who orchestrate ritual events on behalf of clans they represent. I will concentrate on the *kapkap* because it is utilised within the elementary school curriculum for learning about pattern and mathematics; also, it provides an excellent opportunity to understand how, through the patterns on its surface, people use them to participate fully in Nalik society.

Kapkap (Fig. 1) is the name given to a circular composite shell ornament found in the Nalik speaking area. It comprises of two key components: a shiny circular back-plate made from rubbed-down giant clam shell and a turtle-shell plate, into which intricate designs are incised, creating a visual effect not unlike that of a paper doily. The turtle-shell plate rests on top of a clam shell disc, and the two components are fastened together with a length of cord that is threaded through hole in the centre of the clam shell and the turtle-shell. The cord is knotted so that the breast-plate may be hung around the neck like a medal. The most salient features of the *kapkap* are the patterns incised into the turtle-shell disc. They appear as regular incisions creating an optical effect as they appear to radiate out from the centre of the design where a large cross-like motif is carved. At the centre of each pattern is a large quadripartite motif, its design belonging to a clan. No too patterns are alike: rather the arrangement of motifs into combinations of repetitions and transformations means that the *kapkap* patterns look the 'same but different'. This system of style, which combines conservatism with limited innovation, is a hallmark of New Ireland pattern that extends beyond the bounds of style.

The *kapkap* and its pattern denote the rank of a *maimai*. One concentric band refers to his authority to speak for one clan. As a *maimai* attains more rights to speak for clans in the region, the *kapkap* is remade so that each band carved into the design signifies the number of clans a *maimai* can speak for. Thus, the patterned surface area enlarges with each extra concentric band incorporated. With the increased status, the wider his social scope becomes, entitling him to speak in more and more

Fig. 2 Diagram illustrating
the corpus of motifs for all
kapkaps from New Ireland in
the sample

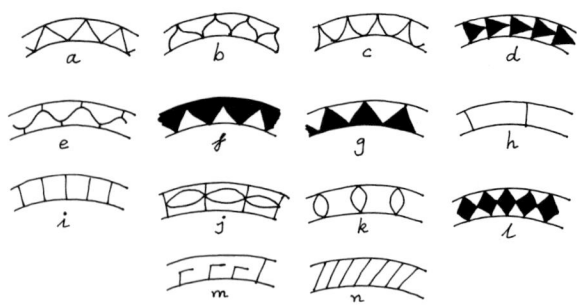

village hamlets with the increased status. The pattern on the *kapkap* thus becomes a means to denote the social power of a *maimai*; its revelation at mortuary feasts allow participants to link *maimai* to place and event as the pattern is easily remembered by the participants.

The *kapkap* is produced by a specialist carver, *aitak*. He knows how to cut distinctive designs for each clan in the region, learning through visual observation and memory. There is no verbal discourse about the design of the pattern; instead he knows that the style of a *kapkap* is based on a standard template that is common to northern New Ireland. This template consists of formal mathematical properties: a series of concentrically arranged circles that emanate out from a large central cross motif. Inside each of the circular bands, motifs are repeated in rows and aligned with motifs in adjacent bands. My analysis of over two hundred New Ireland *kapkaps* reveals that there are fourteen different motifs that make up the stylistic corpus (see Were 2010). However, a carver usually incorporates only two or three classes of motif from within this corpus rather than the full range when making a *kapkap*. Thus, we can begin to see how an idea of 'same but different' emerges: whilst there is an emphasis on conservatism, so too is slight innovation encouraged to create difference.

The basic element of the *kapkap* pattern is an anthropomorphic motif (motif a) that resembles a zigzag (Fig. 2). This motif may be transformed through repetition, rotation, reflection, and scaling to create other motifs in the corpus. Moreover, its straight lines may be transformed into curvilinear forms (motif a → b; motif a → c); with lines added (motif a → e). The curves formed may be reflected along the plane (motifs j and k). Motif (a) may be filled in and rotated (motif a → d); reversed using a figure-ground technique[6] (motif a → f; motif f → g); and reflected (motif g → l). Motif (a) may also be truncated in half to a single line (motifs h, i, and n) as well as bent in the centre (motif m). The result of analysing this stylistic corpus is to show how the anthropomorphic motif can be transformed into all possible forms of motifs within the New Ireland corpus through simple transformations involving symmetry, geometry and coordinate changes[7].

---

[6]A figure-ground relationship is a design where the figure defines the ground and the ground defines the figure; the two elements are inseparable. It forces the viewer to shift from one element to the other, but not both simultaneously. The face/vase illusion is an example of this.

[7]See Were (2003) for a more in-depth analysis of the relation between the various motifs.

The objective of briefly describing how the transformations of the anthropomorphic motif (a) is to demonstrate how individual motifs can be transformed into all others, thus generating the complete range of designs that I call the New Ireland 'corpus of motifs'. Since I have shown how the entire corpus of motifs may be generated by the one anthropomorphic motif (a), then we could say that the motifs within the corpus are relational, as each one relates to any of the others.

The next stage of this analysis is to investigate the implications in conceiving of pattern as a set of relational designs. To answer this, we need to begin by first examining this system of relationality between motifs that govern New Ireland style, so that we can begin to explore how this system could extend to other systematic properties of culture. My formal analysis has showed that a system of transformations that governs the composition of motifs. These are as follows:

- Symmetrical transformations of motifs and part-motifs, including reflection and rotation.
- Scaling of individual motifs, such as halving the proportions so that the transformed motif fits inside the concentric circles.
- Co-ordination transformation from straight line to curved line and vice versa.
- Figure-ground reversal (a motif may be generated from the negative space between the turtle-shell design and clam shell disc beneath).

This list of transformations covers all the possible motivic forms that are generated in New Ireland *kapkap* patterns. The complete set of motivic transformations accounts for all possible stylistic variations in *kapkap* patterns and dictates the principles that govern the creation of a pattern, enabling a *kapkap* to be recognisable as coming from New Ireland. These principles that govern the generation of individual motifs are constraints that are imposed on the possibility of transforming any motif in the corpus to any of the others. In other words, something belongs to New Ireland only if a motif can be transformed into another motif inside the corpus in accordance to the principles. It therefore demonstrates that style is founded by studying the set of connections between the various *kapkaps*, since this list of transformations can only be generated by examining all stylistic variations of *kapkap* pattern, as evidenced in my study of over two hundred *kapkap*.

The patterns are uniform throughout northern New Ireland suggesting that stress is placed on stylistic coherence and accounts for the way patterns are recognised as 'coming from New Ireland'. Minute stylistic variations make individual *kapkaps* recognisable though these differences are difficult to notice. So by formulating a system of parameters that defines New Ireland style, we can begin to account for an abstract principle that governs the generation and application of carved motifs inside the *kapkap* pattern. This structuring principle becomes evident if we see how motivic transformations occur with the least possible modification. Take for example the carving of a pattern with two concentric bands. If the anthropomorphic motif (motif a) is carved in the innermost band, the principle of least difference dictates that the following motif in the adjacent band could either be repeated, or could be a transformation of the first motif by reflection or rotation along the plane or by following a scaling procedure. This system of style—governed by this abstract logic—accounts for slight variations in the composition of the pattern, allowing for small differences between *kapkaps* while at the same time remaining within the New Ireland style. These relations between motifs permit individual patterns to be distinguishable and therefore recognisable, suggesting that the stylistic system is also a way to articulate minute differences between individual *kapkaps*, even when the *kapkap* patterns look almost identical. But this abstract principle also constrains style, which is evident in the uniformity of the *kapkap* patterns, so that the patterns appear the same. Thus, we could say that the system of style is conservative as much as it is innovative.

So far I have discussed how the system of style regulates the types of pattern produced on the *kapkap*. Could this abstract system have any bearing on the system of structures that regulate and maintain order in the social domain?

Let us begin by considering these abstract relational principles from a cultural perspective in the context of the Nalik people. Firstly, principles of 'same but different' can be traced to the Nalik people's origins in the mountain village of Baum. Now abandoned, Naliks once fled the village and moved to the coastal settlements where they now reside, escaping hostilities and open warfare that plagued the region prior to the arrival of Christian missionaries and colonial government. As exogamous groups migrated to various parts of the island, clans fragmented into lineages as they moved through the mountain landscape. Many of the names of the motifs used in the *kapkap* refer to the social history of this particular period of the Nalik past. The result of this outward migration and movement down to the coastal areas is that clan lineages trace their origins from Baum. The system of 'same but different' is captured in the metaphor of the banana tree and its tuberous roots. Naliks say how they all come from the same banana tree and the roots that taper off from the main trunk represent their fragmentation but also emphasise their interconnectedness.

The principles that structure pattern, I suggest, also impinge on the way social relations are structured in the Nalik area. The clan lineage is the most important element of kinship and social organisation in the Nalik area. It is the clan lineage that determines a person's relation to the land and to obligations within the community. The notion of 'the same but different' is expressed through the way Naliks place stress on social difference whilst at the same time maintaining unity. This highly fraught dynamic entails maintaining social distance between kin relations, a process that necessitates avoidance, spatial separation or elements of respect. These rules are enforced so as to avoid possible situations that could shame a person such as coming into contact with avoidance relations. Thus, a Nalik man must avoid his wife's sisters as well as his mother-in-law to avoid shameful contact, deemed too close and disrespectful.

The dynamic of 'same but different' also means that Naliks are acutely aware of the binding elements of social relations, where inter- and intra-clan unity is especially emphasised. Such a dynamic is spectacularly foregrounded at mortuary feasts when clans participate in competitive exchange in acknowledgement of their interrelatedness. It is the corporate effort on the part of each clan expended in the ritual events that articulates one's close relation to the dead and renews the social ties that interconnect the various clans on the island. Failure to recognise one's obligations is tantamount to breaking the unity.

One specific kinship relationship that especially captures this acute tension between separation and proximity is the brother–sister relation. The brother and sister are of the same clan and the same mother, but spatial distance is traditionally maintained between them. Even though the sibling pair maintains distance from each other, a bond still exists between them both, exercised in the brother's protection of his sister's social well being as they once shared the same womb. In turn, the sister has to carefully manage her sexual relations by concealing them to avoid harming her brother's social identity. Thus, the relationship demonstrates the paradox of a concern for difference (through spatial separation) but at the same time, a concern to maintain close ties (through mutual respect and care for one another).

Concerns for homogeneity and difference are made more complex for Naliks as social relations are in constant flux. The complex nature of social relations means that men and women can always appear to trace common ancestors between each other. However, the changing nature of social relations also means that kinship relations are constantly shifting in Nalik society. As people get older, they acquire new affinal relations[8]. This entails that relationships are transformed from respect or joking relations to altogether avoidance. Hence as clan siblings marry, many males and females who may once have brought gifts of betelnut to the house, now strictly avoid particular residences.

It seems plausible therefore, after examining the tensions that structure social relations amongst Naliks that the abstract structuring principles that order the generation of the *kapkap* pattern appear to be lived out in Nalik everyday life as a concern for social difference and homogeneity. So how did this congruence arise? The dynamic generation of style is the result of a long series of initiatives that have developed historically, which have given rise to what we understand today to be New Ireland style. The process of carving each new *kapkap* is a product of a sequence of stylist decisions that New Ireland carvers take. These decisions that generate coherence and stability in New Ireland style, are never enacted without the knowledge of their social outcome and the threats that transgression may bring to clan strength. Thus, the subtle differences in generating the *kapkap* pattern with the prominent uniformity show a concern to establish difference, but concurrently show a recognition of their relation to a background of sameness. The abstract logic that generates the *kapkap* pattern is also lived out in ritual performances. Nalik men and women follow circular and linear pathways as they engage in ceremonial exchanges during mortuary feasts, appearing to mimic the circles and lines that compose the *kapkap*. And it is exactly performances, objects and practices that the elementary cultural mathematics programme seeks to utilise, a time when the *kapkap* is displayed, the drum is beaten and dancers move through the village space.

3 Concluding Comments: Toward the Mathematical Mind

This chapter has established that Papua New Guinea is full of patterns that flow through the lives of the people there. We have seen that there is a culture of pattern; that people internalise them through observation and they appear at key points

---

[8]Affinal relations are based upon marriage e.g. son-in-law.

in the lifecycle. We could argue that the human mind has developed a formal system of thought for recognising, classifying, and employing these patterns and have observed that mathematics is at the heart of this system. By using mathematics to organise and systematise ideas about pattern, we have discovered how people think through pattern as a mode of engaging with the world. It appears that the human mind is intrinsically attracted to the logical properties of pattern: as a composition of unit motifs arranged symmetrically, pattern appeals to our visual sense because it is repetitive and predictable, which also allows it to be easily remembered, transmitted, and reproduced. Mathematics is not simply a means of measurement but indeed also contributes to the way we understand our social world through its logical properties. As Lave shows in her analysis of supermarket shoppers (Lave 1988), people develop cognitive skills and abilities specific to their social environment, and these may be internalised from a very young age. The only approach to pattern is through seeing it and working with it in order to think through it. And in the Pacific at least, we have seen how people there are especially adept at thinking through pattern in order to fully engage in their lives.

This renewed examination of pattern in the Pacific raises critical questions about the nature of intellectual creativity and the human mind. As Gardner (1985) argues, individuals have many intelligences, such as linguistic, naturalistic, musical, logical-mathematical, spatial, bodily-kinaesthetic, intra-personal, and interpersonal. Gardner's theory of multiple intelligence has been fruitfully taken up by the educationalist Rose (2004), who reassesses our understanding of intelligence through a detailed study of undervalued jobs in the US workplace, including occupations such as hairdressing and waiting tables, which are often associated with banal, repetitive, and routine workloads. Taking issue with the stereotyped idea that such careers are intellectually undemanding and disputing that these jobs involve only simple tasks, Rose highlights some of the complex tasks workers perform in the course of their daily work and, in so doing, challenges the orthodoxy of what constitutes intelligence. For example, Rose shows how a waitress employs timesaving routines, calculates serving times, and engages in various interactions according to the demands of each customer. Drawing heavily on the theory of multiple intelligences, he points out that there are many 'ways to know', some of which are little understood or are undervalued. His key point is to show how simple forms or routines mask complex calculations; that is, what appears economic on the surface derives from high-order intellectual activity.

Such an approach gives insight into the workings of the mathematical mind and offers clues as to why pattern has taken on such a pronounced role in the cultural curriculum in Papua New Guinea amongst non-literate societies. It is pattern's spatial and logical-mathematical characteristics, in addition to its kinetic, rhythmic, and musical associations, that enable pattern to flourish and be expressed in a multitude of forms. But we must also consider pattern's economic form. It is easily produced, remembered, and transmitted through the symmetrical transformation of a unit motif, the counting of regular beats, the rhythmic gesturing of the arms in dance, and so forth. Its hallmark is its communicability, and its versatility, and this obviously sustains it. Like Rose's routine tasks, pattern's economic form is based on a logical set of mathematical relations that provides the basis for thinking through complex sets of ideas and actions. Moreover, its abductive quality gives rise to material and visual associations external to it, situating pattern as a form of knowledge expressed across various media as it surfaces in performances, objects, bodies, and so forth. It is from this perspective that we can begin to fully appreciate how the mathematical mind enables people to fully participate in their respective life-worlds.

Given the remarkable prevalence of pattern as a hallmark of Pacific creativity, this chapter has revealed the shortcomings in our perception of pattern in terms of its sophistication. Despite recent work in the field of ethnomathematics that draws attention to complexity in pattern (e.g. Eglash 1999), there remains stubborn resistance towards treating pattern in indigenous cultures as anything other than a basic form of visual communication, part of an economy of thought that lies outside the forms in which pattern is incarnated. On reflecting on pattern's versatility within Nalik society, this chapter has attempted to situate pattern as an expression that is highly mobile and good to think. I have shown in learning about pattern, Nalik people in Papua New Guinea are learning about how to engage fully in their social world. Pattern, as a logical set of calculations, is as much about mathematics, counting and symmetry as it is about a process of socialisation. In taking such an approach, I believe, has profound implications in the way we think, know, and learn, especially in light of recent educational initiatives, as we have seen with the curriculum changes in Papua New Guinea. These reflections on pattern, I hope, will prompt new debates on the workings of the mathematical mind in the Pacific and elsewhere.

References

Ascher, M. (2002). *Mathematics elsewhere: An exploration of ideas across cultures.* Princeton: Princeton University Press.

Ascher, M. & Ascher, R. (1981). *Code of the quipu: A study in media, mathematics, and culture.* Ann Arbor: University of Michigan Press.

Barton, B., Fairhall, U., & Trinick, T. (1998). Tikanga reo tātai: Issues in the development of a Māori mathematics register. *For the Learning of Mathematics, 18*(1), 3–9.

Barton, B. (2008). *The language of mathematics: Telling mathematical tales.* New York: Springer.

Bell, F. L. S. (1935). Geometrical art. *Man, 35,* 16.

Borofsky, R. (1987). *Making history: Pukapukan and anthropological constructions of knowledge.* Cambridge: Cambridge University Press.

Crump, T. (1990). *The anthropology of numbers.* Cambridge: Cambridge University Press.

D'Ambrosio, U. (1989). On ethnomathematics. *Philosophia Mathematica, 2–4*(1), 3–14.

Deacon, A. & Wedgwood, C. (1934). Geometrical drawings from Malekula and other islands of the New Hebrides. *Journal of the Royal Anthropological Institute of Great Britain and Ireland, 64,* 129–175.

Department of Education (1998). *Elementary scope and sequence.* Waigani, Papua New Guinea: Department of Education.

Department of Education (2006). *Good beginnings in mathematics with patterns: Elementary patterns resource book.* Waigani, Papua New Guinea: Department of Education.

Eglash, R. (1999). *African fractals: Modern computing and indigenous design.* New Brunswick: Rutgers University Press.

Gardner, H. (1985). *Frames of mind: The theory of multiple intelligences.* New York: Basic Books.

Goetzfridt, N. (2008). *Pacific ethnomathematics: A bibliographic study*. Honolulu: University of Hawaii Press.

Kuechler, S. (1987). Malangan: Art and memory in a Melanesian society. *Man, 22*(2), 238–255.

Kuechler, S. (1999). Binding in the Pacific: Between loops and knots. *Oceania, 69*(3), 145–156.

Lave, J. (1988). *Cognition in practice: Mind, mathematics and culture in everyday life*. Cambridge: Cambridge University Press.

Mimica, J. (1988). *Intimations of infinity: The mythopoeia of the Iqwaye counting system and number*. Oxford: Berg.

Rose, M. (2004). *The mind at work: Valuing the intelligence of the American worker*. New York: Viking.

Schiralli, M. (2007). The meaning of pattern. In N. Sinclair (Ed.), *Mathematics and the aesthetic: New approaches to an ancient affinity* (pp. 234–279). New York: Springer Books.

Stafford, B. (1999). *Visual analogy: Consciousness as the art of connecting*. Cambridge: The MIT Press.

Toren, C. (1990). *Making sense of hierarchy: Cognition as social process in Fiji. LSE Monograph on Social Anthropology*: Vol. 61. London and Atlantic Highlands: Athlone Press.

Urton, G. (2003). *Signs of the Inka khipu: Binary coding in the Andean knotted-string records*. Austin: University of Texas Press.

Vagi, O. & Green, R. (2004). The challenges in developing a mathematics curriculum for training elementary teachers in Papua New Guinea. *Early Child Development and Care, 174*(4), 313–319.

Washburn, D. & Crowe, D. (1988). *Symmetries of culture: Theory and practice of plane pattern analysis*. Seattle: University of Washington Press.

Wassmann, J. (1991). *The song to the flying fox*. Boroko, Papua New Guinea: National Research Institute.

Were, G. (2003). Objects of learning: An anthropological approach to mathematics learning. *Journal of Material Culture, 8*(1), 25–44.

Were, G. (2010) *Lines that connect: Rethinking pattern and mind in the Pacific*. Honolulu: University of Hawaii Press.

Zaslavsky, C. (1973). *Africa counts: Number and pattern in African culture*. Boston: Prindle, Weber & Schmidt.

Commentary on the Chapter by Graeme Were, "From the Known to the Unknown: Pattern, Mathematics and Learning in Papua New Guinea"

From the Known to the Unknown: What It Means to 'Know' Mathematics

Alan J. Bishop

Papua New Guinea (PNG) is an anthropologist's paradise—several hundred still existing languages, people still living in remote villages with memories of first contact with European civilisations, a country still trying to define itself in the context of the modern world, and many cultures still waiting to be documented and understood. Part of its push to modernisation relates to formal school education and Were's chapter is situated in the context of a PNG Department of Education curriculum development concerning elementary mathematics education. The thrust which interests Were is the plan "to deliver a community-based mathematics education programme in the local vernacular that was sensitive to a child's existing knowledge of local cultural traditions." This was indeed a bold initiative, one which few developing countries have successfully realised due to an array of challenges.

Rather than address these primarily educational challenges, Were explores the thrust of the initiative through analysing the cultural significance of pattern as an example of local knowledge, which could be used to deliver the intended curriculum in the schools. He convincingly demonstrates the power and widespread presence and knowledge of pattern in the culture which he has studied—the Nalik speaking people of New Ireland, an island just east of the New Guinea mainland. He also argues the significance of pattern in the lives of PNG people more generally, and thus appears to give strong support to the Government's curriculum initiative.

However, he does point out two of the many challenges faced by those trying to realise the curriculum initiative, namely

- Restrictions on 'publicising' specific cultural knowledge and practices, in terms of taboos and 'ownership' of the knowledge,
- The often non-verbal ways in which this knowledge is communicated and transmitted within the community.

A.J. Bishop (✉)
Monash University, Melbourne, Australia
e-mail: alan.bishop@monash.edu

H. Forgasz, F. Rivera (eds.), *Towards Equity in Mathematics Education*,
Advances in Mathematics Education,
DOI 10.1007/978-3-642-27702-3_39, © Springer-Verlag Berlin Heidelberg 2012

He quotes others who point out the shortage of time for teacher training to enable teachers to put into practice these ideas, and the difficulties they face in working with local elders to uncover the relevant cultural practices. Not only will the local knowledge not always be known by all in the community, there may be severe restrictions on sharing this knowledge with anyone from outside the culture, in particular with a teacher who is most likely to come from a different cultural group.

No solutions to these problems are offered by Were, but rather the author concentrates on exploring the richness of pattern in cultural life, via objects, practices and customs. Were is an anthropologist and part of his training will have been to try to uncover the richness of any cultural group studied without abstracting that richness from the culture's context. The search is for specific details, to be theorised and analysed, rather than working with vague generalisations, and Were gives us plenty of examples. This kind of work as an anthropologist is important to us in mathematics education, now that there is widespread acceptance of the idea of culture influencing the decisions to be made about mathematics curricula in any societal context.

Having said all that, Were is not a mathematics educator, and his paper loses some of its significance in this educational text by ignoring, or neglecting to reference, the huge amount of relevant mathematics educational literature. Ethnomathematics is far from being an emerging field. It is one that has existed now for nearly 40 years ever since D'Ambrosio (1992) from Brazil raised the concept at the International Congress on Mathematics Education in Adelaide, Australia in 1984.

Since then huge numbers of research studies have been undertaken, in a variety of educational contexts, and with a variety of goals, for example Gerdes (1995) in Mozambique, Barton (1996) in New Zealand, Joseph (1991) in UK, Knijnik (1993) in Brazil, Lipka et al. (2007) in USA, Pinxten et al. (1987) in Belgium, and Zaslavsky (1993) in USA and Africa. In that literature are many ideas of relevance for this discussion. The anthropological paradigm is a hugely important one for those of us who work in the educational world, and it is true that it has only been within the last 40 years that its value has been fully recognised. But this is not to say that this paradigm has made practitioners' and teachers' tasks easier—far from it. When one moves the cultural discourse into the educational sphere, the problems revealed are far more complicated than first realised.

Consider initially the issue of the educational aim. Is it to teach the specific indigenous cultural knowledge? Or is it to use that knowledge as an example to test a theory? Or is it to demonstrate the relevance of culture in education? Or is it to teach the idea of a relationship between one form of cultural knowledge and another? What about teaching mathematics to show the irrelevance of European cultural artefacts in the education of PNG students? What about the teaching of values? What would an anthropologist make of the relevant values behind a subject like mathematics? Aims exist aplenty to which anthropology offers little understanding, advice, or even interest. Perhaps the biggest issue raised by the cultural discourse is, educationally, the danger of colonisation of the local cultural knowledge by the educators. What gives educators the right to choose which aspects of culture to take for the curriculum, and which to ignore?

So Were's paper suffers in that there is little serious discussion of what he might mean by education in this context. Equally it could be said that he offers no definition of what he means by "mathematics". He does, however, make an interesting claim that the kind of knowledge demonstrated by the rich cultural understandings of pattern can be considered as mathematics. This is essentially what concerns him in this chapter. It would have helped his argument had he also revealed his definition of mathematics. The ideas of Wilder (1981) might have offered him a start.

Similarly my own work in the 1980s was based around trying to understand mathematics as a form of cultural knowledge, and what that implies for mathematics education. This has meant analysing the nature of mathematics to reveal its structure, in order to create a conceptual context in which both an indigenous mathematics and Western mathematics can be related. In summary, I identified six contributing activities which meet two criteria, namely that they can be found in every culture studied, and also they give rise to the familiar Western mathematics. In brief the six activities are counting, locating, measuring, designing, playing and explaining (Bishop 1988).

Were's interest is with pattern, which also clearly develops in my interpretation just as any mathematics educator can see the mathematical connections. But Were takes his argument further than merely demonstrating the ways pattern relates to much of the cultural life of the Nalik people. He argues that the manifold ways that pattern structures the lives of the Nalik implies a mathematics of its own. He seems to be taking mathematics to be the fundamental idea structuring the whole of a culture.

As he says in his conclusion: "We have seen that there is a culture of pattern: that people internalise them through observation, and they appear at key points in the lifecycle. We could argue that the human mind has developed a formal system of thought for recognising, classifying, and employing these patterns and have observed that mathematics is at the heart of this system. By using mathematics to organise and systematise ideas about pattern, we have discovered how people think through pattern as a mode of engaging with the world. . . . Mathematics is not simply a means of measurement but indeed also contributes to the way we understand our social world through its logical properties." That is indeed a provocative claim but without his definition of mathematics the claim will have to remain untested.

What is clearer with the knowledge gained from ethnomathematical research is that the position of learners is always going to be one of cultural conflict, between their home culture and the school culture. Now perhaps we can understand that this particular teaching/learning situation is no different to that for children from say a Western farming community, or one where the family is oriented to a skilled trade, or one where the child is physically or mentally handicapped, or from an immigrant worker's family. In each case educators face the issue of the learners' conflicts between the culture of the home or community and that of the school. That interface is a crucial one for us educators, and indeed it can be seen at the heart of any education, including mathematics education.

References

Barton, B. (1996). Making sense of ethnomathematics: Ethnomathematics is making sense. *Educational Studies in Mathematics, 31*, 201–233.

Bishop, A. J. (1988). *Mathematical enculturation*. Dordrecht, Holland: Springer.

D'Ambrosio, U. (1992). Ethnomathematics: A research program on the history and philosophy of mathematics with pedagogical implications. *Notices of the American Mathematical Society, 39*(10), 1183–1184.

Gerdes, P. (1995). *Ethnomathematics and education in Africa*. Sweden: Stockholm University.

Joseph, G. G. (1991). *The crest of the peacock: Non-European roots of mathematics*. London: I.B.Taurus.

Knijnik, G. (1993). An ethnomathematical approach in mathematical education: A matter of political power. *For the Learning of Mathematics, 13*(2), 23–26

Lipka, J., Sharp, N., Adams, B., & Sharp, F. (2007). Creating a third space for authentic biculturalism: Examples from math in a cultural context. *Journal of American Indian Education, 46*(3), 94–105.

Pinxten, R., van Dooren, I., & Soberon, E. (1987). *Towards a Navajo Indian geometry*. Gent: K.K.L. Books.

Wilder, R. L. (1981). *Mathematics as a cultural system*. Oxford: Pergamon.

Zaslavsky, C. (1993). *Multicultural mathematics*. Portland: J. Western Watch.

Commentary on the Chapter by Graeme Were, "From the Known to the Unknown: Pattern, Mathematics and Learning in Papua New Guinea"

Ethnomathematics, Patterns, and Mathematical Minds

Steven K. Khan

> ...while pattern is not spoken about, people are nevertheless especially adept at engaging with it (Were, Chap. 38, p. 415)

In this straightforwardly written ethnomathematical report of geographically and socio-culturally situated mathematical practices—patterning of artefacts and social relations among the Nalik of New Ireland—Were houses a subtle critique of the field of ethnomathematics (EM). However, his critique is rendered less potent by not having demonstrated an attention to the conversational patterns in the recent EM literature. As a reader somewhat familiar with a significant portion of the "basic" literature and the ongoing debates in the field (Khan 2008), I appreciated, however, the author pointing me to texts and manuscripts that I had not encountered. In the hope that this commentary and Were's chapter be of utility to readers I first offer an outline of some complementary patterns in the more recent (ethno)mathematics education literature before returning to the seed that lies dormant within Were's paper, which is to be found via a consideration of his key motif—pattern(ing).

1 Patterns in a Field[1]

EM continues to be productive as evidenced by continued publications in major journals (Doolittle and Glanfield 2007; Khan 2008; Palmer 2010; Wagner and Lunney Borden in press), books (Barton 2009; Nicol 2011), book chapters (Rivera and Becker 2007; Rosa and Orey 2007), conference proceedings and plenaries (Doolit-

---

[1]I wish to acknowledge the extreme partiality of my commentary in that I am most familiar with the literature that has been published or translated into English and that there exists a significant corpus of work in other languages, for example, Portuguese, Spanish, Arabic, Mandarin, etc., as well as in other more geo-politically localised language systems.

S.K. Khan (✉)
University of British Columbia, Vancouver, BC, Canada
e-mail: stkhan@interchange.ubc.ca

H. Forgasz, F. Rivera (eds.), *Towards Equity in Mathematics Education,*
Advances in Mathematics Education,
DOI 10.1007/978-3-642-27702-3_40, © Springer-Verlag Berlin Heidelberg 2012

tle 2007; D'Ambrosio 2010), symposia, and, most importantly, by continued debate and critique among members in (e.g. Pais 2011) and allied with the field itself (e.g.: Denzin et al. 2008; Nasir et al. 2008). Readers with specific interests in equity should also consult Choppin et al. (2012) that adopts a more discursive approach to the question of achieving equity. Gutiérrez (2012) in that volume discusses the way "equity" has been deployed in mathematics education to refer to issues of access, achievement, identity and power, while Wagner and Lunney Borden (2012) also in that volume specifically address the relationship of equity in ethnomathematics research. In working with a Mi'kmaw community in eastern Canada, they foreground the difficulties and problematize any easy road to equity or developing mathematical ideas in any learning situation that involves interactions among multiple agents with complex and entangled historical trajectories, beliefs, goals, motivations and valuation networks.

Ethnomathematical studies and critiques now come from all parts of the globe and not only (previously) colonized nations, though, it is worth noting that, as a response to (ongoing) histories of colonialism and imperialism (Bishop 1990) a significant proportion of research (necessarily) continues to be situated in these societies. Researchers in (former) colonizer nations who grapple with issues of multiculturalism or countries receiving large numbers of immigrants have also turned in recent times to ethnomathematics or culturally responsive mathematics programmes in an attempt to address or redress curricular inequities affecting various groups.

With respect to the latter, Pais (2011) issues an important and timely critique which draws upon Žižek's critique of multiculturalism within capitalism and utilizes the philosopher's concept of a *desubstantialized* Other with respect to ethnomathematical projects in referring to pedagogical practices in which cultural artefacts with mathematical significance are stripped and sanitized of all other cultural (social, political, economic, religious) associations to become "just a[nother] way of teaching mathematics" (Pais 2011, p. 224). Addressing directly those of us who work in or come from privileged WEIRD[2] (Henrich et al. 2010) societies, the pattern that Pais (2011) describes is that of,

> [t]he power of capitalism to produce variety is at work in the educational applications of ethnomathematics... This incorporation of ethnomathematical ideas into capitalist dynamics is made possible through the deployment of an ideological injunction where we are willing to accept the Other deprived of its otherness (Žižek 1992). That is, we are willing to accept the Other as long as it fits into our symbolic order; as long as it is kept at a safe distance, the distance that prevents us from reaching its non-symbolic dimension. I love the Other (the poor, the indigenous) precisely because he is poor, oppressed, and utterly helpless, needing protective care (p. 225).

In calling our attention to a tendency towards desubstantialization in EM projects, Pais reminds us of its major ongoing epistemological debate: viz. the conflation of the term 'ethno' with the species of otherness that is represented by the terms

---

[2]The acronym WEIRD is used by Henrich et al. (2010) to describe the extreme exceptionality of the Western Educated Industrialized Rich Democratic populations which make up the majority of samples in psychology. They are clear to note that they, "do not intend any negative connotations or moral judgments by the acronym" (p. 83).

'ethnic' and 'indigenous.' Consequently some curriculum revitalization projects using an EM philosophical framework become merely an attempt to translate and insert very situated mathematical practices (Palmer 2010) *into* classrooms often from marginalized or forgotten cultures into the dominant culture of school.

Given the context of Were's study and the goals of the Papua New Guinean Department of Education, desubstantialization is a potential issue.[3] Were draws our attention to related difficulties in terms of "seeing culture as a resource" (p. 2) and the situated enactments, or substantiations of cultures. For example in discussing the *kirugu* (p. 2) he raises related issues of taboo and gender restrictions which may conflict with some notions of equity noting that examples like these problematize desires like liberal access and "developing a curriculum that is sensitive to the stipulations placed on traditional knowledge" (p. 2). In this report while he takes care not to desubstantialize the practices of the Nalik people, explicitly juxtaposing his paper with these types of frameworks and situating his descriptions in relation to ongoing concerns makes for a more powerfully resonant and perhaps relevant ethnomathematical critique.

2 The Mathematical Mind: An Issue for Equity

The subtle critique of (ethno)mathematics education is to be found in Were's emphasis on patterns and his speculations as to its relationship to social organization and hierarchies. A pattern, Were suggests, "can be understood as an expression of the mathematical mind at work" (p. 6). In naming the Nalik's practices of patterning as evidence of the mathematical mind at work, he begins to draw closer together the discursive seams that have made a spectacle of Otherness by scribing some groups outside the locus of privilege afforded by being seen or labeled, 'marked' as being Mathematically able (e.g. Damarin 2000, 2008). Studies of 'The Mathematical Mind' (e.g.: Burton 2004; Byers 2007) have, however, with few exceptions, been a WEIRD science. Ethnomathematical studies, among other perspectives (most notably feminist, postcolonial and indigenous critiques), however, have, and continue to pose a direct challenge to these mythologies of who can/can't *do* mathematics and provides an avenue for making the study of (the mathematical) mind less WEIRD and more representative of the diversity of the human.

Later in the paper Were shifts focus to the social realm raising equity concerns in wondering whether or not, "this abstract system [could] have any bearing on the system of structures that regulate and maintain order in the social domain?" (p. 10) and suggests that "[t]he principles that structure pattern... also impinge on the way social relations are structured in the Nalik area" (p. 11). Here Were begins to echo Whitehead's (1941/1951) famous argument in *Mathematics and the Good* in which Whitehead suggested that changes in mathematical thought are related to

---

[3]I know relatively little about the dynamics of the Papua New Guinea educational situation, and what I do know is from having read a small section of the literature published in English.

changes in aesthetic considerations, and perhaps changes in patterns of human behavior (ethics).[4] When this utterance is (temporarily) unmoored from its referential
anchor in this particular paper and this particular context, it can be used to construct
a critique of what continues to constitute the dominant epistemological perception
of ethnomathematics, viz. it is *an Other's* mathematics—thus, as Pais (2011) suggests, robbing it of most of its critical transformative power at the very moment that
it might celebrate its seeming success.

In the concluding paragraph Were articulates his critique most clearly when he
argues,

> ...there remains stubborn resistance towards treating pattern in indigenous cultures as any-
> thing other than a basic form of visual communication, part of an economy of thought that
> lies outside the forms in which pattern is incarnated... Pattern... is as much about mathe-
> matics, counting and symmetry as it is about a process of socialization (p. 14).

I read here in Were's attentiveness to the relation of patterns both at the level of
artifacts and at the level of social structuring in *this* culture, a call some in the field,
including myself, to begin to attend to and engage in examining the very specific
yet diffuse and abstracted mathematical practices in the networks of powerful and
privileged elites in *our* own cultures.

To Shockey's (2006) discussion of heart surgeons for example could now be
added any number of groups in our WEIRD societies. Given recent global events
convincing arguments could be made to examine the patterns of 'situated' mathematical practices and decision making involved in, for example: BP's 2010 Gulf of
Mexico Deepwater Horizon oil spill disaster, the global economic meltdown and
housing bubble collapses and perhaps the patterns of adjunct hiring and tenure promotion models in Universities, as specifically ethnomathematical research projects.
I wonder, for example, how reading recent books such as Patterson's *The Quants*
(2010) or Sorkin's *Too Big to Fail* (2010), or, closer to home, Soifer's, *The Mathematical Coloring Book* (2009), and Hacker and Dreifus' *Higher Education? How
colleges are wasting our money and failing our kids* (2010) through an EM lens
alongside those "traditionally" included in the situated mathematical practices
of teacher preparation courses, graduate seminars, and professional development
workshops which self-describe as ethnomathematically *oriented* (double entendre
intended), might change our understanding of the patterns at the social level of
schooling and perhaps begin the process of making explicit the way in which familiar WEIRD mathematical practices are *also* ethnomathematical?

These are perhaps goals for some of the next generation of ethnomathematicians
who may better help us to see those patterns structuring everyday life which are not
spoken about but with which we are especially adept at engaging and negotiating.

---

[4]"...the cohesion of social systems depends on the maintenance of patterns of behavior; and advances in civilization depend on the fortunate modification of such behavior patterns. Thus the
infusion of pattern into natural occurrences, and the stability of such patterns, and the modification
of such patterns, is the necessary condition for the realization of the Good.... Mathematics is the
most powerful technique for the understanding of pattern and for the analysis of the relationships
of patterns" (Whitehead 1941/1951, pp. 677–678)

Re-socializing the economies of power/knowledge and value in WEIRD societies also holds great promise for interrupting intractable inequities and promoting sustainability perhaps by drawing from a consideration of the patterns of ethnomathematics as a specifically *human* genre (Khan 2010).

References

Barton, B. (2009). *The language of mathematics. Telling mathematical tales*. New York: Springer.

Bishop, A. J. (1990). Western mathematics: The secret weapon of cultural imperialism. *Race & Class, 32*(2), 51–65.

Burton, L. (2004). *Mathematicians as enquirers: Learning about learning mathematics*. Dordrecht, The Netherlands: Kluwer Academic Publishers.

Byers, W. (2007). *How mathematicians think: Using ambiguity, contradiction, and paradox to create mathematics*. New Jersey: Princeton University Press.

Choppin, J., Herbel-Eisenmann, B., Pimm, D., & Wagner, D. (Eds.) (2012). *Equity in discourse for mathematics education: Theories, practices and policies*. New York: Springer.

Damarin, S. (2000). The mathematically able as a marked category. *Gender and Education, 12*(1), 69–85.

Damarin, S. (2008). Toward thinking feminism and mathematics together. *Signs: Journal of Women in Culture and Society, 34*(1), 101–123.

D'Ambrosio, U. (2010). From Ea, through Pythagoras, to Avatar: Different settings for mathematics. In M. F. Pinto & T. F. Kawasaki (Eds.), *Proceeding of the 34th conference of the international group for the psychology of mathematics education* (Vol. *1*, pp. 1–20). Brazil: Belo Horizonte.

Denzin, N. K., Lincoln, Y. S., & Smith, L. T. (Eds.) (2008). *Handbook of critical and indigenous methodologies*. California: Sage Publications Inc.

Doolittle, E. (2007). Mathematics as medicine. In P. Liljedahl (Ed.), *Proceedings of the 2006 CMESG annual meeting* (Vol. *1*, pp. 17–25). Burnaby: Canada.

Doolittle, E., & Glanfield, F. (2007). Balancing equation and culture: Indigenous educators reflect on mathematics education. *For the Learning of Mathematics, 27*(3), 27–30.

Gutiérrez, R. (2012). Context matters: How should we conceptualize equity in mathematics education. In J. Choppin, B. Herbel-Eisenmann, D. Pimm, & D. Wagner (Eds.), *Equity in discourse for mathematics education: Theories, practices and policies* (pp. 17–33). New York: Springer.

Hacker, A., & Dreifus, C. (2010). *Higher education?: How colleges are wasting our money and failing our kids*. New York: Times Books.

Henrich, J., Heine, S. J., & Norenzayan, A. (2010). The weirdest people in the world? *Behavioral and Brain Sciences, 33*, 61–83.

Khan, S. K. (2008). Mathematics and the steel drum. *For the Learning of Mathematics, 28*(2), 33–38.

Khan, S. K. (2010). *After gender: Towards more human genres of mathematics education*. Invited faculty/public research seminar, University of the West Indies, St. Augustine, Trinidad.

Nasir, N. S., Hand, V., & Taylor, E. (2008). Culture and mathematics in school: Boundaries between "cultural" and "domain" knowledge in the mathematics classroom and beyond. *Review of Research in Education, 32*(1), 187–240.

Nicol, C. (Ed.) (2011). *Living culturally responsive mathematics curriculum and pedagogy: Making a difference with/in indigenous communities*. The Netherlands: Sense Publishers.

Pais, A. (2011). Criticisms and contradictions of ethnomathematics. *Educational Studies in Mathematics, 76*(2), 209–230.

Palmer, M. A. (2010). Situated mathematical research: The interaction of academic and non-academic practices. *For the Learning of Mathematics, 30*(2), 32–39.

Patterson, S. (2010). *The quants: How a new Breed of math whizzes conquered Wall Street and nearly destroyed it*. New York: Crown Publishing.

Rosa, M., & Orey, D. C. (2007). Pop: A study of the ethnomathematics of globalization using the sacred Mayan mat pattern. In B. Atweh, M. Borba, R. Vithal, & C. Vistro-Yu (Eds.), *Internationalisation and globalisation in mathematics and science education* (pp. 227–246). New York: Springer.

Rivera, F., & Becker, J. R. (2007). Ethnomathematics in the Global Episteme: Quo Vadis? In B. Atweh, M. Borba, R. Vithal, & C. Vistro-Yu (Eds.), *Internationalisation and globalisation in mathematics and science education* (pp. 209–225). New York: Springer.

Shockey, T. (2006). Left-ventricle reduction through an ethnomathematics lens. *For the Learning of Mathematics*, 26(1), 2–6.

Soifer, A. (2009). *The mathematical coloring book: Mathematics of coloring and the colorful life of its creators*. New York: Springer.

Sorkin, A. (2010). *Too big to fail: The inside story of how Wall Street and Washington fought to save the financial system and themselves*. New York: Penguin.

Wagner, D., & Lunney Borden, L. (2012). Aiming for equity in ethnomathematics education research. In J. Choppin, B. Herbel-Eisenmann, D. Pimm, & D. Wagner (Eds.), *Equity in discourse for mathematics education: Theories, practices, and policies* (pp. 69–87). New York: Springer.

Wagner, D., & Lunney Borden, L. (in press). Common sense and necessity in (ethno)mathematics. *Cultural Studies in Science Education*, special issue on "informal science education".

Whitehead, A. N. (1941/1951). Mathematics and the good. In P. A. Schlipp (Ed.), *Library of living philosophers*, Vol. *III. The philosophy of Alfred North Whitehead* (pp. 666–681). Stanford: Stanford University Press.

Part IV
Equity and Biology

Helen Forgasz and Ferdinand Rivera

This section of the book includes three chapters focusing on equity issues associated with biological factors. The contents are at the cutting edge of research on equity with respect to biological determinants. In earlier research on biological factors, the focus was on trying to explain gender differences in performance by biological differences associated with spatial capabilities (e.g., Benbow and Stanley 1980). At the time, critics of biological explanations (e.g., Meyer and Fennema 1988) pointed to the lack of supportive scientific evidence for the claims. Other detractors (e.g., Selkow 1985) outlined the psychological and social consequences for women if genetic explanations for gender differences were embraced. One of the chapters in this section is consistent with these earlier criticisms of biological determinism. The other chapters included here go beyond gender differences to examine other aspects of biological factors. The work discussed contributes to emerging and important understandings of the interaction of biology and equity considerations in the learning of mathematics education.

In their chapter, *Gender differences in mathematics and science achievement across the distribution: What international variation can tell us about the role of biology and society*, **Andrew Penner** and **Todd CadwalladerOlsker** focus on the potential of biology and/or social factors on mathematics proficiency. They provide an overview of the literature on genetic theories associated with gender differences in mathematics achievement, and the literature on social factors. Using TIMSS 1995 data, variation and gender differences in international mathematics achievement were examined using two analytical approaches: descriptive statistics, and modeling using country level predicators. Patterns were evident that were inconsistent with the implications suggested by biological theories: i. the magnitude of the gender differences varied widely across countries; ii. the variation in scores was not always highest for males; iii. females in some countries scored higher than males in other countries; and iv. contextual, societal factors may impact differentially on the possibilities for high achieving girls and boys.

The chapter by **Marjorie Montague** and **Asha Jitendra**, *Research-based mathematics instruction for students with learning disabilities*, provides valuable insights into an under-researched field affecting between 5% and 8% of children in the general population. The authors describe the difficulties experienced by children with mathematics learning difficulties that put them at risk educationally. An overview of the existing research literature identifying evidence-based practices that are effective in teaching these students is presented. This is accompanied by descriptions of two programs designed to improve the mathematics learning of students with learning difficulties: i. *Solve It!*, founded in cognitive strategy instruction (CSI) designed for secondary students to improve their mathematical problem solving; and ii. a four step schema-based instruction (SBI) intervention to assist children to identify a problem types or problem schema. As children with learning disabilities move into mainstream schooling, the authors recommend that classroom teachers and specialist remedial teachers need to work collaboratively. The success of interventions, they claim, "often depend on how, when, and by whom they are provided to students".

A synopsis of research on brain-related explorations within the general field of neuroscience as it has been applied to understandings of mathematics learning is provided by **Ferdinand Rivera** in his chapter, *Neural correlates of gender, culture, and race and implications to embodied thinking in mathematics*. Rivera believes that information about the "neural mechanisms that support cognitive processes in mathematical thinking and learning" will be derived from such studies. Of particular interest in the chapter is the overview of research that "provide a neural grounding of gender, culture, and race" within and outside mathematics education. Rivera concludes with implications for mathematics education of the neuroscientific approach and, quoting Eliot (2010), notes that "[B]rain differences are indisputably biological, but they are not necessarily hardwired"; experience changes the structure and functioning of the brain.

References

Benbow, C. P., & Stanley, J. (1980). Sex differences in mathematical ability: Fact or artifact? *Science, 210*, 1262–1264.

Eliot, L. (2010). The truth about boys and girls. *Scientific American Mind, 21*(2), 22–29.

Meyer, M. R., & Fennema, E. (1988). Girls, boys, and mathematics. In T. R. Post (Ed.), *Teaching mathematics in grades K-8: Research based methods* (pp. 406–425). Boston: Allyn and Bacon.

Selkow, P. (1985). Male/female differences in mathematical ability: A function of biological sex or perceived gender role? *Psychological Reports, 57*, 551–557.

Gender Differences in Mathematics and Science Achievement Across the Distribution: What International Variation Can Tell Us About the Role of Biology and Society

Andrew M. Penner and Todd CadwalladerOlsker

Abstract This study examines international data on mathematics and science achievement to illuminate the ways in which macrostructural factors influence gender differences in mathematics, particularly among high achievers. Examining the importance of macrostructural factors for gender differences not only helps us better understand the role that larger contexts play in contributing to these differences, but international variation also provides a unique opportunity to present simple and powerful arguments for the continued importance of social factors vis-à-vis biological considerations. We show that there is considerable international variation in gender differences, and that gender differences among high achievers in both mathematics and science literacy are related to gender inequality in the labor market and differences in the overall status of men and women. We conclude by discussing the implications of our findings for mathematics education and suggesting fruitful avenues for future research.

1 Introduction

While the proportion of women in science, technology, engineering and mathematics (STEM) fields is increasing, women are still a minority in these fields. The underrepresentation of women in these fields has generated a considerable body of research, much of which focuses on the roles that a wide variety of social factors have in creating gender differences in achievement. However, there is a persistent notion that biological factors are somehow responsible for the underrepresentation of women in STEM fields, famously expressed in remarks made by former Harvard University president Summers (2005). Recent scholarship has witnessed the resurgence of biological explanations for gender differences in mathematical achievement. Geary (1998), for example, posits an evolutionary explanation, Kimura (1999)

A.M. Penner (✉)
University of California, Irvine, USA
e-mail: andrew.penner@uci.edu

T. CadwalladerOlsker
California State University, Fullerton, USA
e-mail: tcadwall@fullerton.edu

H. Forgasz, F. Rivera (eds.), *Towards Equity in Mathematics Education*,
Advances in Mathematics Education,
DOI 10.1007/978-3-642-27702-3_41, © Springer-Verlag Berlin Heidelberg 2012

proposes a hormonal theory, and Baron-Cohen (2003) argues that there is a cerebral basis. Further, while most researchers examining biological mechanisms for gender differences in mathematics achievement argue that it is important to keep social factors in mind, some seem to regard biology as an immutable destiny (Blum 1997).

Despite the growing literature on the potential importance of biological considerations for gender differences in mathematics achievement, these issues remain difficult to address, as it is impossible to examine people apart from either their biological bodies or their social contexts. Given the inextricable link between the biological and the social, one way to proceed is through examining these differences in a variety of national contexts. An international perspective is useful here because there is no reason to believe that the biological factors involved in determining gender vary across countries, implying that to the degree that gender differences in mathematics result from biological factors, there should be no international variation in these differences. International data can thus provide an intuitive and straightforward way to demonstrate the importance of social factors. There is relatively little nuance involved: If gender differences vary across countries (and they do), then social factors are important.[1]

International variation also provides us with analytical leverage to address the possibility of biological limitations, as we can compare the scores of girls in one country to those of boys in another. That is, if girls in Canada score higher than boys in the United States, it seems implausible that the biological limitations of girls in the United States could account for the gender differences we observe. Further, international variation allows us to examine what characteristics of countries are related to larger and smaller gender differences in mathematics achievement, providing a sense of how macro-level societal factors impact gender differences in mathematics (e.g., Charles and Bradley 2009). In examining the role of these macro-level factors, we address questions highlighted by Hanna's (1989) international research on gender differences in mathematics achievement. Hanna argues that the international variation in gender differences she finds suggests the importance of psychosocial (and not biological) factors, concluding that "to gain an understanding of the psychosocial factors which contribute to girls' lower achievement (in certain countries), one would have to be able to assess the impact of the sociocultural influences transmitted by the environment on the achievement of students of both sexes" (p. 31). We argue that in addition to looking across countries, looking across subject areas can also be informative, as the social and societal pressures that are faced by women in mathematics and the sciences are likely to be very similar, so that the macro-contextual factors related to gender differences in one domain should also be relevant to similar domains.

---

[1]It is important to note that this is not the same thing as saying that biological factors are unimportant. As an example, it could be that diet interacts with hormones so that the gender differences we observe are in some sense biological, but importantly, social factors are still central. International variation thus precludes a deterministic biological account, but is eminently compatible with bio-social interactions.

This paper takes two analytical approaches to examine the international variation in achievement differences. First, we examine descriptive statistics about gender differences across countries. This analytical approach involves not only examining gender differences in basic descriptive statistics like means and variances, but also examining gender differences throughout the distribution. Here we focus on differences in mathematics achievement. Second, we model the cross-national variation in gender differences throughout the distribution with country-level predictors. Here we examine both differences in mathematics and science achievement, as we would expect that many of the macro-social mechanisms responsible for differences in mathematics achievement are also important in shaping science achievement. Looking at differences in science achievement thus functions as a robustness check for the macro-contextual factors that appear to affect mathematics achievement, and allows us to see how widespread the effects of the various macro-contextual factors affecting gender differences in mathematics achievement are.

Where the first approach establishes the importance of societal factors, the second reveals what specific societal factors are related to gender differences in mathematics and science achievement. In both of these approaches, we respond to previous work calling for an understanding of differences in populations that encompasses more than just their central tendencies (e.g., Favreau and Everett 1996; Handcock and Morris 1999). This is important not only because biological theories about gender differences in mathematics make arguments about differences throughout the distribution, but also because the underrepresentation of women in mathematics and science related fields is more closely related to differences at the top of the distribution than differences in the mean.

We proceed by outlining how biological and societal factors could create gender differences in mathematics, and then discuss the data and methods used in this paper. After presenting the results, we conclude by discussing their implications for mathematics education.

Ultimately, the work of identifying the role of biology in creating gender differences in mathematics achievement falls to biologists. But whatever biologists find, there is also no question that social factors have had, and will continue to have major effects. Biological bodies provide us with both opportunities and constraints, and in different societies it is possible for different people to mobilize these opportunities and constraints to different degrees. If, as Dyson (2007) argues, the 21st century is the century of biology, then understanding how social and biological factors interact provides one of the great challenges and opportunities for social scientists. In the dialogue that social scientists are increasingly entering into with biologists, it is thus important to have a firm grasp on how social factors matter, and how they might interact with biological factors. Further, while we discuss biological and social factors separately for heuristic purposes, it is important to remember that far from being independent, they are better thought of as mutually interpenetrating.[2]

---

[2]For an excellent recent review that synthesizes biological and social factors, see Ceci and Williams (2010b).

2 The Biological Production of Gender Differences in Mathematics

Although this study does not examine biological factors by directly modeling them, many biological theories about gender differences in mathematics achievement have clear implications for the international patterns of gender differences that we should observe. While finding patterns of differences consistent with these implications would not establish that gender differences in mathematics achievement are biological, finding patterns that are inconsistent with these implications suggests that the theories examined cannot completely account for the gender differences we observe. As we show below, our findings reveal patterns that are inconsistent with the implications suggested by some of these biological theories.

Against this backdrop, we briefly review the major biological theories relevant to gender differences in mathematics. Biological explanations for gender differences in mathematics achievement tend to focus on measurements of general intelligence and its specific components, especially spatial abilities (Benbow 1988; Casey et al. 1997; Halpern 2000), and can be roughly grouped into three categories: genetic, hormonal, and cerebral.

2.1 Genetic Considerations

Given the role of genes in determining gender, it is not surprising that theories attributing gender differences in mathematics achievement to genetic differences exist. Studies typically find that around 50% of general intelligence is accounted for by heredity (McClearn et al. 1997; Petrill 1997; Plomin et al. 1997), although as Plomin et al. (1997) note, this is an average, and likely varies in extreme environments.[3] While there is considerable evidence that both general intelligence and spatial abilities are inherited genetically, and even some work investigating the genes involved (e.g., Plomin and Craig 2001), in order for genetic influences to account for gender differences in mathematics achievement, gender differences in inheritances are necessary. Although early studies purported to evince the sex-linked recessive gene theory (Bock and Kolakowski 1973), which claims that a recessive gene for high-spatial ability is carried on the X chromosome, later studies could not replicate this finding (see, for example, DeFries et al. 1976). Boles (1980), in a review of the literature, concludes that there is no convincing evidence for the sex-linked linked recessive gene theory, though Thomas and Kail (1991) argue that Boles' dismissal of this theory is premature.

---

[3]This raises an interesting question, as it suggests that most of the estimates that we receive about the relative importance of genetic versus environmental influences are from contexts with relatively homogeneous environments, and thus underestimate the amount of variation attributable to the environment. It is possible, for example, that estimates of the genetic and environmental portions of the variance in cognitive abilities might differ if estimated in a context where international environmental differences were able to be taken into account.

More recent work by Geary (1998) suggests that sexual selection could drive the underrepresentation of women at the top of the distribution by creating greater variability among males than females. Specifically, Geary (1998) posits that sexual selection might have resulted in males being more responsive to environmental conditions than females, so that adverse environmental conditions would result in more boys at the bottom of the distribution, while beneficent environmental conditions would result in more boys at the top. Under this logic, a combination of adverse and beneficent conditions would result in greater male variability.

In general, while research has moved away from studies looking for a spatial abilities gene (e.g., the sex-linked recessive gene theory), genes still play a prominent role in this arena due to their influence on other biological considerations (Blum 1997).

2.2 Hormonal Considerations

The relevance of hormones to gender differences in spatial ability is visible in the literature linking natural variation in hormones to performance on tests of spatial abilities. Hampson (1990), for example, shows that menstrual cycle variation in estradiol impacts spatial abilities, and studies examining testosterone fluctuations across seasons (Kimura and Hampson 1994) and the time of day (Moffat and Hampson 1996) find evidence that testosterone levels also affect spatial abilities. In reviewing this literature, Kimura (1999) suggests that there is an optimal level of testosterone for spatial performance, and that this level falls low in the male range. Thus, we would expect the highest levels of mathematical achievement from females with high levels of testosterone and males with low levels of testosterone. This curvilinear relationship between testosterone and spatial abilities is also supported by research on people with congenital adrenal hyperplasia (CAH), a condition which results in the overproduction of an androgen similar to testosterone. While women with CAH are found to have higher spatial abilities the other women, men with CAH have, if anything, lower spatial abilities than other men (Hampson and Altmann 1998; Resnick and Bouchard 1986).

2.3 Cerebral Considerations

Cerebral theories of gender differences typically focus on brain lateralization. Generally these theories are based on the finding that males have more asymmetrically organized brains than females (Halpern 2000). This means that spatial abilities tend to be processed in both hemispheres in females and in one hemisphere in males. Evidence of this is found in studies showing that in men, damage to the left hemisphere results in decreased verbal abilities, and damage to the right hemisphere in decreased spatial abilities, while in women, damage to the left hemisphere results in equal decreases in verbal and spatial abilities, and damage to the right does not seem to decrease either (Gazzaniga et al. 1998). EEG research on people with normally functioning brains finds that males perform better on a spatial task, and tend to access their right hemisphere more than their left, while females exhibit greater degrees of bilateral activity (Gill and O'Boyle 1997). However, it is not clear that one way of processing is more efficient than the other. Levy (1974) hypothesizes that hemispheric specialization is the optimal neural organization for performing spatial tasks, but Halpern (2000) notes that data from left-handers does not support this. Bryden (1986) and Geschwind and Galaburda (1987), however, find that greater hemisphere specification could contribute to gender differences in spatial and mathematical differences, as they find that right-handers exhibiting greater right hemisphere specialization for non-linguistic tasks tend to have higher spatial abilities.

3 The Social Production of Gender Differences in Mathematics

In contrast to our treatment of the biological theories, where we are constrained to examining implications of the theories and are unable to incorporate biological considerations directly into our models, one of the advantages of using international data is that they allow us to examine how gender differences vary across different national contexts.[4] Thus, while most social theories examine the role of micro-level social considerations like parent, teacher, and peer effects (e.g., Ma 2001), this study explores broader issues of national context (e.g., Charles and Bradley 2009). Although analyses of gender differences in mathematics achievement typically take broader macrostructural settings as given, by exploiting variation in national contexts we are able to systematically explore how differences in this macro-level context are related to the mathematics gender gap in that country. There are at least two ways in which national level factors could play a significant role in creating gender differences in mathematics and science achievement: first, through creating incentive structures; and second, through establishing gender differences in status.

3.1 Incentive Structures

Country-level social inequalities can be thought of as providing different incentive structures within which students decide whether to pursue mathematics and science education. Riegle-Crumb (2005), for example, uses international data to argue that girls in school look to the kinds of opportunities that are available to women in their society in determining whether to pursue mathematics and science education. This suggests that when boys and girls are deciding what courses to take, how hard to work on their homework, and otherwise where to invest their human capital, they do so with knowledge of the kinds of opportunities that are realistically available to

---

[4]It is worth noting, however, that due to the correlational nature of this study, even when we are modeling social factors we are examining necessary (but not sufficient) conditions.

them. From this perspective, we can understand girls' underachievement in mathematics as a rational response to the perceived lack of opportunities for women in fields where mathematics achievement is valued. Frank et al. (2008) argue that this rational choice process includes not only future costs and benefits, but also more immediate social considerations such as popularity. Here too it is easy to imagine that national contexts would affect gender differences in mathematics and science achievement, in this scenario by dictating how popular mathematics is. Thus, regardless of the precise time horizon under which students are performing their cost-benefit analyses, it is reasonable to expect that differences in national contexts will alter students' decisions and result in differing gender gaps in mathematics achievement.

3.2 Status

Findings from the sociological literature on the psychology of status suggest that national contexts are important not just in determining the incentive structure within which decisions are made, but that they also impact achievement through creating general gender differences in status and making gender more or less salient to mathematics and science achievement. The sociological literature on status characteristics theory suggests that beliefs and stereotypes about status characteristics have a broad range of effects. Especially relevant to gender differences in mathematics and science achievement are findings of status effects on cognitive performance (Lovaglia et al. 1998) and self-perceptions of performance (Correll 2004).

The psychological analogue of status characteristics theory focuses on stereotype threat effects, with similar findings. Stereotype threat research finds that when gender is made salient prior to administering a mathematics test women score lower and men score higher than when gender is not made salient (Spencer et al. 1999; Walton and Cohen 2003). Stereotype threat effects in mathematics can be evoked with stimuli as diverse and subtle as a mixed gender setting (Inzlicht and Ben-Zeev 2000), gender stereotyped television commercials (Davies et al. 2002), and by telling subjects that gender differences are due to genetic (as opposed to social) factors (Dar-Nimrod and Heine 2006). These findings suggest that stereotype threat effects could have a substantial scope, and that country-level factors such as the representation of women in the media and beliefs about the biological origin of gender differences could be important. Taken together, research on stereotypes and status characteristics suggests that larger sociocultural factors might impact gender differences in mathematics and science achievement through impacting gender schemas at the national level.

4 Data

Data for this study come from several sources. Mathematics and science achievement scores are obtained from the Third International Mathematics and Sciences Study conducted in 1995. Gender differences in these scores are modeled using country-level data from the World Bank, the United Nations Statistics Department, the International Labour Organization, the World Values Survey, and the International Social Survey Programme.

4.1 Mathematics and Science Data

Mathematics and science achievement data for this paper are from the Third International Mathematics and Science Study (TIMSS), which was conducted by the International Association for the Evaluation of Educational Achievement (IEA) in 1995. TIMSS includes information from over forty countries and measures mathematics and science achievement in elementary, middle, and high school. This study analyzes mathematics and science literacy scores from students in the final year of secondary school using data from the 22 countries with information on these students.[5] A two-stage stratified sample was used within each of the countries, with schools sampled with probability-proportional-to-size at the national level, and a fixed number of students from different categories (e.g., students who have taken advanced mathematics courses as opposed to those who have taken only basic mathematics) sampled from each school. Weights are thus used to obtain nationally representative results. More information about the TIMSS dataset is available in Martin and Kelly (1996) and Mullis et al. (2000).

While international comparisons in education are often complicated by cross-national differences in curriculum and population definitions (Bracey 2000), these concerns are less of an issue for this study as the focus here is not a direct comparison of scores between countries, but rather an international comparison of within country gender differences. Of more concern for this analysis is that TIMSS countries with information on high school students consist primarily of European countries, and the non-European countries that are represented (Australia, Canada, Israel, New Zealand, South Africa, and the United States) are relatively western. Although there is still a considerable amount of variation in national contexts, it is difficult to know how generalizable the findings are, and whether we would find similar patterns in a broader context.

We use TIMSS data from 1995 instead of more recent IEA data (from the Trends in International Mathematics and Science Study) or data from OECD's Programme for International Student Assessment (PISA), because the TIMSS 1995 data allow us to compare students in their final year of secondary school and are not restricted to students who have taken advanced coursework. While there is a considerable body of research examining how early educational inequalities lead to later inequalities

---

[5]Mathematics and science literacy scores are used as they are less susceptible to biases arising from country-level differences in curriculum. The mathematics literacy items in TIMSS are designed to assess students' knowledge, their ability to execute routine and complex procedures, and their problem solving skills in the content areas of number sense, algebraic sense, and measurement and estimation (Martin and Kelly 1996). Science literacy items focus on student's ability to apply scientific knowledge to real-world questions, and cover content in earth science, life science, and physical science (Mullis et al. 2000).

(Farkas 2003), research on gender differences in mathematics achievement generally suggests that differences emerge between middle and high school (e.g., Muller 1998). Table A.1 reports the gender differences for 4th, 8th, and 12th grade students for the 13 countries with information on students at all three grade levels. In general, we see that gender differences are more likely to be statistically significant in 12th grade, that the shape of gender differences across the distribution (the relative size of differences at the 10th and 90th percentiles) changes across grade levels, and that differences are considerably larger in 12th grade. Thus, while it is interesting and important to examine gender differences at various stages in students' academic careers, gender differences in earlier grade levels do not necessarily provide an accurate picture of the gender inequality that we observe later. We focus on 12th grade mathematics achievement because we believe that the final year of secondary school provides a key point of comparison for examining the gender gaps produced by different national educational systems.

4.2 Country-Level Data

While some work analyzing cross-national variation relies on sorting countries into a typology (for example, Esping-Anderson's (1990) three worlds of welfare capitalism), other work seeks to use country-level variables to model international variation. This study takes the latter approach, using a variety of country-level predictors to analyze variance in country-level gender differences in mathematics achievement. Analytically, this is similar to Fuwa's (2004) analysis of how macro-level gender inequality impacts national differences in the gendered division of household labor. However, where Fuwa uses the United Nations Development Programme's Gender Empowerment Measure (GEM) to examine how gender inequality writ large impacts the division of household labor, we are interested in examining what specific aspects of macro-level gender inequality are related to gender differences in mathematics achievement. Thus, while the GEM combines factors like the percentage of women in parliament and the income gap between men and women into one measure, we combine variables like these into five different measures tapping different aspects of macro-level gender inequality.

Data from the World Bank, the United Nations Statistics Department, the International Labour Organization, the World Values Survey, and the International Social Survey Programme are combined to create five variables.[6]

1. Education: measures gender differences in education broadly. This measure is comprised of variables such as differences in secondary enrollment and educational attainment. Higher values indicate that women are more advantaged (or less disadvantaged) in education.
2. Domesticity: measures the salience of children and home life considerations to women. This measure is comprised of variables such as the amount of time

---

[6]These variables were created using information from 1995 wherever possible, and from the nearest adjacent year when no information was available for 1995. Principle components were estimated using an eigen decomposition of the correlation matrix, and are unrotated.

women spend on housework and the importance of children to women's fulfillment. Higher values indicate that women are more likely to be viewed as strongly tied to the home and to children.

3. Status: measures differences between the general status of men and women. This measure is comprised of variables such as the percentage of women in politics and gender differences in the age at first marriage. Higher values indicate that women have higher status in that society.

4. Labor force equality (Equality): measures equality in labor force position. This measure is comprised of variables such as the percentage of managers who are female and women's rights to a job. Higher values indicate that women are more advantaged (or less disadvantaged) in the labor market.

5. Labor force representation (Representation): measures differences in the representation of men and women in the labor force. This measure is comprised of variables such as the percentage of women who are in the labor force and the percentage of the labor force that is female. Higher values indicate that women are more highly represented in the labor force.

By design, the factor scores are on a standardized metric; all have a mean of zero and standard deviations between 1.17 and 1.59. This similarity allows for straightforward comparisons between the coefficients of the various factors.[7] However, while it may be easy to grasp their importance vis-à-vis the other factors, factor scores can be somewhat obtuse. To ameliorate this shortcoming, Table 1 contains information about the scores of countries on the five factors used in this analysis, as well as GEM, providing a sense for how the measures used here compare to GEM. Table 1 also allows us to understand more concretely what it means to be one unit apart on each of the five dimensions. For example, a one unit difference in labor market representation roughly corresponds to the difference between the United States (1.12) and Norway (2.06).

4.3 Modeling Strategy and Precedents

While most research examining gender differences in achievement focuses on average differences, recent work has begun to argue for the importance of examining differences throughout the distribution, and there is some precedent for an extreme-oriented approach. Several studies of gender differences in achievement report not only gender differences in the means and variances, but also the ratios of girls to boys at the extremes of the distribution (e.g., Hedges and Nowell 1995; Nowell and Hedges 1998; Stumpf and Stanley 1996). More recent work on extreme achievement by Xie and Shauman (2003) begins modeling gender differences in representation at the distribution extremes using logistic regression.

---

[7]It is worth noting the that the underlying metrics of the variables that go into these factor scores vary, so that we cannot really compare the metrics. Rather, we can compare the effects of moving one standard deviation in the different distributions.

Table 1 Country-level factor scores and the United Nations Statistics Department's Gender Empowerment Measure (GEM)

	Education	Domesticity	Representation	Equality	Status	GEM
Australia	−.95	−.11	−.40	.84	.82	.664
Austria	−.76	.93	−.37	−1.01	.26	.686
Canada	−.31	−.19	.82	.38	.12	.720
Cyprus	−.33	−.13	−2.34	−1.40	−2.75	.379
Czech Republic	−.94	1.75	−.51	−.36	−1.54	.527
Denmark	1.58	.00	1.75	−1.44	.52	.739
France	−.28	−.42	−.04	.46	.64	.489
Germany	−1.56	.82	−.11	.43	.95	.694
Hungary	−.52	2.51	−1.28	.96	−1.84	.491
Iceland	.82	−1.18	2.54	−1.18	2.37	.723
Israel	.43	−2.27	−.96	.30	−1.57	.484
Italy	−.91	1.56	−3.76	−.99	−1.21	.521
Lithuania	.25	1.39	.67	2.05	−1.13	.517
The Netherlands	−1.64	−.79	−.38	−1.04	1.55	.689
New Zealand	1.58	−1.17	.88	.25	1.15	.725
Norway	.85	−1.42	2.06	−1.04	1.74	.790
Russian Federation	−.31	2.37	.04	2.38	−1.91	.426
Slovenia	.35	1.17	.10	.75	−1.78	.475
South Africa	.63	−2.72	−2.50	−.62	1.23	.531
Sweden	4.60	−.62	2.67	−1.55	2.88	.790
Switzerland	−2.55	.07	.00	−.09	.47	.654
United States	−.03	−1.53	1.12	1.93	−.97	.675
Overall						
Mean	.00	.00	.00	.00	.00	.632
Std. Dev.	1.44	1.43	1.59	1.17	1.56	.108

Note: The Gender Empowerment Measure statistics listed are from 1995 except for Lithuania (1999) and Russia (2000).
Source: The United Nations Statistics Department and authors' calculations based on World Bank, United Nations Statistics Department, International Labour Organization, World Values Survey, and International Social Survey Programme data. All calculations were done in STATA 10.

In addition to modeling differences in the composition of the extremes, it is also possible to measure differences between boys and girls at different percentiles using quantile regression. While quantile regression has a long history in economics, there is little work employing quantile regression to analyze differences in extreme educational achievement (but see Grodsky et al. 2009; Levin 2001; Penner and Paret 2008). Whereas the logistic regression approach examines the likelihood of being above and beneath various cutpoints in the distribution, quantile regression models estimate the size of the differences between boys and girls at various points in the distribution. Thus, in contrast to OLS regression, which provides information about differences in conditional means, quantile regression can provide information regarding conditional differences at the 90th (or any other) percentile.

National models also control for per capita gross national income (GNI), and standard errors for these models are adjusted for country-level clustering using a sandwich estimator. Before proceeding, a word of clarification is in order. Although model results are often used to make causal claims, this is not the emphasis here. Rather, the following cross-national models are better understood as stylized facts, a quantitative thick description designed to explore what kinds of factors covary with gender differences in mathematics achievement. The attempt here is thus not to make causal claims, but rather to examine what kinds of societies tend to have larger and smaller gender differences at the extremes of the distribution.

In order to examine which country-level predictors are related to gender differences in extreme mathematics and science achievement, we estimate quantile regression models including gender, the five country-level predictors, and the interaction of gender and the country-level predictors. The interaction effects reveal the relative impact of that particular factor on gender differences in mathematics and science at the national level.

5 Results

Much of the discussion surrounding genetic theories for gender differences in mathematical achievement revolves around basic descriptive statistics, especially differences in means and variances. We thus begin by revisiting these statistics in an international context, and then proceed to examine statistics that more accurately describe the tails of the mathematics achievement distribution. Finally, we present results from models of cross-national differences across the distribution of mathematics and science achievement

5.1 Basic Descriptive Statistics

The conventional understanding against which we will be framing our findings argues that: (1) Mean differences in mathematics always favor males, evincing a biological basis; and (2) Males have a greater variance in mathematical ability, again evincing a biological basis. These points are significant, as together they suggest that the gender gap is especially large at the top of the distribution.

Table 2 presents basic descriptive information from the countries in the TIMSS data, and lists countries in descending order of the magnitude of their effects size (mean difference in standard deviation units, column 9). Comparing the mean differences (column 7) across countries reveals that mean differences in all of the countries favor males, but also that there is a considerable amount of variation in the size of the mean differences that exist in different countries.[8] Previous research has

---

[8]Other analyses not reported establish that the variation across countries is statistically significant. It is also worth noting that a more recent study of advanced mathematics students in the final year of secondary school finds that in Lebanon girls outscore boys—though in Sweden, Russia, Slovenia, Iran, and the Philippines boys outscore girls, and the Netherlands, Italy, Norway, and Armenia have no statistically significant differences between boys and girls (Mullis et al. 2009).

Table 2 Descriptive statistics and gender differences in the different countries

Country	N	Proportion female	Overall mean	Overall SD	Male mean	Female mean	Dif in mean	Dif in Std. Dev.	Effects size	Male 90%tile	Female 90%tile
	1	2	3	4	5	6	7	8	9	10	11
The Netherlands	1470	.48	562	90	590	533	57*	−11.3	.63	687	649
Denmark	2604	.55	549	85	578	525	53*	.8	.62	685	632
Norway	2518	.49	529	94	557	501	56*	10.3	.59	685	615
Austria	1779	.62	515	75	542	500	42*	8.2	.57	638	593
Czech Republic	1899	.48	462	96	486	435	51*	10.5	.53	630	555
Iceland	1703	.52	535	87	560	514	46*	3.5	.53	671	620
France	1590	.53	525	78	546	506	40*	5.4	.51	645	605
Slovenia	1387	.50	505	80	526	488	38*	.7	.48	618	589
Sweden	2816	.51	551	94	573	530	43*	13.2	.45	699	640
Canada	4832	.53	519	88	537	503	34*	4.2	.39	662	614
Switzerland	2976	.44	541	86	555	522	33*	1.4	.38	667	627
Germany	2182	.44	493	91	508	476	32*	−4.7	.35	617	590
Russian Federation	2289	.62	470	85	487	459	28*	2.0	.33	599	575
Italy	1578	.54	472	84	487	460	27*	6.8	.33	593	561
Australia	1844	.58	522	97	539	510	29*	11.0	.30	670	621
New Zealand	1763	.51	523	98	537	509	29*	9.2	.29	673	627
Lithuania	2887	.65	468	84	484	460	23*	−5.3	.28	589	571
Israel	1045	.48	445	109	460	432	27*	3.3	.25	596	579
South Africa	2757	.51	351	81	361	343	18*	3.4	.22	483	435
Cyprus	473	.55	440	69	447	433	14*	8.3	.20	543	519
United States	5371	.50	460	91	466	454	12*	7.8	.13	599	574
Hungary	5091	.48	483	91	485	480	5	15.4	.05	623	591

Note: Countries are sorted in descending order by their effects size (column 9). For each country, column 1 reports the number of observations; column 2 reports the proportion of the sample that is female; column 3 reports the mean mathematics literacy score; column 4 reports the standard deviation of mathematics literacy; column 5 reports the mean male score; column 6 reports the mean female score; column 7 reports the gender difference (male-female) in the mean of mathematics literacy; column 8 reports the gender difference (male-female) in the standard deviation of mathematics literacy; column 9 reports the effects size (mean difference in standard deviation units); column 10 reports the 90th percentile male score; and column 11 reports the 90th percentile female score.
Source: Authors' calculations based on TIMSS 1995 Population 3 data. All calculations were done in STATA 10.
* $p < .05$.

argued that the prevalence of the male advantage implies that gender differences have an evolutionary origin (e.g., Geary 1996), a claim that assumes that biological factors and their effects are constant while sociological factors and their effects vary. However, by this reasoning, the considerable variation in the size of mean differences suggests that social factors are quite important. Mean differences can thus be used to advance both social and biological theories for the origin of gender differences.

Further, the variation in the size of the mean differences makes it unclear exactly how to interpret the fact that all of the means exhibit a male advantage. That is, it is impossible to tell whether: 1) Biological differences slightly favor boys and sociological factors range from neutrality to strongly favoring boys, or, 2) Biological factors are neutral and sociological factors range from slightly favoring boys to strongly favoring boys. Of course, it is also possible that 3) Biological factors give a sizeable advantage to girls, and it is only the strong social female disadvantage, which varies in size, that creates gender differences favoring boys. Likewise, another little discussed possibility is that 4) Biological factors severely disadvantage girls, and sociological factors actually affect the gender gap in favor of girls, mitigating the gap. While some of these appear more or less plausible given our understanding of current gender regimes (i.e. the social order generally privileges men), from the international comparisons the only conclusion that we can draw with certainty is that social factors have some effect on the gender gap, as evinced by the variation in the gender differences.

Table 2 also shows that the common assumption that males have greater variance in mathematics achievement is not universally true (column 8).[9] In three of the countries examined (Germany, Lithuania, and the Netherlands) women actually have greater variance than men, and among the countries that exhibit greater male variance the differences in variance range considerably in size. Thus, while claims about greater male variability are often used to support evolutionary arguments about gender differences in mathematics achievement (e.g., Geary 1998), these international findings call for caution.

We turn next to the question of representation in the distribution extremes explicitly. This is important because supplemental analyses (not shown) find that while country-level gender differences in the means and standard deviations are not significantly correlated with the proportion of women working in STEM fields, differences in achievement at the top of the distribution are.

5.2 Examining the Distribution Extremes

We begin by examining columns 10 and 11 of Table 2, which show the male and female 90th percentile scores. This allows us to compare the 90th percentile score

---

[9]It is worth mentioning that although males are typically assumed to be more variable in mathematical abilities, Feingold (1994), in a cross-national review, also finds that differences in variability for mathematical and spatial abilities differed across countries, with males being more variable in some countries and females being more variable in others.

of girls in one country to the 90th percentile score of boys in another. Thus, for example, we see that in Germany boys have a 90th percentile score of 617, and in Switzerland girls have a 90th percentile score of 627. If we assume that the biological limitations of girls are the same in Germany and Switzerland, this suggests that there is no biological reason why the 90th percentile girls in Germany could not score as high as the 90th percentile boys in Germany. Using this logic to examine gender differences in mathematics achievement in the United States, we find that 10 of the 21 other countries in this study, including Canada, have female 90th percentile scores that are above the United States male 90th percentile score (599). Further, in 5 of these countries girls score at least 24 points higher than US boys, so that when we compare girls from these countries to boys in the United States, we get a difference that is as big as the gender gap favoring boys in the US. It thus seems reasonable to conclude that gender differences between the 90th percentile boy and girl in the United States are not the result of US girls' biological limitations.

Table 3 builds on this by reporting the unadjusted gender differences at multiple points in the distribution, which allows us to see, for example, the difference between the scores of the 10th percentile boy and the 10th percentile girl. As it can be difficult to interpret the magnitude of differences in raw scores, the results presented here use logged mathematics score as their dependent variable. Logged score coefficients approximate percent differences, so that a gender coefficient of .01 indicates that girls' scores are roughly 1 percent higher than boys' scores at that same percentile. Table 3 reports results from five quantile regression models examining differences at the 10th, 25th, 50th, 75th, and 90th percentiles. As in Table 2, the countries in Table 3 are sorted by their effects size. We include OLS results as a point of comparison; comparing OLS and quantile regression results provides a sense of what is missed by examining only the mean of the distribution.

By way of example, the results in Table 3 indicate that in Hungary: (1) At the 10th and 25th percentiles gender differences are significant and favor girls (columns 2–3); (2) At the median gender differences are insignificant (column 4); and (3) At the 75th and 90th percentiles gender differences are significant and favor boys (columns 5–6). While the OLS coefficient in column 1 reports that the gender difference in Hungary is not statistically significant, the quantile regression results provide a more nuanced view, showing that the statistically significant female advantage at the bottom of the distribution and the statistically significant male advantage at the top of the distribution cancel each other out at the mean. In other countries the findings are less striking, but the differences across the distribution are equally important. In the United States, for example, the results indicate that boys score 2% higher than girls at the mean (column 1), but the difference at the 90th percentile is twice that, at 4% (column 6). Column 7 reports the p-value from an ANOVA test that the different quantiles have the same effects. This shows that of the 22 countries observed here, 10 have coefficients that vary significantly across the five points of the distribution observed, highlighting the importance of observing differences across the distribution, and not looking exclusively at the middle of the distribution.

Internationally, the results in Table 3 can be summarized by four broad patterns of gender differences across the distribution. First, in roughly half of the countries

Table 3 The effect of being female on logged mathematics score at different quantiles

	OLS	.10	.25	.50	.75	.90	p(ANOVA)
	1	2	3	4	5	6	7
The Netherlands	−.11*	−.16*	−.16*	−.11*	−.06*	−.06*	.00
Denmark	−.10*	−.11*	−.11*	−.11*	−.08*	−.08*	.00
Norway	−.10*	−.09*	−.11*	−.12*	−.11*	−.11*	.28
Austria	−.08*	−.06*	−.07*	−.11*	−.09*	−.07*	.01
Czech Republic	−.09*	−.12*	−.10*	−.10*	−.11*	−.13*	.49
Iceland	−.11*	−.08*	−.08*	−.10*	−.09*	−.08*	.36
France	−.08*	−.06*	−.08*	−.09*	−.08*	−.06*	.38
Slovenia	−.08*	−.08*	−.10*	−.09*	−.07*	−.05*	.17
Sweden	−.08*	−.05*	−.07*	−.08*	−.10*	−.09*	.03
Canada	−.07*	−.07*	−.06*	−.06*	−.08*	−.08*	.83
Switzerland	−.06*	−.07*	−.07*	−.06*	−.06*	−.06*	.81
Germany	−.07*	−.12*	−.07*	−.05*	−.06*	−.04*	.01
Russian Federation	−.06*	−.05*	−.04*	−.08*	−.06*	−.04*	.00
Italy	−.06*	−.05*	−.06*	−.05*	−.06*	−.06*	.91
Australia	−.05*	−.04	−.03	−.07*	−.07*	−.08*	.23
New Zealand	−.05*	−.01	−.05*	−.07*	−.05*	−.07*	.20
Lithuania	−.05*	−.11*	−.07*	−.04*	−.03*	−.03*	.00
Israel	−.07*	−.08*	−.08*	−.05*	−.06*	−.03	.29
South Africa	−.05*	−.03*	−.02*	−.04*	−.10*	−.10*	.00
Cyprus	−.03	−.03	−.01	−.04	−.04	−.05*	.53
United States	−.02*	.01	.01	−.02	−.04*	−.04*	.05
Hungary	−.00	.03*	.02*	.00	−.03*	−.05*	.00

Note: Countries are sorted in descending order by their effects size (mean difference in standard deviation units). For each country, column 1 reports the effect of being female on logged mathematics literacy score from an OLS regression, and columns 2–6 report the effect of being female on logged mathematics literacy score from quantile regression models estimated at the 10th percentile (column 2), 25th percentile (column 3), the 50th percentile (column 4), the 75th percentile (column 5), and the 90th percentile (column 6). Column 7 reports the p-value from an ANOVA test that the coefficients at the 10th, 25th, 50th, 75th, and 90th percentiles are equivalent.
Source: Authors' calculations based on TIMSS 1995 Population 3 data. All calculations were done in R.
* $p < .05$.

the male advantage is the same across the distribution. This pattern is exemplified by Switzerland, and in these countries the OLS results provide an accurate summary of gender differences. Second, in a few countries like the Netherlands and Lithuania, gender differences are found to be larger at the lower extreme of the distribution. Third, in several countries differences are larger at the top extreme. In some of these countries, like Sweden, the boys do better throughout the distribution but this advantage increases at the top, while in Hungary and the United States girls are found to do at least as well or better than boys at the bottom of the distribution but worse at the top. Finally, it is interesting to note that in Russia and Austria gender differences are actually larger in the middle of the distribution than at either extreme. That cross-national gender differences vary not only in magnitude but also in the pattern that they take across the distribution suggests that social factors have a substantial influence on gender differences in mathematics achievement.

5.3 Modeling Cross-National Gender Differences Across the Distribution

5.3.1 Mathematics Achievement

Where Table 3 documents that gender differences vary across countries, we now expand on this by modeling how country-level factors are related to gender differences at different points in the distribution. To do this we include an individual level dummy variable for being female, which is interacted with country-level predictors. By looking at the strength of the interaction effects we are able to see which of the country-level factors are most strongly associated with gender differences in mathematics achievement.

Table 4 presents results from quantile regression models of mathematics achievement estimated at various quantiles throughout the distribution. Each column reports the results for a model estimated at a different point in the distribution, and all models contain the same independent variables.

The interaction effects provide a sense of how relevant the various country-level factors are to gender differences. Looking at the interaction effects we find that education and domesticity impact gender differences only at the bottom of the distribution, status has an effect only at the top, and labor force representation and equality are related to gender differences across the distribution. Thus, at the bottom of the distribution there is no effect of status, but all of the other factors are pertinent (though the effect of domesticity is relatively small), while at the top of the distribution labor force factors and general status are germane, and in the middle of the distribution only labor force related factors seem to be relevant.

Looking at the direction of the interaction effects, we find that the effects are generally in the expected direction. Countries with smaller female disadvantages in mathematics achievement tend to be countries where: 1) women are less associated with home and children; 2) there is greater educational gender equality; 3) there is little gender inequality in the labor force; and 4) women have higher general status. Interestingly, the effect of the representation of women in the labor force is such that countries with greater female representation tend to have larger gender differences in mathematics achievement. While this is somewhat counterintuitive, it seems plausible that countries with greater female representation in the labor force offer more opportunities for women in female-dominated non-technical sectors.[10]

How should we interpret the overall findings from Table 4? The first and most obvious point is that labor force considerations (representation and equality) are closely related to gender differences in mathematics. Net of other factors, higher levels of female labor force participation tend to exacerbate the female disadvantage in mathematics, while higher levels of gender equality tend to play a mitigating

---

[10]Other analyses (not shown) find that in this sample countries with greater female representation also tend to have higher degrees of occupational gender segregation in the labor market. It is also worth noting this counterintuitive effect of labor market representation is net of the other factors in the model, and that the sample of countries includes only European and other western countries.

Table 4 Quantile regression models reporting the effects of country-level predictors on logged mathematics score at various percentiles, net of per capita GNI

	.05	.10	.25	.50	.75	.90	.95
Female	−.058	−.048	−.057	−.069	−.073	−.069	−.068
	(22.79)	(11.32)	(9.27)	(8.84)	(13.97)	(23.30)	(40.48)
Education	−.018	−.015	−.012	−.009	−.007	−.006	−.004
	(6.52)	(2.75)	(1.11)	(.66)	(.74)	(1.20)	(1.37)
Domesticity	.052	.053	.055	.049	.027	.013	.008
	(17.81)	(9.40)	(3.65)	(2.08)	(1.33)	(1.07)	(1.09)
Representation	.044	.045	.045	.044	.031	.025	.016
	(15.55)	(7.58)	(3.42)	(2.52)	(2.08)	(2.75)	(2.89)
Equality	−.037	−.039	−.033	−.032	−.027	−.024	−.022
	(8.25)	(4.26)	(1.94)	(1.47)	(1.66)	(2.66)	(3.90)
Status	.019	.014	.013	.005	.007	.006	.007
	(4.74)	(1.75)	(.77)	(.19)	(.31)	(.50)	(.92)
Fem X Educ	.014	.009	.007	.001	−.003	.000	−.001
	(8.17)	(3.25)	(1.80)	(.24)	(.98)	(.14)	(.67)
Fem X Domestic	−.006	−.004	.004	.006	.007	.001	.003
	(2.17)	(.76)	(.52)	(.69)	(1.29)	(.34)	(1.82)
Fem X Rep	−.015	−.013	−.015	−.008	−.002	−.007	−.005
	(8.36)	(4.23)	(3.95)	(2.05)	(.63)	(3.02)	(2.52)
Fem X Equality	.020	.025	.026	.024	.020	.016	.013
	(7.79)	(4.76)	(3.61)	(2.44)	(3.02)	(4.29)	(5.69)
Fem X Status	−.001	.004	.009	.010	.010	.010	.010
	(.30)	(.83)	(1.14)	(.92)	(1.31)	(2.32)	(3.86)
Constant	5.820	5.880	5.987	6.134	6.315	6.446	6.504
	(652.95)	(308.21)	(140.73)	(97.07)	(123.03)	(224.66)	(386.26)
Observations	54,965	54,965	54,965	54,965	54,965	54,965	54,965

Absolute value of t statistics are in parentheses.

Note: Each column reports the results from a model at a different quantile. Female is a dummy variable for being female, and education, domesticity, representation, equality, and status are national level factor scores. The coefficients for the interactions between female and the national level factors are reported as Fem X (the respective factors). Per capita GNI is included as a control variable.

Source: Authors' calculations based on TIMSS 1995 Population 3 data. All calculations were done in STATA 10.

role. The effects of these two factors are visible throughout the distribution. Second, gender equity in education writ large seems to matter most for gender differences in mathematics achievement at the bottom of the distribution. Third, the effect of status is found to matter only at the top of the distribution. Finally, differences in ideologies concerning the importance of home and children for women do not track well with cross-national variation in gender differences in mathematics achievement.

Thinking more specifically about mathematics achievement at the top of the distribution, these results are interesting both for what factors they suggest are important, and also for what factors they suggest are not important. For example, finding that gender differences at the top of the mathematics distribution are associated with labor market factors, but not with broader educational equality, suggests that gender differences in mathematics achievement are more closely linked to the labor market than to these broader educational considerations. Similarly, the finding that differences in the relative status of men and women are linked to gender differences at the top end of the mathematics distribution, but that ideologies regarding women and children are not, suggests that gender differences in status impact mathematics achievement at the top of the distribution more than ideologies about the importance of children to women. The finding that labor market factors matter suggests that students make decisions about their education with some kind of awareness about the types of labor market opportunities available to them, and the findings for status suggest that national level gender differences in status play a role in creating the framework within which status characteristics effects appear. Taken together, the results for labor market factors and the relative status of women in a society highlight how national contexts define the realm of possibilities for boys and girls in high school.

5.3.2 Science Achievement

In thinking about how generalizable the results presented above are, it is useful to compare the results for mathematics achievement to those found in another domain in which similar processes should be at work. Table 5 presents results from models of science achievement that are analogous to the mathematics achievement results presented in Table 4, showing that the findings for science are in many respects similar to those for mathematics.

As with the mathematics results presented in Table 4, the results in Table 5 show that labor market factors (particularly equality) and status matter at the top of the distribution. However, there are also some interesting differences between the mathematics results presented in Table 4 and the science results presented in Table 5. Notably, we see that the factors that impact the bottom of the distribution in mathematics do not appear to matter for science. This suggests that gender differences at the top of the distribution in mathematics and science are driven by the same processes, but that the factors that matter for gender differences at the bottom of the mathematics distribution do not seem to affect differences in science.

It is also worth noting that the main effect of being female is slightly larger in science than in mathematics, as are the significant interaction effects at the top of the distribution. The larger main effect of gender indicates that gender differences in countries with average scores on the five indices are larger in science than in mathematics, and the interaction effects suggest that the magnitude of differences in science varies more as a function of these five indices than mathematics differences do. Thus, for example, if we compare the gender differences in science achievement that we would expect in two hypothetical countries that were at the mean for all of the indices except for equality, we would expect that a country that was one

Table 5 Quantile regression models reporting the effects of country-level predictors on logged science score at various percentiles, net of per capita GNI

	.05	.10	.25	.50	.75	.90	.95
Female	−.077	−.081	−.076	−.082	−.087	−.089	−.084
	(26.63)	(20.64)	(11.86)	(11.98)	(18.18)	(27.80)	(38.37)
Education	−.017	−.020	−.014	−.012	−.007	−.004	−.004
	(5.19)	(3.46)	(1.27)	(.84)	(.68)	(.61)	(1.18)
Domesticity	.064	.056	.052	.033	.013	.005	.000
	(20.05)	(7.70)	(2.91)	(1.23)	(.63)	(.39)	(.01)
Representation	.070	.070	.060	.048	.030	.024	.021
	(15.75)	(8.45)	(3.51)	(2.09)	(1.83)	(2.50)	(3.35)
Equality	−.023	−.028	−.024	−.021	−.019	−.022	−.022
	(4.24)	(3.07)	(1.47)	(1.08)	(1.20)	(2.02)	(3.08)
Status	.017	.011	.007	.006	.005	.000	.000
	(3.31)	(1.20)	(.40)	(.23)	(.27)	(.01)	(.04)
Fem X Educ	.000	.002	−.006	−.004	−.006	−.006	−.003
	(.06)	(.63)	(.90)	(.70)	(1.57)	(2.62)	(1.81)
Fem X Domestic	−.006	−.001	−.001	.005	.004	.001	−.002
	(1.84)	(.33)	(.08)	(.66)	(.96)	(.35)	(1.32)
Fem X Rep	.001	.001	.002	.002	.000	−.005	−.010
	(.39)	(.31)	(.37)	(.25)	(.03)	(1.96)	(5.07)
Fem X Equality	−.004	.004	.010	.013	.013	.017	.019
	(1.08)	(.91)	(1.18)	(1.41)	(2.43)	(3.62)	(5.98)
Fem X Status	−.002	.000	.005	.006	.009	.014	.017
	(.53)	(.07)	(.65)	(.61)	(1.91)	(3.21)	(5.33)
Constant	5.847	5.925	6.023	6.173	6.338	6.452	6.514
	(522.99)	(254.37)	(109.99)	(77.33)	(108.06)	(208.10)	(360.62)
Observations	54,965	54,965	54,965	54,965	54,965	54,965	54,965

Absolute value of t statistics are in parentheses.
Note: Each column reports the results from a model at a different quantile. Female is a dummy variable for being female, and education, domesticity, representation, equality, and status are national level factor scores. The coefficients for the interactions between female and the national level factors are reported as Fem X (the respective factors). Per capita GNI is included as a control variable.
Source: Authors' calculations based on TIMSS 1995 Population 3 data. All calculations were done in STATA 10.

standard deviation above the mean on equality would have a gender difference of 6 percent at the 95 percentile, whereas we would expect that a country that was one standard deviation below the mean on equality would have a difference of 11 percent. That is, in the egalitarian country we would expect the 95th percentile girl to score 6 percent lower than the 95th percentile boy, and in the non-egalitarian country we would expect the 95th percentile girl to score 11 percent lower than the 95th percentile boy. By contrast, if we were to look at mathematics achievement for these same hypothetical countries, we would expect that instead of differences of 6 and 11 percent, we would find differences of 5 and 8 percent respectively, so that differences in science are both higher overall and more closely linked with the social factors we examine.

The results from science reported in Table 5 are perhaps most interesting because they show that social factors have the most impact at the top of the distribution. It is worth considering the implications of these findings for biological and social arguments about gender differences in mathematics and science achievement. While these findings do not necessarily imply that biological factors matter less at the top of the distribution, they do suggest that social factors are particularly important for gender differences in these domains at the upper extreme of the distribution. This runs counter to notions that at the top of the distribution social factors matter less, and that differences among high achievers are primarily a function of biological abilities. Again, it is worth noting that these findings do not rule out that biological factors may be at play, but they do indicate that to the degree that macro-social factors matter for mathematics and science achievement, they matter for high achievers.

6 Discussion

Our key findings can be summarized as follows. First, we find that there is considerable variation in the magnitude of gender differences in mathematics achievement across countries, which we argue is evidence that social factors are important in producing these differences. That is, arguments for a biological underpinning for gender differences in mathematics based on an international male advantage (e.g., Geary 1996) are not particularly useful in accounting for the variation we observe in international gender differences. Second, we find that variation in mathematics achievement is not universally higher among men. This has important implications for many of the biological theories, which posit that men are more variable then women. In particular, this finding contradicts what we would expect from theories of a sex-linked recessive gene or condition dependent variability, as these theories are not well-suited to explaining why France exhibits greater male variability while Germany has greater female variability.[11] Third, women in many countries have considerably higher mathematics achievement than men in the United States, suggesting that the difference between US men and women is not being driven by the biological limitations of US women. Fourth, macro-level contextual factors related to gender differences in the labor market and the overall status of women are associated with gender differences at the top of the achievement distribution in mathematics and science. We also find that not all aspects of national gender equity are related

---

[11] In supplementary analyses (not shown) we use simulations to examine how often one would expect to find countries with greater female variability if students were randomly assigned to countries. This was done to address concerns that there might be a relatively high likelihood of observing greater female variation among any given subgroup in the data. We stopped the simulations when, after over 50,000 simulations, we had still failed to find even one country in which boys had a lower variance than girls. Given that our data contain three such countries, we believe that this is unlikely to have arisen due to chance.

to gender differences in mathematics and science achievement, and in particular, domestic issues seem largely orthogonal to the gender differences between boys and girls at the top of the distribution. These findings highlight the important, and sometimes unexpected, ways in which national contexts shape the realm of possibilities for high achieving girls and boys.

While the TIMSS 1995 data are the most recent data containing information on students in their final year of high school that are not restricted to students of advanced mathematics, they are now 15 years old, so that it is worth considering how things might have changed since then. As trends in the United States suggest that women's empowerment writ large is increasing (United Nations Development Programme 2008), to the degree that gender differences in mathematics and science achievement are less pronounced in contexts where women experience higher levels of empowerment, there is reason to be optimistic for the reduction of gender differences in achievement in the United States. However, while the overall empowerment of women has increased in the United States, changes across domains have been rather uneven, and there is a sense that progress towards gender equality has stalled in many areas (Hochschild and Machung 1989). Thus, given the domain specific nature of our findings, and the fact that considerations like the relative position of women in the labor market have largely remained constant since 1990's (Cohen et al. 2009), it is perhaps not surprising that estimates of the gender gap in mathematics achievement from NAEP suggest that very little has changed for US 12th graders between 1996 and 2008.

Likewise, women continue to be under-represented in employment in STEM fields, making up only 26 percent of employed scientists and engineers in the United States in 2006 (National Science Foundation 2009). This disparity in employment may contribute to gender differences in mathematics and science achievement. While our analysis does not specifically look at the representation of women in STEM professions, perceptions of equality in the workplace are likely to be affected by gender differences in STEM employment. In addition, while our findings suggest that broader national context plays an important role in shaping the pool of potential scientists by influencing gender differences at the top of the achievement distribution, it seems plausible that national contexts might also affect the likelihood of women being choosing to pursue careers in STEM fields, even net of their achievement (see, e.g., Ceci and Williams 2010a).

Research from the United States suggests that this is this case, showing that contextual factors play an important role in creating gender differences in interest in mathematics net of achievement (Correll 2001). We might also expect that this pattern is the result not only of the opportunities available to men and women within STEM fields, but also of the opportunities that they have outside of these fields. Indeed, given that Webb et al. (2002) note that men who excel in mathematics often have poor verbal skills, whereas women who excel in mathematics tend to also have strong verbal skills, we might speculate that these findings are due in part to a rational effort allocation by women away from STEM fields.

Finally, given the range of skills involved in mathematics achievement, it is also instructive to consider the implications of our findings for research on various domains within mathematics. Research suggests that there is considerable heterogeneity in the gender differences found in different kinds of tasks within mathematics, such that girls do better than boys in some areas, but worse in others (e.g., Gibbs 2010). As the first order effects of national context are presumably constant across the various domains within mathematics, this variation in gender differences between domains is difficult to reconcile with our findings. One possible explanation comes from research on gender differences in problem-solving strategies: Fennema et al. (1998) note that men tend to use more exploratory strategies, while women stick to the strategies that they have been taught in class, and Spelke (2005) notes that when men and women are confined to the same strategy, gender differences disappear, suggesting that differences are largely the result of the different strategies used to solve the problems. To the degree that different national contexts encourage more exploration by men (or women) in creating and selecting problem-solving strategies, we might expect that these national contexts can account for the gender differences that we observe not just in international achievement, but also in problem-solving strategies and sub-domains within mathematics.

Hanna (1989) suggested another explanation for how differences across subdomains might interact with international context. Hanna uses data on 13 year-olds from the Second International Mathematics Study to show that girls do as well as boys in the mathematical subdomains of arithmetic, algebra, and statistics, and that the domains that favor boys (measurement and geometry) are those that get little coverage in class. She thus posits that gender differences in these areas arise due to experiences outside of the classroom. Thus, we might expect that macro-level social considerations are important not primarily because of their influence on what happens inside of the classroom, but for how they shape the universe of possibilities and experiences outside of the classroom. This is congruent with our finding that large-scale gender equality in education matters less for differences at the top of the distribution than labor market inequality.

In sum, by showing how five different axes of gender inequality at the national level affect gender differences in mathematics and science achievement, our findings begin to provide some ideas about the impact of sociocultural factors, particularly at the top of the distribution. We believe that international variation provides a unique opportunity to assess the role of broad contextual and sociocultural considerations, and that the insights gained by this analysis represent a starting point for understanding the role of these factors. We hope that future research will expand on the findings presented here by considering not just how macro-level factors shape mathematics and science achievement broadly, but also how the effects of macro-level factors vary across subdomains within mathematics. We also hope that future research will begin to link macro-level sociocultural factors with the micro-level processes through which they are transmitted, such as the strategies employed by teachers and students, as well as out of school experiences.

Acknowledgements Portions of this chapter previously appeared in Penner (2008). We are grateful to Jim Dietz and to the Irvine Comparative Sociology Workshop at the University of California, Irvine for useful comments and discussions.

Appendix

The analyses presented in this study focus on gender differences in the final year of secondary school, allowing us to compare the degree to which national education systems vary in the gender differences that they produce at this key juncture. However, it is also informative to examine how these gender differences diverge from those observed among students earlier in their educational careers. Table A.1 thus augments the results reported above by presenting information on gender differences for 4th, 8th, and 12th grade students for the 13 countries with information on students at all three grade levels. For each grade level we present information on the mean gender difference (OLS), as well as gender differences at the 10th and 90th percentiles, and the p-value from an ANOVA test examining whether the gender differences at the 10th and 90th percentiles are different. Given that the students examined are at the most around eight years apart, we adopt a synthetic cohort approach in discussing these results.

We see that there are substantial differences in both the number of significant coefficients and the magnitude of the coefficients as students progress through school. Looking first at gender differences at the bottom of the distribution, we see that in 4th grade the 10th percentile score of boys was higher than the 10th percentile score of girls in 4 of the countries, and lower in one country (New Zealand). By 8th grade there were no countries where boys had higher 10th percentile scores than girls, and in 2 countries girls had higher 10th percentile scores than boys (Australia and Cyprus). In contrast, by 12th grade, boys had higher 10th percentile scores than girls in 8 of the countries, while girls had higher 10th percentile scores in one country (Hungary). These findings highlight the temporal variability of the results, and suggest that differences at the bottom of the distribution emerge between 8th and 12th grade.

By contrast, differences at the top of the distribution emerge between the 4th and 8th grades—in 4th grade five countries had significant differences favoring boys, by 8th grade there were significant differences favoring boys in 10 countries, and by 12th grade 12 of the 13 countries had gender differences favoring boys at the 90th percentile. However, there were still profound changes in the magnitude of the coefficients between the 8th and 12th grades. Examining the 8th grade results, we see that only one country (Israel) had a male advantage of more than .03 (roughly 3 percent), while by 12th grade, all 12 of the countries with significant differences had male advantages of more than .03. We believe that differences between the results for 4th, 8th, and 12th grade students are important not only for what they reveal about how gender differences emerge over time, but also because they underscore the importance of selecting the appropriate grade level for making international comparisons. In this study, given our interest in comparing the gender differences produced by different educational systems, we believe that it is important to examine students in the final year of secondary school.

Table A.1 The effect of being female on logged mathematics score at the mean, the 10th percentile, and the 90th percentiles for 4th, 8th, and 12th grade students

Country	4th grade				8th grade				12th grade			
	OLS	.10	.90	p(ANOVA)	OLS	.10	.90	p(ANOVA)	OLS	.10	.90	p(ANOVA)
	1	2	3	4	5	6	7	8	9	10	11	12
The Netherlands	-.02*	-.02*	-.02*	.60	-.01*	-.01	-.02*	.57	-.11*	-.16*	-.06*	.00
Norway	-.03*	-.04*	-.02*	.11	-.01	-.02	-.02*	.48	-.10*	-.09*	-.11*	.28
Austria	-.02*	-.01	-.03*	.10	-.01	.00	-.02*	.03	-.08*	-.06*	-.07*	.01
Czech Republic	-.01*	-.01	-.01	.70	-.02*	-.01	-.02*	.32	-.09*	-.12*	-.13*	.49
Iceland	-.02*	-.03*	-.02	.62	.00	.01	-.01	.01	-.11*	-.08*	-.08*	.36
Slovenia	-.01	-.01	.00	.64	-.01*	-.01	-.02*	.41	-.08*	-.08*	-.05*	.17
Canada	-.02*	-.01	-.02*	.58	.00	.00	.00	.53	-.07*	-.07*	-.08*	.83
Australia	-.01	.01	-.01	.10	.01*	.04*	.00	.00	-.05*	-.04	-.08*	.23
New Zealand	.02*	.08*	-.01	.00	-.01	.02	-.02*	.00	-.05*	-.01	-.07*	.20
Israel	-.02*	-.03*	.00	.05	-.06*	-.04	-.05*	.83	-.07*	-.08*	-.03	.29
Cyprus	-.01*	.01	-.02*	.02	.00	.02*	-.02*	.00	-.03	-.03	-.05*	.53
United States	.00	.00	.00	.53	-.01	.01	-.02*	.01	-.02*	.01	-.04*	.05
Hungary	-.01	.00	-.01	.32	.00	.01	-.01*	.07	-.00	.03*	-.05*	.00

Note: Countries are sorted in descending order by their 12th grade effects size (mean difference in standard deviation units). For each country, column 1 reports the effect of being female on logged mathematics score for 4th graders from an OLS regression, and columns 2 and 3 report the effect of being female on logged mathematics score for 4th graders from quantile regression models estimated at the 10th percentile (column 2) and the 90th percentile (column 3). Column 4 reports the p-value from an ANOVA test that the coefficients at the 10th and 90th percentiles for 4th graders are equivalent. Columns 5–8 mirror columns 1–4, but examine 8th grade mathematics achievement, and columns 9–12 examine 12th grade mathematics achievement.

Source: Authors' calculations based on TIMSS 1995 Population 1, 2, and 3 data. All calculations were done in R.

* $p < .05$.

References

Baron-Cohen, S. (2003). *The essential difference*. New York: Basic Books.

Benbow, C. P. (1988). Sex-differences in mathematical reasoning ability in intellectually talented preadolescents: Their nature, effects, and possible causes. *Behavioral and Brain Sciences, 11*(2), 169–183.

Blum, D. (1997). *Sex on the brain: The biological differences between men and women*. New York: Viking.

Bock, R. D., & Kolakowski, D. (1973). Further evidence of sex-linked major-gene influence on human spatial visualizing activity. *American Journal of Human Genetics, 25*, 1–14.

Boles, D. B. (1980). X-linkage of spatial ability: A critical review. *Child Development, 51*, 625–635.

Bracey, G. W. (2000). The TIMSS final year study and report: A critique. *Educational Researcher, 29*, 4–10.

Bryden, M. P. (1986). Dichotic listening performance, cognitive ability, and cerebral organization. *Canadian Journal of Psychology, 40*, 445–456

Casey, M. B., Nuttall, R. L., & Pezaris, E. (1997). Mediators of gender differences in mathematics college entrance test scores: A comparison of spatial skills with internalized beliefs and anxieties. *Developmental Psychology, 33*, 669–680.

Ceci, S. J., & Williams, W. M. (2010a). Sex differences in math-intensive fields. *Current Directions in Psychological Science, 19*(5), 275–279.

Ceci, S. J. & Williams, W. M. (2010b). *The mathematics of sex: How biology and society conspire to limit talented women and girls*. Oxford: Oxford University Press.

Charles, M., & Bradley, K. (2009). Indulging our gendered selves? Sex segregation by field of study in 44 countries. *American Journal of Sociology, 114*, 924–976.

Cohen, P. N., Huffman, M. L., & Knauer, S. (2009). Stalled progress?: Gender segregation and wage inequality among managers, 1980–2000. *Work and Occupations, 36*(4), 318–342.

Correll, S. J. (2001). Gender and the career choice process: The role of biased self-assessments. *American Journal of Sociology, 106*, 1691–1730.

Correll, S. J. (2004). Constraints into preferences: Gender, status, and emerging career aspirations. *American Sociological Review, 69*, 93–113.

Dar-Nimrod, I., & Heine, S. J. (2006). Exposure to scientific theories affects women's math performance. *Science, 314*, 435.

Davies, P. G., Spencer, S. J., Quinn, D. M., & Gerhardstein, R. (2002). Consuming images: How television commercials that elicit stereotype threat can restrain women academically and professionally. *Personality and Social Psychology Bulletin, 28*, 1615–1628.

DeFries, J., Vandenberg, S., & McClearn, G. (1976). Genetics of specific cognitive abilities. *Annual Review of Genetics, 10*, 179–207.

Dyson, F. (2007). Our biotech future. *New York Review of Books, 54*(12), 4–8.

Esping-Anderson, G. (1990). *The three worlds of welfare capitalism*. Princeton: Princeton University Press.

Farkas, G. (2003). Cognitive skills and noncognitive traits and behaviors in stratification processes. *Annual Review of Sociology, 29*, 541–562.

Favreau, O. E., & Everett, J. C. (1996). A tale of two tails. *American Psychologist, 51*(3), 268–269.

Feingold, A. (1994). Gender differences in variability in intellectual abilities: A cross-cultural perspective. *Sex Roles, 30*(1–2), 81–92.

Fennema, E., Carpenter, T. P., Jacobs, V. R., Franke, M. L., & Levi, L. W. (1998). A longitudinal study of gender differences in young children's mathematical thinking. *Educational Researcher, 27*(5), 6–11.

Frank, K. A., Muller, C., Schiller, K., Crosnoe, R., Riegle-Crumb, C., & Mueller, A. S. (2008). The social dynamics of mathematics coursetaking in high school. *American Journal of Sociology, 113*, 1645–1696.

Fuwa, M. (2004). Macro-level gender inequality and the division of household labor in 22 countries. *American Sociological Review, 69*, 751–767.

Gazzaniga, M. S., Ivry, R. B., & Magnum, G. R. (1998). *Cognitive neuroscience: The biology of the mind*. New York: Norton.

Geary, D. C. (1996). Sexual selection and sex differences in mathematical abilities. *Behavioral and Brain Sciences*, *19*, 229–284.

Geary, D. C. (1998). *Male, female: The evolution of human sex differences*. Washington: American Psychological Association.

Geschwind, N., & Galaburda, A. M. (1987). *Cerebral lateralization: Biological mechanisms, associations, and pathology*. Cambridge: MIT Press.

Gibbs, B. (2010). Reversing fortunes or content change? Gender gaps in math-related skill throughout childhood. *Social Science Research*, *39*, 540–569.

Gill, H. S., & O'Boyle, M. W. (1997). Sex differences in matching circles and arcs: A preliminary EEG investigation. *Laterality*, *2*, 33–48.

Grodsky, E., Warren, J. R., & Kalogrides, D. (2009). State high school exit examinations and NAEP long-term trends in reading and mathematics, 1971–2004. *Educational Policy*, *23*, 589–614.

Halpern, D. F. (2000). *Sex differences in cognitive abilities*. Mahwah: Erlbaum.

Hampson, E. (1990). Estrogen-related variations in human spatial and articulatory motor skills. *Psychoneuroendocrinology*, *15*, 97–111.

Hampson, E., & Altmann, D. (1998). Spatial reasoning in children with congenital adrenal hyperplasia due to 21-hydroxylase deficiency. *Developmental Neuropsychology*, *14*, 299–320.

Handcock, M. S., & Morris, M. (1999). *Relative distribution methods in the social sciences*. New York: Springer.

Hanna, G. (1989). Mathematics achievement of girls and boys in grade eight: Results from twenty countries. *Educational Studies in Mathematics*, *20*, 225–232.

Hedges, L. V., & Nowell, A. (1995). Sex-differences in mental test-scores, variability, and numbers of high-scoring individuals. *Science*, *269*(5220), 41–45.

Hochschild, A. R., & Machung, A. (1989). *The second shift*. New York: Avon.

Inzlicht, M., & Ben-Zeev, T. (2000). A threatening environment: Why females are susceptible to experiencing problem-solving deficits in the presence of males. *Psychological Science*, *11*, 365–371.

Kimura, D. (1999). *Sex and cognition*. Cambridge: MIT Press.

Kimura, D., & Hampson, E. (1994). Cognitive pattern in men and women is influenced by fluctuations in sex hormones. *Current Directions in Psychological Science*, *3*, 57–61.

Levin, J. (2001). For whom the reductions count: A quantile regression analysis of class size and peer effects on scholastic achievement. *Empirical Economics*, *26*, 221–246.

Levy, J. (1974). *Hemisphere function in the human brain: Psychobiological implications of bilateral asymmetry*. New York: Wiley.

Lovaglia, M. J., Lucas, J. W., Houser, J. A., Thye, S. R., & Markovsky, B. (1998). Status processes and mental ability test scores. *American Journal of Sociology*, *104*, 195–228.

Ma, X. (2001). Participation in advanced mathematics: Do expectation and influence of students, peers, teachers, and parents matter? *Contemporary Educational Psychology*, *26*, 132–146.

Martin, M. O., & Kelly, D. L. (1996). *Third international mathematics and science study technical report volume 1: Design and development*. Boston College, Chestnut Hill: Center for the Study of Testing, Evaluation, and Educational Policy.

McClearn, G. E., Johansson, B., Berg, S., Pedersen, N. L., Ahern, F., Petrill, S. A., & Plomin, R. (1997). Substantial genetic influence on cognitive abilities in twins 80 or more years old. *Science*, *276*, 1560–1563.

Moffat, S. D., & Hampson, E. (1996). A curvilinear relationship between testosterone and spatial cognition in humans: Possible influence of hand preference. *Psychoneuroendocrinology*, *21*, 323–337.

Muller, C. (1998). Gender differences in parental involvement and adolescents' mathematics achievement. *Sociology of Education*, *71*, 336–356.

Mullis, I., Martin, M., Robitaille, D., & Foy, P. (2009). *TIMSS advanced 2008 international report: Findings from IEA's study of achievement in advanced mathematics and physics in the final year of secondary school*. Chestnut Hill: TIMSS and PIRLS International Study Center.

Mullis, I. V., Martin, M. O., Fierros, E. G., Goldberg, A. L., and Stemler, S. E. (2000). *Gender differences in achievement: IEA's third international mathematics and science study.* Chestnut Hill: TIMSS and PIRLS International Study Center, International Study Center, Lynch School of Education.

National Science Foundation (2009). *Women, minorities, and persons with disabilities in science and engineering: NSF 09-305.* Available at http://www.nsf.gov/statistics/wmpd/.

Nowell, A., & Hedges, L. V. (1998). Trends in gender differences in academic achievement from 1960 to 1994: An analysis of differences in mean, variance, and extreme scores. *Sex Roles, 39*(1–2), 21–43.

Penner, A. M. (2008). Gender differences in extreme mathematical achievement: An international perspective on biological and social factors. *American Journal of Sociology, 114,* S138–S170.

Penner, A. M., & Paret, M. (2008). Gender differences in mathematical achievement: Exploring the early grades and the extremes. *Social Science Research, 37,* 239–253.

Petrill, S. A. (1997). Molarity versus modularity of cognitive functioning? A behavioral genetic perspective. *Current Directions in Psychological Science, 6,* 96–99.

Plomin, R., & Craig, I. (2001). Genetics, environment and cognitive abilities: Review and work in progress towards a genome scan for quantitative trait locus associations using DNA pooling. *The British Journal of Psychiatry, 178,* s41–s48.

Plomin, R., Fulker, D. W., Corley, R., & DeFries, J. C. (1997). Nature, nurture, and cognitive development from 1 to 16 years: A parent-offspring adoption study. *Psychological Science, 8,* 442–447.

Resnick, S. M., & Bouchard, T. J. (1986). Early hormonal influences on cognitive functioning in congenital adrenal hyperplasia. *Developmental Psychology, 22,* 191–198.

Riegle-Crumb, C. (2005). The cross-national context of the gender gap in math and science. In L. Hedges & B. Schneider (Eds.), *The social organization of schooling* (pp. 227–243). New York: Russell Sage Foundation.

Spelke, E. S. (2005). Sex differences in intrinsic aptitude for mathematics and science? A critical review. *American Psychologist, 60*(9), 950–958.

Spencer, S. J., Steele, C. M., & Quinn, D. M. (1999). Stereotype threat and women's math performance. *Journal of Experimental Social Psychology, 35,* 4–28.

Stumpf, H., & Stanley, J. C. (1996). Gender-related differences on the college boards advanced placement and achievement tests, 1982–1992. *Journal of Educational Psychology, 88,* 353–364.

Summers, L. H. (2005). *Remarks at NBER conference on diversifying the science and engineering workforce.* Retrieved June 15, 2005, from http://www.president.harvard.edu/speeches/2005/nber.html

Thomas, H., & Kail, R. (1991). Sex differences in speed of mental rotation and the x-linked genetic hypothesis. *Intelligence, 15,* 17–32.

United Nations Development Programme (2008). *Human development indices: A statistical update 2008.* http://data.un.org/DocumentData.aspx?id=118#15.

Walton, G. M., & Cohen, G. L. (2003). Stereotype lift. *Journal of Experimental Social Psychology, 39,* 456–467.

Webb, R. M., Lubinski, D., & Benbow, C. P. (2002). Mathematically facile adolescents math-science aspirations: New perspectives on their educational and vocational development. *Journal of Educational Psychology, 94*(4), 785–794.

Xie, Y., & Shauman, K. A. (2003). *Women in science.* Cambridge: Harvard University Press.

Commentary on the Chapter by Penner and CadwalladerOlsker, "Gender Differences in Mathematics and Science Achievement Across the Distribution: What International Variation Can Tell Us About the Role of Biology and Society"

James S. Dietz

In their article, Penner and CadwalladerOlsker argue that sex differences in performance are largely due to societal factors, not biology. While they acknowledge several biological differences between males and females, they reach their conclusion based largely on differences in the variance in achievement across countries from the 1995 dataset of the Third International Mathematics and Science Study (TIMSS). Given there is no reason to believe that there are relevant biological differences across large cross-sections of the world, the country differences must be societal and cultural in origin.

The lengthy debate over sex differences in mathematics and science performance, stretching back more than 30 years, matters to society for reasons of equity and diversity and because it is simply counterproductive to allow roadblocks to impede the progress of science and presumably economic growth. In 2009, less than one in three doctorate recipients in the physical sciences were women. In engineering it was about one in five (National Science Foundation 2009). How many female Einsteins have been missed by the system for one reason or another?

But with respect to addressing these gender inequities, student learning may not be the crux of the issue (Lachance and Mazzocco 2006). Penner and CadwalladerOlsker may be turning over the wrong rock or at least not the biggest one. This is because, in terms of student performance, there appear to be no great differences between the sexes (at least on average). The data from our 2007 National Assessment of Education Progress (NAEP) in mathematics show that at the fourth and eighth grades there was a two-point difference in exam scores favoring boys over girls (National Center for Education Statistics (NCES) 2010). If this were a 100-point test *may* warrant some attention, however, the scale is 500 points at these grade levels—leaving us with a distinction without a difference. At the twelfth

J.S. Dietz (✉)
National Science Foundation, Wilson Boulevard, USA
e-mail: jdietz@nsf.gov

The opinions expressed are the authors own and do not necessarily reflect the position of the National Science Foundation (NSF). The author wishes to thank NSF for the support provided.

H. Forgasz, F. Rivera (eds.), *Towards Equity in Mathematics Education*,
Advances in Mathematics Education,
DOI 10.1007/978-3-642-27702-3_42, © Springer-Verlag Berlin Heidelberg 2012

grade, the difference between males and females is really no larger—3 points on a 300-point scale (NCES 2011). A recent meta-analysis of 242 research articles on gender differences in mathematics essentially confirms no difference, showing .05 effect size in favor of males (Lindberg et al. 2010). When these same authors examined national federally funded longitudinal data sets (such as, the National Longitudinal Survey of Youth (NLS), National Educational Longitudinal Study (NELS), the National Assessment of Education Progress (NAEP), and the Longitudinal Study of American Youth (LSAY)), they found a .07 effect size. Again, this is a miniscule effect. On the international front, via the most recent eighth-grade Trends in Mathematics and Science Study (TIMSS) exam (2007), in eight countries boys outscored girls, in 25 countries there was no statistically detectable difference, and in 17 countries girls outscored boys (Mullis et al. 2008).

So if there are no apparent differences in student performance, why the big deal and what is the equity argument all about? Well, the story gets a bit, but not much, more complicated for three reasons and this is where Penner and CadwalladerOlsker's argument comes into play. First, quite consistently the top one, two, three, or five percent of boys score higher than the top one, two, three, or five percent of girls in virtually all major tests (Halpern et al. 2007). Or put differently, there are more boy extreme-high-scorers than girl extreme-high-scorers. The overall mean differences on exams such as the ones discussed above may be masking a phenomenon in the right tail of the distribution (e.g. Olszewski-Kubilius and Lee 2011). Penner and CadwalladerOlsker examine this through quantile regression and confirm that there are indeed such differences. Second, probably related to the first, there may be some reason to believe that boys perform better on more complex problems. Hyde and colleagues, probably the leading authorities on this issue, write in their meta-analytic study, "Overall, we conclude that a small gender difference favoring boys in complex problem solving is still present in high school" (Lindberg et al. 2010, p. 1132). This is underscored by the fact that sex differences are larger (20 point difference) on the Programme for International Student Assessment (PISA) exam—a test of 15-year olds that tests more applied, situated, and perhaps complex problems (Wu 2010)—than on TIMSS. And, third, there *may* be some differences at subfields of mathematics as Penner and CadwalladerOlsker discuss. In TIMSS eighth grade, girls performed better than boys in algebra, whereas boys outperformed girls in number (Mullis et al. 2008). This is disputed, however, and there is evidence to the contrary (Lindberg et al. 2010).

So we are left with the following situation. If you are concerned that future mathematicians and scientists are drawn from this top few percent, then the sex differences in performance should matter to you. Penner and CadwalladerOlsker's findings support this. They conclude that sex differences in student performance at the top of the distribution are correlated with the proportion of women in the science and engineering workforce across countries. So there probably is good reason for some concern. If you are concerned with general equity and access to opportunities to learn mathematics, then the differences shouldn't alarm you greatly at least not when compared to differences between white and black children. The difference in NAEP scores was 26 points at the fourth grade, 32 points at the eighth grade, and 30 at twelfth grade (NCES 2011).

However, let's assume that these sex differences in performance matter (as I personally believe). The question then becomes is society or biology at fault? Penner and CadwalladerOlsker use macro data to conclude that they are largely due to broad societal factors rather than biology. The principal findings in their macrostructural analyses are, first, that the "status" index is positively related to performance but only at the top of the distribution of scorers. Penner and CadwalladerOlsker write, "This measure is comprised of variables such as the percentage of women in politics and gender differences in the age at first marriage." However, one could argue that age at first marriage may be as much a socio-economic indicator as a status one. Second, equality (encompassing variables such as percentage of managers who are female and women's rights to a job) is related positively to performance across the distribution (from students in low achieving quantiles to high achieving ones). Next, educational enrollment indicators relate to those scoring at the bottom of the distribution. And, fourth, surprisingly, representation of women in the labor force is negatively related to performance across the distribution—a true head-scratcher. Finally, it should be noted that all of these analyses use data from 1995.

Upon closer examination, though, I would argue that the data in Table 4 reveal that none of these indices has a substantively significant relationship with the outcomes and the overall story these data present is not clear. The scale of the indices is not shown or interpreted in the paper. However, the coefficient estimates hover around .01 to .02 in quantile regression models. It is almost certainly the case that societal discrimination plays some role, but the effects found by Penner and CadwalladerOlsker's measures are fairly weak. Despite this, their findings are consistent with at least one other study. Else-Quest et al. (2010) explored four of the most commonly used international gender equity indices—such as the Gender Equity Measure (GEM), Gender Equality Index (GEQ), Standard Index of Gender Equality (SIGE), and the Global Gender Gap Index (GGI)—and find that the measures were not strongly related to TIMSS scores but were of PISA scores. Taking broader issues into account, they conclude that "gender equity in school enrollment, women's share of research jobs [not in the indices], and women's parliamentary representation were the most powerful predictors of cross-national variability in math gender scores in the TIMSS and PISA exams" (Else-Quest et al. 2010, p.103). Thus, there is evidence to support Penner and CadwalladerOlsker's finding that macro-level labor force equity factors are weakly associated with variance in test scores across countries. But how much do these aggregate societal contexts matter when compared with more proximal, meso-level social factors? And, is student learning—rather than access and persistence, for example—the right place to look for effects of these factors?

At the meso-level, researchers have identified a wide array of psychological, social psychological, and sociological factors when accounting for gender differences, such as self-efficacy (Else-Quest et al. 2010; Wigfield and Eccles 2000), identity (Kiefer and Sekaquaptewa 2007), attitudes (Else-Quest et al. 2010), motivation (Preckel et al. 2008), societal stratification parental influences (Furnham et al. 2002; Herbert and Stipek 2005; Jacobs et al. 2005), teacher behavior (Tiedemann 2009; Beilock et al. 2010), and stereo-type threat (Spencer et al. 1999). Incidentally, this last one appears, by all accounts in the literature, to be a major factor. This goes well beyond Penner and CadwalladerOlsker's interest in this paper but factors such as these, in the author's opinion, are more likely to be important in locating where the discrimination rubber meets the road, so to speak.

At the level of the individual, as Penner and CadwalladerOlsker summarize, there are biological differences among men and women—brain organization (Keller and Menon 2009), hormonal (e.g., Valla and Ceci 2011), and evolutionary (e.g.: Geary 1996; Joseph 2000) for example—and differences that are often potentially biological, social, or biopsychosocial that may be relevant to the development of future mathematicians and scientists. But the question here, too, is how pertinent are these latter factors? Spelke (2005) concludes that the skills of girls and women exceed those of boys and men in the areas of verbal fluency, arithmetic calculation and spatial locations of objects. NAEP studies find that girls have advantage in verbal skills, reading, and writing (NCES 2008, 2011)—all clearly required of good scientists and engineers. Men tend to excel at verbal analogies, mathematical word problems, and memory for geometric configurations of objects (mental rotation) (cf. Lippa et al. (2010) for international comparisons). Webb et al. (2007) argue convincingly that it is these cognitive-spatial reasoning skills that are key to excelling in science and engineering. Their studies show large sex differences among gifted and talented students on this dimension (Wai et al. 2009). But the differences found by Spelke appear to be small, they also appear to be confounded with differences in problem solving strategies that are favored by men and women. As Penner and CadwalladerOlsker note, when men and women are confined to the same strategies, they tended to perform more similarly (Jay 2006; Spelke 2005).

In sum then the data teach us a few things. First, the vast cross-section of boys and girls perform equally on our tests of national performance in mathematics. International studies teach us there is a high variability in sex differences across countries of the world and that macro social factors appear to explain some part of this difference. How much is not clear from Penner and CadwalladerOlsker's chapter. A large number of studies conclude that males outperform females at the top of the distribution and probably on more complex problems. There are gender-based skill and ability differences that are due to biological or societal factors or both (we don't know) but it is not clear that these affect the achievement of students in the short or long term. And, finally, the search for gender differences that matter in mathematics learning is hampered by the fact that there really aren't many differences with which to be concerned.

References

Beilock, S. L., Gunderson, E. A., Ramirez, G., & Levine, S. C. (2010). Female teachers' math anxiety affects girls' math achievement. *Proceedings of the National Academy of Sciences of the United States of America, 107*, 1860–1863.

Else-Quest, N. M., Hyde, J. S., & Lynn, M. C. (2010). Cross-national patterns of gender differences in mathematics: A meta-analysis. *Psychological Bulletin, 136*, 103–127.

Furnham, A., Reeves, E., & Budhani, S. (2002). Parents think their sons are brighter than their daughters: Sex differences in parental self-estimations and estimations of their children's multiple intelligences. *Journal of Genetic Psychology, 163*, 24–39.

Geary, D. C. (1996). Sexual selection and sex differences in mathematical abilities. *Behavioral and Brain Sciences, 19*(2), 229–284.

Halpern, D. F., Benbow, C. P., Geary, D. C., Gur, R. C., Hyde, J. S., & Gernsbacher, M. A. (2007). The science of sex differences in science and mathematics. *Psychological Science in the Public Interest, 8*, 1–41.

Herbert, J., & Stipek, D. (2005). The emergence of gender differences in children's perceptions of their academic competence. *Journal of Applied Developmental Psychology, 26*, 276–295.

Jacobs, J. E., Davis-Kean, P., Bleeker, M., Eccles, J. S., & Malanchuk, O. (2005). "I can, but I don't want to": The impact of parents, interests, and activities on gender differences in math. In A. M. Gallagher & J. C. Kaufman (Eds.), *Gender differences in mathematics: An integrative psychological approach* (pp. 246–263). New York: Cambridge University Press.

Jay, T. (2006). Gender differences in patterns of strategy use amongst secondary school mathematics students. *Proceedings of the 30th conference of the international group for the psychology of mathematics education* (Vol. 3, pp. 361–368). Greece: PME.

Joseph, R. (2000). The evolution of sex differences in language, sexuality, and visual–spatial skills. *Archives of Sexual Behavior, 29*, 35–66.

Keller, K., & Menon, V. (2009). Gender differences in the functional and structural neuroanatomy of mathematical cognition. *Neuroimage, 47*, 342–352.

Kiefer, A. K., & Sekaquaptewa, D. (2007). Implicit stereotypes, gender identification, and math-related outcomes—A prospective study of female college students. *Psychological Science, 18*, 13–18.

Lachance, J. A. & Mazzocco, M. M. (2006). A longitudinal analysis of sex differences in math and spatial skills in primary school age children. *Learning and Individual Differences, 16*, 195–216.

Lindberg, S. M., Hyde, J. S., Petersen, J. L., & Linn, M. C. (2010). Gender similarities characterize math performance. *Psychological Bulletin, 136*, 1123–1135.

Lippa, R. A., Collaer, M. L., & Peters, M. (2010). Sex differences in mental rotation and line angle judgments are positively associated with gender equality and economic development across 53 nations. *Archives of Sexual Behavior, 39*, 990–997.

Mullis, I. V. S., Martin, M. O., & Foy, P. (with Olson, J. F., Preuschoff, C., Erberber, E., Arora, A., & Galia, J.) (2008). *TIMSS 2007 international mathematics report: Findings from IEA's trends in international mathematics and science study at the fourth and eighth grades.* Boston College, Chestnut Hill: TIMSS & PIRLS International Study Center.

National Center for Education Statistics (2008). *Writing 2007: National assessment of educational progress at grades 8 and 12.* Washington: Institute for Education Sciences, US Department of Education.

National Center for Education Statistics (2010). *Mathematics 2009: National assessment of educational progress at grades 4 and 8.* Washington: Institute for Education Sciences, US Department of Education.

National Center for Education Statistics (2011). *Grade 12: Reading and mathematics 2009 (National and pilot state results from the national assessment of education progress).* Washington: Institute for Education Sciences, US Department of Education.

National Science Foundation (2009). *Characteristics of doctoral scientists and engineers in the United States 2006* (available online at http://nsf.gov/statistics/doctoratework/).

Olszewski-Kubilius, P., & Lee, S. Y. (2011). Gender and other group differences in performance on off-level tests: Changes in the 21st century. *Gifted Child Quarterly, 55*, 54–73.

Preckel, F., Goetz, T., Pekrun, R., & Kleine, M. (2008). Gender differences in gifted and average-ability students—Comparing girls' and boys' achievement, self-concept, interest, and motivation in mathematics. *Gifted Child Quarterly, 52*, 146–159.

Spelke, E. S. (2005). Sex differences in intrinsic aptitude for mathematics and science? A critical review. *American Psychologist, 60*, 950–958.

Spencer, S. J., Steele, C. M., & Quinn, D. M. (1999). Stereotype threat and women's math performance. *Journal of Experimental Social Psychology, 35*, 4–28.

Tiedemann, J. (2009). Gender-related beliefs of teachers in elementary school mathematics. *Educational Studies in Mathematics, 41*, 191–207.

Valla, J. M., & Ceci, S. J. (2011). Can sex differences in science be tied to the long reach of prenatal hormones?: Brain organization theory, digit ratio (2D/4D), and sex differences in preferences and cognition. *Perspectives on Psychological Science, 6*, 134–146.

Wai, J., Lubinski, D., & Benbow, C. P. (2009). Spatial ability for STEM domains: Aligning over 50 years of cumulative: Psychological knowledge solidifies its importance. *Journal of Educational Psychology, 101*, 817–835.

Webb, R. M., Lubinski, D., & Benbow, C. P. (2007). Spatial ability: A neglected dimension in talent searches for intellectually precocious youth. *Journal of Educational Psychology, 99*, 397–420.

Wigfield, A., & Eccles, J. S. (2000). Expectancy–value theory of achievement motivation. *Contemporary Educational Psychology, 25*, 68–81.

Wu, M. (2010). Comparing the similarities and differences of PISA 2003 and TIMSS. *OECD Education Working Papers* (Vol. *32*). Switzerland: OECD Publishing.

Commentary on the Chapter by Penner and CadwalladerOlsker, "Gender Differences in Mathematics and Science Achievement Across the Distribution: What International Variation Can Tell Us About the Role of Biology and Society"

Robert (Bob) Klein

In their chapter, Penner and CadwalladerOlsker combine TIMSS data with country data from the United Nations, World Bank, and other resources to give two primary results related to mathematics performance by sex. First, they cast doubt on the utility of biological causes as an effective explanation of gender differences in mathematics performance on the TIMSS. They do so by demonstrating variability in the gender differences in math achievement measures between countries, corroborating Hanna's (1989) refutation of biology as a primary basis for gender differences in mathematics performance. As there are not significant biological differences between people by country, they suggest that the differences are better explained by differences in social factors within national contexts.

Second, since gender differences in math scores depend on social factors, they investigate which country-level factors matter most in explaining those differences. The authors create five variables calculated from country-level variables from extant data sets and use these to predict gender differences in TIMSS Math scores (and then Science scores). Following the work of Lubinski and Benbow (1992), Favreau and Everett (1996), and others, they look for explanatory factors for differences within quantiles throughout the distribution, and in particular, in the right tail, representing high-achieving students. This approach acknowledges that comparisons of means mask many of the subtleties explaining variation within, say high-achieving males or females or low-achieving males or females. They cite studies reporting equal mean measures of mathematical achievement by gender, suggesting gender equity. Yet Penner and CadwalladerOlsker's analysis demonstrates how the right tail of the achievement distribution evidences significant gender differences favoring men.

Methodologically, the effects of this argument are to offer an analytic treatment that provides far more nuanced findings and implications than simple comparisons of means as has been typical in quantitative research on gender differences. This approach, used more frequently in econometrics, is finding greater embrace by

R. Klein (✉)
Ohio University, Athens, OH, USA
e-mail: kleinr@ohio.edu

H. Forgasz, F. Rivera (eds.), *Towards Equity in Mathematics Education*,
Advances in Mathematics Education,
DOI 10.1007/978-3-642-27702-3_43, © Springer-Verlag Berlin Heidelberg 2012

educational researchers, and deservedly so. Discursively, mathematics education research seems to have grown tired of quantitative studies of gender differences in mathematics—an intellectual ennui resulting from a dearth of quantitative work with results that either motivate action or stimulate curiosity and a plethora of qualitative work on the subject that offers intriguing snapshots that fail to generalize well or speak to foundations and major policy groups hungry for "scientifically based research."

Penner's and CadwalladerOlsker's work could revitalize work on gender and mathematics (using a variety of methodologies) by providing a useful analytic technique focused on understanding a more targeted population (for instance, high-achieving boys and girls). Multi-level modeling, tail analysis, and econometric techniques offer a much broader set of interpretive tools and more fine-grained analysis than has been used commonly in educational research. Many of the questions that were best answered with qualitative methods increasingly may be open to quantitative and mixed methods analyses. This should delight the community of researchers as the more varied methods will likely speak to a greater audience than before and offer more in-depth analyses than were possible before. Even better, these tools may suggest more targeted and more effective curriculum, structural changes, and policies geared toward equitable student opportunities and outcomes.

1 The Problem of Scope and Scale

Focusing on country-level variables is an effective way to refute the power of biological explanations of gender differences, and Penner and CadwalladerOlsker do so effectively while admitting the limitations of doing this for 22 mostly European countries. Others, such as Condron (2011), have conducted similar county-level regressions on mathematics achievement data. Condron's recent work regresses PISA scores from 27 of the 30 OECD nations on the Gini index (a measure of distribution of wealth) and finds significant correlation between high achievement on PISA math scores and egalitarian distribution of wealth. The ability to conduct country-level research is aided by the ubiquity of data sources at that level. Collecting data at the national and international levels is something that national governments do well; indeed it is responsible for the creation of modern mathematics in early civilizations (Grattan-Guinness 2000).

Yet the scale on which research focuses determines to a significant extent the set of available actions justified by that research. Penner and CadwalladerOlsker create five variables (social factors) on which to predict achievement scores: Education, Domesticity, Labor force representation, Labor force equality, and Status differences between men and women. They find that, among these macro-level contextual factors and in the right tail of the distribution, Labor market issues matter most, followed by Status. What set of actions are implied by this finding and, perhaps more importantly, who is best able to implement them? The authors' focus on macro-level factors does not preclude action on the micro-level as in the case of provincial, district, community, school building, or teacher/classroom interventions. Still, educators' history of trying to counter macro-level messages involving gender,

SES, and racial inequities is limited to isolated successes and generally on micro-levels rather than the macro-scale, country levels. Significant changes in labor force representation and equality seem to imply actions at the level of nations though the incentive for nations to move quickly to improve labor markets is not clear. The authors' measure of status includes measures such as the relative participation of women in politics and differences in the age at first marriage. Given its significance as a correlate to high levels of achievement in mathematics, how does one act (including who acts) to improve on this status? Again, the answer to improving this macro-level variable seems to be at the macro-level, involving government or large grassroots organizations to affect such change.

To complicate matters, as the scale of inquiry changes, the relevant variables (not to mention the relevant datasets) must change. The set of available actions implied by analysis of smaller-scale data might be more actionable "on the ground." Could similar methodologies be employed on smaller scales to motivate more immediate and effective efforts to foster gender equity across the distribution? For instance, could revisiting this method of analysis according to region, locale type, or even by funding categorizations at national levels may imply a set of actions that can be taken by communities, school districts, teachers, and parents, instead of by countries?

There is no reason why Penner and CadwalladerOlsker's use of quantile regression to model gender differences could not be used productively on different scales using variables that are appropriate on those scales. Indeed, this work has motivated me to revisit a paper that a colleague and I presented (Klein and Johnson 2010) using a multi-level model to examine individual- and school-level variables for the state of Kentucky (USA) in understanding rural versus non-rural students' (relatively homogeneous categories within Kentucky) math scores on Kentucky's Commonwealth Test of Basic Skills at the ninth grade level. The entire set of population ($n = 49,979$) data was available to regress math scores on SES, ethnicity, school size, and gender. We found that schools in rural locales erased the gender gap, diminished the race and poverty gaps in math achievement for all students combined, and erased the influence of low SES on mathematics achievement for females and minorities (a partial table of these results is found in Table 1). What set of actions are implied by this and who are the relevant actors? Certainly the answer is not in remaking non-rural places into rural ones, but Theobald (1995) and others have argued how bonds among community and kin, as well as ties to the land, are prominent, uniting features of rural communities, which suggests one avenue for further investigation into how educational practices at the level of *locale* may improve instruction for all students.

My colleague and I did not compare data across the distribution as did Penner and CadwalladerOlsker, though it is imminently conceivable that part of how schools, communities, and teachers might address gender, race, and SES gaps could vary according to talent levels (measured by test scores). The within-country variances in scores by gender and between-country variances in gender differences both lead me to wonder what quantile regressions on the Kentucky data set would demonstrate, but it is also interesting to consider the set of possible actions or recommendations that could come from such an analysis. State-level analysis of means masked the

Table 1 Comparison of results for multilevel regression model one analysis for variables predicting student performance on grade 9 CTBS-5 math (from Klein and Johnson 2010)

Variable	All	Rural	Non-rural
	($N_1 = 46,680$) ($N_2 = 334$)	($N_1 = 16,839$) ($N_2 = 126$)	($N_1 = 29,918$) ($N_2 = 208$)
Student Sex (SEX1)	-2.922^{***}	$-.650$	-2.953^{***}
Student Ethnicity (ETH1)	-16.985^{***}	-14.071^{***}	-17.061^{***}
School Percent Female (SEX2)	.222	$-.096$	$.453^{*}$
School Percent Minority (ETH2)	-1.853^{**}	-1.787	-1.923^{*}
School SES (SES2)	$-.783^{***}$	$-.637^{***}$	$-.831^{***}$
School Enrollment Size (SIZ2)	$-.0005$	.002	$-.003$
SEX1 X SEX2 interaction	$-.071$	.131	.096
SEX1 X ETH2 interaction	-2.486^{***}	$-.090$	-3.038^{***}
SEX1 X SES2 interaction	$.071^{*}$	.046	$.122^{**}$
SEX1 X SIZ2 interaction	$-.0003$	$-.002^{*}$	.0009
ETH1 X ETH2 interaction	-9.400^{***}	-7.656	-9.642^{***}
ETH1 X SEX2 interaction	$-.519$	.908	.459
ETH1 X SES2 interaction	$.195^{*}$	.138	$.269*$
ETH1 X SIZ2 interaction	$-.004$	.004	$-.006$
SES2 X SIZ2 interaction	$-.0004^{***}$	$-3.43E\text{-}05$	$-.0008^{***}$

Notes: $N_1 = N$ at level one (students); $N_2 = N$ at level two (schools).
$^{*}p \leq 0.05$; $^{**}p \leq 0.01$; $^{***}p \leq 0.001$.

differences by locale, a disaggregation that embeds significant differences in social factors (Howley 2002; Howley and Howley 1999). This embedding may be productive in that it draws on smaller communities of potential actors where changes impact educational practice more immediately than at the country level.

2 The Pushmi-Pullyu Except This Thing Has TWO Tails

Dr. Dolittle's pushmi-pullyu was an animal with two identical heads that could not move because each head wanted to move the beast in a direction opposite the other. Penner & CadwalladerOlsker's animal is, instead, a two-tailed animal with notably different tails. Their analysis across the distribution focuses a bit more on the high-achieving right tail than the low-achieving left, and while the clear preference of educators would be to have the whole animal move to the right, it is unclear what carrot or what stick will do this. Their analysis reveals that the two tails respond to different variables, with the high-achieving right side responding more to labor-market conditions and the status accorded math-using fields in a given country. The low-achieving left tail wags according to all factors except status. Given the country-level focus of the analysis, the set of actions that would be implied by these correlates are difficult to affect at any level other than through national government policy (e.g., funding schemes).

Still, focusing on the extreme tails further moves the discussion away from essentializing a "female" or "woman's" mathematics performance (located at the mean) to understanding relative performances of high- or low-achieving women's and men's performances, thus grounding analysis in more categorically actionable terms—"What explains high-achieving women's and men's different choices or different levels of performance on TIMSS tests?" "What incentives influence increased participation by high-achieving women in advanced coursework in mathematics?" Questions that arise from understanding left-tail results are potentially different, but no less important. The increased attention on male experiences in mathematics and, in particular, at the left tail of performance (Holland 2003), make it clear that the lessons inherent to the left tail are no less important to insuring equitable outcomes and opportunities for all students. Researchers' focus on equity should move the entire animal toward the right.

3 Conclusion

There is little doubt of the importance of Penner's and CadwalladerOlsker's work in two key areas. First, it offers a model methodology for educators to bring quantitative methods to bear on more focused sections of the achievement distribution. This approach could be put into conversation with qualitative methods to address new and old questions in fruitful ways that both drive constructive interventions promoting equity (not limited to gender) and speak to a wider audience of potential funders and consumers of research. A more robust description of the analytic approach would offer others greater access to interpreting the results presented here but also to adapting the strategy to other data sets and at other scales.

Second, the results of this study suggest that economic incentive structures and gender status factors (and not biology) correlate strongly with gender differences in high-achieving students. The implication is that women and men are actively making decisions in response to these factors, underscoring the importance of agency but also the power of ingrained macro-structural messages. Future work could expand the construct of gender used here (male/female) to more of a continuum (masculine to feminine) so as to understand how social constructions of gender map onto mathematics achievement in relation to other variables. The significance of the Status variable to accounting for gender differences in high achievers underscores the power of socially constructed norms—norms that impact how masculine and feminine impact societal valuations. Though admittedly hard to measure, seeing gender as continuum rather than binary could offer even more fine-grained understanding of the mechanisms of status and how they impact educational outcomes.

Even as the results shared in Penner and CadwalladerOlsker's chapter suggest social factors that matter most, changing these macro-level structures seems to require macro-level strategies that are well out of the reach of educators. Attempts by national governments to engage in social engineering, as Scott points out (1998), fail due to the importance of local practices that large-scale efforts ignore. Methodologically, we have done the same when viewing achievement score distributions, and the excellent work presented here suggests that different neighborhoods within the distribution tell different stories. Applied on a more micro-scale, this strategy could imply actions toward equity that are within range of educators.

References

Condron, D. (2011). Egalitarianism and educational excellence: Compatible goals for affluent countries? *Educational Researcher, 40*(2), 47–55.

Favreau, O. E., & Everett, J. C. (1996). A tale of two tails. *American Psychologist, 51*(3), 268–269.

Grattan-Guinness, I. (2000). *The rainbow of mathematics: A history of the mathematical sciences.* New York: W. W. Norton & Company.

Hanna, G. (1989). Mathematics achievement of girls and boys in grade eight: Results from twenty countries. *Educational Studies in Mathematics, 20*(2), 225–232.

Holland, V. (2003). Underachieving boys: Problems and solutions. *Support for Learning, 13,* 174–178.

Howley, C. (2002). Understanding the circumstance of rural schooling: The parameters of respectful research. In S. Henderson (Ed.), *Understanding achievement in science and mathematics in rural schools* (pp. 11–16). Lexington: Appalachian Rural Systemic Initiative.

Howley, A. A. & Howley, C. B. (1999). The transformative challenge of rural context. *Educational Foundations, 14*(4), 73–85.

Klein, R., & Johnson, J. (2010). On the use of locale in understanding the mathematics achievement gap. In P. Brosnan, D. B. Erchick, & L. Flevares (Eds.), *Proceedings of the 32nd annual meeting of the North American chapter of the international group for the psychology of mathematics education* (pp. 489–496), Columbus, Ohio, USA, October 28–31, 2010. The Ohio State University.

Lubinski, D., & Benbow, C. (1992). Gender differences in abilities and preferences among the gifted: Implications for the math-science pipeline. *Current Directions in Psychological Science, 1,* 61–66.

Scott, J. (1998). *Seeing like a state: How certain schemes to improve the human condition have failed.* New Haven: Yale University Press.

Theobald, P. (1995). *Teaching the commons: Place, pride, and the renewal of community.* Boulder: Westview Press.

Research-Based Mathematics Instruction for Students with Learning Disabilities

Marjorie Montague and Asha K. Jitendra

Abstract The purpose of this chapter is to describe several advances and trends in teaching mathematics to students with learning disabilities. Mathematics disabilities affect between 5% and 8% of students in the general population. These students generally experience serious difficulties in developing requisite concepts, skills, and strategies for success in mathematics and, consequently, are at risk for school failure and poor post-secondary outcomes. Research in special education has identified a number of evidence-based practices that have proven effective for teaching students with learning, attention, and/or behavioral disorders that interfere with success in mathematics. Following a discussion of student characteristics and evidence-based practices, two examples of research programs focusing on mathematical problem-solving interventions for students with learning disabilities are highlighted.

Students in our nation's schools have consistently performed poorly on state (e.g., Florida Comprehensive Assessment Test; FCAT), national (e.g., National Assessment of Educational Progress; NAEP 2003, 2007), and international mathematics tests (e.g., The International Mathematics and Science Study 2003, 2007). In 1989, the National Council of Teachers of Mathematics (NCTM) responded to this poor performance with the publication of *Curriculum and evaluation of NCTM standards for school mathematics* (NCTM 1989) and again in 2000 with a revision titled *Principles and NCTM standards for school mathematics* (NCTM 2000). NCTM called attention to the dismal mathematics performance of students and recommended a more meaningful approach to teaching and learning mathematics. While these standards have helped educators to improve students' mathematics learning in many ways, students, particularly those with special learning needs, continue to fare poorly.

M. Montague (✉)
Department of Teaching and Learning, University of Miami, 5202 University Dr., Merrick Bldg. 321, Coral Gables, FL 33146, USA
e-mail: mmontague@aol.com

A.K. Jitendra
Department of Educational Psychology, University of Minnesota, 245 Education Sciences Building, 56 E. River Road, Minneapolis, MN 55455, USA
e-mail: jiten001@umn.edu

H. Forgasz, F. Rivera (eds.), *Towards Equity in Mathematics Education*, 481
Advances in Mathematics Education,
DOI 10.1007/978-3-642-27702-3_44, © Springer-Verlag Berlin Heidelberg 2012

NAEP 2007 results indicated that while 32% of our students scored at or above the proficient level in grade 8, only 23% were proficient at grade 12. The National Mathematics Advisory Panel's report (NMAP 2008) suggested this decline in performance may be traced to middle school students' difficulties with algebra, which is typically introduced in middle school and, according to the NMAP, is a "demonstrable gateway to later achievement" (p. xiii). In 2004, 26 states required at least one year of algebra for high school graduation, and 13 states required two years of algebra (American Diploma Project 2004). By 2015, nearly all states will require all students to successfully complete at least one year of algebra for graduation. The challenge, then, is to prepare students for algebra by ensuring that they have the critical foundational concepts and skills needed to enter and be successful in algebra and then by explicitly teaching strategies needed to apply those concepts and skills at a higher level in a real world context. Success in algebra will provide more academic and, ultimately, more post-secondary opportunities for students. Research has shown that students completing algebra II are twice as likely to graduate from college and earn more from employment than students who have less math preparation (NMAP 2008).

Compounding the problem is that students in urban schools, particularly students with disabilities, often fare much worse, putting them at even greater risk for poor academic outcomes and school dropout. There remains a persistent performance gap in mathematics between poor students from diverse cultural backgrounds and their white, middle-class counterparts and students with and without disabilities (NAEP 2003, 2007). For example, NAEP 2003 results indicated that white students scored 40 points higher, on average, than African American students, and 71% of students with disabilities contrasted with 27% of students without disabilities scored below the basic level. NAEP 2007 indicated that 40% of participating grade 4 students with disabilities scored below the basic level compared to 15% of grade 4 students without disabilities. The gap widens in grade 8 and again in grade 12 with 83% of students with disabilities scoring below the basic level compared to 36% of peers without disabilities. In keeping with the No Child Left Behind Act (NCLB 2001), all students, including students with disabilities, must meet high standards in mathematics as measured by state-administered achievement tests. To meet these standards, low performing students and students with disabilities, who vary considerably in ability, achievement, and motivation, must develop the necessary mathematical concepts, skills, and strategies needed not only to perform well on mathematics assessments but also to apply these skills successfully in real-world settings. Thus, improving students' mathematical abilities and performance should have a positive impact on students' overall math performance, high stakes test performance, graduation rate, and post-secondary outcomes.

The purpose of this chapter is to describe several advances and trends in teaching mathematics to students with learning disabilities (LD). First, the characteristics of students with mathematical learning disabilities (MLD) that appear to impede success are described. Second, an overview of research-based practices designed to address these student characteristics is provided. Next, we present two cognitively based intervention research programs designed to improve math problem solving for students with MLD–Montague's (2003) *Solve It!* and Jitendra's (2007) schema-based instruction. Finally, we present several recommendations for delivering instruction for students with MLD.

1 Characteristics of Children and Adolescents with MLD

Children and adolescents with MLD display a variety of cognitive and behavioral characteristics that interfere with successful performance in mathematics. Geary (2003) estimated that between 5% and 8% of children have MLD and proposed that they may have one or more of three subtypes of math disability. The first subtype, procedural deficits, is characterized by difficulties in retaining information in working memory and monitoring counting processes. These children often rely on immature counting strategies such as "counting all" and make more errors in executing mathematical procedures, particularly complex procedures, for example, carrying or borrowing from one column to the next when regrouping is required. He attributes these behaviors to developmental delay as these children often perform similarly to younger typical learners but improve with age. Children with the second subtype, semantic memory deficits, have considerable difficulty retrieving mathematical facts or answers to simple arithmetic problems (e.g., memorizing the multiplication tables when they are introduced in third grade), make more errors when they do, and vary considerably on reaction time for correct retrieval. This subtype seems to represent a developmental difference among children as these children appear cognitively different from younger typical peers and generally do not improve with age. The third subtype is visuo-spatial MLD, which is characterized by difficulties in representing numerical relationships spatially and interpreting and understanding spatially represented information. Children with visuo-spatial deficits have difficulties visually and spatially representing and interpreting numerical information and relationships in math problems. Visuo-spatial deficits are manifested in problems with measurement, place value, geometry, and aligning and rotating numbers.

Johnson et al. (2010) reviewed nine studies of children with MLD and found that these students, despite average intelligence, were significantly different from typically achieving students in several cognitive categories. Specifically, they were significantly lower on tests of verbal and visual working memory, executive function, processing speed, and short-term memory. Difficulties in retrieving arithmetic facts from long-term memory and storing numbers in working memory affect the ability to compute, learn algorithmic procedures, and solve mathematical word problems. Children with MLD characteristically do not remember certain combinations and patterns of numbers, which is most evident when they fail to learn the multiplication tables or struggle with division.

Students with MLD also have considerable difficulty manipulating numerical and linguistic information in mathematical word problems (Montague and Jitendra 2006). Textbooks rarely provide sufficient instruction in how to solve mathematical word problems or how to accommodate diverse student learning needs. Instruction usually is restricted to textbook models that give a sequenced list of activities (i.e., read, decide what to do, solve, and check the answer). These instructions are not very helpful for students who have few resources for "deciding what to do." Montague and Applegate (1993) found significant differences between average- and high-achievers and students with MLD in their ability to represent the information in math word problems. As a consequence, students with MLD have no systematic approach for solving the problems. Problem representation is a prerequisite for problem solution. As Johnson et al. (2010) and Melia de Alba and Montague (2010) found, students with MLD also have problems with executive functions (i.e., metacognition or self-regulation). Metacognitive or self-regulation strategies are important for monitoring one's thought processes and task performance. Students who are effective self-regulators self-instruct by telling themselves what to do when they are engaged in a task, self-question by asking themselves questions, and self-monitor through consistent error detection and correction and self-evaluation. Ineffective self-regulators are disorganized, do not know where or how to begin, lack perseverance, and do not evaluate what they do. Self-regulation is essential for successful academic performance across domains (Montague 1998).

Behavioral problems (e.g., inattention and impulsivity) also interfere with learning (Wolraich et al. 1996). Children with attention problems are careless, distractible, and forgetful. They have difficulty listening, following directions, finding their place, staying on task, and completing tasks. Impulsive children lack self-control and generally do not think before they act. They typically operate in the present without regard for past experiences or consequences. These children frequently fidget and squirm and have difficulty staying in their seats and completing their school work. They appear unmotivated and indecisive. By early adolescence, they are characterized by an inability to self-regulate. Also, they are notoriously poor at generalizing their learning across situations and settings. For example, students may learn a strategy for solving a certain type of problem and use it with confidence and success in a small-group instructional setting. However, application does not transfer to different but similar types of problems or to the general classroom setting. Remedial programs must include instructional techniques such as cues and prompts or a self-monitoring checklist to assist students in generalizing their learning across different settings (e.g., classroom, home) and different tasks (e.g., class assignments, homework, tests).

2 Research-Based Math Instruction for Students with MLD

Several theoretical frameworks have influenced research focusing on interventions to improve mathematics performance of students with MLD. These frameworks are the foundation for the instructional approaches and generally are based on behavioral theory, cognitive theory (i.e., developmental theory, information processing theory, sociocultural theory), or a combination (Adams and Carnine 2003; Case 1985; Flavell et al. 1993; Sternberg 1985; Vygotsky 1989). Behavioral theory is the basis for direct instruction whereas a combination of behavioral and cognitive theory underlies most cognitively based instructional approaches such as cognitive strategy instruction and schema-based instruction, two instructional approaches for improving math problem solving that are described later in this chapter.

In several meta-analyses of intervention research with students with LD (e.g., Kroesbergen and van Luit 2003; Swanson 1999), direct instruction and cognitive strategy instruction were found to be the most effective interventions for these students. Both instructional approaches are highly structured and organized and include similar research-based instructional practices and procedures such as cueing, modeling, verbal rehearsal, and feedback. Direct instruction is more teacher-directed, didactic, and scripted compared with cognitive strategy instruction and has been used extensively for improving basic skills in mathematics. Cognitive strategy instruction is more interactive and uses explicit instruction that focuses on teaching students the processes involved in the application of skills (i.e., solving mathematical problems). Swanson (1999) advocated a *"Combined Direct Instruction and Strategy Instruction Model"* for teaching students with LD that includes 10 components found to be most predictive of treatment effectiveness:

1. Sequencing instruction (e.g., breaking down the task and then sequencing short activities that build on one another, fading prompts and cues).
2. Drill-repetition and practice-review (e.g., daily skill testing, sequenced review, distributed practice, daily feedback).
3. Segmentation (e.g., breaking the task into small units and then synthesizing the parts into a whole).
4. Directed questioning and response (e.g., process or content questions, teacher/student dialogue).
5. Controlling task difficulty or processing demands of the task (e.g., sequencing from easy to difficult, prompts and cues that fade over time).
6. Technology (e.g., computers, calculators, flow charts, pictorial representations, structured materials, graphing performance over time).
7. Modeling problem-solving processes (e.g., "think-aloud" demonstrations of how successful problem solvers think and behave).
8. Group instruction (e.g., small groups, teacher/student interaction).
9. Supplements to instruction (e.g., tutors, homework).
10. Strategy cues (e.g., reminders, mnemonics, think-aloud models).

Several examples of research in mathematics with students with LD that have utilized direct instruction as the instructional approach are described in the next section. Following that, an overview of cognitively based instruction is presented and two examples for improving math problem solving of students with LD are described in detail.

3 Direct Instruction Research in Mathematics

Research using direct instruction to improve mathematics has focused primarily on drill and practice for improving math fact recall, computation skills, and algorithmic procedures for students with MLD because it provides challenge, appropriate time on task, and numerous response opportunities (e.g., Fuchs and Fuchs 2001). For example, Burns' (2005) incremental rehearsal was used to teach multiplication facts to third-grade students with LD. Like other drill and rehearsal programs, he used a gradually increasing ratio of known to unknown items until the students achieved at least 90% accuracy. He introduced 10 new facts each time until students met a preset mastery criterion of known facts. A caveat associated with drill and practice routines is that students usually do not generalize what they have learned to similar but unfamiliar math facts, which may be attributable to poor number sense as Gersten and Chard (1999) suggested. Woodward (2006) noted that students need a fundamental understanding of number before they can sufficiently learn facts and procedures needed to perform the mathematical operations. Concrete manipulatives have been used to develop number sense (Funkhouser 1995), that is, before memorizing basic facts, students with LD in kindergarten and grade one were taught to recognize the number of objects in a set without counting them. They were given vertical displays of rectangles divided into equal squares with 0 to 5 dots or jellybeans placed within the squares. Students counted the dots or jellybeans and then identified the number represented using different combinations. Gradually, students were introduced to the addition symbol and the basic addition facts.

Direct instruction is also the basis for C-R-A instruction, which is an acronym for a routine that progresses from introducing math concepts at a concrete level and then moving to a representational level and, lastly, to the abstract level. C-R-A instruction has been used to teach students with LD place value (Peterson 1988), coin sums (Miller et al. 1992), basic facts (Mercer and Miller 1992), multiplication (Harris et al. 1995; Miller et al. 1998), algebraic reasoning (Witzel 2005; Witzel et al. 2003), and geometry (Cass et al. 2003). Three levels of instruction are used beginning with the concrete level, in which three-dimensional manipulatives are used to demonstrate specific mathematical concepts. Then, at the representation level, students use visual representations of the mathematical concept to develop conceptual understanding. Finally, when conceptual understanding is demonstrated, instruction moves to the abstract level where students solve problems using only number symbols. The CRA teaching sequence uses a four-step direct instruction lesson format: (a) introduce the lesson, (b) model the new procedure, (c) guide students through the procedure, and (d) move students toward working independently. Witzel (2005) developed a C-R-A algebra model to teach concepts and solution strategies for variables with multiple coefficients, fractions, and exponents. Figure 1 presents his example of a C-R-A teaching sequence for reducing algebraic expressions. The concrete level uses manipulative objects to teach the math concept, the representation level uses pictures, and, finally, Arabic symbols are used to teach the abstract lessons.

4 Cognitively Based Math Problem Solving Research

Two cognitively based approaches for improving math problem solving for students with LD and supporting efficacy research are described in the next two sections. Both Montague's *Solve It!*, a cognitive strategy instructional program, and Jitendra's schema-based instructional program are strategic in their approach, were developed specifically for students with LD, and have been researched extensively

Concrete, Representational, and Abstract Examples of a Reducing Expressions Problem

Step 1: A **concrete** representation of 5 – 2X – 6 must use manipulative objects. For this problem it would appear in this order: five small sticks, a minus sign, one coefficient marker, an X, a plus symbol, a large stick, an equal line, and three small sticks. Manipulating the objects leaves the answer as a minus sign with one stick remaining and a minus sign followed by two cups of X.

Step 2: A pictorial **representation** would closely resemble the concrete objects but could be drawn exactly as it appears here. A student solves representational problems exactly as she would solve them concretely. For example,

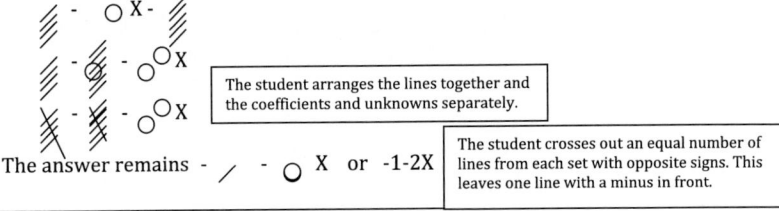

The student arranges the lines together and the coefficients and unknowns separately.

The student crosses out an equal number of lines from each set with opposite signs. This leaves one line with a minus in front.

Step 3: An **abstract** problem is written using Arabic symbols as displayed in most textbooks and standardized exams. Students in the comparison group used this format for problem solving during each lesson. The Multisensory group only used this format after concrete and representational manipulation.
To solve abstract problems, students write each step to solving this problem. For example,

$$5 - 2X - 6$$
$$^{+}5 - 6 - 2X$$
$$^{-}1 - 2X$$

From Witzel (2005). Copyright by Learning Disabilities: A Contemporary Journal. Reprinted with permission.

Fig. 1 *Concrete, representational, and abstract examples of a reducing expressions problem* (from Witzel (2005). Copyright by Learning Disabilities: A Contemporary Journal. Reprinted with permission)

using both single-subject and group research designs. They also incorporate many instructional procedures associated with cognitive strategy instruction. As noted by Swanson (1999), direct instruction and cognitive strategy instruction share several instructional procedures such as corrective and positive feedback, reinforcement, rehearsal, and prompts and cues. However, strategy instruction has been described as more explicit than direct in its methodology (Montague 2003). Like direct instruction, however, explicit instruction is characterized by structured, organized lessons; appropriate cues and prompts; guided and distributed practice; immediate and corrective feedback on learner performance; positive reinforcement; overlearning; and mastery (Montague et al. 2000). Explicit instruction is more flexible than direct instruction and promotes ongoing interaction between teachers and students. Teachers are encouraged to adapt the teaching routine and tailor instruction to accommodate the specific learning needs of students. Explicit instruction as exemplified in *Solve It!* (Montague 2003) and Jitendra's schema-based instruction includes the following proven effective instructional strategies.

1. Active student participation and interaction is encouraged among students and between students and teachers (Peterson 1988). A guided discussion technique ensures that all students participate. Role reversals and problem solving teams/groups engage all students in the process (Jenkins and O'Connor 2003).
2. Verbal rehearsal, a mnemonic technique, helps students learn the processes and strategies until they can recall the problem-solving routine from memory (Scruggs and Mastropieri 2003). Cue cards and wall charts are phased out as students become more proficient problem solvers (Hutchinson 1993).
3. Cognitive modeling, or thinking aloud while performing an activity to demonstrate how successful problem solvers solve math problems, is used repeatedly during initial learning and practice sessions. Teachers and then students model problem solving demonstrating both correct and incorrect behaviors (Wong et al. 2003). The class is then engaged in discussion about alternative solutions to each problem.
4. Distributed practice and performance feedback provide students with ongoing evaluation and correction as they become more competent. Students learn how to evaluate their own performance and set learning goals from the outset (Adams and Carnine 2003). Students also graph their performance over time to demonstrate progress and provide feedback to the teacher and student to determine if additional review is needed.
5. Reinforcement is essential for students as they make progress towards mastery. They need specific information on when and why they are successful. Labeled praise indicates precisely the behaviors that contribute to effective problem solving. As students gain confidence, they can begin to reinforce one another and themselves (Eisenberger and Cameron 1996).

4.1 Cognitive Strategy Instruction (CSI)

CSI is the foundation of *Solve It!*, an instructional program that was designed specifically to improve mathematical problem solving of middle and secondary school students with LD (Montague 2003). *Solve It!* incorporates essential cognitive processes and metacognitive strategies that are integral to problem representation and problem execution and underlie successful problem solving (Mayer 1985). Problem representation, generally speaking, is a combination of concrete representations (e.g. manipulatives), something written, and/or a carefully constructed mental image (Janvier 1987; Mayer 1985). Problem representation processes and strategies are necessary when comprehending the problem, that is, integrating problem information, maintaining mental images of the problem in working memory, and developing a solution plan, frequently by using alternative solutions to the problem (Silver 1985). Problem representation requires the ability to translate and transform linguistic and numerical information into verbal, graphic, symbolic, and quantitative representations that show the schemata or relationships among the information in the problem (Mayer 1985; Montague and Applegate 1993; van Garderen and Montague 2003) before generating the mathematical solutions (i.e., the equations or algorithms). After the solution plan is reviewed and found acceptable, the problem solver proceeds with problem execution strategies. Problem execution requires the problem solver to work forward and backward constantly returning to the problem information. *Solve It!* places particular emphasis on teaching students how to represent mathematical problems by paraphrasing problems, using visualization strategies such as diagram drawing or mental imaging, and hypothesizing or setting up a plan without resorting to trial-and-error. The ultimate goal of *Solve It!* is to have students internalize the cognitive and metacognitive routine so that the problem-solving processes and strategies become automatic during problem solving. *Solve It!* incorporates the following cognitive processes and activities:

1. Reading the problem for understanding includes reading and rereading the problem or parts of the problem to identify relevant/irrelevant information.
2. Paraphrasing means translating the linguistic information by putting the problem into one's own words without changing the meaning of the "story" or "situation."
3. Visualizing is the ability to transform the linguistic and numerical information to form internal representations in memory by developing either a graphic representation or mental image that shows the relationships among the components of the problem.
4. Hypothesizing requires the problem solver to establish a goal, think about the outcome, and develop a plan to solve the problem by deciding on the operations that are needed, selecting and ordering the operations, and transforming the information into correct equations and algorithms.
5. Estimating the outcome or answer means predicting the outcome based on the question/goal and the information presented.
6. Computing the outcome or answer requires the problem solver to recall the correct procedures for the basic operations. Calculator skills are taught and reinforced.
7. Checking is important for determining the accuracy of the solution by checking understanding and representation as well as the process, procedures, and computation.

Cognitive processes are proactive in nature. They are the "to do" strategies. In contrast, metacognitive strategies are more reflective and require problem solvers to tell themselves what processes to use and then question what they are doing and what they have done. Problem solvers use metacognitive strategies to recall what they know, locate and correct errors, and monitor performance throughout the problem-solving process. These metacognitive or self-regulation strategies help problem solvers gain access to strategic knowledge, guide their application, and regulate use of strategies and overall performance as they solve problems. They are

READ (for understanding)
Say: Read the problem. If I don't understand, read it again.
Ask: Have I read and understood the problem?
Check: For understanding as I solve the problem.

PARAPHRASE (your own words)
Say: Underline the important information. Put the problem in my own words.
Ask: Have I underlined the important information? What is the question? What am I looking for?
Check: That the information goes with the question.

VISUALIZE (a picture or a diagram)
Say: Make a drawing or a diagram. Show the relationships among the problem parts.
Ask: Does the picture fit the problem? Did I show the relationships?
Check: The picture against the problem information.

HYPOTHESIZE (a plan to solve the problem)
Say: Decide how many steps and operations are needed. Write the operation symbols (+, -, x, and /).
Ask: If I ..., what will I get? If I ..., then what do I need to do next? How many steps are needed?
Check: That the plan makes sense.

ESTIMATE (predict the answer)
Say: Round the numbers, do the problem in my head, and write the estimate.
Ask: Did I round up and down? Did I write the estimate?
Check: That I used the important information.

COMPUTE (do the arithmetic)
Say: Do the operations in the right order.
Ask: How does my answer compare with my estimate? Does my answer make sense? Are the decimals or money signs in the right places?
Check: That all the operations were done in the right order.

CHECK (make sure everything is right)
Say: Check the plan to make sure it is right. Check the computation.
Ask: Have I checked every step? Have I checked the computation? Is my answer right?
Check: That everything is right. If not, go back. Ask for help if I need it.

Fig. 2 *Solve it!* Math problem solving cognitive routine (from Montague (2003). Copyright by Exceptional Innovations. Permission to photocopy this figure is granted for personal use only)

used overtly by talking out loud or whispering to oneself or covertly through internal self-talk. Metacognitive strategies included in *Solve It!* include self-instruction (SAY what to do before and while performing actions), self-questioning (ASK questions of oneself while engaged in an activity to stay on task, regulate performance, and verify accuracy), and self-monitoring (CHECK to make certain that everything is done correctly throughout the problem solving process). Figure 2 presents the entire *Solve It!* cognitive routine.

4.2 Research Evidence in Support of CSI for Math Problem Solving

All of the research studies that have focused on CSI for improving math problem solving used Montague's (2003) *Solve It!* or used a cognitive routine similar to *Solve It!* (Case et al. 1992; Cassel and Reid 1996; Chung and Tam 2005; Hutchinson 1993; Montague 1992; Montague et al. 1993; Montague and Bos 1986). A total of 142 students with LD or mild intellectual disabilities ranging in age from 8–4 to 16–7 years participated in these studies. For a detailed review, see Montague and Dietz (2009). *Solve It!* was designed to teach students how to understand, analyze, solve, and evaluate mathematical problems by developing the processes and strategies that effective problem solvers use. Montague's research focused primarily on middle and high school students with LD and, most recently, also on middle school students in general education classes who were identified as at risk for mathematics failure (Montague et al. 2011).

The first study was a single-subject, multiple-baseline design with six secondary students with LD (Montague and Bos 1986) who met participation criteria (e.g., average ability, minimum reading level, computation competence). Following intervention, all students improved to criterion (at least 70% correct on at least three of four consecutive tests of math problem solving) and also generalized strategy use to more difficult problems. Additional validation studies using both single-subject and group designs were then conducted as part of a federally funded grant to improve mathematical problem solving for middle school students with LD (Montague 1992; Montague et al. 1993). In the group study, 72 middle school students with LD were given the *Solve It!* intervention and, following intervention, performed at the same level as average-achieving students on math problem-solving tests and maintained performance over a four-month period. Effect sizes based on pretest performance for students with LD were 1.09 (posttest), 1.08 (maintenance 1), .74 (maintenance 2 at two months), and 1.25 (following a booster session), all moderate to large effect sizes. Furthermore, students were able to generalize the problem-solving routine to more difficult problems. This study also indicated that the combination of cognitive and metacognitive components of instruction was more effective than either component alone. Daniel (2003) used *Solve It!* and also found significant improvement in math problem solving for her group of middle school students with LD compared with a control group. She also noted that their knowledge and awareness of strategies improved to the level of average-achieving students following *Solve It!* instruction. Thus, the program was validated in four studies with students with LD, students who struggle most with mathematics (Geary 1994; Jitendra and Xin 1997). These students characteristically have limited cognitive strategies, particularly problem representation strategies, poor self-regulation, and low motivation and self-efficacy.

The potential for addressing students with other types of disabilities has also been established. Coughlin and Montague (2010) successfully adapted *Solve It!* for three adolescents with spina bifida. As a result of a pilot study, the researchers eliminated the estimation process, provided manual support for visual representation, and built in a graduated program whereby students demonstrated mastery with one-step problems before they advanced to two-step problems. All students improved to criterion on both types of problems. Whitby (2009) adapted *Solve It!* for three middle school students with high functioning autism/Asperger's Syndrome by adding a video modeling component and gradually increasing the difficulty level of the problems.

Solve It! was the focus of a federally funded efficacy study (2007–2010) that provided general education math teachers with professional development to incorporate *Solve It!* instruction into the prescribed curriculum. The results from the first year of the study conducted in a large, urban school district (Montague et al. 2011) with students in grades 7–8 indicated that average-achieving students as well as low-achieving students and students with LD ($n = 185$) who received *Solve It!* instruction made significantly greater progress in math problem solving over the school year than students in the comparison group ($n = 127$) as measured by curriculum-based measures of textbook-type problems ($d = .44$). They also made significantly greater growth in math problem-solving self-efficacy and math confidence over the school year than students in the comparison group ($d = .37$). As expected, students in grade 8 had higher scores initially than grade 7 students. However, the rate of growth was the same for both grades. In the second year (Montague et al. 2011), 24 middle schools (matched on performance level and SES) participated. *Solve It!* was implemented for seven months in general education math classrooms and periodic progress monitoring was conducted across the school year. One eighth-grade math teacher participated at each school. A cluster-randomized design was used, and the data were consistent with a 3-level model where repeated measures were nested within students and students were nested within schools. The results indicated that students who received the intervention ($n = 319$) showed significantly greater growth in math problem solving over the school year than students in the comparison group ($n = 460$) who received typical classroom instruction ($d = .91$, a large effect size). The intervention had the same impact for students with LD, low-achieving students, and average-achieving students. By the end of the school year, the students with LD significantly outperformed even the average-achieving students in the comparison group. Thus, the findings were positive and support the efficacy of the intervention for students with LD in both special education and inclusive classroom settings. In sum, the research conducted thus far suggests that CSI is a promising approach for teaching students how to solve math problems.

4.3 Schema-Based Instruction (SBI)

Another well researched and documented approach to improving mathematical problem solving for children and adolescents with LD is schema-based instruction (see Fuchs et al. 2008; Jitendra et al. 2007; Jitendra et al. 1998; Xin and Jitendra 2006). The work of Marshall (1995), Mayer (1999), and Riley et al. (1983) serves as the foundation for Jitendra and colleagues' SBI approach, which emphasizes schema knowledge (i.e., identifying the problem schema underlying a given text), elaboration knowledge (e.g., using the schema as a vehicle to elaborate on and translate information in the text into a semantic representation), strategic knowledge (i.e., planning to represent the problem as a math equation), and execution knowledge (e.g., carrying out single or chains of calculations). Identifying the problem schema entails recognizing the situation in simple or complex problems from among several problem schemata. The set of schemata described in the literature for arithmetic word problems include change, group, compare, restate, and vary (Marshall 1995). These schemata are separated into two problem categories, additive and multiplicative structures. Change, group, and compare problems are additive structures, because the solution operation is either addition or subtraction. Restate (i.e., multiplicative compare) and vary (i.e., equal groups, proportion) problems belong to the multiplicative field, because the solution operation is either multiplication or division (Christou and Philippou 1999, p. 269).

In recent research, SBI has included metacognitive strategy knowledge as a salient component to encourage student "think-alouds" for monitoring and directing problem-solving behavior along the following dimensions: (a) problem comprehension (e.g., "Did I read and retell the problem to understand what is given and what must be solved?" "Why is this a compare problem?" "How is this problem similar to or different from ones I already solved?"), (b) problem representation (e.g., "What diagram can help me represent information in the problem to show the relation between quantities?"), (c) planning (e.g., "How can I set up the math equation?"), and (d) problem solution (e.g., "Does the answer make sense?" "How can I verify the solution?")

The SBI intervention described here entails a four-step strategy (FOPS; F—Find the problem type, O—Organize the information in the problem using the diagram, P—Plan to solve the problem, S—Solve the problem) that a teacher uses to scaffold the cognitive and metacognitive processes as he/she thinks aloud to solve word problems (see Fig. 3). For example, consider the following problem: *Lisa made 12 cupcakes for the Annual Teacher Appreciation week at her school. She made $\frac{1}{3}$ as many cupcakes as her friend Mary. How many cupcakes did Mary make?* (Jitendra 2007, p. 199). The primary focus of SBI is on helping youngsters identify the problem type or problem schema (i.e., "a general description of a group of problems that share a common underlying structure") (Xin and Jitendra 2006, p. 53). The teacher identifies the problem type using Step 1 of the strategy by reading, retelling, and examining information in the problem to recognize it as a multiplicative compare problem via self-instructions (e.g., Are there compare words in the problem that tell me about a comparison? Does the *comparison* statement tell what the problem is comparing?). SBI emphasizes that compare words such as "$\frac{1}{3}$ as many as" in this problem can cue the learner to multiplicative compare situation. In addition, the teacher makes the connection between previously solved compare problems by noting that the multiplicative compare problem is similar to the compare problem in that they compare two quantities. However, they are different because multiplicative compare problems describe the comparison of one quantity as a multiple or part (e.g., one-third) of the other quantity and involve the operation of multiplication or division whereas compare problems (e.g., A redwood tree can grow to be 85 meters tall. A Douglas fir can grow to be 15 meters taller. How tall can the Douglas fir grow?) describe the comparison of an additive structure.

MULTIPLICATIVE COMPARE (MC) PROBLEM CHECKLIST

Step 1.　**Find the problem type**

☐ Did I read and retell the problem?

☐ Did I ask if it is a MC problem? Did I look for the multiplicative compare words, such as "n times as many/much as," "n^{th} of," to see whether there is a comparison sentence that tells about a multiple (e.g., 2 times) or partial (e.g., $\frac{2}{3}$) relation?

Step 2.　**Organize information using the compare diagram**

☐ Did I underline the comparison sentence, circle the two things compared (i.e., compared and referent), and write them in the diagram? Did I write the relation between the compared and referent in the diagram?

☐ Did I underline the compared and referent, and circle numbers and labels for the compared and referent, and write them in the diagram?

☐ Did I write a "?" for what must be solved? (Did I find the question sentence?)

Step 3.　**Plan to solve the problem**

☐ Did I translate the information in the diagram into a math equation?

Step 4.　**Solve the problem**

☐ Did I solve the math sentence?

☐ Did I write the complete answer?

☐ Did I check if the answer makes sense?

Fig. 3 Compare problem checklist

Fig. 4 Multiplicative compare schematic diagram

In Step 2, students learn to represent that problem schema (i.e., multiplicative compare) using a schematic diagram (see Fig. 4). Unlike pictorial representations of problems that include concrete but irrelevant details, which "are superfluous to solution of the math problem" (Edens and Potter 2006, p. 186), a schematic diagram depicts the spatial relations between objects in the problem text (Hegarty and Kozhevnikov 1999, p. 685). Given that many students with LD lack problem translation skills, SBI focuses on comprehending sentences that express a relation between two quantities that is critical to solving multiplicative compare word problems. Using the problem described earlier, the teacher elaborates on and translates information in the problem by first identifying the comparison or relational sentence, for example, *"She (Jill) made $\frac{1}{3}$ as many cupcakes as her friend Mary. How many cupcakes did Mary make?"* Further, SBI teaches that the comparison sentence can help figure out the two things (number of cupcakes Jill made and number of cupcakes Mary made) compared and identify the compared and referent quantities. SBI demonstrates that the compared quantity is the number of cupcakes Jill made because this is the quantity that is compared to the number of cupcakes made by Mary (i.e., referent or base). In addition, instruction focuses on the relation ($\frac{1}{3}$) between the two things compared in the comparison sentence. Finally, information in the remaining verbal text specifies the known and unknown quantities for the two things compared. In sum, for Step 2 the teacher demonstrates how to organize information using the schematic diagram via self-instructions. For example, questions such as "What does this problem compare?" (e.g., number of cupcakes Jill made and number of cupcakes Mary made) "What is the relation between Jill and Mary with regard to the number of cupcakes? Is it a multiple or partial relation" (e.g., $\frac{1}{3}$), are used to make sense of the problem schema. Specifically, the questioning makes clear the relational value (i.e., $\frac{1}{3}$), which is first represented in the diagram, by reading the comparison sentence to find the compared and referent sets and writing the terms associated with those sets (i.e., Jill's cupcakes is the compared set and Mary's cupcakes is the referent set). Next, the quantities for each of the two sets are identified by reading the problem text and writing the given amount(s) or a "?" for what must be solved in the diagram. The teacher then analyzes the problem situation using the completed diagram in Fig. 4 as follows: "The number of cupcakes Jill made to the number of cupcakes Mary made is $\frac{1}{3}$. Jill made 12 cupcakes. We need to solve for the number of cupcakes Mary made."

Step 3 involves translating the information in the diagram into a math equation (e.g., $\frac{12}{x} = \frac{1}{3}$). Finally, for Step 4 students solve the problem using an appropriate solution method (e.g., equivalent fractions or cross-product algorithm). Students also learn to justify the derived solution using the schema features as anchors for explanations and elaborations and check the accuracy of not only the computation but also the representation.

4.4 Research Evidence in Support of SBI for Math Problem Solving

The research studies that have focused on SBI for improving student learning have emphasized (a) modeling the semantic structure of problem types as in SBI (Fuchs et al. 2008; Hutchinson 1993; Jaspers and Van Lieshout 1994; Jitendra et al. 2002; Jitendra et al. 1998; Jitendra and Hoff 1996; Jitendra et al. 1999; Jitendra et al. 2007; Xin et al. 2005; Xin 2008; Xin et al. 2008; Xin and Zhang 2009; Zawaiza and Gerber 1993) or (b) teaching for transfer (e.g., searching novel problems for familiar problem types that may seem different based on different format, different key vocabulary, additional or different question, and irrelevant information) as in schema-broadening instruction (SBI-T) (Fuchs et al. 2008; Fuchs et al. 2002; Fuchs et al. 2004; Owen and Fuchs 2002). A total of 758 students, with LD and mild intellectual disabilities as well as English language learners (ELL) and nondisabled students struggling in mathematics, ranging in age from 8–10 to 26–7 years, participated in the 20 studies on SBI and SBI-T. For the purpose of this chapter, we describe in detail Jitendra and colleagues' research on SBI that addresses the semantic structure of problems as the instructional focus given that this type of instruction of broad problem types is less tied to surface features within problems and therefore may not require direct transfer instruction as in SBI-T.

SBI that emphasizes modeling the semantic structure of the word problem using schematic diagrams was studied extensively in a series of single subject and group design studies. Students with LD in two single subject studies (Jitendra and Hoff 1996; Jitendra et al. 1999) who received SBI to solve change, group, and compare word problems not only improved word problem-solving performance (mean increase ranged from 47% to 69%) but also maintained their problem-solving improvement 2 to 4 weeks following the intervention. Three of the four middle school students in the Jitendra et al. (1999) study also generalized strategy use from one-step to two-step problems. Interestingly, although all four students made greater progress on one-step problems (mean increase = 29%) compared to a normative sample of third graders (mean increase = 36%), the mean performance on two-step word problems for students with LD (71% correct) was substantially higher than the normative sample (28% correct). The findings of these two studies showed SBI to be a promising approach for improving the word problem solving of students with LD.

Two other studies, (i.e., Jitendra et al. 1998; Jitendra et al. 2007) used randomized control trials to validate the effectiveness of SBI for elementary school students with and without disabilities. Following intervention in the Jitendra et al. (1998) study, students (e.g., students with LD, intellectual disabilities, and/or emotional/behavior disorders and students at risk for poor problem-solving outcomes) in the SBI group not only outperformed the control group, but performed at the same level as average-achieving students on math problem-solving posttests. Further, the SBI group maintained and generalized problem-solving skills to novel problems. Effect sizes comparing SBI with the control condition were .65 at posttest, .81 at delayed posttest, and .74 for transfer, all medium to large effect sizes.

Jitendra et al. (2007) then extended this SBI approach by embedding metacognitive strategy instruction in SBI and then providing general education teachers with professional development to implement SBI during their math classes. The results of the study conducted in five third-grade classrooms in a high poverty elementary school with a sample of mostly low achieving students, including students with LD, English language learners (ELL), and Title I math students, indicated that SBI was more effective than typical classroom instruction in improving students' mathematical problem-solving skills at posttest ($d = .52$) and delayed posttest ($d = .69$). Further, the SBI group outperformed the comparison group on the state assessment of mathematics performance ($d = .65$).

Jitendra and colleagues also extended their work on SBI to the domain of multiplication and division word problem solving involving multiplicative compare and proportion problems. Result of a single-subject, multiple baseline design study (Jitendra et al. 2002) showed that SBI improved the learning of multiplicative compare and proportion word problems (mean improvement ranged from 50% to 71%) by four middle school students with LD. Further, all four students maintained their performance 2 to 10 weeks following the intervention and generalized to more complex problems including multistep problems. Xin et al. (2005) and Jitendra et al. (2009) further validated the effects of SBI using randomized control trials. Xin et al. (2005) worked with 22 students with LD, behavior disorders, and students at-risk for poor problem solving to teach multiplicative compare and proportion problems. Results indicated that students receiving SBI significantly outperformed a comparison group on an immediate posttest ($d = 1.69$), delayed posttests ($d > 2.50$), and a transfer test ($d = .89$).

In a series of single subject design studies, Xin (2008), Xin et al. (2008), and Xin and Zhang (2009) modified SBI in terms of the schematic diagrams and employed word problem story grammar (metacognitive strategy instruction) to examine the effects of their conceptual model-based problem solving in improving student learning. Xin (2008) found that all four students' mean improvement on word problems involving equal groups and multiplicative compare ranged from 75% to 92%, which was further maintained 4 to 10 days later. Results also showed transfer effects ($>50\%$ correct) to novel word problems for two students. Similarly, Xin et al. (2008) found that the three participants who learned to solve part-part-whole and additive compare problems improved following the intervention (mean improvement range $= 58\%$ to 63%). Results also indicated substantial mean improvement (97% and 100%) from baseline to post-intervention for the two participants who received SBI to solve multiplicative compare and equal group problems. In addition, all five participants showed transfer effects (mean improvement range $= 50\%$ to 100%) to a standardized math test and prealgebra concepts and skills that ranged from 34% to 100%. In a third study, Xin and Zhang (2009) found that three fourth- and fifth-grade students in their study improved from baseline to post-intervention and also showed transfer effects.

Jitendra et al. (2009) extended the work of Xin and colleagues by providing seventh-grade general education teachers with professional development to incorporate SBI into their prescribed curriculum to teach ratios and proportions. The SBI curriculum not only addressed ratio and proportion word problem solving but also emphasized the foundational concepts (e.g., ratios, rates) and multiple solution methods (e.g., unit rate, equivalent fractions, cross multiplication) to develop conceptual and procedural knowledge. Participants included 148 seventh graders from 8 classrooms in one middle school in a large, urban school district. Results revealed that students in the SBI condition outperformed students in the "business as usual" control classrooms on the problem-solving posttest ($d = .45$) and a 4-month delayed posttest ($d = .56$). In sum, the research conducted by Jitendra and colleagues suggests that SBI is effective particularly for students with LD by developing their schemata for a variety of math problem types.

5 Recommendations for Instructing Students with LD

With the move toward inclusion of students with disabilities in general education, more and more students with LD will be receiving math instruction in general education math classrooms. As a result, general education math teachers need to become knowledgeable about the characteristics of these students and the interventions that have been effective for improving their achievement in math. We offer three important recommendations to facilitate research-based instruction for these students. First, to maximize instruction, it seems essential that special education teachers work collaboratively with general education math teachers such that instruction becomes a shared responsibility. It should be noted that not all students in general education will need the interventions described in this chapter because they have developed the necessary concepts, skills, and strategies to be successful. However, students with LD and even some students who have not been identified with disabilities may need specialized instruction such as CSI and SBI. Consequently, it is important that students be assessed to determine who will benefit from the intervention.

Our second recommendation is that instruction be provided by specialists (i.e., expert remedial teachers) who are familiar with the characteristics of these students and also are skilled in implementing research-based interventions that been found effective for improving math performance of students with special learning needs. Our third recommendation is based on assessment results in that the specialized instruction is provided to small groups of students (e.g., 8 to 10 students) who have similar needs and would benefit from instruction. Since these interventions are intense and time-limited, it may be more beneficial to remove students from the classroom for a brief period for initial instruction or booster sessions to review previous learning. For example, for *Solve It!* instruction, students would receive the three days of intensive instruction in the cognitive routine in small groups in a setting outside the general education classroom. Following this intensive instruction, students would return to the classroom and be prompted to use the routine whenever they solve math word problems. That is, the general education teacher would be responsible for reinforcing what students have learned during small-group instruction in the context of whole class problem-solving activities. Students who do not make adequate progress may need booster sessions or additional small-group instruction, again provided by the remedial specialist. In conclusion, there are research-based interventions that have proven effective for students with LD. However, the success of these interventions often depend on how, when, and by whom they are provided to students. These are important considerations if we expect these students to profit from what we have learned about effective instruction in mathematics.

References

Adams, G., & Carnine, D. (2003). Direct instruction. In H. L. Swanson, K. R. Harris, & S. Graham (Eds.), *Handbook of learning disabilities* (pp. 403–416). New York: Guilford.

American Diploma Project (2004). *Ready or not: Creating a high school diploma that counts.* Washington: American Diploma Project (available online at www.achieve.org).

Burns, M. K. (2005). Using incremental rehearsal to increase fluency of single-digit multiplication facts with children identified as learning disabled in mathematics computation. *Education and Treatment of Children, 28,* 237–249.

Case, R. (1985). *Intellectual development: Birth to adulthood.* San Diego: Academic Press.

Case, L. P., Harris, K. R., & Graham, S. (1992). Improving the mathematical problem-solving skills of students with learning disabilities: Self-regulated strategy development. *The Journal of Special Education, 26,* 1–19.

Cass, M., Cates, D., Smith, M., & Jackson, C. (2003). Effects of manipulative instruction on solving area and perimeter problems by students with learning disabilities. *Learning Disabilities Research and Practice, 18,* 112–120.

Cassel, J., & Reid, R. (1996). Use of a self-regulated strategy intervention to improve word problem-solving skills of students with mild disabilities. *Journal of Behavioral Education, 6,* 153–172.

Christou, C., & Philippou, G. (1999). Role of schemas in one-step word problems. *Educational Research and Evaluation, 5,* 269–289.

Chung, K. H., & Tam, Y. H. (2005). Effects of cognitive-based instruction on mathematical problem solving by learners with mild intellectual disabilities. *Journal of Intellectual and Developmental Disability, 30,* 207–216.

Coughlin, J., & Montague, M. (2010). The effects of cognitive strategy instruction on the mathematical problem solving of students with spina bifida. *Journal of Special Education,* doi:10.1177/0022466910363913.

Daniel, G. E. (2003). *Effects of cognitive strategy instruction on the mathematical problem solving of middle school students with learning disabilities.* Unpublished doctoral dissertation, Ohio State University.

Edens, K., & Potter, E. (2006). How students "unpack" the structure of a word problem: Graphic representations and problem solving. *School Science and Mathematics, 108,* 184–196.

Eisenberger, R., & Cameron, J. (1996). Detrimental effects of reward: Reality or myth? *American Psychologist, 51,* 1153–1166.

Flavell, J. H., Miller, P. H., & Miller, S. A. (1993). *Cognitive development.* Englewood: Prentice Hall.

Fuchs, L. S., & Fuchs, D. (2001). Principles for the prevention and intervention of mathematics difficulties. *Learning Disabilities Research and Practice, 16,* 85–95.

Fuchs, L. S., Fuchs, D., Hamlett, C. L., & Appleton, A. C. (2002). Explicitly teaching for transfer: Effects on the mathematical problem-solving performance of students with mathematics disabilities. *Learning Disabilities Research & Practice, 17,* 90–106.

Fuchs, L. S., Fuchs, D., Finelli, R., Courey, S. J., & Hamlett, C. L. (2004). Expanding schema-based transfer instruction to help third graders solve real-life mathematical problems. *American Educational Research Journal, 41,* 419–445.

Fuchs, L. S., Seethaler, P. M., Powell, S. R., Fuchs, D., Hamlett, C. L., & Fletcher, J. M. (2008). Effects of preventative tutoring on the mathematical problem solving of third-grade students with math and reading difficulties. *Exceptional Children, 74,* 155–173.

Funkhouser, C. (1995). Developing number sense and basic computational skills in students with special needs. *School Science and Mathematics, 95,* 236–239.

Geary, D. (1994). *Children's mathematical development.* Washington: American Psychological Association.

Geary, D. (2003). Math disabilities. In H. L. Swanson, K. R. Harris, & S. Graham (Eds.), *Handbook of learning disabilities* (pp. 199–212). New York: Guilford.

Gersten, R., & Chard, D. (1999). Number sense: Rethinking arithmetic instruction for students with mathematical disabilities. *The Journal of Special Education, 44,* 18–28.

Harris, C. A., Miller, S. P., & Mercer, C. D. (1995). Teaching initial multiplication skills to students with disabilities in general education classrooms. *Learning Disabilities Research and Practice, 10,* 180–195.

Hegarty, M., & Kozhevnikov, M. (1999). Types of visual-spatial representations and mathematical problem solving. *Journal of Educational Psychology, 91,* 684–689.

Hutchinson, N. L. (1993). Effects of cognitive strategy instruction on algebra problem solving of adolescents with learning disabilities. *Learning Disability Quarterly, 16*, 34–63.

Janvier, C. (1987). *Problems of representation in the teaching and learning of mathematics.* Mahwah: Laurence Erlbaum.

Jaspers, M. W. M., & Van Lieshout, E. C. D. M. (1994). The evaluation of two computerized instruction programs for arithmetic word-problem solving by educable mentally retarded children. *Learning and Instruction, 4*, 193–215.

Jenkins, J. R., & O'Connor, R. E. (2003). Cooperative learning for students with learning disabilities: Evidence from experiments, observations, and interviews. In H. L. Swanson, K. R. Harris, & S. Graham (Eds.), *Handbook of learning disabilities* (pp. 417–430). New York: Guilford.

Jitendra, A. K. (2007). *Solving math word problems: Teaching students with learning disabilities using schema-based instruction.* Austin: Pro-Ed.

Jitendra, A. K., DiPipi, C. M., & Perron-Jones, N. (2002). An exploratory study of word problem-solving instruction for middle school students with learning disabilities: An emphasis on conceptual and procedural understanding. *Journal of Special Education, 36*, 23–38.

Jitendra, A. K., & Hoff, K. (1996). The effects of schema-based instruction on mathematical word problem solving performance of students with learning disabilities. *Journal of Learning Disabilities, 29*, 422–431.

Jitendra, A. K., Hoff, K., & Beck, M. (1999). Teaching middle school students with learning disabilities to solve multistep word problems using a schema-based approach. *Remedial and Special Education, 20*, 50–64.

Jitendra, A. K., Griffin, C., McGoey, K., Gardill, C, Bhat, P., & Riley, T. (1998). Effects of mathematical word problem solving by students at risk or with mild disabilities. *Journal of Educational Research, 91*, 345–356.

Jitendra, A. K., Griffin, C., Haria, P., Leh, J., Adams, A., & Kaduvetoor, A. (2007). A comparison of single and multiple strategy instruction on third grade students' mathematical problem solving. *Journal of Educational Psychology, 99*, 115–127.

Jitendra, A. K., Star, J., Starosta, K., Leh, J., Sood, S., Caskie, G., Hughes, C., & Mack, T. (2009). Improving students' learning of ratio and proportion problem solving: The role of schema-based instruction. *Contemporary Educational Psychology, 34*, 250–264.

Jitendra, A., & Xin, A. (1997). Mathematical word problem solving instruction for students with mild disabilities and students at risk for failure: A research synthesis. *Journal of Special Education, 30*(4), 412–438.

Johnson, E. S., Humphrey, M., Mellard, D. F., Woods, K., & Swanson, H. L. (2010). Cognitive processing deficits and students with specific learning disabilities: A selective meta-analysis of the literature. *Learning Disability Quarterly, 33*, 3–18.

Kroesbergen, E. H., & van Luit, J. E. H. (2003). Mathematics interventions for children with special educational needs. *Remedial and Special Education, 24*, 97–114.

Marshall, S. P. (1995). *Schemas in problem solving.* New York: Cambridge University Press.

Mayer, R. E. (1985). Mathematical ability. In R. J. Sternberg (Ed.), *Human abilities: Information processing approach* (pp. 127–150). San Francisco: Freeman.

Mayer, R. E. (1999). *The promise of educational psychology, Vol. 1: Learning in the content areas.* Upper Saddle River: Merrill Prentice Hall.

Melia de Alba, A., & Montague, M. (2010). Mathematical problem-solving processes and strategies of middle school students: Results of structured interview analysis. Unpublished raw data.

Mercer, C. D., & Miller, S. P. (1992). Teaching students with learning problems in math to acquire, understand, and apply basic math facts. *Remedial and Special Education, 13*, 19–35.

Miller, S. P., Harris, C., Strawser, S., Jones, W. P., & Mercer, C. (1998). Teaching multiplication to second graders in inclusive settings. *Focus of Learning Problems in Mathematics, 20*, 50–70.

Miller, S. P., Mercer, C. D., & Dillon, A. (1992). Acquiring and retaining math skills. *Intervention, 28*, 105–110.

Montague, M. (1992). The effects of cognitive and metacognitive strategy instruction on mathematical problem solving of middle school students with learning disabilities. *Journal of Learning Disabilities, 25*, 230–248.

Montague, M. (1998). Research on metacognition in special education. In T. Scruggs & M. Mastropieri (Eds.), *Advances in learning and behavioral disabilities* (Vol. *12*, pp. 151–183). Greenwich: JAI Press.

Montague, M. (2003). *Solve it: A mathematical problem-solving instructional program.* Reston: Exceptional Innovations.

Montague, M., & Applegate, B. (1993). Mathematical problem-solving characteristics of middle school students with learning disabilities. *The Journal of Special Education, 27*, 175–201.

Montague, M., Applegate, B., & Marquard, K. (1993). Cognitive strategy instruction and mathematical problem-solving performance of students with learning disabilities. *Learning Disabilities Research and Practice, 29*, 251–261.

Montague, M., & Bos, C. (1986). The effect of cognitive strategy training on verbal math problem solving performance of learning disabled adolescents. *Journal of Learning Disabilities, 19*, 26–33.

Montague, M., & Dietz, S. (2009). Evaluating the evidence base for cognitive strategy instruction and mathematical problem solving. *Exceptional Children, 75*, 285–382.

Montague, M., Enders, C., & Dietz, S. (2011). Effects of cognitive strategy instruction on math problem solving of middle school students with learning disabilities. *Learning Disability Quarterly, 34*, 262–272.

Montague, M., & Jitendra, A. K. (Eds.) (2006). *Teaching mathematics to middle school students with learning difficulties.* New York: Guilford.

Montague, M., Warger, C., & Morgan, H. (2000). Solve it! Strategy instruction to improve mathematical problem solving. *Learning Disabilities Research and Practice, 15*, 110–116.

National Assessment of Educational Progress (2003). *NAEP 2003 mathematics report card for the nation and the states.* Princeton: Educational Testing Service.

National Assessment of Educational Progress (2007). *NAEP 2007 mathematics report card for the nation and the states.* Princeton: Educational Testing Service.

National Council of Teachers of Mathematics (1989). *Curriculum and evaluation NCTM standards for school mathematics.* Reston: The National Council of Teachers of Mathematics, Inc.

National Council of Teachers of Mathematics (2000). *Principles and NCTM standards for school mathematics.* Reston: The National Council of Teachers of Mathematics, Inc.

National Mathematics Advisory Panel Report (2008). Washington, DC: USDOE (available online at http://www.ed.gov/about/bdscomm/list/mathpanel/report).

No Child Left Behind Act (2001). *Reauthorization of the elementary and secondary education act.* Pub. L. 107-110 2102(4).

Owen, R. L., & Fuchs, L. S. (2002). Mathematical problem-solving strategy instruction for third-grade students with learning disabilities. *Remedial and Special Education, 23*(5), 268–278.

Peterson, P. (1988). Teachers' and students' cognitional knowledge for classroom teaching and learning. *Educational Researcher, 17*, 5–14.

Riley, M. S., Greeno, J. G., & Heller, J. I. (1983). Development of children's problem-solving ability in arithmetic. In H. P. Ginsberg (Ed.), *The development of mathematical thinking* (pp. 153–196). New York: Academic Press.

Scruggs, T. E., & Mastropieri, M. A. (2003). Science and social studies. In H. L. Swanson, K. R. Harris, & S. Graham (Eds.). *Handbook of learning disabilities* (pp. 364–381). New York: Guilford.

Silver, E. A. (1985). Foundations of cognitive theory research for mathematics problem solving. In A. Schoenfeld (Ed.), *Cognitive science and mathematics education* (pp. 33–60). Mahwah: Laurence Erlbaum.

Sternberg, R. (1985). *Beyond IQ: A triarchic theory of human intelligence.* Cambridge: Cambridge University Press.

Swanson, H. L. (1999). Instructional components that predict treatment outcomes for students with learning disabilities: Support for a combined strategy and direct instruction model. *Learning Disabilities Research & Practice, 16*, 109–119.

Trends in International Mathematics and Science Study (2003). Washington: U.S. Department of Education.

Trends in International Mathematics and Science Study (2007). Washington: U.S. Department of Education.

van Garderen, D., & Montague, M. (2003). Visual-spatial representation, mathematical problem solving, and students of varying abilities. *Learning Disabilities Research & Practice, 18*, 246–254.

Vygotsky, L. S. (1989). *Thought and language*. Cambridge: MIT Press.

Whitby, P. (2009). *The effects of a modified learning strategy on the multiple step mathematical word problem solving ability of middle school students with high-functioning autism or Aspergers' disorder*. Unpublished doctoral dissertation, University of Central Florida.

Witzel, B. S. (2005). Using CRA to teach algebra to students with math difficulties in inclusive settings. *Learning Disabilities: A Contemporary Journal, 3*, 49–60.

Witzel, B. S., Mercer, C. D., & Miller, M. D. (2003). Teaching algebra to students with learning difficulties: An investigation of an explicit instruction model. *Learning Disabilities Research and Practice, 18*, 121–131.

Wolraich, M. L., Felice, M. E., & Drotar, D. (Eds.) (1996). *The classification of child and adolescent mental diagnoses in primary care: Diagnostic and statistical manual for primary care (DSM-PC) child and adolescent version*. Elk Grove Village: American Academy of Pediatrics.

Wong, B. Y., Harris, K. R., Graham, S., & Butler, D. L. (2003). Cognitive strategies instruction research in learning disabilities. In H. L. Swanson, K. R. Harris, & S. Graham (Eds.), *Handbook of learning disabilities* (pp. 273–292). New York: Guilford.

Woodward, J. (2006). Making reform-based mathematics work for academically low-achieving middle school students. In M. Montague & A. Jitendra (Eds.), *Teaching mathematics to middle school students with learning difficulties* (pp. 29–50). New York: Guilford Press.

Xin, Y. P. (2008). The effects of schema-based instruction in solving mathematics word problems: An emphasis on prealgebraic conceptualization of multiplicative relations. *Journal for Research in Mathematics Education, 39*, 526–551.

Xin, Y. P., & Jitendra, A. K. (2006). Teaching problem-solving skills to middle school students with mathematics difficulties. In M. Montague & A. K. Jitendra (Eds.), *Teaching mathematics to middle school students with learning difficulties* (pp. 51–71). New York: Guilford.

Xin, Y. P., Jitendra, A. K., & Deatline-Buchman, A. (2005). Effects of mathematical word problem solving instruction on students with learning problems. *Journal of Special Education, 39*, 181–192.

Xin, Y. P., Wiles, B., & Lin, Y-Y. (2008). Teaching conceptual model based word problem story grammar to enhance mathematics problem solving. *Journal of Special Education, 42*, 163–178.

Xin, Y. P., & Zhang, D. (2009). Exploring a conceptual model based approach to teaching situated word problems. *Journal of Educational Research, 102*, 427–441.

Zawaiza, T. B. W., & Gerber, M. M. (1993). Effects of explicit instruction on community college students with learning disabilities. *Learning Disabilities Quarterly, 16*, 64–79.

Commentary on the Chapter by Marjorie Montague and Asha Jitendra, "Research-Based Mathematics Instruction for Students with Learning Disabilities"

How do Cognitive Strategy Limitations Relate to Other Aspects of Mathematical Difficulties?

Ann Dowker

This is an extremely interesting chapter that describes children's difficulties in mathematics. It discusses the problem of learning difficulties in mathematics. As the authors point out, mathematical difficulties are indeed very common. This has been found to be so internationally (Butterworth 2005; Bzufka et al. 2000; Gross-Tsur et al. 1996), though there can be problems in defining the point where they are severe, specific and persistent enough to be diagnosed as a specific disability (Mazzocco and Myers 2003). Studies in the UK (e.g. Bynner and Parsons 1997) have shown that persistent numeracy difficulties in adulthood are much commoner and at least as disabling as literacy difficulties.

Montague and Jitendra explore the interesting issue of weaknesses in organizing cognitive strategies and report very interesting studies suggesting that training in cognitive strategy selection and use, for example in using schemas in word—problem solving, can lead to significant improvement in mathematical performance. This is a most important result, which will help to guide instruction for children with, and possibly without, mathematical difficulties.

Questions arise about the reasons for the findings. There is the 'chicken and egg' problem: are the problems with strategy formation and use a cause or an effect of mathematical difficulties? Do children start by having difficulties with strategy selection and planning in general, perhaps because of weaknesses in working memory or executive function, and therefore have problems in selecting and planning mathematical strategies in particular? Or do they fail to plan and use strategies effectively because of their general lack of mathematical knowledge or understanding? They may not have access to some important strategies, or to knowledge of arithmetical properties and relationships that may inform such strategies; they may respond randomly as a result of failing to comprehend what the mathematical problems involve; or they may be so insecure about their ability to get the right answer—or perhaps even any answer—if they stray from known paths that they stick to a small range

A. Dowker (✉)
University of Oxford, Oxford, UK
e-mail: ann.dowker@psy.ox.ac.uk

H. Forgasz, F. Rivera (eds.), *Towards Equity in Mathematics Education,*
Advances in Mathematics Education,
DOI 10.1007/978-3-642-27702-3_45, © Springer-Verlag Berlin Heidelberg 2012

of strategies, even when these are inefficient or even completely inappropriate for a given task. It is known that children with mathematical difficulties often stick to a restricted range of counting-based strategies for longer than other children do; and at the other end of the scale, expertise in mathematics, e.g. in professional mathematicians, is associated with the use of a particularly wide range of appropriate strategies (Dowker et al. 1996).

As regards further research that needs to be done, the two main areas that I would propose are (1) the relationship of the findings to the componential nature of arithmetical thinking and (2) the importance of emotion in arithmetical performance and cognition.

There is by now a great deal of evidence that arithmetical ability is not a single entity but is made up of multiple components—for example, memory for number facts; estimation derived fact strategies; word problem comprehension solving; counting concepts and procedures; etc. This evidence comes from many converging sources, including experimental, educational and factor analytic studies of typically developing children and adults (e.g. Dowker 1998, 2005; Geary and Widaman 1992; Jordan et al. 2009; Siegler 1988); studies of children with mathematical difficulties (Butterworth 2005; Dowker 2005; Geary and Hoard 2005; Jordan and Hanich 2000; Russell and Ginsburg 1984); and functional brain imaging studies (e.g. Ansari et al. 2005; Castelli et al. 2006; Dehaene et al. 1999; Kaufmann et al. 2009). Moreover, though the different components often correlate with one another, weaknesses in any one of them can occur relatively independently of weaknesses in the others. The components do not necessarily function *as a hierarchy*. A child may perform well at an apparently difficult task (e.g. word problem solving) while performing poorly at an apparently easier component (e.g. remembering the counting word sequence). The componential nature of arithmetic seems to have very early roots: 4-year-olds can show marked discrepancies, in both directions, between counting procedures and concepts, and between different counting concepts (Dowker 2008).

There is recent evidence that intervention programmes that involve assessments of the different components and are targeted to children's specific strengths and weaknesses are particularly effective in improving the performance of primary school children with mathematical difficulties (Dowker and Sigley 2010; Dunn et al. 2010; Gross 2007; Wright et al. 2005).

There should now be more research on the extent to which cognitive strategy instruction is particularly important to different components. Is it particularly important with regard to components that specifically involve applying arithmetical principles to devising effective strategies (notably derived fact strategy use), or comprehending the meaning of a mathematical problem (e.g. word problem solving, as in the schema-based instruction approaches discussed in the article), or would it be equally beneficial to all components?

The second important issue is that of emotions, especially negative emotions. Emotion as well as cognition is important to mathematical performance. Dehaene (1997) points out that even research on the neural bases of mathematics must take emotion into account: "cerebral function is not confined to the cold transformation of information according to logical rules" (p. 235).

Most notably, some people have extremely negative attitudes to mathematics, amounting to *mathematics anxiety* (Hembree 1990) or even a true phobia of mathematics. This can affect mathematical performance both by causing avoidance of mathematical activities and reducing opportunities for learning in the medium and long term and possibly by overloading working memory during mathematics tasks and thus inhibiting performance in the short term. Ashcraft et al. (2007, p. 345) have proposed that math anxiety operates in some ways 'like a genuine math learning disability', and affects a larger proportion of the population than most other mathematical learning disabilities. Even among people who do not have very marked degrees of mathematics anxiety, attitudes to mathematics tend to become more negative with age during the primary school years, and to decline markedly from primary to secondary school age (Blatchford 1996; Dowker 2005).

It should be examined whether anxiety has a particularly deleterious effect on effective strategy choice, either through the overloading of working memory resources, or because the child become desperate to give *some* sort of answer and therefore does not take the time to develop a strategy (Holt 1966). There should also be studies of the extent to which cognitive strategy instruction—and other forms of intervention—relieve anxiety as well as improving performance.

Montague and Jitendra's final statement is particularly crucial: "In conclusion, there are research-based interventions that have been proven effective for students with LD. However, the success of these interventions often depends on how, when, and by whom they are provided to students." This is indeed a very important issue both for practice and for further research.

References

Ashcraft, M. H., Krause, J. A., & Hopko, D. R. (2007). Is math anxiety a mathematical learning disability? In D. Berch & M. Mazzocco (Eds.), *Why is math so hard for some children? The nature and origins of mathematical learning difficulties and disabilities* (pp. 329–348). Baltimore: Paul Brookes.

Ansari, D., Garcia, N., Lucas, E., Hamon, K., & Dhilil, B. (2005). Neural correlates of symbolic number processing in children and adults. *Neuroreport, 16*, 1769–1775.

Blatchford, P. (1996). Pupils' views on school work and school from 7 to 16 years. *Research Papers in Education, 11*(3), 263–288.

Butterworth, B. (2005). Developmental dyscalculia. In J. I. D. Campbell (Ed.), *Handbook of mathematical cognition* (pp. 455–467). Hove: Psychology Press.

Bynner, J. & Parsons, S. (1997). *Does numeracy matter? Evidence from the national child development study on the impact of poor numeracy on adult life.* London: Basic Skills Agency.

Bzufka, M. W., Hein, J., & Neumärker, K. J., 2000. Neuropsychological differentiation of subnormal arithmetic abilities in children. *European Child and Adolescent Psychiatry, 9*, 65–76.

Castelli, F., Glaser, D. E., & Butterworth, B. (2006). Discrete and analogue quantity processing in the parietal lobe: A functional MRI study. *Proceedings of the National Academy of Sciences, 103*, 4693–4698.

Dehaene, S. (1997). *The number sense.* London: Macmillan.

Dehaene, S., Spelke, E., Pinel, P., Stanescu, R., & Tsivkin, S. (1999). Sources of mathematical thinking: behavioural and brain evidence. *Science, 284*, 970–974.

Dowker, A. (1998). Individual differences in normal arithmetical development. In C. Donlan (Ed.), *The development of mathematical skills* (pp. 275–302). Hove: Psychology Press.

Dowker, A. (2005). *Individual differences in arithmetic: Implications for psychology, neuroscience and education.* Hove: Psychology Press.

Dowker, A. (2008). Individual differences in numerical abilities in preschoolers. *Developmental Science, 11,* 650–654.

Dowker, A. & Sigley, G. (2010). Targeted interventions for children with mathematical difficulties. *British Journal of Educational Psychology, Monograph Series II, 7,* 65–81.

Dowker, A. D., Flood, A., Griffiths, H., Harriss, L., & Hook, L. (1996). Estimation strategies of four groups. *Mathematical Cognition, 2,* 113–135.

Dunn, S., Matthews, L., & Dowrick, N. (2010). Numbers count: Developing a national approach to intervention. In I. Thompson (Ed.), *Issues in teaching numeracy in primary schools.* Maidenhead, Berks: Open University Press.

Geary, D. C. & Hoard, M. K (2005). Learning disabilities in arithmetic and mathematics. In *Handbook of mathematical cognition* (pp. 253–267). Hove: Psychology Press.

Geary D. C. & Widaman, K. E. (1992). On the convergence of componential and psychometric models. *Intelligence, 16,* 47–80.

Gross, J. (2007). Supporting children with gaps in their mathematical understanding. *Educational and Child Psychology, 24,* 146–156.

Gross-Tsur, V., Manor, O., & Shalev, R. S. (1996). Developmental dyscalculia: Prevalence and demographic features. *Developmental Medicine and Child Neurology, 38,* 25–33.

Hembree, R. (1990). The nature, effects and relief of mathematics anxiety. *Journal for Research in Mathematics Education, 21,* 33–46.

Holt, J. (1966). *How children fail.* New York: Pitman.

Jordan, N. C. & Hanich, L. B. (2000). Mathematical thinking in second grade children with different forms of LD. *Journal of Learning Disabilities, 33,* 567–578.

Jordan, J. A., Mulhern, G. & Wylie, J. (2009). Individual differences in trajectories of arithmetical development in typically achieving 5- to 7-year-olds. *Journal of Experimental Child Psychology, 103,* 455–468.

Kaufmann, L., Vogel, S. E., & Starke, M. (2009). Numerical and non-numerical ordinality processing in children with and without developmental dyscalculia: Evidence from fMRI. *Cognitive Development, 24,* 486–494.

Mazzocco, M. & Myers, G. F. (2003). Complexities in identifying and defining mathematics learning disability in the primary school age years. *Annals of Dyslexia, 53,* 218–253.

Russell, R. L. & Ginsburg, H. P. (1984). Cognitive analysis of children's mathematics difficulties. *Cognition and Instruction, 1,* 217–244.

Siegler, R. S. (1988). Individual differences in strategy choice: Good students, not-so-good students and perfectionists. *Child Development, 59,* 833–851.

Wright, R., Martland, J. & Stafford, L. (2005). *Early numeracy: Assessment for teaching and intervention* (2nd ed.). London: Sage.

Commentary on the Chapter by Marjorie Montague and Asha Jitendra, "Research-Based Mathematics Instruction for Students with Learning Disabilities"

Delinda van Garderen

In their chapter, Montague and Jitendra highlight that many students with disabilities are not making adequate progress in their mathematical performance. Further, both cognitive and behavioral characteristics appear to interfere with their performance and development in mathematics. In response to these concerns, Montague and Jitendra highlight two instructional approaches—*Solve It!* and Schema-Based Instruction (SBI), that can be used to help students with disabilities improve their mathematical performance.

Both *Solve It!* and SBI draw on research that suggests many students with learning disabilities (LD) have poor 'strategic competence' (e.g. the ability to use representations while solving a problem as well as a systematic approach to solve problems) and executive functions (e.g. the self-regulating of problem solving performance through the use of metacognitive strategies). For this reason, Montague and Jitendra both engage the students in the use of cognitive (e.g., representation via a diagram) and metacognitive strategies (e.g. self-question and self-check), through explicit instruction, in order to learn how to solve mathematical word problems in general or within specific content or schema such as additive and multiplicative structures. The outcomes of their research clearly suggest that these approaches are effective practices for improving the mathematical performance for solving word problems of students with LD. Building on the effectiveness of these research-based interventions, they provide three excellent recommendations (i.e., collaboration between general and special educators, increased content specialization of special educators, and the use of specialized small group instruction for those student who need it) to promote the use of interventions such as *Solve It!* and SBI by classroom teachers.

For this commentary, I am going to focus my response on two main thoughts. First, the continued need to build on and extend the research on cognitively based

D. van Garderen (✉)
Department of Special Education, University of Missouri, 303 Townsend, Columbia, MO 65211, USA
e-mail: vangarderend@missouri.edu

H. Forgasz, F. Rivera (eds.), *Towards Equity in Mathematics Education,*
Advances in Mathematics Education,
DOI 10.1007/978-3-642-27702-3_46, © Springer-Verlag Berlin Heidelberg 2012

mathematics problem solving. Second, the implications their recommendations for use of interventions, such as *Solve It* and SBI, have on factors beyond the classroom.

1 Building on and Extending the Cognitively Based Research in Math Problem Solving

According to the National Research Council ("NRC" 2001), there are five strands that are involved in being mathematically proficient. Those strands included: (1) conceptual understanding, (2) procedural fluency, (3) strategic competence, (4) adaptive reasoning, and (5) productive disposition. Although each strand is dependent on the other in the development of mathematical proficiency, it is understood that each also has a unique contribution that plays a critical role in the acquisition of mathematical proficiency (e.g. NRC 2001). Arguably, the research-based practices *Solve It!* and SBI have a primary focus on developing the strategic competence of students in order to use strategies such as representation and cognitive/metacognitive steps to solve mathematical word problems. Despite the solid research suggesting these intervention strategies, and others like it (e.g.: Hutchinson 1993; van Garderen 2007) can improve mathematical word problem performance, what is unclear is the extent to which these interventions impact mathematical proficiency. Further, it is also unclear whether sustained or even increased growth in mathematics performance can occur following intervention instruction. Therefore, one of the challenges I believe needs further attention is the development of interventions that not only incorporate a strategic focus, but also address the other strands (i.e., conceptual understanding, adaptive reasoning, productive disposition, and procedural fluency), which contribute to mathematical proficiency.

To elaborate, it has been established that representations such as a diagram can play a pivotal role for solving word problems (e.g.: Hegarty and Kozhevnikov 1999; van Garderen and Montague 2003). However, in order to accurately solve the problem, the quality and information depicted within those representations is crucial (e.g.: Diezmann 2000; Dufour-Janvier et al. 1987; Hegarty and Kozhevnikov 1999; van Garderen et al. under review). In particular, it has been found that students who draw diagrams that are primarily relational (i.e. images that depict the spatial relations described in a problem) are more likely to solve the problem correctly than those who draw primarily pictorial (i.e. an image of the persons or objects referred to in the problem) diagrams (van Garderen and Montague 2003). Yet, in order to draw a diagram that is primarily relational and to effectively use that diagram in the solution process, an understanding of the mathematical concepts presented in the word problem is necessary (e.g.: Brown and Presmeg 1993; Diezmann 2000; van Garderen et al. under review). For example, consider the following problem: "Adams Middle School students made an Italian submarine sandwich that was 12 ¾ feet long. After making it, they decided to divide the sandwich into smaller portions to share with other students. If each portion was ¾ of a foot long, how many students would get a portion?" In order to draw a diagram that is relational, the problem solver would need to recognize that this problem involves knowledge of division and fractions. This knowledge can then be used to generate an image that goes beyond a depiction of a sandwich, to a sandwich divided into "equal" sized portions as a way to determine the number of portions that can be generated from a given whole. Therefore, it may be necessary to develop interventions that not only focus on strategic competence, but also incorporate the conceptual understanding of the mathematical concepts presented.

2 Promoting Instruction for Students with Disabilities: Implications Beyond the Classroom

To conclude their chapter, Montague and Jitendra propose three recommendations (i.e. collaboration between general and special educators, increased content specialization of special educators, and the use of more intensive small group instruction) that are important and, I believe, timely. With these recommendations in mind, I would like to highlight some challenges and additional considerations that go beyond the classroom setting that may also be important to the implementation of these recommendations.

Like Montague and Jitendra, I believe that there is a need to promote collaboration among general educators and special educators to increase a shared responsibility and expertise for instructing students with disabilities in mathematics. Although collaboration is often promoted as "best practice" (Hang and Rabren 2009; Scruggs et al. 2007; Winn and Blanton 2005), the extent it truly happens within the schools is unclear (Murawski and Swanson 2001). What is also disconcerting is that when collaboration is reported to occur within the schools, it is often fraught with tension and challenges that frequently undermine its effectiveness (e.g. Kloo and Zigmond 2008; Parmar and DeSimone 2006; Scruggs et al. 2007). To help address these concerns, a common recommendation is for teacher preparation programs to provide strong programs that help prepare teachers for collaboration across the disciplines (Baroody 2011; Brownell et al. 2010). However, there is even less information on what teacher preparation programs are doing to help teachers prepare for collaboration than what teachers are doing in the classroom (Brownell et al. 2010). Certainly, many special education teachers feel unprepared to teach mathematics and often lack mathematical content knowledge, whereas general education teachers have limited knowledge of specialized practices that can assist students struggling with mathematics including those with disabilities (Maccini and Gagnon 2002, 2006). What is clear is that there is an important need for targeted research to identify effective practices for preparing all teachers to teach mathematics to students with disabilities.

Interestingly, recent research has begun to suggest that teacher preparation alone may be insufficient for improving student achievement. It has been suggested that the teachers' subject matter knowledge may be a critical component for improving student achievement (Wayne and Young 2003). Further, effective teachers not only have content knowledge but possess a domain expertise that is, "…knowledge of how the discipline is structured and how students build knowledge within it" (Brownell et al. 2010, p. 368). This domain expertise has been found to be an essential element of effective instruction (Ball et al. 2008; Hill et al. 2008). The implications of these findings for special educators are huge. Not only do they need disability-specific knowledge, they must also be knowledge-able of intervention strategies that address student needs in mathematics and have an excellent understanding of the content in mathematics in order to recognize how a disability may affect or impact learning in mathematics, (Brownell et al. 2010). To have this knowledge and expertise in one content area such as mathematics let alone the numerous content areas special educators are required to manage is a daunting task. Clearly, this would support Montague and Jitendra's contention for the need of specialists in mathematics to work with students with special needs. However, it has been suggested that teacher preparation no longer focuses on specific disability-related needs or even content areas rather, the emphasis of teacher preparation programs is on general behavioral principles and classroom management practices across the disabilities and content areas (Brownell et al. 2010).

Along with the need for specialists in the content areas there is a need to reconceptualize how instruction for students with disabilities within schools is provided. Montague and Jitendra recommend the use of specialized, small group instruction for those that need intense and focused instruction outside the general education classroom. While this recommendation makes logical sense given the needs of students with disabilities in mathematics, it is, however, counter to what is happening in schools. In response to policy, research, and school practice, the practice of special education service delivery has increasingly moved into the general education classroom rather than outside it (Brownell et al. 2010; Kloo and Zigmond 2008). Further, the majority of intervention research currently being conducted in special education appears to be related to improving instruction in mathematics for students with special needs within the general education classroom or multi-tier based systems (e.g. research on screening measures [Bryant et al. 2008; Jordan et al. 2010; Lembke et al. 2008; Seethaler and Fuchs 2010], or classroom-based interventions for tiered instruction [Clarke et al. 2008; Fuchs et al. 2010; Powell and Fuchs 2010]). Although this research is extremely valuable (e.g., to identify and support those struggling in mathematics as early as possible), the development and further refinement of interventions specifically targeted for students identified as having a disability in mathematics is critical to support these students in all areas of mathematical proficiency. In addition, the preponderance of current intervention research for students with disabilities is primarily focused on one content strand of mathematics, numbers and operation (van Garderen et al. 2009). While this is a critical and foundational strand for mathematical understanding it is only one of many strands of mathematics that a student is expected to be proficient in throughout their educational career. Thus, in order for specialized instruction to be effective the research field needs to continuing widening its scope of exploration and address the specific needs of students with disabilities across all of the mathematical strands both in and outside of the general education setting.

Montague and Jitendra's recommendations, to promote stronger more effective collaboration among teachers, to develop content specific specialists who provide are able to provide intense and specialized instruction in mathematics for students with disabilities, and to shift how and where the student is receiving instruction to more intensive small group settings, provide the reader with some school based directions and implications. I have chosen to extend those recommendations beyond the school setting with two general suggestions. First, there is a need to re-think special education teacher preparation. Like Brownell et al. (2010) note, "reconceptualizing the preparation of special educations teachers... is a perplexing but necessary undertaking" (p. 366). Specifically, requiring a solid foundation in educational practices in both general and special education may not be enough. Rather it may be necessary for special education programs to consider adding advanced or alternative preparation courses focused on training special educators to work within a specific content area in order to provide services directly to students with special needs (Brownell et al. 2010). Second, in order for intervention research to occur and schools be able to implement various models of instruction that best meet the needs of students with disabilities, support for the actual implementation is crucial in both legislature policy and public funding. Unfortunately, many school districts are struggling to fund additional services and supports with budget cuts and cutbacks particularly when it comes to providing additional yet necessary supports for students with special needs (e.g. Santangelo 2009). However, without targeted, intense instruction for students with special needs, these students will continue to perform more poorly than their peers without disabilities (e.g., National Assessment of Educational Progress 2007; National Mathematics Advisory Panel 2008) making little if any progress over time.

Over the past few years the research focused on mathematics and students with disabilities has increased. This is an encouraging sign. Further, as Montague and Jitendra have demonstrated, there are research-based practices that demonstrate improved mathematical performance of students with disabilities. However, there is still a lot to do. Clearly, there is a great need to keep working towards identifying, refining and extending interventions designed to improve mathematical proficiency for students with disabilities, along with developing rigorous programs that prepare special educators to work with students with disabilities in mathematics.

References

Ball, D. L., Thames, M. H., & Phelps, G. (2008). Content knowledge for teaching: What makes is special? *Journal of Teacher Education, 59*, 389–407.

Baroody, A. J. (2011). Learning: A framework. In F. Fennell (Ed.), *Achieving fluency: Special education and mathematics* (pp. 15–58). Reston: National Council of Teachers of Mathematics.

Brown, D. L., & Presmeg, N. C. (1993). Types of imagery used by elementary and secondary school students in mathematical reasoning. In *Proceedings of the 17th annual meeting of the international group for the psychology of mathematics education* (Vol. *II*, pp. 137–144). Tsukuba, Japan: PME.

Brownell, M., Sindelar, P. T., Kiely, M. T., & Danielson, L. C. (2010). Special education teacher quality and preparation: Exposing foundations, constructing a new model. *Exceptional Children, 76*(3), 357–377.

Bryant, D. P., Bryant, B. R., Gersten, R. M., Scammacca, N. N., Funk, C., Winter, A., Shih, M. & Pool, C. (2008). The effects of tier 2 intervention on the mathematics performance of first-grade students who are at-risk for mathematics difficulties. *Learning Disability Quarterly, 31,* 47–63.

Clarke, B., Baker, S., Smolkowski, K., & Chard, D. J. (2008). An analysis of early numeracy curriculum-based measurement. *Remedial and Special Education, 29,* 46–57. doi:10.1177/0741932507309694.

Diezmann, C. (2000). Making sense with diagrams: Students' difficulties with feature-similar problems. In *Proceedings of the 23rd annual conference of mathematics education research group of Australasia* (pp. 228–234). Freemantle, Australia: MERGA.

Dufour-Janvier, B., Bednarz, N., & Belanger, M. (1987). Pedagogical considerations concerning the problem of representation. In C. Janvier (Ed.), *Problems of representation in the teaching and learning of mathematics* (pp. 109–122). Hillsdale: Lawrence Erlbaum Associates.

Fuchs, L. S., Powell, S. R., Seethaler, P. M., Cirino, P. T., Fletcher, J. M., Fuchs, D., & Hamlett, C. L. (2010). The effects of strategic counting instruction, with and without deliberate practice, on number combination skill among students with mathematics difficulties. *Learning and Individual Differences, 20,* 89–100.

Hang, Q. & Rabren, K. (2009). An examination of co-teaching: Perspectives and efficacy indicators. *Remedial and Special Education, 30*(5), 259–268.

Hegarty, M. & Kozhevnikov, M. (1999). Types of visual-spatial representations and mathematical problem solving. *Journal of Educational Psychology, 91,* 684–689.

Hill, H. C., Blunk, M. L., Charalambos, C. Y., Lewis, J. M., Phelps, G. C., Sleep, L., & Ball, D. L. (2008). Mathematical knowledge for teaching and the mathematical quality of instruction: An exploratory study. *Cognition and Instruction, 26,* 1–81.

Hutchinson, N. L. (1993). Effects of cognitive strategy instruction on algebra problem solving with adolescents. *Learning Disability Quarterly, 16,* 34–63.

Jordan, N. C., Glutting, J., & Ramineni, C. (2010). The importance of number sense to mathematics achievement in first and third grades. *Learning and Individual Differences, 20,* 82–88. doi:10.1016/j.lindif.2009.07.004.

Kloo, A. & Zigmond, N. (2008). Coteaching revisited: Redrawing the blueprint. *Preventing School Failure, 52,* 12–20. doi:10.3200/PSFL.52.2.12-20.

Lembke, E. S., Foegen, A., Whittaker, T. A., & Hampton, D. (2008). Establishing technically adequate measures of progress in early number. *Assessment for Effective Intervention, 33,* 206–214.

Maccini, P. & Gagnon, J. C. (2002). Perceptions and application of NCTM standards by special and general education teachers. *Exceptional Children, 68,* 325–344.

Maccini, P. & Gagnon, J. C. (2006). Mathematics instructional practices and assessment accommodations by secondary special and general educators. *Exceptional Children, 72,* 217–234.

Murawski, W. W. & Swanson, H. L. (2001). A meta-analysis of co-teaching research. Where are the data? *Remedial and Special Education, 22,* 258–267.

National Assessment of Educational Progress. (2007). *NAEP 2007 mathematics report card for the nation and the states.* Princeton: Educational Testing Service.

National Mathematics Advisory Panel. (2008). *Foundations for success: The final report of the National Mathematics Advisory Panel.* Washington: U.S. Department of Education.

National Research Council. (2001). *Adding it up: Helping children learn mathematics.* J. Kilpatrick, J. Swafford, & B. Findell (Eds.). Washington: National Academy Press.

Parmar, R. S. & DeSimone, J. R. (2006). Facilitating teacher collaboration in middle school mathematics classrooms with special-needs students. In M. Montague & A. Jitendra (Eds.) *Middle school students with mathematics difficulties* (pp. 154–174). New York: Guildford Press.

Powell, S. R. & Fuchs, L. S. (2010). Contribution of equal-sign instruction beyond word problem tutoring for third-grade students with math difficulty. *Journal of Educational Psychology, 102*(2) 381–394.

Santangelo, T. (2009). Collaborative problem solving effectively implements, but not sustained: A case for aligning the sun, the moon, and the starts. *Exceptional Children, 75*(2), 185–209.

Scruggs, T. E., Mastropieri, M. A., & McDuffie, K. A. (2007). Co-teaching in inclusive classrooms: A metasynthesis of qualitative research. *Exceptional Children, 73*(4), 392–416.

Seethaler, P. M. & Fuchs, L. S. (2010). The predictive utility of kindergarten screening for math difficulty. *Exceptional Children, 77*, 37–59.

van Garderen, D. (2007). Teaching students with learning disabilities to use diagrams to solve mathematical word problems. *Journal of Learning Disabilities, 40*(6), 540–553.

van Garderen, D., & Montague, M. (2003). Visual-spatial representation, mathematical problem solving, and students of varying abilities. *Learning Disabilities Research and Practice, 18*(4), 246–254.

van Garderen, D., Scheuermann, A., & Jackson, C. (in press). Examining how students with diverse abilities use diagrams to solve mathematics word problem. *Learning Disability Quarterly.*

van Garderen, D., Scheuermann, A., Jackson, C., & Hampton, D. (2009). Supporting collaboration between general educators and special educators to teach students who struggle with mathematics: A review of recent empirical literature. *Psychology in the Schools, 46*(1), 56–77.

Wayne, A. J. & Young, P. (2003). Teacher characteristics and student achievement gains: A review. *Review of Educational Research, 73*, 89–122.

Winn, J. & Blanton, L. (2005). The call for collaboration in teacher education. *Focus on Exceptional Children, 38*(2), 1–10.

Neural Correlates of Gender, Culture, and Race and Implications to Embodied Thinking in Mathematics

Ferdinand Rivera

Abstract In this chapter, I discuss neuroscience research and selected findings that are relevant to mathematics education. What does it mean, for example, to engage in a neuroscientific analysis of symbol reference? I also discuss various research programs in neuroscience that have useful implications in mathematics education research. Further, I provide samples of studies conducted within and outside mathematics education that provide a neural grounding of gender, culture, and race. The chapter closes with three brief implications of neuroscientific work in mathematics education research, in general, and in individual- and intentional-embodied cognition in mathematical thinking and learning, in particular.

1 Introduction

If one has to hazard a guess, the science of the early 21st century will be driven by brain research. ...At the core of the new brain science is an astounding mix of technologies adapted from other sciences. None has been developed with the brain in mind, but they have radically transformed neuroscience. (Hacking 2004, p. 26)

Functional neuroimaging techniques pick up on signals indicating brain activity. These signals, by themselves, do not specify a behavior. Only by linking these brain signals with behavior do they have psychological meaning. (Phelps and Thomas 2003, p. 755)

Broadly speaking, the aim of Cognitive Neuroscience is to elucidate how the brain enables the mind... to constrain cognitive, psychological theories with neuroscientific data, thereby shaping such theories to be more biologically plausible. (Ansari et al. 2011, p. 1)

While there have been significant advances in neuroscientific methods, tools and techniques, and findings in the last decade, neuroscience, a term coined in the 1960s, as a scientific field that studies brain structure and functioning is still in its emergent state. Hacking (2004) points out that current interests and investments in neu-

F. Rivera (✉)
San Jose State University, San Jose, CA, USA
e-mail: ferdinand.rivera@sjsu.edu

Writing of this chapter has been supported by a grant from the National Science Foundation (DRL Grant #0448649) awarded to the author. The ideas explored and expressed in this chapter are those of the author and do not reflect the views of the foundation.

roscience in fields outside of education[1] basically seek the resolution of medical issues such as the negative effects of aging and related diseases and disabilities that we all will experience at some point in our lives. Certainly, there is as well a "narcissistic" expectation that "as we learn more about the brain we shall learn more about human nature, about ourselves and our kind" (Hacking 2004, p. 27), especially "our own mental processes" and "higher cognitive functions" (Editorial 2003, p. 1239) in both individual and collective contexts, which could be potentially controversial in many cases.

Despite and amidst healthy skepticism and productive critique in various theoretical and methodological components of neuroscience research in nonmedical contexts (e.g., Anderson and Reid 2009; Coch and Ansari 2009; Fuson 2009; Geertz 2000; Hardcastle and Stewart 2002; Kaufmann 2008; Willingham 2009), I share the view of Ansari et al. (2011) in the opening epigraph, that the impressive eruptive findings in this field will consequently provide valuable descriptive (versus prescriptive) information—cautionary tales, perhaps (Goswami 2005)—about neural mechanisms that support cognitive processes in mathematical thinking and learning. 'Descriptive' means that the neurally drawn information is not meant to be interpreted as a recipe manual for optimal learning but as providing knowledge or a level of explanation that sees mathematical thinking as also being about mind/brain functioning and relationships (Anderson and Reid 2009; Ansari 2005). Unfortunately, current mathematics education research knowledge, practice, and policy appear to dawdle through scientific endeavors that address the material or biological components in both cognitive and affective analyses relevant to, say, the learning of concepts, skills, and other processes (Campbell 2006; Grabner et al. 2010; Schlöglmann 2003).

The title to this chapter involves the frequently used term "neural correlates" in matters relevant to neuroscientific data to convey the current content of available empirical evidence. Various findings drawn from neuroscientific experiments in both nonmathematical and mathematical contexts offer, at least for the time being, supporting evidence that is primarily correlational (versus causational) in nature. Correlations between a target behavior and an activated region in the brain do not mean the latter is involved in the former. "[T]here is some relationship," Phelps

---

[1]Certainly motives behind interests in neuroscience outside education depend on stakeholder contexts. Hacking (2004) articulated medical interests as an example. Neuroscientific findings and programs in the in nonmedical issues that bear on national security (National Research Council 2008) are also of interest to federal and military agencies in the USA. In 2008, the NRC published the document, *Emerging Cognitive Neurosciences and Related Technologies*, in which an attempt is made to address ways in which neuroscientific knowledge could be used to eventually develop usable "future warfighting applications" (p. 14) for the intelligence community. Such applications would have neuroscience associated with the following tasks: (1) "read" the "cognitive states and intentions of persons of interest;" (2) "enhance" the "cognitive capacities" of soldiers (how to make them learn faster and process information more quickly and precisely than usual, how to help them make correct decisions when engaged in battle); (3) "control" the "states and intentions" of oneself (e.g., pain, fear) and others (e.g., "disrupt" an "enemy's motivation to fight"), and: (4) "drive devices" via "cognitive states" (e.g., using white noise to impair senses, using neuropsychopharmacology to develop drugs that "target specific sensory receptors") (pp. 16–17).

and Thomas (2003) note, but "an activation response does not inform us as to what, exactly, a brain region does in the generation of a behavior" (p. 753). Nevertheless, the correlational findings should help "deepen our understanding of causal mechanisms" (Goswami 2009, p. 176) underlying, say, skills and representations involved when learners engage in mathematical knowledge construction. Also, consistent and strong correlations may be used to infer neural markers (or what Goswami 2009, calls biomarkers), neural signatures, or neural specificities (Cantlon et al. 2009) with more data that could then provide useful information in constructing relevant and more reliable cognitive measures, assessments, and diagnostic tools.

My interest in neuroscience, in particular those studies that directly tackle issues relevant to mathematical cognition, has been spurred by Thagard's (2010) thoughts about the role of neural processes in making sense of embodied thinking. Embodied action is an emerging area of research interest among mathematics educators around the globe. Two instances of this kind of work involve understanding functions of representational gestures in conveying mathematical meaning, for example, the special *Educational Studies in Mathematics* (*ESM*) issue on gestures and multimodality in mathematical contexts (Radford et al. 2009), and the influence of emotions and other affective factors in sustaining interest in mathematical knowledge construction, for example, the special *ESM* issue on affect in mathematics education (Zan et al. 2006). Thagard (2010) points out, and rightly so, the mutually determining relationships between neural and psychological accounts of human actions, that is, that our "cognitive capacities" could be seen as a complex of "representational/computational abilities that outstrip embodied action" (p. 9). The purported gendered/cultural/racialized nature of surprise, insight, perception, abduction, creativity, emotion, inference making, meaning construction and signifying practices, and so on, that all bear on mathematical thinking processes might "be illuminated by consideration of neural mechanisms" (Thagard 2010, p. 449). Current models of mathematical thinking are, in fact, based on representations that are both physically available (e.g., gestures) and linguistic, which can also be analyzed in computational and neuroscientific terms.

This chapter is organized in five sections. Section 2 clarifies the different (but overlapping) contexts and purposes of neuroscience that are pertinent to issues in mathematics education. What does it mean, for example, to engage in a neuroscientific analysis of symbol reference? Also briefly discussed are various research programs in neuroscience that have useful implications in mathematics education research. Section 3 provides selective examples of work in which a neuroscientific analysis in understanding basic mathematical processes was employed. This section is meant to showcase recent exciting investigations in which attempts are made to ground, albeit not entirely, mathematical thinking in neurophysiological terms. I also point out constraints and limitations of such investigations so that a cautionary habit is developed of seeing where the science ends and the speculation begins, a disposition that Bruer (1999) would have readers acquire, given the allure of developing misleading brain-based educational implications (e.g., mixing correlation and causation; overgeneralizing) or "neuromyths" (Organization for Economic Cooperation and Development [OECD] 2007) on very limited and targeted experiments.

Section 4 provides samples of studies within and outside mathematics education that provide a neural grounding of gender, culture, and race. Section 5 includes three brief implications of neuroscientific work in mathematics education research in general, and in individual- and intentional-embodied cognition in mathematical thinking and learning in particular.

2 Analyzing Issues in Mathematics Education from a Neuroscience Perspective

Modules in the brain or distributed patchworks? (Ansari 2008, p. 279)

Basically, neuroscience involves studies of brain functioning and development or, more generally, the human nervous system (Szücs and Goswami 2007). One implication of this characterization for educational research involves situating the brain in a mediating function so that changes and developments in psychological or behavioral processes can be explained in material terms by understanding the constraints and connections that emerge when brain cells interact with one another (see Fig. 1; Ansari et al. 2011; Pennington et al. 2007). The connections influence either individual structures or pathways between two or more structures. Thus, brain development is seen to be driving developmental changes in various aspects of behavior in both individual and collective contexts. Acquired experiences also exert influence in brain development. Neuroscientists aptly refer to this as synaptic plasticity (Howard-Jones 2008), and learning is one purposeful tool in which experiences are acquired. In school mathematical contexts, for example, children learn mathematical concepts and processes through "targeted experiences" (Szücs and Goswami 2007, p. 114) of the exact nature, which come into contact with their "approximate number sense" that has been neurologically established to be a characteristic of both human and animal brains. In nonschool mathematical contexts such as the home, the manner in which young children experience being cared for early in their development is correlated with adult behavior. From a neuroscientific perspective this means that their experiences help produce brain cells that affect their memory functions and how they cope with stress in the long term (Eliot 2010).

Learning is a central issue in any study involving educational phenomena. In neuroscience, concerns about learning are routed through studies involving memory, that is, sensory[2], short term, long term, and working memory (Howard-Jones

---

[2]Sensory memory lasts for a few seconds and quickly keeps and discards copies of immediately acquired visual and auditory information. Short-term memory (STM) is a short-term storage of information transferred from sensory memory and does not manipulate the acquired information. STM provides a space for engaging in quick calculations and holds visual and auditory information. Working memory (WM) is the active operational component in STM. It actively processes information acquired in STM and is central in the development of language, reading, mathematics, and problem solving. WM also deals with attentional resources in STM such as the ability to concentrate on one aspect of a target object and shutting off others. Long-term memory (LTM) stores information over periods of time and is organized via schemes that join together to form

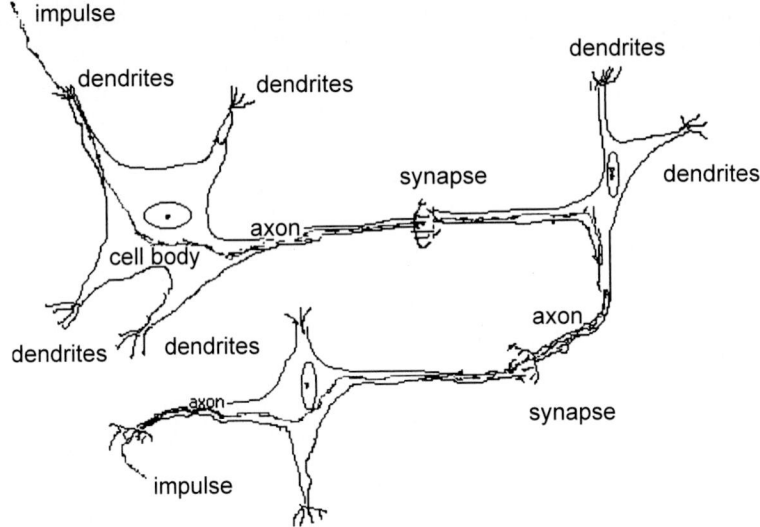

Fig. 1 Brain cell networking (reprinted with permission from http://www.duboislc.org/ EducationWatch/JCameron/01_09_02_HowWeLearn.html)

2008). How the human brain functions and neurally correlates with various parts of the body, and the physical and linguistic operations that come with acting and thinking are all indications of a neuroscientific approach to the study of learning. Aside from learning, the development of one's identity as a gendered, cultured, and racialized being is also a significant educational issue. In this chapter I focus on issues surrounding learning and identity; other concerns in education such as curriculum, instruction, and assessment are not discussed. Certainly, any discussion surrounding equity issues in education involves understanding how these five elements can be aligned well together; this is not of concern here. But I should point out that neuroscientific findings in learning and identity could be used to develop useful implications in the effective design and delivery of curriculum, instruction, and assessment. Space prohibits an exploration of all these aspects in full detail, but I recommend readers to access the most recent references on this matter which are provided in the bibliography.

Goswami (2004) notes that understanding learning at the neuroscientific level involves determining ways in which synapses (i.e. junctions between nerve cells) work in neural (or neuronal) functioning (see Fig. 1). The human nervous system consists of neurons or cells that process and transmit information via synaptic signaling. This signaling occurs in a structure that enables neurons to connect to each other eventually forming a network. The nervous system also includes the brain, the

---

new knowledge structures. Readers are referred to Menon (2010) for an extended discussion of the neuroanatomical correlates of working memory and other relevant cognitive processes relevant in the development of mathematical thinking and skills.

spinal cord, and the peripheral ganglia. Ganglia are tissue masses of bundled nerve cells; they mediate and serve as relay points between two or more neuronal systems such as the peripheral (nerves) and central nervous (brain and spinal cord) systems. Neurons come in several types depending on function. There are sensory neurons that react to sound, touch, and light, and which affect the sensory organ. There are also motor neurons that affect muscle movement, and neural patterns exist that are associated with specific mental representations or mental states. Thus, from a neuroscientific perspective, (successful) learning constitutes (effectively) understanding how changes occur in the neural connections in an individual or among individuals in a social context. Bechtel's (2002) thoughts concerning the complementary analytic approach needed in neuroscientific work echoes the following important point initially raised by Petersen and Fiez (1993) about being wary of the difference between brain localization of a task (i.e. the neuromyth of brain modules) and brain localization of an information processing operation:

> [E]lementary operations, defined on the basis of information processing analyses of task performance, are localized in different regions of the brain. Because many such elementary operations are involved in any cognitive task, a set of distributed functional areas must be orchestrated in the performance of even simple cognitive tasks. ... A functional area of the brain is not a task area; there is no "tennis forehand area" to be discovered. Likewise, no area of the brain is devoted to a very complex function; "attention" or "language" is not localized in a particular Brodmann area or lobe. Any task or "function" utilizes a complex and distributed set of brain areas. (Petersen and Fiez 1993, p. 513)

Figure 2 illustrates the different major subdivisions of the cerebral cortex in a human brain. The cerebral cortex consists of two mirror halves that are often referred to in terms of the left and the right sides of the brain (i.e. brain laterality). The hippocampus and the amygdala are located in the midbrain, under the cerebral cortex. The cortex itself is the largest part of the brain that deals with higher brain functioning such as thinking and acting. The hippocampus is associated with long-term memory and affects spatial orientation performance. Four lobes or sections divide the cerebral cortex and engage in different activities, as follows:

- The *frontal lobe* is engaged in activities that involve planning, reasoning, problem solving, and controlling speech, movement, and emotions;
- The *temporal lobe* is engaged in activities that pertain to memory, language, and recognition of auditory stimulus;
- The *parietal lobe* is engaged in activities that involve the use of spatial processing, orientation, perception, recognition, movement, and touch;
- The *occipital lobe* is engaged in activities that tap vision and visual processing.

A simple example of a neuroscientific investigation is instructive at this stage. Neuroscience research has produced an interesting finding concerning gender differences in spatial processing that is important because it is correlated with mathematical ability. Males and females have been neuroscientifically assessed to employ different neural patterns when thrown in an unfamiliar environment, with the males' left hippocampus showing increased activation compared to females' (Editorial 2005; Grön et al. 2000). Of significance here are sample size and numerical

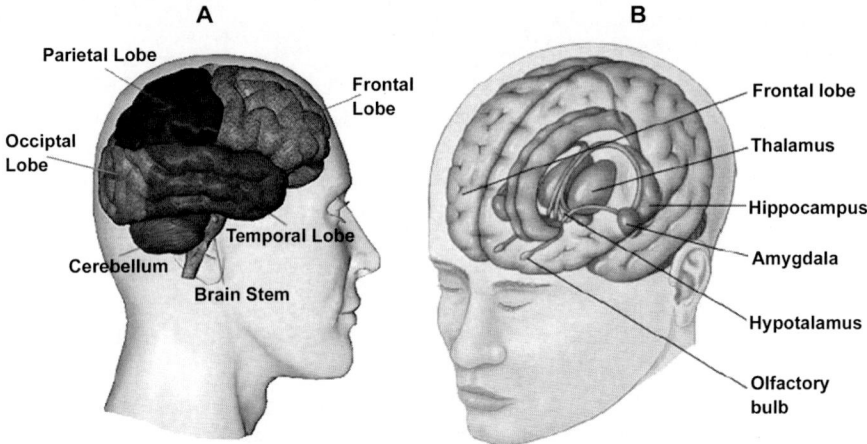

Fig. 2 The human cerebral cortex and the major lobes (reprinted with permission from: (A) http://www.neuroskills.com/edu/ceufunction1.shtml; (B) http://cwx.prenhall.com/bookbind/pubbooks/morris5/chapter2.hml)

amounts of differences in neuroscientific experiments; these values need to be considered in context and analyzed based on the particular tasks presented to the participants. Small differences, for instance, may have neuroscientific value but carry little to no educational impact. Even when there are differences, they may only make sense relative to the tasks used in the study (Bruer 1999). Of concern is the temptation to infer causes on a single part (or parts) of the brain (e.g., the very misleading implications of brain modules and laterality localizations) when the available neuroscientific evidence is basically concerned with establishing correlations. As Szücs and Goswami (2007) have noted, like Petersen and Fiez (1993) before them, "no complex representation can be localized in a single part of the brain" and that "complex phenomena are coded by the interplay of various interconnected neural networks" (p. 115; cf. NRC 2008). Further, we need to know when neuroscientific results end and psychological explanations begin. For example, the observable finding that females in unfamiliar settings employ landmarks while their male counterparts use Euclidean properties of space (e.g., land shape, distances between walls in a room) is a psychological observation. It is, however, reasonable to assume that there is a mutually determining relationship between particular neural and psychological functions[3].

---

[3]In particular, it is worth noting the interesting methodological reflections of Poldrack (2006) and Henson (2006) concerning ways functional neuroimaging data are employed in developing arguments that they term as *forward* and *reverse inferences*, which involve establishing relationships between cognitive functioning and brain activation; for ethical issues involving neural-based reverse inferences, see Poldrack 2008. Forward inferences are deductively valid, and proceeds from assessing neural activity on the basis of performing certain cognitive tasks. Reverse inferences are deductively invalid since they involve making conclusions about cognitive functioning on the basis of brain activation. For Poldrack (2006),

To measure neural activity in the brain, blood flows are actually monitored. A positron emission tomography is not used in educational neuroscientific contexts since it is highly invasive and requires injecting participants with radioactive tracers. Instead, functional magnetic resonance imaging (fMRI), which emerged in 1991, is used to measure neural activity throughout the brain in a non-invasive manner. Participants, oftentimes 5-year-old children and older[4], in recent fMRI experiments wear a head cap that allows a brain scanner to monitor and measure changes in blood flows in the relevant brain regions. What happens is that increased neural activity activates a demand for oxygen, which is then delivered to neurons by hemoglobin. Consequently, increased blood flows occur in the appropriate brain regions. Differences in the magnetic resonance signals of blood (oxygenation) are then used to detect brain activity (i.e. what is oftentimes referred to as the "blood oxygenation level dependent" imaging technique). There are other safe, noninvasive, and indirect neuroimaging techniques (e.g., multiple electrode recording that measures the interactions of brain cells in different cortical lobes, and near-infrared spectroscopy that scans cortical tissue and measures changes in blood hemoglobin concentration) and those that are used to overcome weaknesses (such as time delay) in an fMRI machine (e.g., using a multimodal technique of fMRI and (scalp-based) electroencephalography, which yields gamma and alpha rhythm data waves that measure traces and electrical signals drawn from activation of neurons), but they are not central to the discussion in this chapter (see Bandettini 2009 for an impressive methodological reflection of fMRI technology in neuroscientific work). Suffice it to say, there are indirect ways of measuring neural processes and patterns underlying cognitive activity. Neural patterns that represent sequences of neural firings are indications that some information is being transmitted in a particular way. For example, the 18 adults in an fMRI study by Delazer et al. (2003) have been shown to initially activate their frontal cortical areas associated with working memory in performing long multiplication. With continued practice, activation then shifted to the parietal areas (including shifts within the parietal areas) which are associated

---

cognitive neuroscience is generally interested in a mechanistic understanding of the neural processes that support cognition rather than the formulation of deductive laws. To this end, reverse inference might be useful in the discovery of interesting new facts about the underlying mechanisms. Indeed, philosophers have argued that this kind of reasoning (termed 'abductive inference' by Peirce), is an essential tool for scientific discovery" (p. 60).

However, Henson's (2006) point below is a reminder about being mindful of neuroscientific claims:

[I]t is important to think carefully about the type of inferences that can be made from functional neuroimaging data... only by making these caveats and assumptions explicit, and criticizing them, will we be able to assess the real value of functional neuroimaging for cognitive science" (p. 68).

[4]Kaufmann (2008) points out that current fMRI experiments are restricted to 5-year-old children and older because "fMRI technique requires participants to be awake and respond to stimuli presented in the (narrow and very noisy) scanner environment while simultaneously task-processing related changes in the blood oxygen consumption in different brain regions are recorded" (pp. 2–3).

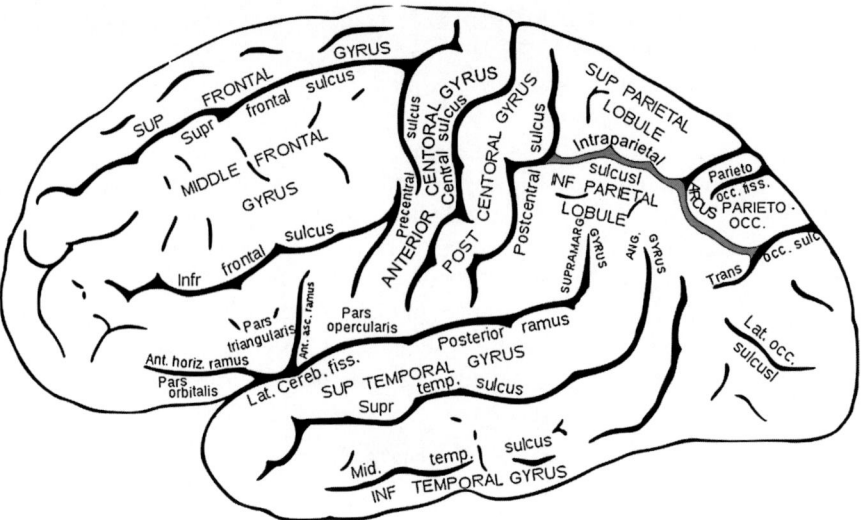

Fig. 3 Lateral surface of the left cerebral hemisphere (http://en.wikipedia.org/wiki/Intraparietal_sulcus)

with automatic processing (Fig. 3). Thus, neural processing of long multiplication in skilled individuals over time tends to shift from quantity-based processing to automatic retrieval.

There are at least three different kinds of neuroscience research programs. While there are clear overlaps insofar as methodology, tools, and analyses, knowing them by label helps identify their primary and fundamental points of interest.

Cognitive neuroscience emerged in the mid-1980s due to dissatisfaction with cognitive psychological theories and findings in scientifically and empirically resolving foundational issues in cognition such as the content of mental images (Pennington et al. 2007). For example, initial studies in neuroscience on the perceptual versus linguistic content of mental images have materially established a neural basis favoring perceptual processing. The methodology in the related cognitive neuroscience experiments adopted connectionist modeling and electrochemical activity, both of which try to give an account of mental states in terms of networks of neural patterns that give rise to learning and development (Goswami 2008). *Developmental cognitive neuroscience* emerged at about the same time. It is basically interested in establishing neural markers of development among individuals relative to some symbol system (e.g., language). While cognitive neuroscience is fundamentally concerned with brain-behavior relations in young and older children, developmental cognitive neuroscience is interested in understanding how such relations evolve over time. Going against the static "hardwired-at-birth" view of cognition, this research program empirically established the existence of developing neural circuits, which then led to the view of mental models as resulting from probabilistic epigenesis and neural constructivism. Here the psychological theories of Piaget have been given a material basis in terms of neural patterns and processes. Further progress in connec-

tionism and molecular genetics has spawned studies in typical and atypical developments and developmental neurobiology (the ontogenetic emergence of the nervous
system) and evolutionary developmental biology (the evolutionary emergence of
different forms of species).

Social/cultural cognitive affective neuroscience is also a recent field of research
interest among (initially) developmental and (later) social and cultural psycholo-
gists. Social neuroscience began in 1992, the year that marked the beginning of the
Decade of the Brain (Adolphs 2010). Initially, it was concerned with the "neurobiology of social behavior" (Adolphs 2010, p. 752). Currently, it involves establishing neural correlates of social cognition, higher-order processes such as moral reasoning, social coordination and cooperation, and perceptions such as stereotyping
and prejudice (about race, social status, etc.). Research concerns in cultural neuroscience, Chiao (2011) points out, are

> motivated by two intriguing questions about human nature: How does culture (that is, val-
> ues, beliefs, practices) shape neurobiology and behavior, and how do neurobiological mech-
> anisms (that is, genetic and neural processes) facilitate the emergence and transmission of
> culture? (Chiao 2011, p. 240)

There is no longer any interest in comparing brain sizes and shapes for the purpose of ascertaining biological primacy and permanent differences between and
among different cultures and race (Eberhardt 2005; Todorov et al. 2006). In this
field, the fundamental aim is "to understand human diversity" (Ames and Fiske
2010, p. 78) that is "perhaps our most precious ability" (Chiao 2011, p. 247). For
example, neuroscientific techniques were used in investigating the powerful social
phenomenon called *theory of mind* in young children in developmental psychology
experiments. The findings indicated that children at about the age of 4 years already understand the intent and mental states of those others around them. Todorov
et al. (2006) underscore the complex and intertwined neural connections between
affect and cognition, which explains why they introduced the term social/cultural
cognitive affective neuroscience as a research field that "entails cognition, emotion, motivation, and readiness for behavior" (p. 82). For example, brain-imaging
experiments have shown greater activity in the amygdala and insular lobe (located between the frontal and temporal lobes, which deals with emotions and interpersonal experiences—see Fig. 4) in both black and white participants when
they were shown black and white faces. However, the activation decreased over
time with more exposure to the colored faces. A study by Sanfey, Rilling, Aronson, Nystrom, and Cohen (2003) had a group of 19 strangers initially meeting
with an experimenter prior to a scanning session. The activity involved evaluating their emotional reactions on both fair and unfair proposals in a simple game
that involved two players (with one assigned the proposer, the other the respon-
der) split a sum of money. Sanfey et al. (2003) also saw significant neural activity in the anterior insular lobe. Apparently, the participants manifested their
disgust neurally (and behaviorally) via a strongly activated anterior insular lobe
in situations when a proposer made an unfair proposal. However, the neural response was not strong in situations when a computer program was used to make

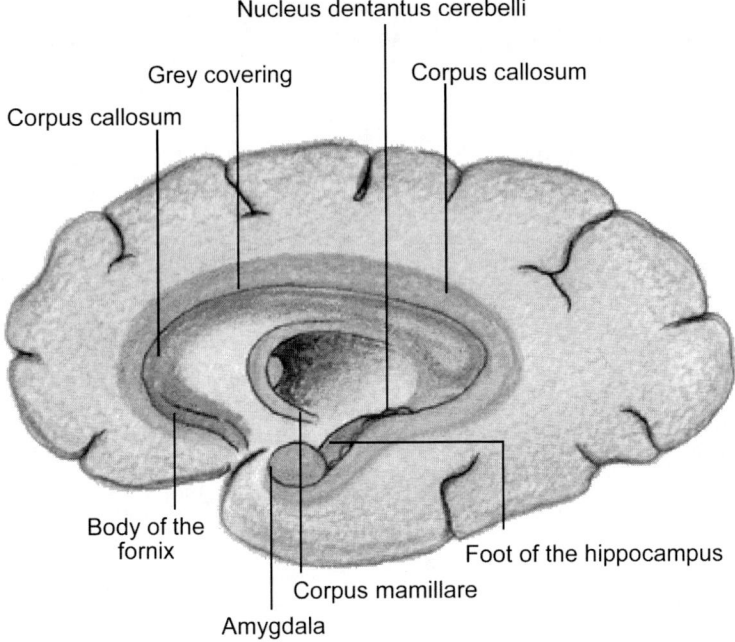

Fig. 4 Amygdala and the insural lobe (reprinted with permission from http://www.hpssandiego. com/VB73a2.jpg)

the same unfair proposal. In a constrained fMRI environment, social neuroscientists basically create psychologically meaningful situations (Todorov et al. 2006, p. 77) that symbolize particular social situations in order to assess participants' neural responses on tasks presented to them in an activity, which may then be correlated with, and could predict, an expected (and perhaps unexpected) behavioral response.

Educational neuroscience is a term recently introduced by Geake (2005), which refers to programs that use educational, neuroscientific, and cognitive psychological methods in understanding mental representational structures. Such structures pertain to how the brain codes information through electrochemical activity, and structural changes could be materially analyzed in terms of cortical changes that occur in individuals. Where a neuroscience approach could complement typical approaches used, cognitive psychology involves establishing a relationship between neural and symbolic activity. Thus, educational neuroscience offers a unifying framework that brings together the analysis of high-level descriptions of the mind (symbolic representations, psychological theories) and lower level data and theories (neuronal activity and function); education is seen as developing "optimal ways of shaping and enriching" individual learners' cognitive systems and mental representations (Szücs and Goswami 2007; Goswami 2008, 2004). Once again, the intent is not about identifying or localizing particular mental functions in the brain, since the functions develop and emerge on the basis of dynamic and distributed neural and cortical networks and pathways that change over time with more experiences, and are recruited to fulfill other relevant cognitive functions. Szűcs and Goswami (2007) clearly articulate one central finding of neuroscience research that should caution educators about laterality recommendations in mathematics education theory and practice, as follows:

> The idea that there is no all-knowing, inner central executive that governs what is known and that orchestrates cognitive development is very important for education. It means that education must deal with the "vast parallel coalition of more-or-less influential forces whose... unfolding makes each of us the thinking beings that we are" (Clark 2006, p. 373). (Szűcs and Goswami 2007, p. 116)

Mental representation of numbers, for example, involves a coordinated activity that taps the parietal lobe, which codes our "approximate sense" of magnitude, and the angular gyrus (Fig. 3) and language-processing parts of the brain, which store memorized arithmetical facts.

In closing this section, I briefly discuss Nieder's (2009) research investigation concerning the neurobiological evolution of symbolic thinking and reasoning in humans and nonhuman primates, which demonstrates the usefulness of neuroscientific findings in developing a case for a material or neural approach to understanding possible biomarkers of abstract thinking in humans. Further details are provided in the next section in which selective findings from the neuroscience of mathematical cognition are presented; for now it simply makes sense to say that humans' evolved number sense ability actually represents a symbol system that does not merely reflect "isolated sign-object associations" (Nieder 2009, p. 99). Some animals have been shown to be capable of limited numerical competence, but it is for the most part nonsymbolic. This means that they operate only at the indexical level (i.e. associations). For example, monkeys and pigeons can be trained to perform simple single-digit and approximate addition, and subtraction limited to very small cardinalities. In humans, number sense competence transcends the indexical status of such associations to include the ability to understand the necessary abstract relationships between signs and the objects they represent, made possible by language. Where neuroscientific methods are useful is in laying neural foundations that support behavioral or psychological explanations and, thus, allows an understanding of how humans and nonhuman primates produce (semantic) meanings to (object) symbols. For example, on the basis of current neurobiological studies with animals and humans, it appears that the dorsolateral granular prefrontal lobe may be neurally responsible for the indexical skill (e.g., macaque monkeys performing shape-to-numerical value associations). The prefrontal lobe is the anterior part (forehead side) of the frontal lobe (see Fig. 3) that does not affect movement when electronically stimulated; one of its psychological roles involves executive function, which deals with processes that pertain to abstract thinking and rule development. The neural correlates of symbolic competence in humans over time (i.e. childhood to adolescence to adulthood) can be initially explained via prefrontal lobe activation. This then shifts to parietal lobe and temporal lobe activations in later development as competence in language improves.

3 Neural Correlates of Mathematical Concepts and Processes

> A central question in cognitive science is whether natural language provides combinatorial operations that are essential to diverse domains of thought. ... We find that only linguistic reasoning excites the known neural substrate of language comprehension, whereas algebra recruits bilateral parietal regions previously implicated in number and magnitude representation. This double dissociation suggests that, at least in the mature brain, the manipulation of algebraic expressions does not rest on the neural machinery of natural language. (Monti et al. 2011, p. 1)

In the psychology of mathematics education research, studies involving the triad of gender, culture, and race (and, more generally, equity) are pursued in terms of their relationships to mathematical learning and thinking. Hence, in this section, the neuroscientific ramp of mathematical cognition is skidded to keep in mind ways in which such findings might inform, change, and advance recent and emerging mathematics-education related theories involving the equity triad. Due space constraints, I engage in a selective reading of exemplar work in the neuroscience of mathematics cognition that addresses the following three basic themes relevant to a study of mathematical thinking and learning in schools, namely: (1) mathematical processing, which is a central skill; (2) linguistic processing, which is a basis for exact mathematical understanding, and; (3) visuospatial processing, which is necessary in visual and spatial (geometric, diagrammatic) thinking. Readers are referred to the special *ZDM* issue on cognitive neuroscience and mathematics learning (Grabner et al. 2010) for further details, but I note one important point raised by Obersteiner et al. (2010) whose work appears in the special issue addressed to the mathematics education community in particular. I echo their view that mathematics education-driven neuroscientific research investigations have the potential to increase the relevance and translational dimension of current cognitive neuroscientific results in mathematical cognition beyond findings drawn from basic numerical tasks presented to participants. Expertise and experience decisions to be made about which complex tasks to use and subject to a meaningful analysis on the basis of their relevance to school mathematical practices and content. Further, since factors such as age, task characteristics, and mathematical competence are often taken into serious consideration in experiments, it is possible to produce a more relevant psychological-neural account of mathematical thinking and learning.

3.1 Mathematical Processing

There has been a tremendous amount of research activity in the area that deals with the neural correlates of arithmetical thinking and skills (e.g., Delazer et al. 2003; Fehr et al. 2007; Kadosh et al. 2008; Rosenberg-Lee et al. 2009; Zamarian et al. 2009; Rocha et al. 2005). Drawn primarily from the influential work of Dehaene (1997, 1992), there is strong and converging neural evidence from various neuroscientific experiments across several countries in different contexts that indicate "number and arithmetic" as being "more than cultural conventions and may have their ultimate roots in brain evolution" (Dehaene et al. 2003, p. 487). Meta-analytic results show that there are regions in human (and animals') brains that support nonverbal or language-independent mental magnitude processing, which enable engagement in simple and complex arithmetical comparison and operation tasks[5]. Dehaene's (1992, 1997) neural network model of triple coding shows that humans possess: (1) an auditory verbal code (left temporal lobe), which is used to recall automatized arithmetical facts such as the multiplication table; (2) a visual code (occipital lobe) for the Hindu-Arabic notation, which is used to perform arithmetical and symbolic operations and make parity decisions, and; (3) a language-independent analog magnitude code (parietal lobe), the famous "mental number line", that allows comparisons to be made among very small numbers, as well as engagement in approximate arithmetic and other spatial judgments of relative sizes. Approximate number processing[6] has been neurally documented to occur *not* in any one particular region but in the parietal, prefrontal, and cingulated regions of the brain with the horizontal segment of the bilateral intraparietal sulcus ("HIPS"; see Fig. 3) as being primarily responsible for the representation and manipulation of numerical quantities and the other regions fulfilling a supportive function in working memory (Dehaene et al. 2004; also see Ansari 2008; Varga et al. 2010 for extended discussions of the HIPS). Jacob and Nieder (2009) also found the same activation pattern network in the case of fraction representation, that is, the "fronto-parietal cortex is tuned to preferred fractions, generalizing across the format of representation" (p. 4652).

Delazer et al. (2003) note the effects of training in significantly modifying patterns in neural activity (e.g., shifts from frontal to parietal activation, or from the intraparietal sulcus to the angular gyrus, or from quantity-based processing to automatic retrieval networks). Dehaene et al. (2004) also point out that arithmetical tasks such as counting and multiplying compared to addition and subtraction might show greater activation in the other regions of the brain, perhaps because they rely on language-based fact retrieval systems or rote verbal memory. For more updated

---

[5]See Varga et al. (2010) and Nieder (2005) for syntheses of research comparing human and animal competence involving concepts of counting, cardinality (numerical quantity), and order (rank) from neuroscientific and neurobiological perspectives. For fMRI comparisons between children and adults involving different aspects of arithmetical processing, see Rocha et al. (2005) and Kawashima et al. (2004). See Ansari (2009) for a review analysis of results drawn from various neuroscientific studies that focused on developmental disorders and difficulties involving numerical cognition and relevant mathematical processes.

[6]Current interests in the implications of approximate number sense are linked to its possible reverse-inferential relationship to school mathematics achievement for both children and adults. For example, based on a longitudinal assessment of 64 14-year-old children with normal development that started in kindergarten, Halberda et al. (2008) established a strong correlation between individual children's approximate number sense and their past scores on standardized school mathematics achievement tests. Mazzocco et al. (in press) established a strong correlation between domain-specific deficits in approximate number processing and persistently deficient mathematics achievement among children with mathematical learning disabilities (i.e. those who scored below the 10th percentile in a mathematics achievement test).

findings, see: Fehr et al. (2007) about the role of working memory capacity and relevant neural networks in complex arithmetical calculations; and Rosenberg-Lee et al. (2009) about different cortical activations that employ the same neural regions due to differing applications of arithmetical strategies such as multiplying from left to right versus right to left and different ordered count-on strategies for adding whole numbers.

Recently, a few studies have focused on other mathematical content beyond arithmetic such as algebra (e.g., Anderson et al. 2003; Lee et al. 2007; Monti et al. 2011; Sohn et al. 2004; Terao et al. 2004). Lee et al. (2007) assessed 18 participants' (all right-handed; 10 males and 8 females whose ages ranged from 20 to 25 years) problem-solving strategies involving typical algebra word problems. Lee et al. (2007) compared and contrasted, at the neural level, the impact of schematic diagrams and alphanumeric symbols in algebra problem solving in working memory. Constructing diagrams taps visual processing while constructing algebraic expressions relies on relevant numerical processing. Results of their fMRI study indicate that no extensive neural differences were found that favored the use of one method over the other, however, there were differences in terms of differential engagement within similar neural processes. The HIPS was actively engaged in both diagrammatic and variable conditions, perhaps as a result of having to compare magnitudes. But what appears to be a more interesting result deals with the neurally drawn finding that using alphanumeric representations actually demands more working memory resources than using diagrams to solve algebra word problems. Monti et al. (2011) neurally assessed 21 right-handed healthy adults when they reasoned about the equivalence and grammatical well-formedness of pairs of linguistic (e.g. "x gave y to z" and "z was given y by x") and algebraic (e.g. "y is greater than z divided by x" and "x times y is greater than z") statements. Results of their fMRI experiment indicate that linguistic equivalence primarily recruited the left fronto-temporal perisylvian (linguistic) regions, while "algebraic equivalence evoked no more activity in these regions than is necessary for simple reading... [but] recruited areas previously reported for number cognition," the bilateral portions of the HIPS (Monti et al. 2011, p. 3).

Developing and assessing inductive and deductive arguments play a central role in constructing and understanding proofs and, more generally, in developing sophisticated mathematical understanding. While there is converging neuroscientific evidence showing activation of the bilaterial fronto-parietal network regions in arithmetical (and algebraic) processing, several recent studies in which the neural correlates of human reasoning were investigated have demonstrated empirically the dominance of the left hemisphere in adult participants who were tested on problems involving inductive and deductive arguments (Goel and Dolan 2004; Goel et al. 1997, 1998, 2000). Both inductive and deductive items activated the left prefrontal lobe. Further, while not evident in deductive processing, inductive processing significantly activated the left dorsolateral prefrontal cortex, and this may be "due to the use of world knowledge in the generation and evaluation of hypotheses" (Goel and Dolan 2004, p. B120). Also, while not evident in inductive processing, deductive processing significantly activated the linguistic neural network and Broca's Area on the basis of significant activation of the left fronto-temporal regions, indicating engagement in syntactical processing (due to the logical form of the propositions) and the use of working memory resources.

3.2 Linguistic Processing

Linguistic processing has been neurally observed to activate the left hemispheric brain regions regardless of ability or disability. What is worth noting is the emerging body of neuroscientific evidence that indicates the neural structures of mathematical cognition as not being linguistic-mediated (or notation-mediated) despite being mediated by numeric symbols (Butterworth et al. 2008; Cantlon et al. 2006; Monti et al. 2011), a view that runs contrary to the Vygotskian thesis that language is necessary for thinking (Brannon 2005), including the well-accepted view that "language forms the basis of structured thought across cognitive domains" (Monti et al. 2011, p. 6). Supporting behavioral evidence has been observed among aphasic (i.e. language-impaired) individuals who have been found incapable of processing simple and complex grammatical relationships such as "The man killed the lion," "The lion killed the man," and "This is the dog that worried the cat that ate the rat that ate the malt that lay in the house that Jack built" but capable of manipulating simple and complex numerical expressions such as $52 - 11$, $11 - 52$, and $(3 + 17) \times 3$ (Brannon 2005, p. 3177; Varley et al. 2005, p. 3519). Certainly, computational recursion is a principle common to both language and mathematics. However, neuroscientific evidence shows that each tends to operate and function on its own, following a distinct syntactical structure. There is also recent work that neurally links the activation of the frontal and parietal lobes (see Fig. 1) in cases when numerical tasks involve the use of linguistic quantifier terms (e.g., some, all, most, more, at least; cf. Hubbard et al. 2008).

3.3 Visuospatial Processing

A recent review of research studies by de Hevia, Vallar, and Girelli (2008) on the role of visuospatial processing in various aspects that matter to arithmetical computing presents converging evidence indicating a "close relationship between numerical abilities and visuospatial processes" (p. 1361). As noted in the previous section on mathematical processing, individuals appear to rely on a spatial mental number line in performing approximate arithmetical tasks. de Hevia et al. (2008) also noted activations of the visuospatial working memory distributed structures, including various regions in the parietal lobe and the supramarginal and angular gyri (see Fig. 3). Interests relevant to possible cultural differences in arithmetical processing have consequently highlighted the salience of visual processing in this domain. For example, a recent brain imaging study by Tang et al. (2006) has shown that neural arithmetical processing appears to be influenced by cultural practices. When the thinking processes of 12 native Western adult participants and 12 native Chinese university students were compared in relation to a simple visually-presented arithmetical task asking them to determine whether a third digit was greater than the bigger one of the first two in a triplet of Arabic numbers, they actually found cortical differences in the way the two groups engaged in number processing. In this particular task, the Chinese participants processed visually, while the Western participants processed verbally, indicated by the increased activation of their left perisylvian lobe language region. Tang et al. (2006) then hypothesized that the visual dominance in number processing among the Chinese participants, evidenced by the activation of their visuo-premotor regions, could be psychologically explained by their reading experiences in school which involves repeatedly learning Chinese characters, and their early experiences in using an Abacus which activated the production of mental images that are visual forms.

4 Neural Correlates of Gender, Culture, and Race

In this section, I provide additional examples of neurocorrelational studies conducted within and outside mathematics education in the areas of gender, culture, and race, with the view that possible neural differences might be used as a scientific basis in designing appropriate and meaningful instruction- and classroom-related strategies or programs that encourage the development of positive findings and reduce the negative effects (e.g., negative emotions, fear, and other stereotyped effects associated with mathematical learning; cf. Hinton et al. 2008; Schlöglmann 2003). Fiske (2007) identified that the neural determinants of social behavior in people can be changed over time, that is, "[p]eople will always gravitate toward the familiar and similar, but they can expand their boundaries, if sufficiently motivated. And this is the substance of social science married to neuroscience" (p. 159). It should be noted that the selected sample of studies in this section excludes those that focus on visual correlates of behavior using eye-tracking methodology.

4.1 Gender

Documenting possible gender differences in spatial thinking with the help of fMRI technology appears to be a major focus of productivity in research in this area. While most studies have established gender differences at the neural level, many of them have proven to be controversial due to constraints in, and the nature of, the tasks presented to participants (see, Cela-Conde et al. 2009, and Kaiser et al. 2008 for brief surveys of relevant studies). One study, in particular, conducted by Kaiser et al. (2008) illustrates a combined neural-psychological analysis relevant to this field of research. The authors investigated the effect of gender on the neural correlates of spatial perspective taking in 12 educated, right-handed adult males and 12 adult females. They found that while both groups activated similar neural patterns in solving a presented task, there were gender differences, albeit small, in strategy use. The women consistently used an egocentric (self) perspective approach and the men employed object-based strategies that neurally activated their precuneus and right inferior frontal gyrus (see Fig. 2).

A study by Keller and Menon (2009) provides an example of a neuroanatomical analysis which involves investigating gender differences in cortical activation (brain functioning) and in gray matter density and volume (brain structure). Keller and Menon (2009) assessed whether there were gender differences between 24 male and 25 female (right-handed) educated adults (aged 18 to 36 years) in various aspects of their brain regions and processing on a relatively simple mental arithmetical task that asked them to determine whether presented addition and subtraction equations (three single-digit whole-number terms on one side and the result on the other) were true or false. The mental arithmetical task has already been shown to activate the appropriate neural regions in the brain (see mathematical processing section above), which then allowed the authors to focus on possible gender-related variations. Their results show that, at the functional level, while overlaps were found in the neural substrates between the male and female participants, gender differences surfaced with greater activation in males than in females around the two right hemisphere regions encompassing the dorsal (right IPS and right angular gyrus) and ventral (right parahippocampal gyrus and right lingual gyrus) visuospatial streams (see Fig. 5). These regions are activated whenever tasks involve number, space, and visual information with the ventral stream proceeding to the temporal lobe and the dorsal stream toward the parietal lobe. Further, at the anatomic or structural level, gender differences were found with the female participants having greater gray matter density and volume than the males in those brain regions that showed activation differences. Isolated from any psychological observation, the authors conjectured that male and female adults might be employing different cognitive strategies (e.g., mental versus overt techniques) despite producing similar performance results. Also, the authors used their structural finding in inferring that differences in the amount of gray matter might explain the functional finding concerning differences in cortical activation.

Two recent lines of gender-related research that use functional imaging techniques and which matter to mathematics education establish neural correlates of: (1) visual aesthetic preference and appreciation with additional interpretive analyses drawn from evolutionary models of symbol use, and parietal differences between humans and animals which could also explain the location versus Euclidean approaches on spatial tasks noted in women and men, respectively (Cela-Conde et al. 2009), and (2) empathic ability, which seems to indicate that males and females process emotional tasks differently on the basis of different neural processing strategies, with males activating the cortical and cognitive-related regions and females the amygdala, inferior frontal regions, and emotion-regulated regions (Derntl et al. 2010).

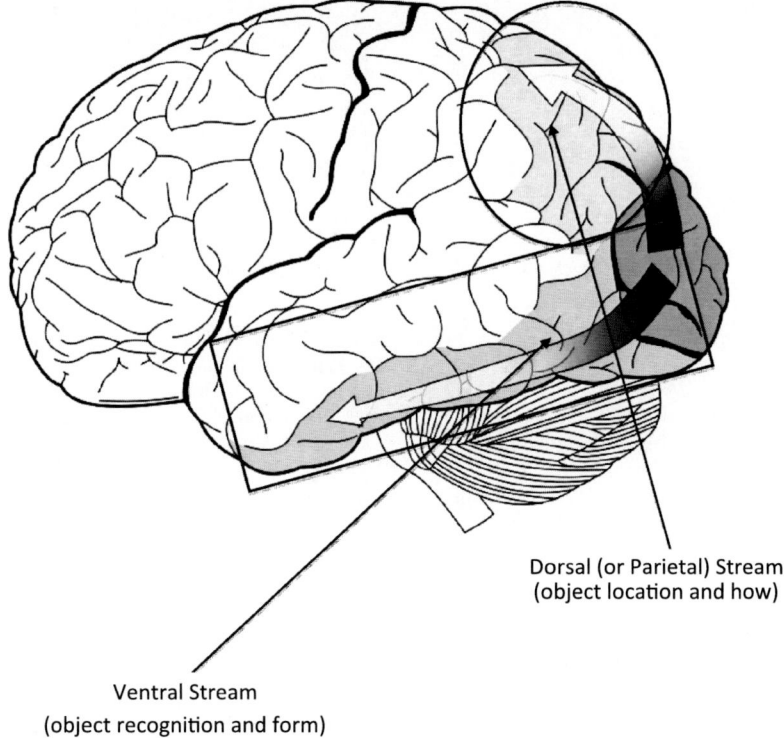

Dorsal (or Parietal) Stream
(object location and how)

Ventral Stream
(object recognition and form)

Fig. 5 Dorsal and ventral streams (http://en.wikipedia.org/wiki/File:Ventral-dorsal_streams.org)

4.2 Race

Two recent syntheses of research by Ito and Bartholow (2009) and Phelps and Thomas (2003) on the neural correlates of race in nonmathematical contexts implicate the regions involving the fusiform gyrus and posterior cingulated cortex (see Fig. 3) in race-based face and familiar face perceptions. Individuals significantly activate the region within their fusiform gyrus when shown faces that belong to the same race as they do. This has led to the phenomena of same-race superiority (or same-race advantage (Phelps and Thomas 2003) and relevant in-group bias. The superiority condition benefits individuals who relate "more naturally" with members that come from the same group. The increased activation might also be explained in terms of an acquired visual expertise that evolves over time and through social practices that encourage individuals to categorize and classify (Van Bavel et al. 2008). Racial categorizing consequently involves activating stereotype beliefs and prejudice that also influence one's personal and perceptual judgments and attitudes toward others. The distributed neural networks involving the posterior cingulated cortex are activated in familiar face contexts. Identifying a person we know, that is, a familiar face, goes beyond visual familiarity to include person knowledge and emotional response (Gobbini and Haxby 2007). Person knowledge involves the subjective characteristics (personal traits, intentions, attitudes, mental states, and objective information of a familiar person) and recruits the anterior paracingulate cortex, the posterior superior temporal sulcus/tempoproparietal junction, and the prucuneus. Emotional responses are tied to activation of the amygdala (tied to attitude towards a person) and the insula (tied to evoked responses towards a face, or faces, that yield an intense emotional effect).

Ito and Bartholow (2009) offer the following bleak implications below concerning how the complexity of race relations and race processing from a neural perspective might be understood:

> [A]ttempts to get people to not "see" race will be relatively ineffective. ... [C]hange occurring at the single level of stereotypical or evaluative associations is unlikely to eliminate racially biased behavior because biased responses could still occur through processes mediated by other parts of the neural network. ... [I]nterventions that seek to improve behavior regulation capabilities might be effective in at least reducing the expression of bias. ... [A]lthough race relations will be affected by race-specific beliefs and feelings, the expression of bias will also be determined by an individual's general regulatory abilities. (Ito and Bartholow 2009, pp. 529–530)

Consequently, the above implications suggest that consideration be given to the possibility of dissociations between intentional and unintentional biases (Phelps and Thomas 2003). For example, several studies have consistently shown that while white Americans explicitly claim they are not biased toward black Americans, implicit measures indicate a negative bias. Further, Phelps and Thomas (2003) recommend "combining the psychological and neural approaches is the best way to advance our understanding of these complex human behaviors more rapidly and with more clarity than could be achieved using either approach in isolation" (p. 754).

4.3 Culture

There are at least two major lines of neurocorrelational studies involving self-other and other-other relations that have implications in the development of understanding of social learning in mathematical contexts. Other people's intentions apart from one's own play a significant role in knowledge acquisition relevant to institutional knowledge. Culture-driven ways of seeing and processing also influence various aspects of basic cognitive processes such as perception, attention, number, language, etc. Due to space constraints, findings from only two recent studies are highlighted. Readers are referred to the research synthesis of Ames and Fiske (2010) who discuss the neural bases of cultural differences in major aspects of cultural cognition.

Concerning self-other relations, in psychology what is known as the *theory of mind* (TOM) involves understanding how young children and adults come to understand people's intentions and mental states other than their own. Neural correlates of TOM have produced mixed results. Some studies with adults implicate the bilateral ventro-medial prefrontal lobe and temporo-parietal junction (see Fig. 3), while others see activation in the medial prefrontal lobe and anterior cingulated regions.

Kobayashi et al. (2007) note that the mixed results perhaps "indicate that some of the neural basis of TOM may be universal whereas others may vary depending upon the person's cultural or linguistic background" (p. 97). However, regardless of cultural and linguistic backgrounds, the "universal" activation of the ventro-medial prefrontal lobe in individuals might be an indication that they are also engaged in reading and conceptualizing other people's emotional behavior, which is central to TOM processing (Kobayashi et al. 2007). Gilbert and Burgess (2008) draw on several other neuroscientific studies in which the medial (rostral) prefrontal lobe is seen as "a region playing an important role in social cognition" (p. 150), especially its role in making judgments about oneself and others and in reflecting on people's emotions and mental states and one's own, the latter being "an important precursor to metacognitive knowledge conducive to efficient learning" (Gilbert and Burgess 2008, p. 150). Concerning other-other relations, neuroscientists have begun to explore possible neural foundations of cultural differences in processing everyday and mathematical phenomena. For example, the fMRI study of Tang et al. (2006; see Visuospatial Processing section) provided neural correlates of Western versus Eastern ways of processing numbers, number relationships, and number operations with a small sample of 12 Western and 12 Eastern native adults. It should also be noted that the two groups activated the occipitoparietal regions, which indicates that some common neural processing across individuals is manifest.

5 Implications for Mathematics Education Research

> Knowledge about the brain... can be relevant in both designing sound educational programs and evaluating existing educational programs, but neuroscience must be considered as just one source of evidence that can contribute to evidence-based practices in education...—it should not be considered alone, out of context from theory or behavioral evidence or the classroom. (Coch and Ansari 2009, p. 547)

> Usable knowledge from [educational neuroscience] is already making important contributions to the field of education. ...The research brings a powerful capability to directly intervene in children's biological makeup, stirring ethical questions about the very nature of child rearing, and the role of education in this process. We argue that there is a key distinction between *raising children* and *designing children*, and the ethical application of neuroscience research to education critically depends upon ensuring that we are *raising* children. (Stein et al. 2011, p. 803)

In this concluding section, I briefly note three implications of neuroscience methodology to mathematics education research.

First, Anderson and Reid (2009) have pointed out the need to clearly articulate three different levels of explanation in any research study that combines neuroscience and education research, namely, biological, cognitive, and behavioral. As demonstrated in the preceding sections, the biological or neural accounts are descriptive in nature, but cognitive and behavioral accounts tend to be more prescriptive and normative (cf. Christodoulou and Gaab 2009; Willingham 2009). Current methodological conditions are still not capable of using all the levels at the same time, hence there is work to be done. Where the methods and interests of educational researchers may differ from typical neuroscientific researchers lie in assessing how proposed cognitive theories and models can be used to understand the resulting behavioral and neural outputs that emerge in activity. The algebra research of Lee et al. (2007; see Mathematical Processing section above) provides an exemplar of this three-level approach.

Second, any analytic discussion concerning the nature, theory, and practice of embodied cognition in mathematics has to take into account the neural dimension in thinking and learning (cf. Campbell 2006; Schlöglmann 2003). Recent characterizations of embodiment tend to dwell on the sensuous or nonrational aspect, such as the role of representational gestures and artifacts in the environment in conveying and developing mathematical meaning. What is not articulated and discussed are the complex neural mechanisms that support and constrain those behaviors within and outside the self who thinks and learns[7]. Concerned stakeholders are, of course, wary of the return of information processing models in mathematical cognition, but this chapter is clear about the psychological-neural or mind-brain nexus. Neuroscience researchers basically establish neural correlates of human behavior and performance on tasks and those involved in mathematics education research possess the requisites and experiences, to borrow Cerulo's (2010) words, "to tell the rest of the story" (p. 120) by situating neuroscientific findings and their implications in issues that concern embodied thinking, acting, and learning in (school) mathematical contexts. In practical terms, it is unlikely that neuroscience will provide usable knowledge for teaching effectively. Hence, the task of mathematics education researchers is to work hard, experiment, and develop and test new design research techniques and hypotheses (Tommerdahl 2010) based on currently available neuroscientific knowledge that is correlational in nature.

From a developmental cognitive neuroscientific perspective, a neural understanding of embodied cognition involves seeing how, say, the frontal-to-parietal lobe mechanism takes place in activity. In such a mechanism, initial learning involves the use of working memory in the frontal lobe. With more training (through learning and experiences), a shift to the parietal lobe occurs as evidenced by automaticity. However, from a social/cultural affective neuroscience perspective, embodied cognition involves seeing how social and cultural (i.e. the triad of race, gender, culture) construction could also be influenced by neural mechanisms that affect emotions, including one's and other people's views (and stereotypes) of the self in relation to others. The main point being addressed here is the complexity of contexts that matter in any account of embodied cognition which should not be limited to psychological or directly observable behaviors. For example, the following neural findings influence embodied thinking: (1) synaptic plasticity; (2) changes in brain tissue development that affect cognitive ability; and (3) symbol-dependent structures that reflect

---

[7]See Gentilucci and Corballis (2006) for an interesting and though-provoking neuroscientific-based account of the evolution of speech and language from manual gestures to vocal communications (i.e. "gestural-origins theory;" an account that differs from the typical sound-to-language perspective).

differing cultural practices which influence neural performance (e.g., visual versus verbal activation in calculating). What is learnt from plasticity in neuroscience is the significant role of experiences in enhancing or constraining, say, cognitive similarities and differences in people. Also learnt is that some personal experiences, which drive those similarities and differences, may not be cultural and linguistic but neural.

Two neurally drawn psychological findings that matter to embodied cognition involve understanding the impact of representations in mathematical learning, as follows: (1) differing mathematical practices place stress and greater demand on working memory if not developed appropriately (e.g., alphanumeric versus visual approaches in solving word problems; right-to-left versus left-to-right—formal mathematical algorithms—approaches in adding and multiplying whole numbers); and (2) levels of representational precision and fluency do not start from the verbal to the symbolic but, especially among young children, from nonverbal and approximate to nonverbal and exact to verbal and counting-based (Clements and Sarama 2007) which have clear implications in the development of algebraic processes such as generalization and abstraction.

What I term *intentionally-embodied cognition* takes as given the crucial role of social and cultural values, beliefs, intentions, and practices in mathematical thinking, learning, and relevant affective processing, which involve both psychological and neurophysiological processes (Chiao et al. 2008). For example, findings drawn from research studies in social cultural affective neuroscience that deal with self-construal style—that is, individualism and collectivism—have much to inform current sociocultural models of embodiment in mathematical processing. Based on converging evidence from several sources, Chiao et al. (2008) note the influence of cultural beliefs on brain-behavior relations involving visual perception and visual experiences. In psychological studies, the notion of culturally preferred style refers to the finding that Caucasian-Americans, trained to live independently, consistently engage in analytic perception (e.g., changes in individual objects independent of context), while East Asians, trained to live interdependently, consistently employ holistic perception (e.g., changes influenced by context).

A modified version of the famous Framed Line Test (FLT) was used in an fMRI study that helped establish, in neural terms, the correlation between visual perception and cultural views of the self. Hedden et al. (2008) recruited twenty adults (10 native European-Americans and 10 East-Asians recently residing in the USA) to participate in a study that measured their blood-oxygenated level-dependent responses using fMRI on a matching FLT experiment. In the experiment, they were asked to judge (easy or difficult) vertical segment lengths in either absolute (ignoring context) or relative (attending to context) conditions. They were initially shown a square frame with a printed segment drawn vertically from the center of the top edge of the square. They were then asked to judge whether a succeeding square frame of a different size had either the same vertical segment shown in the first frame (an absolute task) or a vertical segment whose proportion relative to the size of the succeeding frame reflected the same proportion as the segment-to-frame in the first square (a relative task). In psychological studies that replicated the FLT task

and used other simple visual tasks, culturally preferred styles were noted; Western-
ers consistently performed better than Easterners on absolute conditions, while the
latter consistently performed better than the former on relative conditions. In the
fMRI study of Hedden et al. (2008), the fronto-parietal network, which is generally
associated with working memory, cognitive control, and attention, and is thus employed in demanding tasks, was activated over the temporo-occipital regions in both
groups on culturally nonpreferred judgments, including the left inferior parietal lobe
and right precentral gyrus in situations involving culturally preferred judgments. Increased activation of the fronto-parietal regions was observed in all participants on
difficult tasks that were judged to be incongruous with their cultural preferred style.
Also, East Asians who reported more association with the American culture than
their own did not show increased activation on absolute tasks; this shows the influence of experience in reshaping neural components. Hedden et al. (2008) concluded
that "the cultural background of an individual and the degree to which the individual
endorses cultural values moderate activation in brain networks engaged during even
simple visual and attentional tasks" (p. 12). This finding suggests that the effects
of cultural interaction in visual processing occur in the fronto-parietal lobe regions
rather than the temporo-occipital lobe network that has always been implicated in
early- or primary-stage perceptual processing.

The above finding of Hedden et al. (2008) in regard to the fronto-parietal over
the temporo-occipital network activation and the critical synthesis offered by Chiao
et al. (2008) in relation to their goal of establishing a cultural basis to theories of
consciousness allow further enrichment of understandings of the neural correlates
of intentional embodied cognition in mathematical learning. *Contra* radical constructivist arguments characterize individuals in an embodied, primary all-knowing
cognitive capacity, and as having the "ability to represent one's own thoughts, feelings, and intentions as distinct from another" (Chiao et al. 2008, p. 65). Intentional
embodiment takes as given the role of culture

> in shaping the very nature of conscious experience, such as a person's conceptualization
> and experience of themselves and their relations to others. That is, an individualist has
> a psychological experience and neural representation of themselves that is distinct from
> another whereas a collectivist has a psychological experience and neural representation of
> self-knowledge that overlaps with others. (Chiao et al. 2008, p. 65)

Due to technological advances, the nature of school mathematical knowledge
is slowly experiencing a visual turn. Neural and neuropsychological findings by
Hedden et al. (2008) and many others whose results are periodically synthesized
(e.g., Ames and Fiske 2010; Chiao et al. 2008; Ito and Bartholow 2009; Phelps and
Thomas 2003) further support the significant role of the sociocultural context in
shaping experiences and identifications on visual task performances.

Third, the dark side of intentional embodied cognition in mathematical thinking
and learning involves those neural findings relevant to negative attitudes of individuals toward others. Brain-behavior patterns of prejudice and stereotype will always
complicate conversations involving equity in mathematics education, which consequently translate into inequitable sociocultural practices. Hyde and Mertz (2009),
for example, list the following factors as contributing to why fewer females than
males across several countries excel in mathematics at the high and highest levels:

[D]ynamics in school classrooms leading teachers to provide more attention to boys; guidance counselors, biased by stereotypes, advising females against taking engineering courses; mathematically gifted girls not being identified and nurtured; scarcity of women role models in math-intensive careers leading girls to believe they do not belong in them; unconscious bias against females in hiring decisions; and hostile work environments leading qualified women to drop out in favor of friendlier climes. (Hyde and Mertz 2009, p. 8806)

What we do learn from converging reflections on neuroscientific findings regarding the nature of negative differences which matter significantly in conversations about equity and the proposed and implemented interventions, however, involve developing purposeful, targeted experiences, more inclusive instruction (Hinton et al. 2008), better programs that can effectively nurture (Hyde and Mertz 2009, p. 8806), and those that are able to "improve behavior regulation capabilities" despite the possibility of not being able to fully eliminate them (Ito and Bartholow 2009, p. 259). Eliot's (2010) point about "experience" being a primary interventional tool provides a good provisional closure as follows:

Brain differences are indisputably biological, but they are not necessarily hardwired. The crucial, often overlooked fact is that experience itself changes brain structure and function. Neuroscientists call this shaping plasticity, and it is the basis of all learning and much of children's mental development. Even something as simple as the act of seeing depends on normal visual experiences in early life, without which a baby's visual brain fails to wire up properly and his or her vision is permanently impaired. (Eliot 2010, p. 22)

References

Adolphs, R. (2010). Conceptual challenges and directions for social neuroscience. *Neuron, 65*(6), 752–767.

Ames, D. & Fiske, S. (2010). Cultural neuroscience. *Asian Journal of Social Psychology, 13,* 72–82.

Anderson, J., Qin, Y., Sohn, M., Stenger, V., & Carter, C. (2003). An information-processing model of the BOLD response in symbol manipulation tasks. *Psychonomic Bulletin and Review, 10,* 241–261.

Anderson, M. & Reid, C. (2009). Don't forget about levels of explanation. *Cortex, 45,* 560–561.

Ansari, D. (2005). Paving the way towards meaningful interactions between neuroscience and education. *Developmental Science, 8*(6), 466–467.

Ansari, D. (2008). Effects of development and enculturation on number representation in the brain. *Nature Reviews Neuroscience, 9,* 278–291.

Ansari, D. (2009). Neurocognitive approaches to developmental disorders of numerical and mathematical cognition: The perils of neglecting the role of development. *Learning and Individual Differences, 20,* 123–129.

Ansari, D., Coch, D., & De Smedt, B. (2011). Connecting education and cognitive neuroscience: Where will the journey take us? *Educational Philosophy and Theory, 43*(1), 1–6.

Bandettini, P. (2009). Seven topics in functional magnetic resonance imaging. *Journal of Integrative Neuroscience, 8*(3), 371–403.

Bechtel, W. (2002). Aligning multiple research techniques in cognitive neuroscience: Why is it important? *Philosophy of Science, 69,* 48–58.

Brannon, E. (2005). The independence of language and mathematical reasoning. *Proceedings of the National Academy of Sciences, 102*(9), 3177–3178.

Bruer, J. (1999). In search of... brain-based education. *Phi Delta Kappan, 80*(9), 649–657.

Butterworth, B., Reeve, R., Reynolds, F., & Lloyd, D. (2008). Numerical thought with and without words: Evidence from indigenous Australian children. *Proceedings of the National Academy of Sciences*, *105*(35), 13179–13184.

Campbell, S. (2006). Defining mathematics educational neuroscience. In *Proceedings of the 28th annual meeting of the North American chapter of the international group for the psychology of mathematics education* (Vol. 2, 442–449). Merida, Mexico: Universidad Pedagogica Nacional.

Cantlon, J., Brannon, E., Carter, E., & Pelphrey, K. (2006). Functional imaging of numerical processing in adults and 4-year old children. *Public Library of Science Biology*, *4*(5), 844–854.

Cantlon, J., Platt, M., & Brannon, E. (2009). Beyond the number domain. *Trends in Cognitive Science*, *13*(2), 83–91.

Cela-Conde, C., Ayala, F., Munar, E., Maestu, F., Nadal, M., Capo, M., del Rio, D., Lopez-Ibor, J., Ortiz, T., Mirasso, C., & Mary, G. (2009). Sex-related similarities and differences in the neural correlates of beauty. *Proceedings of the National Academy of Sciences*, *106*(10), 3847–3852.

Cerulo, K. (2010). Mining the intersections of cognitive sociology and neuroscience. *Poetics*, *38*, 115–132.

Chiao, J. (2011). Where does human diversity come from? In M. Brockman (Ed.), *Future science: Essays from the cutting edge* (pp. 236–248). New York: Vintage Books, Random House, Inc.

Chiao, J., Li, Z., & Harada, T. (2008). Cultural neuroscience of consciousness: From visual perception to self-awareness. *Journal of Consciousness Studies*, *15*(10), 58–69.

Christodoulou, J., & Gaab, N. (2009). Using and misusing neuroscience in education-related research. *Cortex*, *45*, 555–557.

Clements, D., & Sarama, J. (2007). Early childhood mathematics learning. In F. Lester, Jr. (Ed.), *Second handbook of research on mathematics teaching and learning* (pp. 461–556). Charlotte: Information Age Publishing Inc.

Coch, D., & Ansari, D. (2009). Thinking about mechanisms is crucial to connecting neuroscience and education. *Cortex*, *45*, 546–547.

De Hevia, M. D., Vallar, G., & Girelli, L. (2008). Visualizing numbers in the mind's eye: The role of visuo-spatial processes in numerical abilities. *Neuroscience and Biobehavioral Reviews*, *32*, 1361–1372.

Dehaene, S. (1992). Varieties of numerical abilities. *Cognition*, *44*, 1–42.

Dehaene, S. (1997). *The number sense: How the mind creates mathematics*. Oxford: Oxford University Press.

Dehaene, S., Molko, N., Cohen, L., & Wilson, A. (2004). Arithmetic and the brain. *Current Opinion in Neurobiology*, *14*, 218–224.

Delazer, M., Domahs, F., Bartha, L., Brenneis, C., Lochy, A., Trieb, T., & Benke, T. (2003). Learning complex arithmetic—An fMRI study. *Cognitive Brain Research*, *18*, 76–88.

Derntl, B., Finkelmeyer, A., Eickhoff, S., Kellermann, T., Falkenberg, D., Schneider, F., & Habel, U. (2010). Multidimensional assessment of emphatic abilities: Neural correlates and gender differences. *Psychoneuroendocrinology*, *35*, 67–82.

Eberhardt, J. (2005). Imaging race. *American Psychologist*, *60*(2), 181–190.

Editorial (2003). Scanning the social brain. *Nature Neuroscience*, *6*(12), 1239.

Editorial (2005). Separating science from stereotype. *Nature Neuroscience*, *8*(3), 253.

Eliot, L. (2010). The truth about boys and girls. *Scientific American Mind*, *21*(2), 22–29.

Fehr, T., Code, C., & Herrmann, M. (2007). Common brain regions underlying different arithmetic operations as revealed by conjunct fMRI-BOLD activation. *Brain Research*, *1172*, 93–102.

Fiske, S. (2007). On prejudice & the brain. *Daedalus*, *136*(1), 156–159.

Fuson, K. (2009). Avoiding misinterpretations of Piaget and Vygotsky: Mathematical teaching without learning, learning without teaching, or helpful learning-path teaching? *Cognitive Development*, *24*, 343–361.

Geake, J. (2005). Educational neuroscience and neuroscientific education: In search of a mutual middle-way. *Research Intelligence*, *92*(3), 10–13.

Geertz, C. (2000). *Available light: Anthropological reflections on philosophical topics*. Princeton: Princeton University Press.

Gentilucci, M. & Corballis, M. (2006). From manual gesture to speech: A gradual transition. *Neuroscience and Biobehavioral Reviews*, *30*, 949–960.

Gilbert, S., & Burgess, P. (2008). Social and nonsocial functions of rostral prefrontal cortex: Implications for education. *Mind, Brain, and Education, 2*(3), 148–156.

Gobbini, M., & Haxby, J. (2007). Neural systems for recognition of familiar faces. *Neuropsychologia, 45*, 32–41.

Goel, V., & Dolan, R. (2004). Differential involvement of left prefrontal cortex in inductive and deductive reasoning. *Cognition, 93*, B109–B121.

Goel, V., Buchel, C., Frith, C., & Dolan, R. (2000). Dissociation of mechanisms underlying syllogistic reasoning. *NeuroImage, 12*, 504–514.

Goel, V., Gold, B., Kapur, S., & Houle, S. (1997). The seats of reason? An imaging study of deductive and inductive reasoning. *NeuroReport, 8*, 1305–1310.

Goel, V., Gold, B., Kapur, S., & Houle, S. (1998). Neuroanatomical correlates of human reasoning. *Journal of Cognitive Neuroscience, 10*(3), 293–302.

Goswami, U. (2004). Neuroscience and education. *British Journal of Educational Psychology, 74*, 1–14.

Goswami, U. (2005). The brain in the classroom? The state of the art. *Developmental Science, 8*(6), 467–469.

Goswami, U. (2008). Principles of learning, implications for teaching: A cognitive neuroscience perspective. *Journal of Philosophy of Education, 42*(3–4), 381–399.

Goswami, U. (2009). Mind, brain, and literacy: Biomarkers as usable knowledge for education. *Mind, Brain, and Education, 3*(3), 176–184.

Grabner, R., Ansari, D., Schneider, M., De Smedt, B., Hannula, M., & Stern, E. (Eds.) (2010). Cognitive neuroscience and mathematics learning. *ZDM—The International Journal on Mathematics Education, 42*(6), 511–663.

Grön, G., Wunderlich, A., Spitzer, M., Tomczak, R., & Riepe, M. (2000). Brain activation during human navigation: Gender-different neural networks as substrate of performance. *Nature Neuroscience, 3*(4), 404–408.

Hacking, I. (2004). The race against time. *New Statesman, 133*(4698), 26–27.

Halberda, J., Mazzocco, J., & Feigenson, L. (2008). Individual differences in nonverbal number acuity correlated with maths achievement. *Nature, 455*, 665–668.

Hardcastle, V., & Stewart, C. (2002). What do brain data really show? *Philosophy of Science, 69*, 72–82.

Hedden, T., Ketay, S., Aron, A., Markus, H., & Gabrieli, J. (2008). Cultural influences on neural substrates of attentional control. *Psychological Science, 19*(1), 12–17.

Henson, R. (2006). Forward inference using functional neuroimaging: Dissociations versus associations. *Trends in Cognitive Sciences, 10*(2), 64–69.

Hinton, C., Miyamoto, K., & Della-Chiesa, B. (2008). Brain research, learning, and emotions: Implications for education research, policy, and practice. *European Journal of Education, 43*(1), 87–103.

Howard-Jones, P. (2008). Philosophical challenges for researchers at the interface between neuroscience and education. *Journal of Philosophy of Education, 42*(3–4), 361–380.

Hubbard, E., Diester, I., Cantlon, J., Ansari, D., van Opstal, F., & Troiani, V. (2008). The evolution of numerical cognition: From number neurons to linguistic quantifiers. *Journal of Neuroscience, 28*(46), 11819–11824.

Hyde, J., & Mertz, J. (2009). Gender, culture, and mathematics performance. *Proceedings of the National Academy of Sciences, 106*(22), 8801–8807.

Ito, T., & Bartholow, B. (2009). The neural correlates of race. *Trends in Cognitive Sciences, 13*(12), 524–531.

Jacob, S., & Nieder, A. (2009). Notation-independent representation of fractions in the human parietal cortex. *Journal of Neuroscience, 29*(14), 4652–4657.

Kadosh, R., Lammertyn, J., & Izard, V. (2008). Are numbers special? An overview of chronometric, neuroimaging, developmental, and comparative studies of magnitude representation. *Progress in Nuerobiology, 84*, 132–147.

Kaiser, S., Walther, S., Nennig, E., Kronmuller, K., Mundt, C., Weisbrod, M., Stippich, C., & Vogeley, K. (2008). Gender-specific strategy use and neural correlates in a spatial perspective taking task. *Neuropsychologia, 46*, 2524–2531.

Kaufmann, L. (2008). Neural correlates of number processing and calculation: Developmental trajectories and educational implications. In A. Dowker (Ed.), *Mathematical difficulties: Psychology and intervention* (pp. 1–12). San Diego: Academic Press.

Kawashima, R., Taira, M., Okita, K., Inoue, K., Tajima, N., Yoshida, H., Sasaki, T., Sugiura, M., Watanabe, J., & Fukuda, H. (2004). A functional fMRI study of simple arithmetic—A comparison between children and adults. *Cognitive Brain Research, 18*, 225–231.

Keller, K., & Menon, V. (2009). Gender differences in the functional and structural neuroanatomy of mathematical cognition. *NeuroImage, 47*, 342–352.

Kobayashi, C., Glover, G., & Temple, E. (2007). Cultural and linguistic effects on neural bases of "Theory of Mind" in American and Japanese children. *Brain Research, 1164*, 95–107.

Lee, K., Lim, Z. Y., Yeong, S., Ng, S. F., Venkatraman, V., & Chee, M. (2007). Strategic differences in algebraic problem solving: Neuroanatomical correlates. *Brain Research, 1155*, 163–171.

Mazzocco, M., Feigenson, L., & Halberda, J. (in press). Impaired acuity of the approximate number system underlies mathematical learning disability. *Child Development.*

Menon, V. (2010). Developmental cognitive neuroscience of arithmetic: Implications for learning and education. *ZDM—The International Journal on Mathematics Education, 42*(6), 515–525.

Monti, M., Parsons, L., & Osherson, D. (2011). *Thought beyond language: Neural dissociation of algebra and natural language.* Retrieved August 8, 2011, online at http://www.princeton.edu/~osherson/papers/Monti9.pdf.

National Research Council (2008). *Emerging cognitive neuroscience and related technologies: Report of the committee on military and intelligence methodology for emergent neurophysiological and cognitive/neural science research in the next two decades.* Washington: National Research Council.

Nieder, A. (2005). Counting on neurons: The neurobiology of numerical competence. *Nature Reviews Neuroscience, 6*, 177–190.

Nieder, A. (2009). Prefrontal cortex and the evolution of symbolic reference. *Current Opinion in Neurobiology, 19*, 99–108.

Obersteiner, A., Dresler, T., Reiss, K., Vogel, A. C., Pekrun, R., & Fallgatter, A. (2010). Bringing brain imaging to the school to assess arithmetical problem solving: Chances and limitations in combining educational and neuroscientific research. *ZDM—The International Journal on Mathematics Education, 42*(6), 541–554.

Organization for Economic Cooperation and Development [OECD] (2007). *Understanding the brain: The birth of a learning science.* Paris, France: OECD.

Pennington, B., Snyder, K., & Roberts, Jr., R., (2007). Developmental cognitive neuroscience: Origins, issues, and prospects. *Developmental Review, 27*, 428–441.

Petersen, S., & Fiez, J. (1993). The processing of single words studied with positron emission tomography. *Annual Review of Neuroscience, 16*, 661–682.

Phelps, E., & Thomas, L. (2003). Race, behavior, and the brain: The role of neuroimaging in understanding complex social behaviors. *Political Psychology, 24*(4), 747–758.

Poldrack, R. (2006). Can cognitive processes be inferred from neuroimaging data? *Trends in Cognitive Sciences, 10*(2), 59–63.

Poldrack, R. (2008). The role of fMRI in cognitive neuroscience: Where do we stand? *Current Opinion in Neurobiology, 18*, 223–227.

Radford, L., Edwards, L., & Arzarello, F. (Eds.) (2009). Special issue on gestures and multimodality in the construction of mathematical meaning. *Educational Studies in Mathematics, 70*(2), 91–215.

Rocha, F., Rocha, A., Massad, E., & Menezes, R. (2005). Brain mappings of the arithmetic processing in children and adults. *Cognitive Brain Research, 22*, 359–372.

Rosenberg-Lee, M., Lovett, M., & Anderson, J. (2009). Neural correlates of arithmetic calculation strategies. *Cognitive, Affective, & Behavioral Neurosciences, 9*(3), 270–285.

Sanfey, A., Rilling, J., Aronson, J., Nystrom, L., & Cohen, J. (2003). The neural basis of economic decision-making in the ultimatum game. *Science, 300*(5626), 1755–1758.

Schlöglmann, W. (2003). Can neuroscience help us better understand affective reactions in mathematics learning? In M. A. Mariotti (Ed.), *CERME 3: Third conference of the European society for research in mathematics education* (pp. 1–10). Bellaria, Italy: CERME.

Sohn, M., Goode, A., Koedinger, K., Stenger, V., Fissell, K., & Carter, C. (2004). Behavioral equivalence, but not neural equivalence— Neural evidence of alternative strategies in mathematical thinking. *Nature Neuroscience, 7,* 1193–1194.

Stein, Z., della Chiesa, B., Hinton, C., & Fischer, K. (2011). Ethical issues in educational neuroscience: Raising children in a brave new world. In J. Illes & B. Sahakian (Eds.), *Oxford handbook of neuroethics* (pp. 803–822). Oxford, UK: Oxford University Press.

Szücs, D., & Goswami, U. (2007). Educational neuroscience: Defining a new discipline for the study of mental representations. *Mind, Brain, and Education, 1*(3), 114–127.

Tang, Y., Zhang, W., Chen, K., Feng, S., Ji, Y., Shen, J., Reiman, E. & Liu, Y. (2006). Arithmetic processing in the brain shaped by cultures. *Proceedings of the National Academy of Sciences of the United States of America, 103*(28), 10755–10780.

Terao, A., Koedinger, K., Sohn, M., Qin, Y., Anderson, J., & Carter, C. (2004). An fMRI study of the interplay of symbolic and visuo-spatial systems in mathematical reasoning. In *Proceedings of the twenty-sixth annual conference of the cognitive science society.* Mahwah: Lawrence Erlbaum Associates.

Thagard, P. (2010). How brains make mental models. In L. Magnani, W. Carnielli, & C. Pizzi (Eds.), *Model-based reasoning in science and technology: Abduction, logic, and computational discovery* (pp. 447–461). Berlin: Springer.

Todorov, A., Harris, L., & Fiske, S. (2006). Toward socially inspired social neuroscience. *Brain Research, 1079,* 76–85.

Tommerdahl, J. (2010). A model for bridging the gap between neuroscience and education. *Oxford Review of Education, 36*(1), 97–109.

Van Bavel, J., Packer, J., & Cunningham, W. (2008). The neural substrates of in-group bias. *Psychological Science, 19*(11), 1131–1139.

Varga, M., Pavlova, O., & Nosova, S. (2010). The counting function and its representation in the parietal cortex in humans and animals. *Neuroscience and Behavioral Physiology, 40*(2), 185–196.

Varley, R., Klessinger, N., Romanowski, C., & Siegal, M. (2005). Agrammatic but numerate. *Proceedings of the National Academy of Sciences, 102*(9), 3519–3524.

Willingham, D. (2009). Three problems in the marriage of neuroscience and education. *Cortex, 45,* 544–545.

Zamarian, L., Ischebeck, A., & Delazer, M. (2009). Neuroscience of learning arithmetic— Evidence from brain imaging studies. *Neuroscience and Biobehavioral Reviews, 33,* 909–925.

Zan, R., Brown, L., Evans, J., & Hannula, M. (Eds.) (2006). Special issue on affect in mathematics education. *Educational Studies in Mathematics, 63*(2), 113–234.

Commentary on the Chapter by Ferdinand Rivera, "Neural Correlates of Gender, Culture, and Race and Implications to Embodied Thinking in Mathematics"

Connecting the Neural Correlates of Gender, Culture and Race with Mathematics Education

Bert De Smedt and Lieven Verschaffel

In his chapter *Neural correlates of gender, culture, and race and implications to embodied thinking in mathematics*, Rivera discusses the fascinating and vivid, yet also hotly debated, association between research in neuroscience and mathematics education, by sketching various neuroscientific studies conducted within and outside mathematics education. Against the background of this volume on gender, culture, and diversity, Rivera pays specific attention to findings from the fields of cognitive and social-affective neuroscience, the latter of which is also gradually, but slowly, shedding further light on the neural correlates of gender, culture, and race. Echoing a more general movement to connect research in (mathematics) education and neuroscience (e.g., Ansari and Coch 2006; Howard-Jones 2008; Stern 2005), Rivera suggests that these neuroscientific findings may offer new perspectives on mathematics education research in general, and on individual and intentional embodied cognition in mathematical thinking and learning in particular.

We congratulate Rivera for this nice contribution. It provides a rich and timely overview of neuroscientific findings that may be relevant for our understanding of gender, culture, and race in relation to mathematics education—issues that have been rarely addressed in most existing reviews on neuroscience and (mathematics) educational research. But more importantly, Rivera provides a nuanced view of the reviewed neuroscientific findings and highlights the dynamic complexity of applying neuroscientific findings to the field of mathematics education, thereby illustrating the promises and pitfalls of exploiting such applications. Against this background, his chapter provides an excellent introduction into the topic to those who are relatively new in the field, while at the same time it is informative and thought-provoking for more informed and experienced readers.

B. De Smedt (✉) · L. Verschaffel
Katholieke Universiteit Leuven, Leuven, Belgium
e-mail: Bert.DeSmedt@ped.kuleuven.be

L. Verschaffel
e-mail: lieven.verschaffel@ped.kuleuven.be

H. Forgasz, F. Rivera (eds.), *Towards Equity in Mathematics Education,*
Advances in Mathematics Education,
DOI 10.1007/978-3-642-27702-3_48, © Springer-Verlag Berlin Heidelberg 2012

The gist of Rivera's chapter deals with the discussion of findings from neuroscientific studies that may be relevant to (mathematics) educational theory and practice, emphasizing that an integration of these neuroscientific findings with theories of mathematics education may yield a better and deeper understanding of children's mathematical learning. This is probably the most-often cited way in which neuroscience can contribute to educational research (De Smedt et al. 2010).

Together with Rivera, we contend that neuroscience, in particular cognitive and social-affective neuroscience, offers social scientists a series of tools, methods, and theories that allow them to constrain and extend the knowledge that has already been accumulated through decades of behavioral research (De Smedt and Verschaffel 2010; Lieberman et al. 2003). Data from neuroscience can confirm what is already known on the basis of cognitive and behavioral research, but yet they add another level of explanation, the biological, to our psychological and educational theories. In other words, the convergence of findings from different research methodologies provides a more solid empirical ground for a given hypothesis or theory. Rivera nicely illustrates this issue by pointing to the work of de Hevia et al. (2008) on the role of visuospatial processing in mathematical thinking and learning, thereby providing a multilevel analysis of how such learning takes place. In addition to assisting in methodological triangulation, neuroscientific research is able to produce findings that could not be accessed by behavioral data alone. In this context, Rivera discusses an interesting study by Keller and Menon (2009) which revealed gender differences in brain activity during a simple mental arithmetic task, despite producing similar performance results. This might suggest that males and females are using different cognitive strategies to solve simple arithmetic tasks which may be difficult to access via behavioral measurements. Although the inferences of cognitive strategies from activity in a particular brain region should be treated with great caution (see Poldrack 2006 for a theoretical discussion), such data can at least be useful for suggesting novel hypotheses that should be tested in subsequent psychological or educational research. Used in this way, neuroscience might set the stage for new educational and psychological research and it can, albeit indirectly, enhance our understanding of learning.

It is important to point out—and we are happy to note that Rivera starts the chapter with highlighting these ideas—that measurements in cognitive and social-affective neuroscience—signals indicating brain activity or structure—can only be interpreted by linking those measurements to behavioral data (e.g., Cacioppo et al. 2008; Phelps and Thomas 2003). Related to this, the scope of a biological explanation of a behavior should be clear. Such a biological explanation does not indicate that a behavior, for example, race perception, is innate, hardwired, or unchangeable. As highlighted various times throughout Rivera's chapter, the human brain indeed shows remarkable plasticity and is shaped by experience (Jäncke 2009). It is also inappropriate to claim that neuroscientific data of a given behavior are more informative than behavioral data, even though there may be a belief, both in public and among scientists, that a biological explanation of behavior is more informative or valid than a non-biological explanation (Racine et al. 2005). Neuroscientific and behavioral data should be considered on a level playing field. There is no knowledge hierarchy, but an appreciation of multiple sources of data at different levels of description is essential to understand better a phenomenon under investigation (see also Stern and Schneider 2010).

One of the strengths of Rivera's contribution is its emphasis on the role of the sociocultural context in shaping experiences and brain activity. Knowledge about the brain should not be considered out of the context of the social reality in the classroom, yet most neuroscientific studies in the field of mathematics, and also in other curricular areas, have investigated academic performance and learning in relative isolation from the classroom context (see also De Smedt and Verschaffel 2010). This is not surprising, given that neuroscientific studies require very restricted and highly controlled environments (e.g., the magnet bore or the EEG lab) which limit the external validity of neuroscientific measurements. This problem might be addressed to some extent by correlating measures inside and outside the scanner or EEG lab or, in other words, by using measures from the laboratory to predict real-word mathematical behavior in the classroom and vice versa (De Smedt et al. 2010). While gender, culture, and race are surely important factors to consider, sufficient attention should also be paid to characteristics of the learning environment in which mathematical learning and problem solving take place, and to how this all interacts with these three factors. Nevertheless, Rivera's focus on the neural correlates of sociocultural factors clearly illustrates that (social-affective) neuroscience has the potential to contribute to socio-cultural and situated perspectives on mathematical thinking and learning (e.g., Greeno et al. 1996), thereby going beyond a purely cognitive perspective which views mathematical learning as a highly individual and mental process occurring in the individual learner, a stance that is often taken when connections between research in neuroscience and education are discussed (De Smedt et al. 2010).

Rivera further suggests that findings from neurocorrelational studies—indeed, most of the existing neuroscientific data are primarily correlational and not causal in nature—might be used as a scientific basis in designing appropriate and meaningful instruction or classroom related strategies, for example, strategies to reduce the negative effects of gender, culture, and race (e.g., stereotypes) associated with mathematical learning. However, merely understanding the cognitive and neural correlates of mental and social processes and their interactions with gender, culture, and race will not suffice to determine how (mathematics) education should take place. The latter requires additional steps of research in which these neuroscientific insights are reconciled with educational theories that are also significantly informed by disciplines other than psychology and neuroscience, for example, mathematics, sociology, philosophy, technology (De Corte et al. 1996). This should then lead to the design and construction of learning environments, the effectiveness of which should be evaluated by educational research. This might again feed into neuroscience research, in which the effect of such learning environments on brain activity can be investigated. Such research might reveal new insights into brain plasticity or into how brain activity changes as a function of schooling, or it might even unravel learning and transfer effects that are hard to detect by behavioral methods.

Against this background, to the best of our knowledge, there currently are no direct applications yet available of neuroscientific findings on what and how to teach in the classroom that go beyond what we already know from decades of educational research, at least not in the field of mathematics education. In a recent review on dyscalculia, Butterworth et al. (2011) concluded that research in cognitive and developmental neuroscience reveals that dyscalculia is due to a core deficit in understanding number, and that remediation should build on foundational number concepts, for example, understanding the meaning of number, rather than addressing isolated conceptual gaps—an implication for teaching that would hardly be considered to be innovative by researchers in the field of mathematics education. By contrast, more progress has been made in other domains of education, such as reading development (McCandliss 2010), where a recent study by Hoeft et al. (2011) showed that brain measures reliably predicted future reading gains in dyslexia, whereas no single behavioral measure, including widely used standardized tests, was able to do so. In particular, activity in those areas of the brain that are not associated with typical reading (the right inferior frontal gyrus) predicted change over time, suggesting compensatory mechanisms which, if better understood, could inform remediation techniques in truly innovative ways.

Without doubt, the input of mathematics education researchers in connecting research on neuroscience and education is crucial. To quote Rivera, "*Neuroscience researchers basically establish neural correlates of human behavior and performance on tasks and those of us involved in mathematics education research possess the requisites and experiences, to borrow Cerulo's (2010) words, "to tell the rest of the story" by situating neuroscientific findings and their implications to issues that concern embodied thinking, acting and learning in (school) mathematical contexts*". However, we would even like to go further in this connection between research in neuroscience and (mathematics) education. Elsewhere, we have argued that this connection should not be restricted to a one-way view in which findings from neuroscience are merely applied to educational theory. In our view, the connection between research in neuroscience and education should rather be conceived as a two-way street with multiple bi-directional interactions between these two fields of research (De Smedt et al. 2010). It will be crucial for (mathematics) educational researchers to also impact on neuroscience, much more than is currently the case by, for example, i. pinpointing the (educational) variables that need to be included (either as control or as experimental variables) in neuroscientific studies, such as participants' learning and teaching histories, ii. collaborating on research that examines the effects of different learning environments on brain activity and, most crucially, iii. by critically interrogating neuroscientists on their research. Although this is something that has not been achieved so far, we hope that this will become a priority in the years to come, and this is exactly one of the aims of a recently founded special interest group of the European Association for Research on Learning and Instruction on Neuroscience and Education (EARLI SIG 22). This being said, a genuine connection between neuroscience and education will be about building a new interdisciplinary field, rather than thinking about two disciplines informing each other. In this context, it is promising to note that there is a new generation of forthcoming researchers who consider themselves as not belonging to either neuroscience or education, but identifying themselves as actors in the new interdisciplinary research area of neuroscience and education[1].

A crucial mechanism to realize this connection between research in mathematics education and research in neuroscience lies in the training of interdisciplinary researchers—this is witnessed by the foundation of undergraduate and graduate programs in Mind, Brain and Education or Educational Neuroscience, for example at the Graduate School of Education at Harvard University (see http://www.gse.harvard.edu/academics/masters/mbe/)—and in the education of educational researchers and neuroscientists (e.g., Ansari and Coch 2006; Szücs and Goswami 2007). Some neuroscientists should receive basic training in educational research, including issues related to gender, culture, and identity, which should allow them to promote the quality of their research on educationally relevant issues, and which might enable them to make qualified educational interpretations of their findings (see, also, Racine et al. 2005). On the other hand, basic training in (cognitive and social-affective) neuroscience is important for some educational researchers to allow them to recurrently evaluate the usefulness of neuroscientific methods and theories for their research questions, an endeavor to which Rivera is already contributing with his chapter in this volume.

The eye-opening view expressed by Rivera is quite optimistic, yet some cautious realism is needed when thinking about the scope of the contribution of neuroscience to research in mathematics education. It needs to be emphasized that only some, but surely not all, research questions in mathematics education might benefit from the use of neuroscientific tools and theories. The use of these should be guided by the research question at hand. Stern and Schneider (2010) described this issue by means of a digital roadmap analogy. In a digital roadmap, the appropriate resolution of the map depends on what the map viewer is looking for, with the viewer being able to zoom in and out between different levels of resolution. While traditional educational research corresponds to an intermediate resolution of the roadmap, neuroscience is on a higher zoom level, allowing researchers to examine cognitive and social-affective processes in a very detailed way, a goal in some but certainly not in all fields of mathematics educational research. In addition to this, the existing body of neuroscientific data on mathematical learning focuses largely, if not solely, on very elementary numerical skills, such as the processing of number or simple arithmetic. Very little is known about more complex mathematical skills that are taught in school, even though researchers are starting to address these issues by investigating, for example, the neural correlates of algebra (Lee et al. 2007). Such research that focuses explicitly on school-taught mathematical contents that go beyond elementary number processing and arithmetic is destined to facilitate broader connections between mathematics education and neuroscience, and it will be exciting to investigate interactions between these more complex mathematical skills and sociocultural variables such as gender, culture, and race.

---

[1] We prefer to use "neuroscience and education" above "educational neuroscience", because "neuroscience and education" indicates more the equality of both neuroscience and education in this new interdisciplinary research field.

References

Ansari, D., & Coch, D. (2006). Bridges over troubled waters: Education and cognitive neuro-science. *Trends in Cognitive Sciences, 10*, 146–151.

Butterworth, B., Varma, S., & Laurillard, D. (2011). Dyscalculia: From brain to education. *Science, 332*, 1049–1053.

Cacioppo, J. T., Berntson, G. G., & Nusbaum, H. C. (2008). Neuroimaging as a new tool in the toolbox of psychological science. *Current Directions in Psychological Science, 17*, 62–67.

De Corte, E., Greer, B., & Verschaffel, L. (1996). Learning and teaching mathematics. In D. C. Berliner, & R. C. Calfee (Eds.), *Handbook of educational psychology* (pp. 491–549). New York: MacMillan.

De Hevia, M. D., Vallar, G., & Girelli, L. (2008). Visualizing numbers in the mind's eye: The role of visuo-spatial processes in numerical abilities. *Neuroscience and Biobehavioral Reviews, 32*, 1361–1372.

De Smedt, B., & Verschaffel, L. (2010). Traveling down the road: From cognitive neuroscience to mathematics education. . . and back. *ZDM—The International Journal on Mathematics Education, 42*, 649–654.

De Smedt, B., Ansari, D., Grabner, R. H., Hannula, M. M., Schneider, M., & Verschaffel, L. (2010). Cognitive neuroscience meets mathematics education. *Educational Research Review, 5*, 97–105.

Greeno, J. G., Collins, A. M., & Resnick, L. B. (1996). Cognition and learning. In D. C. Berliner & R. C. Calfee (Eds.), *Handbook of educational psychology* (pp. 15–46). New York: MacMillan.

Hoeft, F., McCandliss, B. D., Black, J. M., Gantman, A., Zakerani, N., Hulme, C., Lyytinen, H., Whitfield-Gabrieli, S., Glover, G. H., Reiss, A. L., & Gabrieli, J. D. E. (2011). Neural systems predicting long-term outcome in dyslexia. *Proceedings of the National Academy of Sciences of the United States of America, 108*, 361–366.

Howard-Jones, P. (2008). Education and neuroscience [special issue]. *Educational Research, 50*(2) 119–122.

Jäncke, L. (2009). The plastic human brain. *Restorative Neurology and Neuroscience, 27*, 521–538.

Keller, K., & Menon, V. (2009). Gender differences in the functional and structural neuroanatomy of mathematical cognition. *Neuroimage, 47*, 342–352.

Lee, K., Lim, Z. Y., Yeong, S. H. M., Ng, S. F., Venkatraman, V., & Chee, M. W. L. (2007). Strategic differences in algebraic problem solving: Neuroanatomical correlates. *Brain Research, 1155*, 163–171.

Lieberman, M. D., Schreiber, D., & Ochsner, K. N. (2003). Is political cognition like riding a bicycle? How cognitive neuroscience can inform research on political thinking. *Political Psychology, 24*, 681–704.

McCandliss, B. D. (2010). Educational neuroscience: The early years. *Proceedings of the National Academy of Sciences of the United States of America, 107*, 8049–8050.

Phelps, E. A., & Thomas, L. A. (2003). Race, behavior, and the brain: The role of neuroimaging in understanding complex social behaviors. *Political Psychology, 24*, 747–758.

Poldrack, R. A. (2006). Can cognitive processes be inferred from neuroimaging data? *Trends in Cognitive Sciences, 10*, 59–63.

Racine, E., Bar-Ilan, O., & Illes, J. (2005). fMRI in the public eye. *Nature Reviews Neuroscience, 6*, 159–164.

Stern, E., & Schneider, M. (2010). A digital road map analogy of the relationship between neuro-science and educational research. *ZDM—The International Journal on Mathematics Education, 42*, 511–514.

Stern, E. (2005). Pedagogy meets neuroscience. *Science, 310*, 745.

Szücs, D., & Goswami, U. (2007). Educational neuroscience: Defining a new discipline for the study of mental representations. *Mind, Brain and Education, 1*, 114–127.

Commentary on the Chapter by Ferdinand Rivera, "Neural Correlates of Gender, Culture, and Race and Implications to Embodied Thinking in Mathematics"

Stephen R. Campbell

What differences in gender, culture, and race can be attributed to the biological evolution of the species, and what differences in gender, culture, and race can be attributed to social interaction? Such a polarized question is well posed only if these areas of attribution, biology and sociology, are independent of each other. A moment's reflection may suggest that biological evolution and social interaction in the human species are to a large extent co-dependent. There are many distinctions that are but distinctions in thought and perception, and not independent in fact—and there are those who take the co-dependent arising of all things as a matter of fact. Be this as it may, some differences may predate others. For instance, sexual differences in humans may predate social interaction. However, the biological emergence of sexual differences in species may also be seen as an important affordance, perhaps even a *sine qua non*, for social interaction. Beyond and even within the constraints of kinship, social interaction, for better or for worse, must be predicated, for the most part of our evolutionary history, on biological differences of various kinds, covering wide spectra of attractions and repulsions. Through the means of agriculture and industry, and with continued advances in science, mathematics, engineering, and technology, humans have evolved to a point where social interaction can take place in purely semiotic and ideational terms, independently of biological characteristics. Whereas symbols and signs are visible, ideas are not (Merleau-Ponty 1968). Or are they?

Growing realization and acceptance that human cognition is embodied, coupled with lesion studies and advances in neuroimaging, indicate that ideas are based upon and emerge from gestural and sensorimotor activities—and that ideas themselves are

S.R. Campbell (✉)
Simon Fraser University, Metro Vancouver, Canada
e-mail: Stephen_Campbell@sfu.ca

Writing of this commentary was supported by a Research Development Initiative Grant from the *Social Sciences and Humanities Research Council of Canada*. The author assumes responsibility for the manner in which the ideas and perspectives are presented and discussed herein, which may not reflect the views of the Council.

manifest in some way in the structure and behavior of our organ of experience and thought, the brain. That is to say, there are neural correlates of experience, ideas, and, hence, notions associated with gender, culture, and race, within brain and brain behavior. If this is so, what are the implications for embodied thinking in mathematics? This is the main question that comes to mind in reading Rivera's chapter, and, as I confess to being part of the choir (e.g., Campbell 2002, 2003, 2010, 2011; Campbell and Dawson 1995), I am particularly interested in the ways in which cognitive and educational neuroscience can potentially shed light on it.

Rivera knits together a wide variety of literature pertaining to this topic. Careful to qualify the relevance and applicability of findings to this point, he notes that research in cognitive and educational neuroscience at this time is predominantly correlative, not causal (hence the emphasis on neural *correlates*), and descriptive, not prescriptive (hence the emphasis on cognitive measures, assessments, and diagnostics). He also rightly cautions that many studies that have neuroscientific value may hold little in terms of educational value, due to the isolated and restrictive context of the experimental tasks involved. These distinctions give rise to three important and somewhat interrelated problems, which I shall refer to as the causal, prescriptive, and contextual problems. Despite these problems, Rivera notes that it is "reasonable to assume that there is a mutually determining relationship between particular neural and psychological functions" (p. 521). I concur with this statement, as, I am sure, do many of our educational colleagues. Nevertheless, the causal, prescriptive, and contextual problems regarding neuroscientific results remain, leading many educationists[1] to wonder, "so what?" After all, what good is it to simply describe neural phenomena correlated with cognition and learning in contexts that have little or no ecological validity when it comes to the education of human beings?

These are legitimate questions. It is unlikely that initiatives in educational neuroscience or neuroeducation (e.g., Campbell 2011; Patten and Campbell 2011) will catch the attention of too many educationists beyond the most enthusiastic until problems, issues, and concerns underlying these questions are adequately addressed. I believe it is important that they are adequately addressed so that mainstream educationists are more motivated to participate in an educational neuroscientific revolution that will most certainly occur, with or without our participation. If mainstream educationists do not start paying more attention, the field of education will be transformed before our very eyes by the neurosciences, while most of us stand agape on the sidelines.

Paying attention to results and advances in the neurosciences that may have educational implications can be painful for educationists, especially the more humanistically inclined amongst us. It requires learning, or at least becoming familiar with what virtually amounts to a foreign language, replete with unfamiliar customs, norms, values, and mores, not to mention potentially questionable and objectionable philosophical commitments. Rivera attends to this need with some basic introductory notions pertaining to neuroanatomy, neurophysiology, and neuroimaging.

---

[1] By evoking the term "educationist," I mean to include educational scholars, theorists, researchers, practitioners, and policy makers.

His treatment of these notions, as wide-ranging as they are, is necessarily sparse, thereby, I anticipate, requiring some readers wishing to follow along to augment their understandings of these matters of their own accord.

Educationists may variously be baffled, encouraged, or take solace with entice-ments Rivera provides, such as "(h)ere the psychological theories of Piaget have been given a material basis in terms of neural patterns and processes" (p. 523). Oth-ers may wonder what good is it to know that there are "... complex and intertwined neural connections between affect and cognition" (p. 524). Would it not suffice just to know that there are complex and intertwined connections between affect and cog-nition? Such connections are challenging enough to grasp purely in psychological terms. Why bother complicating the matter or confusing educationists further with talk of neural connections between different areas of the brain and brain-imaging results showing different levels of activity between them?

Perhaps *the* crucial point to be made here is that psychological theories of cogni-tion are inherently ideational and speculative in nature (Campbell 2010). They are comprised of constructs that cannot be directly observed, but at best only inferred from introspection and/or behavior. Hence, if valid and reliable links can be estab-lished between psychological states and brain activity, the former can be understood, and more importantly, observed and measured, in terms of the latter. Of course, this is a big "if." Some skeptics may object that such connections are not possible to establish at all, others may admit to the possibility, but remain skeptical that any connections of educational significance can be made in practice, at least anytime soon, the complexities involved being what they are (cf., Churchland 1980).

Alas, the complexities involved regarding understanding connections between psychological states, or cognitive functioning, and brain activity are yet subtler still. Rivera is quite right to raise issues pertaining to whether mapping brain activity to psychological constructs is a one to one correspondence. It is evident that some cognitive functions, such as those pertaining to visual perceptions, for instance, are more localized within the brain, whereas other functions such as those pertain-ing to attention, for instance, can be much more distributed. Indeed, most tasks performed by humans, including the most ordinary and mundane, are comprised of many widely distributed interacting neuronal assemblages working in tandem. Moreover, individual neurons may belong to different neuronal assemblages, just as human beings may be associated with different organizations.

Rivera notes that it is easier to study and infer brain activity associated with different cognitive tasks than it is to proceed in the other direction of inferring cog-nitive functioning on the basis of brain activity. Perhaps an analogy can help convey this point: It would be easier to ascertain which group of people typically watches a given television show than it would be to ascertain what television show a group of people may be watching from the mere fact that that group of people has been previ-ously associated with watching a given show. Connecting cognitive functioning with brain activity, as central as this program of research may be in helping cognitive neu-roscientists ascertain mechanistic foundations underlying cognition and learning, is of little intrinsic interest for educationists. Inferring cognition and learning from brain activity would be much more relevant and instrumental from an educational perspective. Are we there yet? Can we ever be? There are many that would lean toward the negative on both counts.

One way of addressing the skeptics is to address the open-minded, and there are many skeptics who are open-minded to empirical evidence and theoretical rationale supporting the relevance of the neurosciences with regard to educational concerns. Rivera approaches his topic in this spirit. Keeping all due cautions and qualifications in mind, he presents results indicating differences in brain activity associated with gender, culture, and race, while also noting that "(a)side from learning, the development of one's identity as a gendered, cultured, and racialized being is also a significant educational issue" (p. 519). Also aside from learning, there is mounting evidence and rationale that some aspects of mathematical cognition are innate (De Cruz and De Smedt 2010; Kinzler and Spelke 2007). Moreover, there is evidence that geometrical aspects of mathematical cognition transcend both culture (Dehaene et al. 2006) and even species (Landau and Lakusta 2009; Vallortigara et al. 2009), let alone race. Such innate aspects of mathematical cognition are not the end of the story. Otherwise, there would be no need for mathematics education, and clearly that need exists. That is to say, there are clearly aspects of mathematical cognition that are both innate and learned.

From a perspective of embodied cognition, it should not come as a surprise that differences in gender, culture, and race are manifested in the brain activity of individuals of differing gender, culture, and race. Nor should it be much of a surprise that these differences in turn can affect mathematical cognition and learning. Of greater significance, from an educational perspective, is to determine to what extent such differences are innate and to what extent they result from learning and experience. The relatively recent discovery of neural plasticity has given rise to increased interest in developmental cognitive neuroscience (Johnson 1997). Hence, debate regarding what is learned versus what is innate (e.g., Spencer et al. 2009a, 2009b; Spelke and Kinzler 2009) should not be geared toward any misguided kind of either/or resolution. It is important to understand the nature and extent of both, with respect to different populations of individuals, and eventually with respect to individuals themselves. The neurosciences are particularly well-suited and well-poised to help determine these differences. Empowered with this knowledge, a new breed of *neuroeducationists* should eventually be in a better position to design curriculum, instruction, and assessment in ways in which various aspects of mathematical thinking that are innate can be dovetailed with various aspects thereof that can be learned.

In the meantime, it is important that educational researchers not sit on the sidelines. Cognitive neuroscientists to this point have been investigating mathematical cognition and learning "from the ground up." That is, their models to this point do not typically incorporate aspects of mathematical thinking and learning involving higher cognitive functions. Researchers in mathematics education, however, have spent many decades formulating models and theories of mathematical thinking and learning that do involve various aspects of higher cognitive function. A very fertile space for both exists in between, and it is not futile to suggest that this space will be filled in very short order, with or without the participation of researchers in mathematics education. Rivera's chapter helps draw attention to the relevance, importance, complexity, and overall difficulty of this task.

References

Campbell, S. R. (2002). Constructivism and the limits of reason: Revisiting the Kantian problematic. *Studies in Philosophy and Education, 21*(6), 421–445.

Campbell, S. R. (2003). Reconnecting mind and world: Enacting a (new) way of life. In S. J. Lamon, W. A. Parker, & S. K. Houston (Eds.), *Mathematical modelling: A way of life* (pp. 245–253). Chichester: Horwood Publishing.

Campbell, S. R. (2010). Embodied minds and dancing brains: New opportunities for research in mathematics education. In B. Sriraman & L. English (Eds.), *Theories of mathematics education: Seeking new frontiers* (pp. 359–404). Heidelberg: Springer.

Campbell, S. R. (2011). Educational neuroscience: Motivation, methodology, and implications. *Educational Philosophy and Theory, 43*(1), 7–16.

Campbell, S. R., & Dawson, A. J. (1995). Learning as embodied action. In R. Sutherland & J. Mason (Eds.), *Exploiting mental imagery with computers in mathematics education* (pp. 233–249). Berlin: Springer.

Churchland, P. S. (1980). A perspective on mind-brain research. *The Journal of Philosophy, 77*(4), 185–207.

De Cruz, H., & De Smedt, J. (2010). The innateness hypothesis and mathematical concepts. *Topoi, 29*, 3–13.

Dehaene, S., Izard, V., Pica, P., & Spelke, E. (2006). Core knowledge of geometry in an Amazonian indigene group. *Science, 311*, 381–384.

Johnson, M. H. (1997). *Developmental cognitive neuroscience: An introduction.* Cambridge: Blackwell Publishers.

Kinzler, K. D., & Spelke, E. S. (2007). Core systems in human cognition. *Progress in Brain Research, 164*, 257–264.

Landau, B., & Lakusta, L. (2009). Spatial representation across species: Geometry, language, and maps. *Current Opinion in Neurobiology, 19*, 12–19.

Merleau-Ponty, M. (1968). *The visible and the invisible.* Evanston: Northwestern University Press.

Patten, K. E., & Campbell, S. R. (2011). Introduction: Educational neuroscience. *Educational Philosophy and Theory, 43*(1), 1–6.

Spelke, E. S., & Kinzler, K. D. (2009). Innateness, learning, and rationality. *Child Development Perspectives, 3*(2), 96–98.

Spencer, J. P., Blumberg, M. S., McMurray, B., Robinson, S. R., Samuelson, L. K., & Tomblin, J. B. (2009a). Short arms and talking eggs: Why we should no longer abide the nativist—Empiricist debate. *Child Development Perspectives, 3*(2), 79–87.

Spencer, J. P., Samuelson, L. K., Blumberg, M. S., McMurray, B., Robinson, S. R., & Tomblin, J. B. (2009b). Seeing the world through a third eye: Developmental systems theory looks beyond the nativist-empiricist debate. *Child Development Perspectives, 3*(2), 103–105.

Vallortigara, G., Sovrano, V. A., & Chiandetti, C. (2009). Doing Socrates experiment right: Controlled rearing studies of geometrical knowledge in animals. *Current Opinion in Neurobiology, 19*, 20–26.

Editors and Contributors

Bill Atweh is an associate professor of mathematics education at Curtin University of Technology, Australia. His research interests include sociocultural aspects of mathematics education and globalization, the use of action research for capacity building and critical and socially responsible mathematics education. His publications include: Action Research in Practice (Routledge), Research and Supervision in Mathematics and Science Education (Erlbaum), Sociocultural Research on Mathematics Education (Erlbaum), Internationalisation and Globalisation in Mathematics and Science Education (Springer), and Mapping Equity and Quality in Mathematics Education (Springer). His email address is b.atweh@curtin.edu.au.

Anastasios (Tasos) Barkatsas has taught and held leadership positions in secondary schools in Melbourne, Australia. He also held consultancy and research fellowships in Athens, Greece. He was a research project officer before accepting a lectureship at Monash University, Australia. At Monash University, he teaches primary and secondary mathematics education courses. His research interests are: mathematics teachers' content knowledge; students' attitudes, motivation, engagement, and achievement; ICT in mathematics education; students' mathematics task preferences; and multivariate statistical analyzes. He has published numerous conference papers and articles in mathematics education journals and books. His email address is Tasos.Barkatsas@monash.edu.

Bill Barton. In his early career, Bill worked in secondary schools, including a bilingual Maori-English unit, and also produced mathematics television programmes in the 1980s. He then entered the Mathematics Department at The University of Auckland as a mathematics educator specializing in ethnomathematics, specifically mathematics and language. His current main research is university-level mathematics education. In 2010 he became president of ICMI. He currently leads the ICMI/IMU Klein Project, which is attempting to represent the research field of the mathematical sciences to secondary teachers. He is the Associate Dean International for the Faculty of Science at The University of Auckland. His email address is barton@math.auckland.ac.nz.

H. Forgasz, F. Rivera (eds.), *Towards Equity in Mathematics Education*, Advances in Mathematics Education, DOI 10.1007/978-3-642-27702-3, © Springer-Verlag Berlin Heidelberg 2012

Richard Barwell is an associate professor in the Faculty of Education at the University of Ottawa, Canada. His research is located in the intersection of mathematics education and applied linguistics with a particular focus on multilingualism/ bilingualism in the teaching and learning of mathematics. His research interests include mathematics classroom discourse, mathematics learning in multilingual settings, and the relationship between learning language and learning curriculum content. His work has been published in peer-reviewed journals in mathematics education, applied linguistics, and general education. Prior to his academic career, Dr. Barwell taught mathematics in the UK and in Pakistan, where his interest in language and mathematics first arose. His email address is richard.barwell@uottawa.ca.

Alan Bishop (Emeritus Professor) was Professor of Education and Associate Dean at Monash University between 1992–2002 after spending the earlier part of his life in the UK. In 1969 after completing his PhD, he was appointed Lecturer in the Department of Education at Cambridge University, UK where he worked for 23 years. He edited (from 1978 to 1990) the international research journal Educational Studies in Mathematics, published by Kluwer (now Springer), and he has been an Advisory Editor since 1990. He is Managing Editor of the research series Mathematics Education Library, also published by Kluwer/Springer (1980–present). He is also the Chief Editor of the two International Handbooks of Mathematics Education (1996 and 2002) published by Kluwer/Springer, and editor of a handbook on Mathematics Education, published by Routledge. His email address is alan.bishop@monash.edu.

Jeremy Burke was a mathematics teacher for 17 years in four London comprehensive schools. He joined London Metropolitan University to work on a teacher-training programme and then moved to King's College London. At Kings College, he has been the director of the PGCE (teacher training programme) and is now the director of the MA in mathematics education. His research interest is in sociology focusing on strategies deployed in mathematics teaching, in textbooks, and in state level policy interventions in the curriculum and the practice of teaching. His email address is jeremy.burke@kcl.ac.uk.

Leone Burton (1934–2007) passed away in 2007. Her final academic position was as Professor of Education in mathematics and science at the University of Birmingham, UK. She had a diverse range of interests in mathematics education, and her major contributions were in the areas of creative thinking, social justice, and gender equity. What she advocated in theory, she worked fiercely to bring into practice. She was convenor of the *International Organization of Women and Mathematics Education* from 1984–1988. Her contributions leave a strong and lasting legacy. She was influential, inspirational, and a generous friend.

Todd CadwalladerOlsker is an assistant professor in the Department of Mathematics at California State University, Fullerton, California, USA. His research focuses on undergraduate mathematics education and on the mathematical content knowledge of teachers. He has published papers on the understanding of mathematical proof and probability and the use of writing assignments in mathematics. His email address is tcadwall@fullerton.edu.

Rosemary Callingham is an associate professor in mathematics education at the University of Tasmania, Australia. She has an extensive background in mathematics education at all levels in Australia. Her professional experiences include classroom teaching, mathematics curriculum development and implementation, large-scale testing, and pre-service teacher education in two universities. She has worked on projects in Hong Kong and North Korea and has conducted studies in many parts in Australia covering a variety of mathematics education topics, including gifted and talented students. Her research interests include statistical literacy, mental computation and assessment of mathematics and numeracy, and teachers' pedagogical content knowledge. Her email address is Rosemary.Callingham@utas.edu.au.

Stephen R. Campbell is an associate professor in the Faculty of Education at Simon Fraser University in Metro Vancouver, British Columbia, Canada. His scholarly focus is on the historical and psychological development of mathematical thinking from an embodied perspective informed by Kant, Husserl, and Merleau-Ponty. Inspired in part by the pioneering work of Francisco Varela and colleagues, his research incorporates methods of psychophysics and cognitive neuroscience as a means for operationalizing affective and cognitive models of math anxiety and mathematical concept formation. His email address is Stephen_Campbell@sfu.ca.

Marta Civil, formerly at The University of Arizona, is the Frank A. Daniels Jr. Distinguished Professor in Mathematics Education at the University of North Carolina, Chapel Hill. Her research focuses on cultural and social aspects in the teaching and learning of mathematics, equity, linking in-school and out-of-school mathematics, and parental engagement in mathematics. Her work is located primarily in working class, Latino/a communities. She has directed several initiatives in mathematics education, engaging parents, children, and teachers. She is the Principal Investigator for NSF-funded CEMELA (Center for the Mathematics Education of Latinos/as). Her email address is civil@math.arizona.edu.

Claudio José de Oliveira holds a bachelors degree in mathematics at Porto-Alegrense School of Education, Sciences and Languages (FAPA, Brazil). He earned both his masteral and doctoral degrees in education at the University of the Sinos River Valley, Brazil. Presently, he is an associate professor at the University of Santa Cruz do Sul, Brazil. His current research interests are in the area of identity and difference in education among students in the postgraduate education program (masters level). He engages in teaching and research in the following subjects: ethnomathematics; education; mathematics education; pedagogical practices; and teacher education. His email address is coliveir@unisc.br.

Bert De Smedt obtained his PhD in Educational Sciences in 2006 at the Katholieke Universiteit Leuven (Belgium). Until 2010 he was a postdoctoral research fellow of the Research Foundation Flanders, Belgium (FWO). Currently he is an assistant professor in the Faculty of Psychology and Educational Sciences at the Katholieke Universiteit Leuven, Belgium. His primary interest is in individual differences in children's mathematical skills. He uses both behavioral and brain imaging methods to understand how children develop arithmetical skills and what neurocognitive mechanisms underlie this development. He is coordinator of the EARLI Special Interest Group on Neuroscience and Education. His email address is Bert.DeSmedt@ped.kuleuven.be.

James Dietz is a program director in the Division of Research on Learning at the National Science Foundation, USA. His work focuses on education and learning research chiefly in the context of evaluation, policy, and sociology of education and science. He holds a PhD from the Georgia Institute of Technology in Science and Technology Policy and a Master's in Public Administration degree in science policy from Syracuse University's Maxwell School for Citizenship and Public Affairs. His research interests focus on the sociology of science, research evaluation, diffusion of innovation and knowledge transfer, social and human capital theory, evaluation and research methods, organizational theory, and education and human resources. His email address is jdietz@nsf.gov.

Jaguthsing Dindyal is an Assistant Professor of mathematics education at the National Institute of Education Singapore (NIE), an institute of the Nanyang Technological University. He teaches mathematics education courses to both pre-service and in-service schoolteachers and is the programme coordinator of MEd (mathematics) at NIE. His recent work has focused on teacher education and problem solving. He has collaborated in the Teacher Education and Development Study in Mathematics (TEDS-M), an international study by IEA and in the Mathematical Problem Solving for Everyone (M-ProSE) project. His other interests include geometry and proof, algebraic thinking, international studies, and the mathematics curriculum. His email address is jaguthsing.dindyal@nie.edu.sg.

Ann Dowker is a university research lecturer in the Department of Experimental Psychology at the University of Oxford, UK. She has carried out extensive research on developmental psychology and individual differences, especially with regard to mathematical learning. She has a special interest in the development of intervention programs for children with mathematical difficulties. She is the lead researcher on the Catch Up Numeracy intervention project. She is the author of *Individual Differences in Arithmetic: Implications for Psychology, Neuroscience and Education* (2005), co-editor with Arthur Baroody of *The Development of Arithmetic Concepts and Skills* (2003) and editor of *Mathematical Difficulties: Psychology and Intervention* (2008). Her email address is ann.dowker@psy.ox.ac.uk.

Paul Dowling is a full professor of education at the Institute of Education, University of London. He is a sociologist whose principal work has been the introduction and development of Social Activity Method (SAM), an organizational language for the sociological description of pedagogic texts, sites, technologies and institutions. SAM has been applied by Dowling and others in mathematics education as well as in areas as diverse as literary studies, online communities, and the production of academic knowledge in education and medicine. Dowling is the author of *The Sociology of Mathematics Education* (1998, Falmer), and his recent publications include *Sociology as Method* (2009, Sense) and *Doing Research/Reading Research* (2010, Routledge, with Andrew Brown, second edition). His email address is p.dowling@ioe.ac.uk.

Diana B. Erchick is an associate professor at the Ohio State University in Newark and is active nationally and internationally in the mathematics education community. Her work focuses on the intersection of mathematics pedagogy and social constructs such as race, gender, and class, with particular experience around gender and mathematics education and attention to social justice. Her research informs work in a range of contexts, from leading the Gender and Mathematics Working Group of the Psychology of Mathematics Education—North America from 1999–2006, to the development and delivery of Matherscize, a summer mathematics camp for middle grades girls, from 1999–2010. Her email address is erchick.1@osu.edu.

Helen Forgasz is an associate professor and Associate Dean in the Faculty of Education at Monash University, Australia. Her main research interests include: gender and other equity issues, the affective domain, and the non-cognitive effects of using technology for mathematics learning. She is also a member of the editorial boards of several academic journals and a book series. She writes for the scholarly and professional communities, and publishes widely. Her most recent, co-edited monograph is entitled *International Perspectives on Gender and Mathematics Education*. Her email address is helen.forgasz@monash.edu.

Maren Hoffstall graduated from the University of Hamburg with a master degree for secondary schools in 2009. In her study she predominantly concentrated on mathematical modeling, particularly with regard to the prospects and specific problems in school implementation. She also examined gender issues with a focus on mathematics education. Her master thesis, carried out jointly with Anna Barbara Orschulik, was based on a study conducted with two different age groups and examined gender stereotypes in mathematics. Since 2010 she has been teaching mathematics and geography in a secondary school in Hamburg. Her email address is marenhoffstall@web.de.

Laura Jacobsen (formerly Spielman) is as Associate Professor of Mathematics Education in the School of Teacher Education and Leadership at Radford University (RU), recently transitioning from her former position in the Department of Mathematics and Statistics at RU. She serves as coordinator of a new distance education master's program in secondary mathematics education. Dr. Jacobsen's research interests focus on equity and social justice issues in mathematics education, recruitment and retention of under-represented populations in STEM, and critical educational theory. She has published in journals including *ZDM—The International Journal on Mathematics Education*, *Philosophy of Mathematics Education Journal*, and *Democracy and Education*. Her email address is ljacobsen@radford.edu.

Asha K. Jitendra is currently the Rodney Wallace Professor for Advancement of Teaching and Learning at the University of Minnesota. Her research interests include academic and curricular strategies in mathematics and reading for students with learning disabilities, assessment practices to inform instruction, and instructional design and textbook analysis. Her email address is jiten001@umn.edu.

Steven Khan is a doctoral candidate and Vanier Scholar in the Department of Curriculum and Pedagogy at the University of British Columbia. He has taught mathematics and biology at the secondary school level and was a lecturer in mathematics education at the University of the West Indies, St. Augustine. He is a co-founder of the Mindful Mathematics Educators' Research Institute of Trinidad & Tobago. Interests in mathematics education are generally related to intersections among mathematics, aesthetics and ethics and specifically include ethnomathematics, mathematics-for-teaching, mathematics popularization and mindfulness based approaches to mathematics teaching and learning through enacting a philosophy of intervulnerability. His email address is stkhan@interchange.ubc.ca.

Gabriele Kaiser holds a master degree as a teacher of mathematics and humanities, which she completed at the University of Kassel in 1978. She completed her doctorate in mathematics education in 1986 on applications and modeling, and her post-doctoral study in pedagogy on international comparative studies in 1997, both at the University of Kassel. Since 1998, she holds a full professorship for mathematics education at the Faculty of Education, University of Hamburg. Her areas of research include modeling and applications in schools, international comparative studies, gender and cultural aspects in mathematics education, and empirical research on teacher education. Her email address is gabriele.kaiser@uni-hamburg.de.

Robert (Bob) Klein is an associate professor of mathematics at Ohio University in Athens, Ohio. His research integrates critical perspectives on gender and technology with a focus on the impacts of sociocultural issues on mathematics teaching and learning at all levels. Klein, as part of a large research team, recently completed a national cross-case comparison of how rural K-12 schools in the United States connected mathematics instruction to place and community. That work is informing a survey of rural mathematics instructors across the United States. His email address is kleinr@ohio.edu.

Gelsa Knijnik is a full professor in the Graduate Program on Education at Unisinos. She holds a MSc Degree in Mathematics and a PhD in Education from UFRGS. She did postdoctoral work at the Universidad Complutense de Madrid. She is a researcher at the Brazilian National Research Council and coordinates the *Interinstitutional Research Group Mathematics Education and Society*. Her major research interests are related to the education of peasant social movements, politics of knowledge, and ethnomathematics. She is the editor of the Unisinos Journal on Education and has published books, book chapters, and papers in Brazilian journals and abroad. Her email address is gelsak@uol.com.br.

Boris Koichu is a senior lecturer at the Technion—Israel Institute of Technology and is the head of the mathematics education group at the Department of Education in Technology and Science. He received his PhD from the Technion in 2004 and was a postdoctoral fellow at the University of California, San Diego, between 2004 and 2006. His research focuses on mathematical problem solving and problem posing, with special reference to learning and teaching mathematics at secondary school and university. Some of his work concerns the question of how insights from research on mathematically gifted students can be used in favor of wider groups of students. His email address is bkoichu@technion.ac.il.

Gilah Leder is an Adjunct Professor at Monash University and Professor Emerita at La Trobe University, Australia. Her research has focused particularly on gender issues in mathematics education, on exceptionality—predominantly high achievement, and on assessment. She has published widely in each of these areas. She is also a Past President of the Mathematics Research Group of Australasia [MERGA] and of the International Group for the Psychology of Mathematics Education [PME], a Fellow of the Academy of the Social Sciences in Australia, and the recipient of the Felix Klein medal awarded for outstanding lifetime achievements in mathematics education research and development. Her email address is gilah.leder@monash.edu.

Jae Hoon Lim is an assistant professor of research methods at the University of North Carolina at Charlotte. She teaches introductory and advanced research method courses in the College of Education. Her research interests include sociocultural issues in mathematics education and various equity topics in the STEM fields. She has served as a PI or a Co-PI on multiple research and evaluation projects, including a National Science Foundation-funded STEP grant project. She has published over twenty-five articles in scholarly and professional journals worldwide and has authored seven book- and monograph-chapters. Her email address is jhlim@uncc.edu.

Sarah Theule Lubienski is a professor of mathematics education at the University of Illinois at Urbana-Champaign. Her research focuses on inequities in students' mathematics outcomes, and in the policies and practices that shape those outcomes. She has examined social class, race/ethnicity, and gender in large-scale studies using national datasets, as well as in smaller, classroom-based studies. Dr. Lubienski has served as Chairperson of both the National Assessment of Educational Progress (NAEP) Research Special Interest Group of AERA and the Editorial Panel of the *Journal for Research in Mathematics Education*. She is a member of the AERA Grants Governing Board and recently completed a Fulbright-funded study of mathematics education in Ireland. Her email address is Stl@illinois.edu.

David Mittelberg is a senior lecturer in the Graduate Faculty and former head of the Department of Sociology at Oranim Academic College of Education, Israel. He is now a senior research fellow at the Institute for Kibbutz Research, Haifa University, Israel, where he formerly served as its director. When researching issues related to mathematics learning outcomes, his main focus is on the interaction of ethnicity and gender. He has published articles on ethnicity, gender and mathematics learning, migration, tourism, kibbutz education, and the sociology of American Jewry. His email address is davidm@oranim.ac.il.

Marjorie Montague is a full professor at the University of Miami. She specializes in learning disabilities, emotional and behavior disorders, and attention deficit hyperactivity disorder. Dr. Montague is well known for her research in cognitive strategy instruction for students with learning disabilities and has written extensively in the related areas of learning, behavior, and attention problems in children and adolescents. She has had numerous federally funded research and personnel preparation grants and has over 70 publications including journal articles, chapters, books, and curricular materials. Her email address is mmontague@aol.com.

Anna Barbara Orschulik graduated from the University of Hamburg with a master degree for secondary schools in 2009. In her study she was particularly interested in mathematical modeling and its implementation in schools. She also explored gender issues, with a focus on mathematics education. Her master thesis carried out jointly with Maren Hoffstall was based on a study conducted with two different age groups and examined gender stereotypes in mathematics. Since 2010 she has been teaching mathematics and geography in a secondary school in Hamburg. Her email address is annaborschulik@web.de.

Andrew Penner is an assistant professor of sociology at the University of California, Irvine. His research focuses on gender and race inequality in the labor market and educational system. He is currently involved in projects that examine the impact of marriage and children on gender wage inequality in Norway, the organization of gender inequality in transition economies, and the implications of racial fluidity for inequality in the United States. His email address is andrew.penner@uci.edu.

Núria Planas is an associate professor at the University Autonoma of Barcelona, Spain. She was a secondary school teacher of mathematics in several schools before moving into higher education. Her research interests are discourse, language, and social interaction in the mathematics classroom. She has an ongoing project funded by the Spanish Ministry of Science and Innovation. She has written various scientific articles and book chapters in her research area. Her email address is nuria.planas@uab.cat.

Beatriz Quintos is a lecturer in the Department of Curriculum and Instruction at the University of Maryland, USA. She draws on sociocultural and critical perspectives in exploring student learning at the elementary level. She was a fellow at the Center for Mathematics Education of Latinas/os that is housed at the University of Arizona. Her doctoral work explored the influence of a teacher and teaching practice with a vision of social justice on Latina/o students' learning of mathematics and the participation of parents in that context. Her research interests focus on education as a tool for a democratic society that includes the knowledge and perspectives of minority students and their communities. Her email address is bquintos@umd.edu.

Luis Radford is a full professor at Laurentian University in Sudbury Ontario, Canada. He teaches at École des science de l'éducation in the pre-service teachers' training program and conducts classroom research with teachers from Kindergarten to Grade 12. His research interests include the development of algebraic thinking, the relationship between culture and thought, the epistemology of mathematics, and semiotics. He has been co-editor of three special issues of *Educational Studies in Mathematics*. He co-edited the book *Semiotics in Mathematics Education: Epistemology, History, Classroom, and Culture* (Sense Publishers, 2008) and co-authored the book *A Cultural-Historical Perspective on Mathematics Teaching and Learning* (2011, Sense Publishers). He received the Laurentian University 2004–05 Research Excellence Award. His email address is lradford@laurentian.ca.

Ferdinand Rivera is an associate professor in the Department of Mathematics at San Jose State University, California, USA. He conducts research in algebraic thinking from Grades 1 through 14. His general research interest is mathematical cognition with a particular focus on students in urban classrooms. He has published journal articles, conference proceedings, and book chapters on matters involving visualization and pattern generalization. He is the author of the book, *Toward a Visually Oriented School Mathematics Curriculum: Research, Theory, and Practice*, a volume in the Springer Mathematics Education library series. His email address is ferdinand.rivera@sjsu.edu.

Joanne Rossi Becker is a full professor of mathematics at San José State University, specializing in mathematics education. She teaches courses for pre-service secondary mathematics teachers, including methods and capstone problem solving courses. She directs several professional development programs for in-service elementary and secondary mathematics teachers. She is also involved in a 40-district collaboration providing performance assessment for students in grades 2–10. Her research interests include gender and mathematics, teacher professional development, and generalization in algebra. She was one of the founders of Women and Mathematics Education, a professional organization dedicated to improving the mathematics education of girls and women in the U.S. Her email address is joanne.rossibecker@sjsu.edu.

Kenneth Ruthven taught in schools in Scotland and England before joining the Faculty of Education at Cambridge where he is currently Director of Research, recently Chair of Science, Technology and Mathematics Education. He studies issues of curriculum, pedagogy and assessment, especially in mathematics, and particularly in the light of technological change. His current research focuses on designing research-informed classroom interventions for implementation at scale, and on analyzing how to bridge the important gaps that this highlights between educational scholarship and professional needs. His email address is kr18@cam.ac.uk.

Wolfgang Schlöglmann is an emeritus professor in mathematics education at the University of Linz (UL), Austria. Before his retirement, he spent 40 years as a teacher educator to gymnasium teachers at the UL. For 20 years, he also taught mathematics courses for adults. His research interests include: applied mathematics in secondary schools; adult education in mathematics; industrial mathematics and society; affect and mathematics learning; and mathematics teacher education. His email address is wolfgang.schloeglmann@jku.at.

Ole Skovsmose has a special interest in critical mathematics education. He has investigated the notions of landscape of investigation, mathematics in action, students' foreground, and ghettoizing. He was a full professor in the Department of Learning and Philosophy at Aalborg University, Denmark, but he is now retired and spends most of his time in Brazil. He has published several books, including *Towards a Philosophy of Critical Mathematics Education*, *Dialogue and Learning in Mathematics Education* (together with Helle Alrø), *Travelling Through Education*, *In Doubt*, and *An Invitation to Critical Mathematics Education*. His email address is osk@learning.aau.dk.

Hazel Tan. Before commencing doctoral studies, Hazel taught senior secondary mathematics for many years and was the Head of Mathematics in a Singaporean school. She also worked in Singapore's Ministry of Education, spearheading the pedagogical use of technologies in education. Her research interests are in the areas of mathematics teaching and learning with technologies, cross-cultural comparisons, and related teacher education. Her current institutional affiliation is Monash University, Australia. She is researching senior secondary students' learning preferences and their ways of using handheld calculators, and comparing findings from students in Singapore and Australia. Her email address is Hazel.Tan@monash.edu.

Colleen Vale is an associate professor in mathematics education at Victoria University, Australia. She teaches in primary and secondary pre-service teacher programs, and is co-author of the recently published and award winning text, *Teaching Secondary School Mathematics: Research and Practice for the 21st century*. Her research interests include equity and social justice, digital technologies, pre-service teacher education, school improvement, and teacher professional learning, with gender as a common focus in all fields. Recently she conducted a large project reviewing school improvement strategies to progress the mathematics learning of students from low socio-economic school communities. Recent publications have centered on the pedagogical content knowledge and mathematics knowledge for teaching of pre-service teachers. Her email address is Colleen.vale@vu.edu.au.

Paola Valero is a full professor in the Department of Learning and Philosophy at Aalborg University, Denmark. She is the leader of the Science and Mathematics Education Research Group and is the director of the doctoral program, *Technology and Science: Education and Philosophy*. She has been researching the significance of mathematics as a school subject in modern Western societies and its constitution as a field, where power relations are central in generating inclusion and exclusion of different types of students. She focuses on the development of theoretical understandings of mathematics education linking learning to the broad social and political levels of educational practice. Her email address is paola@learning.aau.dk.

Delinda van Garderen is an associate professor in the Department of Special Education at the University of Missouri, Columbia. Her current interests include students with learning disabilities, intervention research focused primarily in mathematics and the use of representations to solve word problems, understanding the characteristics of K-2 struggling learners and their development in number and operations, and developing ways to foster and encourage collaboration among special and general education teachers particularly in the areas of science and mathematics. Her email address is vangarderend@missouri.edu.

Lieven Verschaffel. From 1979–2000, Lieven held research positions under the Fund for Scientific Research, Flanders. He then gained a full professorship in the Faculty of Psychology and Educational Sciences, Katholieke Universiteit Leuven. His major research interest is the psychology of mathematics education, in particular the teaching and learning of elementary arithmetic concepts and skills, and modeling and applied problem solving. He is coordinator of the international scientific network on "Stimulating critical and flexible thinking" (Fund for Scientific

Research–Flanders), and of the Concerted Research Action "Number sense: Analysis and improvement" (Research Fund, University of Leuven). His email address is lieven.verschaffel@ped.kuleuven.be.

Fernanda Wanderer is a full professor at the Universidade Federal do Rio Grande do Sul (UFRGS). She holds a MSc Degree and a PhD Degree on Education from Universidade do Vale do Rio dos Sinos (Unisinos) in the field of mathematics education. Her work is focused on mathematics teaching education of primary and secondary pre-service and in-service teachers and ethnomathematics. Her recent research projects have been developed in different time-space German descendent rural communities of Southern Brazil. She published chapters and papers in national and international books and journals. Her email address is fwanderer@certelnet.com.br.

Tine Wedege is a full professor in mathematics education at Malmö University, Sweden. In 2005–2010 she was also Professor II in the Department of Mathematical Sciences, Norwegian University of Science and Technology, Trondheim, Norway. Until 2005, Wedege was an associate professor at Roskilde University, Denmark, where she defended her doctoral thesis in 2000. She is a board member of the Nordic Society for Research in Mathematics Education and is a member of the editorial committee of Nordic Studies in Mathematics Education. Internationally, Wedege regularly participates in the research forum, Adults Learning Mathematics (ALM). She is also as a member of the editorial board of ALM, the international journal. Her email address is tine.wedege@mah.se.

Graeme Were has a PhD in Anthropology from the University College London and directs the Museum Studies programme at the University of Queensland. He specializes in material culture and museum studies. His current interests include: ethnomathematics and material culture; digital heritage and source community engagement; and object-based learning within the university museum. He has a regional specialization in the Pacific and his publications include: *Lines that Connect: Rethinking Pattern and Mind in the Pacific* (University of Hawaii Press, 2010) and *Pacific Pattern* (Thames & Hudson, 2005 with S. Kuechler). His email address is g.were@uq.edu.au.

List of Reviewers

Alan Bishop
Allan Bernardo
Nerida Ellerton
Margaret Flores
John Francisco
Peter Galbraith
Vasilis Gialamas
Merrilyn Goos
Steven Khan
Richard Kitchen
Robert M. Klein
Evelyn Kroesbergen
Gilah Leder
Roza Leikin
Stephen Lerman
Swapna Mukhopadhyay
Swee Fong Ng
Nuria Planas
Joanne Rossi Becker
Wolfgang Schlöglmann
Bharath Sriraman
Olof Bjorg Steinthorsdottir
Catherine Vistro-Yu
David Wagner

H. Forgasz, F. Rivera (eds.), *Towards Equity in Mathematics Education*, 569
Advances in Mathematics Education,
DOI 10.1007/978-3-642-27702-3, © Springer-Verlag Berlin Heidelberg 2012

Author Index

H. Forgasz, F. Rivera (eds.), *Towards Equity in Mathematics Education*,   571
Advances in Mathematics Education,
DOI 10.1007/978-3-642-27702-3, © Springer-Verlag Berlin Heidelberg 2012

Subject Index

H. Forgasz, F. Rivera (eds.), *Towards Equity in Mathematics Education,*
Advances in Mathematics Education,
DOI 10.1007/978-3-642-27702-3, © Springer-Verlag Berlin Heidelberg 2012

Printed by Books on Demand, Germany